Climatology and Weather Forecasting:
An Integrated Approach

Climatology and Weather Forecasting: An Integrated Approach

Edited by Loren Gilbert

SYRAWOOD
PUBLISHING HOUSE

New York

Published by Syrawood Publishing House,
750 Third Avenue, 9th Floor,
New York, NY 10017, USA
www.syrawoodpublishinghouse.com

Climatology and Weather Forecasting: An Integrated Approach
Edited by Loren Gilbert

International Standard Book Number: 978-1-68286-649-8 (Hardback)

Cataloging-in-Publication Data

Climatology and weather forecasting : an integrated approach / edited by Loren Gilbert.
    p. cm.
Includes bibliographical references and index.
ISBN 978-1-68286-649-8
1. Climatology. 2. Weather forecasting. 3. Climatic changes. 4. Meteorology. I. Gilbert, Loren.
QC981 .C55 2019
551.6--dc23

TABLE OF CONTENTS

Permissions

List of Contributors

Index

PREFACE

Over the recent decade, advancements and applications have progressed exponentially. This has led to the increased interest in this field and projects are being conducted to enhance knowledge. The main objective of this book is to present some of the critical challenges and provide insights into possible solutions. This book will answer the varied questions that arise in the field and also provide an increased scope for furthering studies.

Climatology or climate science is the study of climate, especially weather conditions over a long period of time. The concepts of climatology are used to analyze accumulated data for pattern recognition. Such recognized patterns are used to forecast weather. Recent advances in this field of science have given rise to a number of modern approaches in related areas of research such as paleoclimatology, historical climatology, etc. This book is a compilation of chapters that discuss the most vital concepts and emerging trends in the field of climatology and their application in weather forecasting. It is a complete source of knowledge on the present status of this important field. Easy to understand language and extensive use of examples makes this book an ideal reference text for students and researchers alike.

I hope that this book, with its visionary approach, will be a valuable addition and will promote interest among readers. Each of the authors has provided their extraordinary competence in their specific fields by providing different perspectives as they come from diverse nations and regions. I thank them for their contributions.

Editor

Assessment of Faecal Cortisol Levels in Free-Ranging Nilgiri Tahrs (*Nilgiritragus hylocrius*) in Correlation with Meteorological Parameters: A Non-Invasive Study

Boon Allwin[1*], Nishit S Gokarn[1], Serma S Pandian[1], Stalin Vedamanickam[2], Sathish Gopal[2], Manoj K[3] and Bharath Jothi S[2]

[1]*Madras Veterinary College, Chennai, TANUVAS, Tamil Nadu, India*

[2]*Department of Animal Husbandry, Tamil Nadu, India*

[3]*Forest College and Research institute, TNAU, Tamil Nadu, India*

[*]**Corresponding author:** Boon Allwin, Madras Veterinary College, Chennai, TANUVAS, Tamil Nadu, India, E-mail: boonallwin@gmail.com

Abstract

The faecal glucocorticoid metabolites of a free-ranging small Nilgiri tahr population of Western Ghats, Tamil Nadu was studied to investigate contributing confounding influences of season, ambient temperature, rainfall and water level on the annual secretion pattern. The was done for a period of one year Oct 2013-Sep 2014. Individuals may cope with environmental challenges through the secretion of stress hormones (glucocorticoids) which allows the integration of environmental change as essential life events such as predator stress, food and water availability, resting cover, influence of tourists and life history events such as birth, death, maintenance of an essential population size by means of an adaptive feedback mechanism. Adaptation and eventually acclimatization to cyclic day-to-day activities, short-term environmental stressors or long-term ecological pressures have been observed with these animals. However, being a highly limited population the animals maintained an effective population size. A clear cut seasonal pattern of glucocorticoid metabolites excretion was detected, with increasing levels in summer and winter. The confounding factors such temperature, rainfall, relative humidity, solar radiation, soil temperature were recorded throughout the study period and did not have any correlation with the stress the animals exhibited. The observed pattern might be due to lack of feed availability both during summer and winter, a declining nutritional intake and reduction of metabolism during winter, clearly the animals were not in their "Thermo comfort Zone". However, broad retrospective studies are essential to identify potential contingent environmental stressors. This study reports the baseline cortisol level in Nilgiri Tahrs, with the relevant confounding factors correlating with their annual variation level.

Keywords: Faecal cortisol metabolites; Stress; Nilgiri Tahr; Meteorological parameters

Introduction

The Nilgiri Tahr (*Nilgiritragus hylocrius*) is an endangered mountain ungulate endemic to the southern part of the Western Ghats. The species is found in a roughly 400 km stretch in the Western Ghats which falls in the states of Kerala and Tamil Nadu. The local distribution of the species is attributed to the animal's preference for the habitat with grasslands with steep rocky cliff shelters. Owing to the disturbances in habitat and their degradation, fragmentation, predator pressure and co-inhabitation of other prey species sharing the same ecosystem. However, the factors that qualifies a particular habitat and owing the bioavailability of resources. It is an endangered mountain ungulate listed in Schedule-I of the Indian Wildlife Protection Act 1972. The IUCN lists Nilgiri Tahr as 'endangered' in the Red List 2010. Natural habitats to native animals are acclimatized environments where several situations, either predictable or capricious, may trigger an evident adaptive response in animals through behavioural, morphological or physiological modifications. On exposure to a stressful event, the adrenal cortex releases glucocorticoids into circulation, and their concentrations in the blood increase as part of the stress response that is mediated by an endocrine pathway and the glucocorticoid regulation level is dependent on either acute or chronic exposures. Glucocorticoids are also involved in metabolic regulations and may vary according to reproductive state and seasonal fluctuations adapting the organism to changing conditions and also govern these functions in a specific population. Acute stress enables animals to cope with unforeseen stress events which are favourable for the species survival, on the contrary chronic stress may lead to reduced survivability and reproductive success [1]. Glucocorticoids-either cortisol or corticosterone (glucocorticoid metabolites) are released during stressful situations, they can serve as an index of the stress response, and the development of non-invasive techniques to measure glucocorticoid metabolites in feces or urine has received increasing attention in field research. Such a technique has the advantage of keeping subjects undisturbed during collection of samples that helps in fixing baseline values and also exploiting the non-invasiveness of the technique. Glucocorticoids have been used as physiological indicators of stress in different species and prove as an index [2]. Our objective was to investigate seasonal variations in the Faecal Cortisol Metabolites (FCM) secretion of a free ranging population of Nilgiri Tahrs in response to potential sources of stress such as variations in temperature, rainfall, relative humidity, solar radiation, soil temperature.

Materials and Methods

The study area Valparai is a Taluk and hill station in the Coimbatore district of Tamil Nadu, India. It is located 3,500 feet (1,100 m) above sea level on the Anaimalai hills range of the Western Ghats at co-ordinates 10°22′12″N, 76°58′12″E. It has an average elevation of 3,914 feet (1,193 m). The study population of Nilgiri Tahr that is endemic to the Anaimalai hill range showed a density of about 10 individuals/1000 ha. Forestry activities like timber logging, firewood collection, social forestry are no longer carried out within this range other than the human disturbance that potentiated by tourism and the road being the only source of connecting route. Glucocorticoid metabolites can be measured as a parameter of adrenal activity in faecal samples, which offer the advantage of being easily collected and feedback free [2,3]. The study was carried out from December 2013 to November 2014, 10 fresh faecal samples were collected randomly each month from the group containing 30 animals and immediately stored in 80% ethanol to initiate steroid extraction immediately after collection [4]. All samples at the fringe areas and near roadways where these animals were usually sighted. Post collection, well-mixed wet faeces (0.6 g) was placed in a capped tube, containing 2.00 mL 80% methanol, vortexed for 30 min and then the tubes were carefully centrifuged for 20 min at 2500 rpm. The supernatant material was diluted in Phosphate Buffer Saline and stored at -80°C for subsequent use. Cortisol estimation was done using the ELISA KIT-DSI-EIA. The calibration curve with the mean absorbance on Y-axis and the calibrator concentration on X-axis was obtained using a 4-parameter curve by immuno assay software. The value of cortisol concentration of the unknowns was read directly from the calibration curve. The data for the meteorological parameters were obtained TNAU weather portal and monthly averages were ascertained for temperature, rainfall, relative humidity, solar radiation, soil temperature and wind speed throughout the study period. The monthly variations of glucocorticoids with these predictor variables were compared and their intra monthly, inter monthly variations were subjected statistical analysis using SPSS. The water sources for this particular region were also noted as these animals basically migrated based on their needs for establishing a permanent contact with the water source.

Results

The results suggest a significant variation in the cortisol level between the months and these variations were attributed to the temperature, relative humidity, wind speed, soil temperature, rainfall and solar radiation (Table 1). The maximum cortisol level was recorded 255.02 ng/g of faeces during May and the minimum was 169.84 ng/g of faces during July. The highest temperature was recorded during May and the least was during December. The relative humidity percentage was highest during December and lowest during May. Highest wind speed was recorded in the month of April and the lowest in the month of December. Rainfall was highest during the month of July. The solar radiation and soil temperature recorded was highest in the month of May and the lowest in January (Table 1). There about eight descriptive water sources in the study area namely, Sholayar, Azhiyar, Parambikulam, upper Nirar, Lower Nirar, Kadamparai and Upper Azhiyar.

Season	Months	Cortisol	Temperature	Relative humidity (%)	Wind speed (kmph)	Soil temperature (°c)	Rainfall (mm)	Solar radiation (cal/cm^2)
Winter	Nov	200.51	11.2	85.65	2.2	8.2	0	221.65
Winter	Dec	205.23	2.1	91.25	1.8	0	0	197.45
Winter	Jan	195.87	5.3	88.21	3.3	0	2	159.22
Winter	Feb	180.45	15.25	85.66	2.5	11.33	3	190.22
Summer	March	175.54	20.32	83.25	4.3	18.03	6.08	412.84
Summer	April	219.87	28.02	69.38	6.9	22.22	0	535.05
Summer	May	255.02	37.81	61.11	6.6	35.96	2.9	590.22
Summer	June	210.38	24.45	89.23	4.1	18.5	13.57	530.06
Rainy	July	169.84	27.54	87.56	2.5	20.16	15.9	380.25
Rainy	Aug	180.78	28.66	81.37	4.2	19.65	12.96	280.45
Rainy	Sep	175.13	22.21	70.65	5.1	19.12	6.21	370.16
Rainy	Oct	179.03	18.58	79.15	5.7	13.98	7.54	419.55

Table 1: Results variation in the cortisol level between temperature, relative humidity, wind speed, soil temperature, rainfall and solar radiation.

The statistical analysis by Pearson's correlation revealed no significant co-relations between the cortisol level and the abiotic factors ($p < 0.001$, $p < 0.05$) such as temperature, relative humidity, rainfall and revealed significant co-relations for wind speed, soil temperature, and solar radiation. However there was a significant correlation ($p < 0.05$) between temperature and relative humidity, wind speed and highly significant correlations ($p < 0.001$) between temperature and soil temperature, solar radiation.

Discussion

Huber et al., [5] found a clear seasonal pattern of Glucocorticoid metabolites secretion in captive red deer (*Cervus elaphus*) population, with higher level in winter and lower level in summer. The same variation has been reported by other studies on deer species in temperate climates, like white-tailed deer *Odocoileus virginianus*, [6] and mule deer *Odocoileus hemionus*, [7], this was not coinciding with our findings. We detected a clear pattern in the seasonal level of FCM, with highest concentrations in May and lowest concentration in July, (Figure 1). The data analysis shows no significant correlation between cortisol and other corresponding variable factors that were taken into consideration during the period of study. The high level of secretion of glucocorticoids in summer might be due to drastic change in the temperature and also adding to increased anthropogenic pressures that is because of tourism [8-10]. However individual animal variations and physiology also plays a major role contributing to the cumulative stress quotient, the increased in stress during summer might be due to the non-availability of feed and grazing grounds rising a physiological concern. The temperature, rainfall, relative humidity, solar radiation, soil temperature and wind speed recorded established no influence in the secretion of glucocorticoids. The observed pattern of secretion might be due to various factors including nutritional intake, predator density, tourism [7,11,12]. The meteorological variables like temperature, soil temperature and solar radiation were also high during the month of May but had no contribution to the cumulative stress quotients that increase glucocorticoid secretion additively. The environmental conditions have seen to have played a greater portion in the variations in stress in this small group of animals.

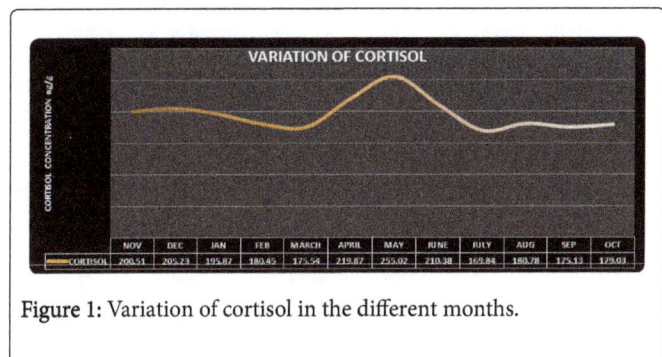

Figure 1: Variation of cortisol in the different months.

In this regard, it becomes essential to mention the report furnished by Pride [13], who quoted that glucocorticoid measures could be useful predictors of individual survival probabilities in the wild populations and existence of high glucocorticoid levels indicated the lowered individual fitness or even population variability. Mateo [14], mentioned that elevation of cortisol observed at emergence might facilitate the acquisition of anti-predator behaviors, with a conclusion that the minimum level of stress operating on the species. The encountering of elevated level of faecal cortisol concentrations in the population of Nilgiri Tahr could be directly attributed to the physiological status of the particular animals. However, it might be impossible to conclude whether it was acute stress or chronic stress that operated in these animals under study.

Schwarzenberger et al., [15] stated that the delayed between the circulation of steroids and their appearance in urine samples was rather short but the lag time of faecal steroids was about 12-24 h in ruminants. This is a baseline data that provide the level of quantifiable stress that prevail in these animals and also owing to the non-invasive tools of its assessment and the ease of sampling. Lesser disturbances in terms of number of visitors might be, however, assigned as the reason for the encountering of lesser faecal cortisol concentration level during the other months of July, August and September. This might be due to

increased availability of feed materials including water for drinking, adequate environmental conditions, absence of various species of predators and most importantly decrease in tourism and no visitors reduce the social challenges that they have meet out, may be putting them in a comfort zone. How the environmental conditions (Meteorological parameters) did not impacted this rhythm still remains unclear. Cavigelli [16] stated that fecal cortisol levels were relatively high corresponding with the end of dry season when high intensity anti predatory behaviour and estimates of feeding effort were high, which is coinciding with the study. Harper et al., [17] observed a clear temporal pattern variation in fecal glucocorticoid levels and it was lowest during October, which was coincident with shortening day length and decreasing ambient temperatures.

The faecal glucocorticoid excretion varied seasonally with a response to cold stress and the parameters such as minimum ambient temperature and snow proved to be the only factors exerting significant effects on fecal glucocorticoid excretion, also the mean daily cortisol concentrations were not significantly different between seasons, but cortisol displayed a circadian rhythm only during the summer. The absence of a circadian rhythm of cortisol during the winter might have been a result of the limited amount of daylight as well as the continual need to produce metabolic heat as a by-product of gluconeogenesis [3,5,18,19]. The increase in the concentration of fecal cortisol was influenced by the days of rainfall and temperature, season and humidity index [20-24].

The mean minimum ambient temperature and mean temperature humidity index values had a significant positive correlation with mean fecal cortisol values [25], however, these values had no correlation with stress quotient of the animal and were individually a physiological response in our study. So in conclusion it is the adrenocortical activity that plays an important role in the seasonal and daily regulation of their physiological states, individually dependant on the animal and not the parameters [26].

Interestingly all the animals are in the same area, with the constant meteorological factors, so the stress acting upon each of the subjects should that are aided by these parameters are constant. The variations in the cortisol level that was observed with the animals may be due to the individual physiological states of the subjects. The quotients of the attributed stress by the meteorological parameters are present but the individual variations observed may be due to, predator pressure, dominance, aggression, reproductive status, competition from co-existing herbivores and finally anthropogenic activities. The conclusion of this study projects a baseline data on glucocorticoid metabolites and their variation and statistically provides proof that the meteorological factors show no correlation with the cortisol, which provides reasons of stress, may be of the physiological conditions of the individual animals such as starvation, pregnancy aggression both intra specific and inter specific. Intense individual animal studies are further required to come to concrete conclusions.

References

1. Sapolsky RM, Romero LM, Munck AU (2000) How do glucocorticoids influence stress response? Integrating permissive, suppressive, stimulatory and preparative actions. Endocrine Reviews 21: 55-89.

2. Mostl E, Palme R (2002) Hormones as indicators of stress. Dom Anim Endocrinol 23: 67-74.

3. Touma C, Palme R (2005) Measuring fecal glucocorticoid metabolites in mammals and birds: the importance of validation. Ann the New York Acad Sci 1046: 54-74.

4. Allwin B, Jayathangaraj MG, Palanivelrajan M, Raman M (2015a). Enumerating endogenous faecal glucocorticoid metabolites as indicators of stress in wild pigs interfering with agriculture adjoining forest regions correlating with conflict and meteorological factors - A Non invasive approach. Int J Adv Multidiscip Res 2: 63-76.

5. Huber S, Palme R, Arnold W (2003) Effects of season, sex, and sample collection on concentration of fecal cortisol metabolites in red deer (*Cervus elaphus*). General Comp Endocrinol 130: 48-54.

6. Bubenik GA, Bubenik AB, Schams D, Leatherland JF (1983) Circadian and circannual rhythms of LH, FSH, testosterone (T), prolactin, cortisol, T3 and T4 in plasma of mature, male white tailed deer. Comp Biochem Physiol 76: 37-45.

7. Saltz D, White GC (1991) Urinary cortisol and urea nitrogen responses to winter stress in mule deer. J Wildlife Manage 55: 1-16.

8. Yousef MK, Cameron RD, Luick JR (1971) Seasonal changes in hydrocortisone secretion rate in reindeer, Rangifer tarandus. Comp Biochem Physiol 40: 495-501.

9. Dantzer R, Mormede P (1983) Stress in farm animals: a need for reevaluation. J Anim Sci 57: 6-17.

10. Huber S, Palme R, Zenker W, Mostl E (2003b) Non-invasive monitoring of the adrenocortical response in red deer. J Wildlife Manage 67: 258-266.

11. DelGiudice GD, Mech LD, Kunkel KE, Gese EM, Seal US (1992) Seasonal patterns of weight, hematology, and serum characteristics of free- ranging female white-tailed deer in Minnesota. Can J Zool 70: 974-983.

12. Tsuma VT, Einarsson S, Madej A, Kindahl H, Lundheim N (1996) Effect of food deprivation during early pregnancy on endocrine changes in primiparous sows. Anim Reprod Sci 41: 267-278.

13. Pride RE (2005) High fecal glucocorticoid levels predict mortality in ring-tailed lemurs (*Lemur catta*). Biol lett 1: 60-63.

14. Mateo MJ (2006) Development and geographic variation in stress hormones in wild Belding's ground squirrels. Horm Beha 50: 718-725.

15. Schwarzenberger F, Mostl E, Palme R, Bamberg E (1996) Faecal steroid analysis for non-invasive monitoring of reproductive status in farm, wild and zoo animals. Anim Reprod Sci 42: 515-526.

16. Cavigelli SA (1999) Behavioural patterns associated with faecal cortisol levels in free ranging female ring-tailed lemurs, Lemur catta Anim Behav 57: 935-944.

17. Harper JM, Austad SN (2001) Effect of Capture and Season on Fecal Glucocorticoid Levels in Deer Mice (*Peromyscus maniculatus*) and Red-Blacked Voles (*Clethrionomys gapperi*). Gen Comp Endocrinol 123: 337-344.

18. Washburn BE, Millspaugh JJ (2002) Effects of simulated environmental conditions on glucocorticoid metabolite measurements in white-tailed deer feces. Gen Comp Endocrinol 127: 217-222.

19. Oki C, Atkinson S (2004) Diurnal patterns of cortisol and thyroid hormones in the Harbor seal (*Phoca vitulina*) during summer and winter seasons. Gen Comp Endocrinol 136: 289-297.

20. Romero LM (2002) Seasonal changes in plasma glucocorticoids concentrations in free-living vertebrates. Gen Comp Endocrinol 128: 1-24.

21. Petrauskas LR, Atkinson SK (2006) Variation of fecal corticosterone concentrations in captive stellar sea lions (*Eumetapias jubatus*) in relation to season and behavior. Aquatic Mammals 32: 168-174.

22. Dalmau A, Ferret A, Chacon G, Manteca X (2007) Seasonal changes in fecal cortisol metabolites in Pyrenean chamois. J Wildlife Manage 71: 190-194.

23. Alejandro CI, Nava CR, Lang CGR, Hernandez DMB (2008) Effect of environmental and meteorological conditions on levels of fecal cortisol in two captive species of carnivorous. J Anim Vet Adv 7: 759-764.

24. Rangel-Negrin A, Alfaro JL, Valdez RA, Romano MC, Serio JC (2009) Stress in Yucatan spider monkeys effects of environmental conditions on fecal cortisol in wild and captive populations. Anim Conserv 12: 496-502.

25. Smitha S, Kannan A, George S, Mercy KA (2011) Radio immuno assay of fecal cortisol to Evaluate climatic stress in New Zealand White rabbits reared at tropical summer. J Veterin Anim Sci 7: 290-294.

26. Smith JE, Monclús R, Wantuck D, Florant GL, Blumstein DT (2012) Fecal glucocorticoid metabolites in wild yellow-bellied marmots: Experimental validation, individual differences and ecological correlates. Gen Comp Endocrinol 178: 417-426.

Kriging Infill of Missing Data and Temporal Analysis of Rainfall in North Central Region of Bangladesh

Nazia Hassan Choudhury*, Ataur Rahman and Sara Ferdousi

Department of Water Resources Engineering, Bangladesh University of Engineering and Technology, Dhaka-1000, Bangladesh

Abstract

The north-central region of Bangladesh is subjected to rapid land use pattern changes as a result of continuous unplanned urbanization, encroachment of water bodies and agricultural land, filling up of designated wet lands and flood flow zones by real estate companies. There is a growing need to consider rainfall regimes an important factor in assessing drainage network design, flood control work, soil and water conservation planning, watershed management and likes. The historic data of daily rainfall of 7 climate stations in the area comprising 4 districts of Dhaka (Banani, Savar and Dhaka_PBO), Narayanganj (Shimrail), Narsingdi (Narsingdi) and Gazipur (Joydebpur and Maona), 4 of which are for 30 years (1984-2013), 1 is for 21 years (1993-2013) and 2 are for 18 years (1996-2013). An assessment of Kriging infill is conducted and appreciable results confirmed for a missing station from moderate values of nearby observed rainfall. The missing volume of data is identified as 7.95% and infilled by Geostatistical approach Kriging in Arc-GIS 10.1. The Kriging generated data and the observed set of data are both then analyzed station-wise for temporal variation on yearly, monthly and seasonal basis. Kriging yields most significant results for monsoon and pre-monsoon seasons. Localized trends are observed at all stations. Kriging Average Annual Rainfall (AAR) and total rainfall values are higher than observed ones; highest AAR of 2047 mm at Dhaka_PBO and lowest 1322 mm at Joydebpur; the highest monsoon rainfall of 1343 mm is at Maona, the highest pre-monsoon rainfall of 465 mm is at Dhaka_PBO and the highest post-monsoon rainfall of 208 mm is at Shimrail.

Keywords: Multi-temporal analysis; Geostatistics; Infill method; Kriging

Background

Introduction

Bangladesh receives some of the heaviest rainfall in the world [1]. Annual Average rainfall in Bangladesh ranges from a low of 1,200 mm in the west to almost 6,000 mm in the east where huge uplifting effects of Meghalaya Plateau compose the highest rainfalls of the world just beyond the border with India [2]. IPCC has termed Bangladesh as one of the most vulnerable countries in the world due to climate change. Hydrological changes are the most significant impacts of climate change in Bangladesh and resultant altered precipitation patterns cause frequent extreme weather events, such as floods, droughts and rainstorms etc., [3-5].

The variability of rainfall and pattern of extreme high or low precipitation are important for the agriculture and hence economy of the country. The study of rainfall patterns and intensities of heavy rainfall events which constitute the hydrologic design and analysis is the primary prioritized requirement in design of development works and study of impact assessments alike.

The variation of rainfall over an area can be studied on spatial or temporal basis. The several spatial variation approaches are by station, basin, sub-basin, watershed, etc.,

Temporal studies can be short term basis or long term basis. Short term variations are concerned with intensities of rainfall and are obtained either from recorded short interval data or from daily data by the application of empirical formulae in 15 minutes, 30 minutes, hourly or three-hourly intervals. Long term variations are usually based on climate normals and are further studied on yearly, monthly or seasonal basis. Four distinct seasons can be recognized in Bangladesh from the climatic point of view:

i. Dry Winter season (December, January, February; DJF).

ii. Pre-monsoon Hot Summer season (March, April, May; MAM).

iii. Rainy Monsoon season (June, July, August, September; JJAS).

iv. Post-monsoon Autumn season (October, November; ON).

Objectives

The study attempts to initially infill the missing daily data within the observed data set to generate a Kriging set of data and then proceed analysis with both sets of data alike. The key objectives are:

i. Assessment of Kriging infill.

ii. Infilling missing daily data of 7 climate stations by Kriging.

iii. Station-wise comparative temporal analysis between Kriging generated data and observed data on the following basis:

a. Yearly

b. Monthly

c. Seasonal

Literature review

Geostatistics: By its introductory definition, Geostatistics offers a way of describing the spatial continuity of natural phenomena

---

*Corresponding author: Nazia Hassan Choudhury, Department of Water Resources Engineering, Bangladesh University of Engineering and Technology, Dhaka-1000, Bangladesh, E-mail: kainazia24@gmail.com

and provides adaptations of classical regression techniques to take advantage of this continuity [6].

Geostatistics is the result of introducing spatial dependence into statistical basics. Disregarding spatial dependence can invalidate methods for analyzing cross-sectional and panel data [7]. It was also established along with the already common statistical analysis of the variables, there must be an assessment of how well the models describe the spatial features of the data [8].

Geostatistical kriging approach

Kriging is a geostatistical method wherein a variable of interest $Z(\mathbf{s})$ is decomposed into a deterministic trend $\mu(\mathbf{s})$ and a random, auto correlated errors form, $\varepsilon(\mathbf{s})$.

$$z(s) = \mu(s) + \varepsilon(s)$$

The symbol $\mathbf{s}$ simply indicates the location; it contains the spatial x (longitude) and y (latitude) coordinates. Deterministic trends are essentially results of math techniques based on the concept that future behavior can be predicted precisely from the past behavior of a set of data. These techniques ignore the existence of disturbances or external 'shocks' that may alter the data's future pattern. In geostatistics, the information on spatial locations allows the computation of distance between observations and thus model autocorrelation as a function of distance.

Geostatistical approach is applicable and methods are optimal when data are:

- normally distributed

- stationary (mean and variance do not vary significantly in space)

The different Kriging methods vary in degrees of complexity and in their underlying assumptions. Ordinary Kriging is one of the simplest forms of Kriging. It assumes that the data points demonstrate local stationarity, i.e., they contain no significant trends over the reasonably homogeneous smaller regions designated by the interpolation search neighborhood. As a result, it is a fairly accepted belief that moderate trends in the data do not significantly affect ordinary Kriging interpolations [9].

Studies on kriging

The pioneer study on Kriging [6] was conducted by Isaaks and Srivastava, 1989. Their book Applied Geostatistics is the introductory text wherein they explain how various forms of the estimation method called Kriging can be employed for specific problems. The book highlights an instructive case study of a simulated deposit.

A recent study performs evaluation of 17 infilling methods [10] including Kriging for time series of daily precipitation and temperature. Results revealed Kriging, among other stochastic methods, performs better infill than the set of deterministic methods. The linear regression residuals between precipitation and elevation incorporated into an ordinary Kriging model gave the best results. Use of elevation in stochastic methods to produce better results in the interpolation of climate variables has been highly recommended.

A case study was conducted on four forms of Kriging [11] and three forms of thin plate splines to predict monthly maximum temperature and monthly mean precipitation in Jalisco State of Mexico. Results show that techniques using elevation as additional information improve the prediction results considerably. From these techniques,

trivariate regression-kriging and trivariate thin plate splines performed best.

Study Area

The study area is composed of the 4 districts Dhaka, Narayanganj, Narshingdi and Gazipur and falls within the North Central Region of Bangladesh as defined by FAP, 1989. The districts span over latitudes of 24.008° to 24.028° and longitudes of 90.006° to 90.989°. A map is illustrated in Figure 1.

Status quo of study area

In these regions, flooding due to rainfall is pronounced and drainage congestion is becoming increasingly uncontrollable [12]. Drainage congestion occurs due to inadequacy of drainage system, heavy rainfall, high water level of peripheral rivers, unplanned development and encroachment. Numerous khals which fed water retention areas are being continuously filled up by new settlements, low-lands and paddy fields are readily converted into build-up areas by the housing estates and natural depressions and wetlands have been filled by solid wastes to reduce the filling costs in many parts. Unplanned urban development sprawls in the area and rivers are subjected to increasing encroachments as a bi-product of development. Water bodies in such settlements are fatally polluted if not filled up already.

BWDB maintains 7 well dispersed climate stations within the study area as per the distribution given in Table 1.

Methodology

Data collection

The station locations and corresponding available data is depicted in Table 2.

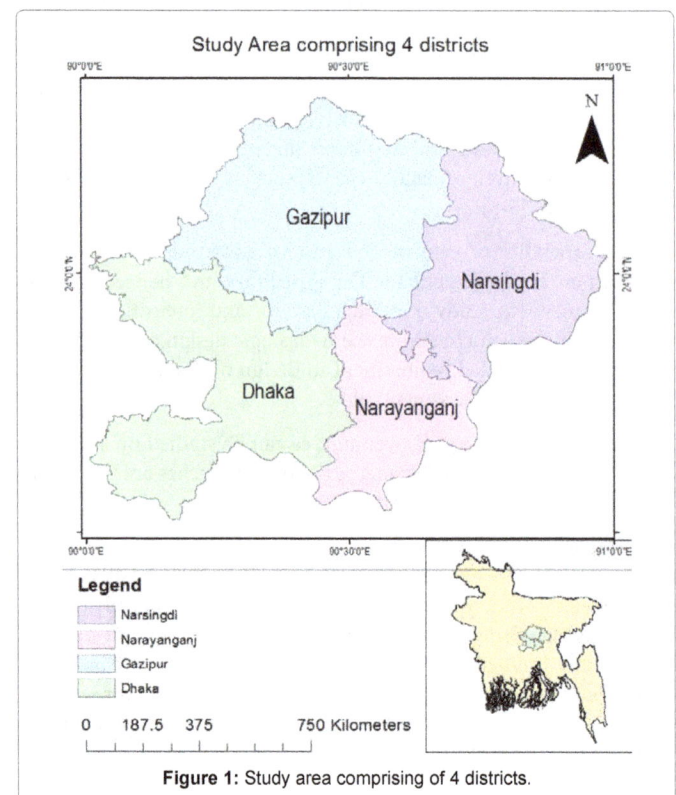

Figure 1: Study area comprising of 4 districts.

District	Number of Stations	Station Name
Dhaka	3	Banani, Savar, Dhaka_PBO
Narayanganj	1	Shimrail
Narsingdi	1	Narsingdi
Gazipur	2	Joydebpur, Maona

Table 1: District wise distribution of climate stations.

Station Name	Station ID	District	Latitude	Longitude	Available Data	No. of Years	Percentage of missing data
Shimrail	CL519	Narayanganj	23.64	90.62	1984-2013	30	2.53
Narsingdi	CL76	Narsingdi	23.94	90.75	1984-2013	30	17.19
Banani	CL42	Dhaka	23.8	90.4	1993-2013	21	5.91
Savar	CL31	Dhaka	23.83	90.26	1984-2013	30	12.35
Joydebpur	CL17	Gazipur	24.01	90.42	1996-2013	18	15.91
Dhaka_PBO	CL9	Dhaka	23.7	90.48	1984-2013	30	6.85
Maona	CL43	Gazipur	24.15	90.4	1996-2013	18	3.09

Table 2: Stations Locations and Available Data.

Data processing

The amount of missing data each station contributes to the total missing volume of 7.95% which is generated by Kriging. Within the missing volume, Joydebpur constitutes the largest volume of missing data at 27% and Dhaka_PBO constitutes the smallest at 4%. Kriging is executed via Arc-GIS 10.1 for each missing day individually with Ordinary Kriging and Prediction Type Surface via the Geostatistical Wizard. The results obtained from data generation of April, May and August for Joydebpur station in 2013 are depicted in Figures 2 and 3.

Temporal analysis

Firstly, for each year in each station calculations for both sets of data are executed to obtain:

- Total Annual Rainfall (TAR).
- Average Daily Rainfall(ADR).
- Total no. of wet days, Total no. of dry days.
- Maximum Daily Rainfall.
- Total Rainfall for each season and each month.

Secondly, for each station the Average Annual Rainfall (AAR) and finally, for all stations, values of AAR, maximum daily rainfall and seasonal rainfall distribution calculations are executed.

Data Analysis, Results and Discussions

Data analysis

Kriging assessment for the reference year 1999 is conducted for Joydebpur station and the months May from pre-monsoon season, August from monsoon season and October from post monsoon season. The generated Kriging data is then compared to actual observed data to reveal appreciable results via similar values and distribution.

Results

The Temporal Analysis results of Maona station are depicted in Figures 4 and 5. The station-wise results are tabulated in Table 3. The trends observed for TAR, ADR, number of wet days, number of dry days and maximum daily rainfall, the appreciable Kriging years observed per season and highest rainfall month for each station is shown.

For all stations, AAR, maximum daily rainfall and seasonal rainfall distribution results are depicted in graphs of (Figures 6-8) respectively.

Discussions

Kriging results are observed to be non-appreciable for post monsoon while it is significant and appreciably deviant from observed data for monsoon and pre-monsoon. The results are observed to be most effective for moderate rainfall values. Kriging for monsoon is also useful for the purpose of obtaining and studying expected moderate values although it is liable to uncertainties due to characteristic dynamism of Bangladesh's monsoon.

Kriging results are seen to be appreciably deviant from observed data for all stations and mildly for Dhaka_PBO. It is seen to be most effective for Joydebpur which had the highest amount of missing data.

However, anomalies and limitations exist. No results were obtained regardless of moderate observed rainfalls for May 2013 at Shimrail. The only observed values were at Maona and in spite of two fold Kriging, no

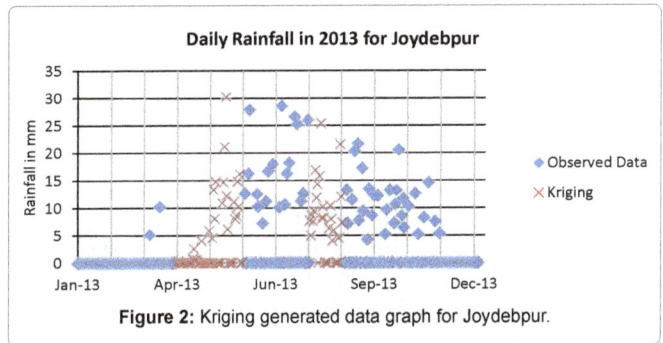

Figure 2: Kriging generated data graph for Joydebpur.

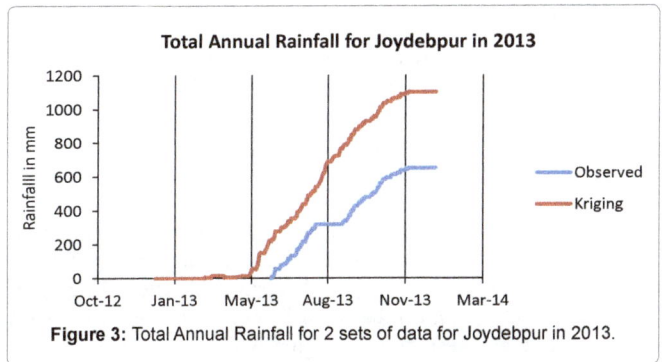

Figure 3: Total Annual Rainfall for 2 sets of data for Joydebpur in 2013.

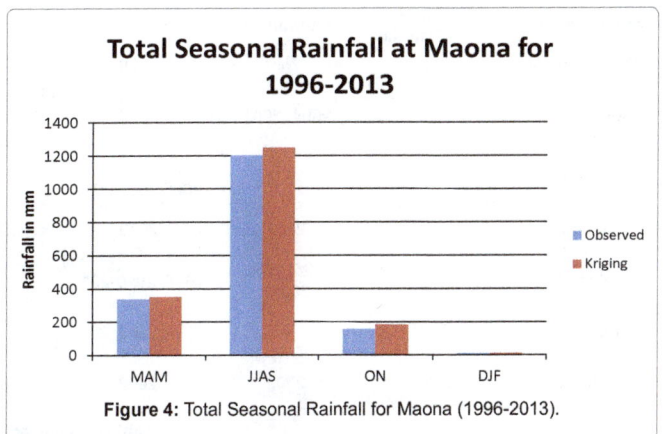

Figure 4: Total Seasonal Rainfall for Maona (1996-2013).

Total Montly Rainfall at Maona for 1996-2013

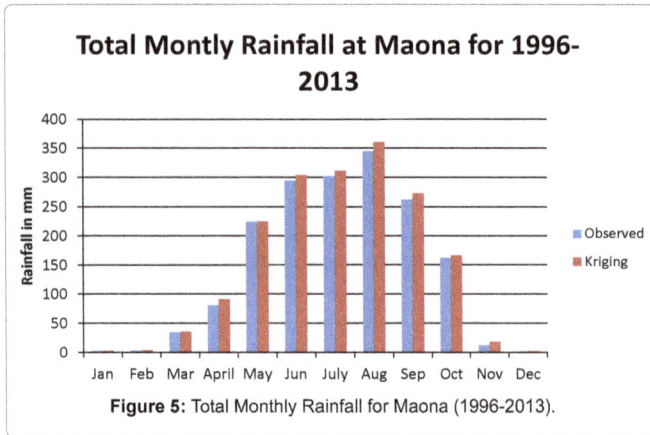

Figure 5: Total Monthly Rainfall for Maona (1996-2013).

generation was possible. Furthermore, missing data can be calculated only for days which have observed rainfall in another location. For low rainfall values, this may not necessarily mean rain in the station of interest.

Trend interpretations

The number of wet days is increased in any year due to Kriging but the overall trend over the years is observed to be same for both Kriging and observed data.

As depicted in results by trend lines, there is a localized effect in number of wet days and in number of dry days over the years. This can be cited as a result of global warming and as a sign of Bangladesh's North Central Zone's vulnerability to the phenomena. This further confirms the impacts of global warming to be continuously effective over the study area.

Conclusions

The assessment of Kriging reveals it to be a suitable method for infilling missing rainfall data and successful data processing for stations with missing data. Minimum errors and appreciable values are obtained for a missing station from moderate values of nearby observed rainfall. The values generated only for the highest or lowest observed rainfall values depict errors and deviant results. Most significant results are obtained for pre-monsoon and monsoon seasons; the volume of expected rainfall in these seasons can be deduced using Kriging. Kriging for post-monsoon has no or negligibly appreciable effect. Kriging in winter is redundant and can be anomalous. The number of wet days is increased in any year due to Kriging but the overall trend over

the years is observed to be same for both Kriging and observed data. Maximum daily rainfall is unaffected by Kriging infill for all stations except Joydebpur when the observed value of 83.7mm is generated to be 104.12 mm.

The limitation lies in the fact that missing data can be calculated only for days which have observed rainfall in at least one other location. For low rainfall values, this may not necessarily mean rain in the station of interest. Applications of Kriging can be resourceful in studying the rainfall over ungauged vulnerable areas, locations of importance and likes. Kriging for monsoon is also useful for the purpose of obtaining and studying expected moderate values although it is liable to uncertainties due to characteristic dynamism of Bangladesh's monsoon.

The temporal analysis firstly as per station spanning over corresponding available years of data, for both Kriging and observed data illustrates total seasonal rainfalls and total monthly rainfalls to follow a normal distribution trend with Monsoon season giving the peaks; July is the rainiest month for all stations except Maona wherein August is both observed and Kriging generated as the rainiest month. Trend line visualizations used in all graph plots of number of wet days, number of dry days and daily maximum rainfall show increasing or decreasing trends. There is a localized effect in number of wet days and in number of dry days over the years. The number of wet days is decreasing at 5 stations (Dhaka_PBO, Joydebpur, Savar, Narsingdi and Shimrail) and increasing at 2 stations (Banani and Maona). The number of dry days follows no trend at 2 stations (Dhaka_PBO and Savar) and is increasing for 5 stations (Joydebpur, Narsingdi, Shimrail, Banani and Maona). This can be cited as a result of global warming and as a sign of Bangladesh's North Central Zone's vulnerability to the phenomena. This further confirms the impacts of global warming to be continuously effective over the study area. Localized changes in maximum daily rainfall are observed as it is decreasing for 4 stations (Joydebpur, Savar, Narsingdi and Shimrail) but increasing for 3 stations (Dhaka_PBO, Banani and Maona).

The temporal analysis secondly for all stations combined, deduced values of average annual rainfall, maximum daily rainfall and seasonal rainfall distribution. The average annual rainfall values are highest and lowest at Dhaka_PBO and Joydebpur respectively for both observed dataset and Kriging dataset; the highest average annual rainfall values are observed to be 2034 mm and generated by Kriging to be 2047 mm.; the lowest average annual rainfall values are observed to be 1083 mm and generated to be 1322 mm. The highest value of observed maximum daily rainfall is 448 mm in 2009 at Banani and lowest of 158.2 mm in 1996 at Joydebpur. From seasonal distributions at all stations,

Station Name	Month of highest rainfall	Post-monsoon	Monsoon	Pre-monsoon	Maximum Daily Rainfall	No. of Dry days	No. of Wet days	TAR, ADR
Dhaka_PBO	July	Nil	1998	1996	Increasing	No trend	Decreasing	Decreasing
Joydebpur	July	2004, 2005	1997, 2002-2005, 2012, 2013	1996,1997, 2002, 2003, 20012, 2013	Decreasing	Increasing	Decreasing	Decreasing
Savar	July	1993	1993, 1997, 2005,2009,2013	1997	Decreasing	No trend	Decreasing	Decreasing
Banani	July	2002	2013	2004,2013	Increasing	Increasing	Increasing	Slightly Decreasing
Maona	Aug	1996, 2002, 20013	2007, 2010-2013	1996, 1997, 2003,2013	Increasing	Increasing	Increasing	Increasing
Narsingdi	July	Nil	1997, 2003, 2013	1997, 2003, 2004, 2006, 2013	Decreasing	Increasing	Decreasing	Significantly Decreasing
Shimrail	July	2002	2002	2005,2007	Decreasing	Increasing	Decreasing	Significantly Decreasing

Table 3: Summary of station-wise results.

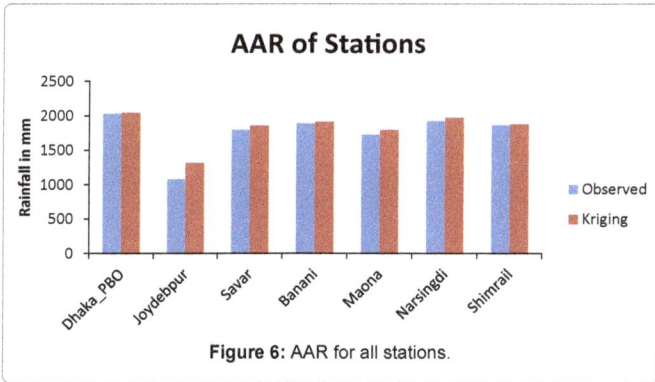

Figure 6: AAR for all stations.

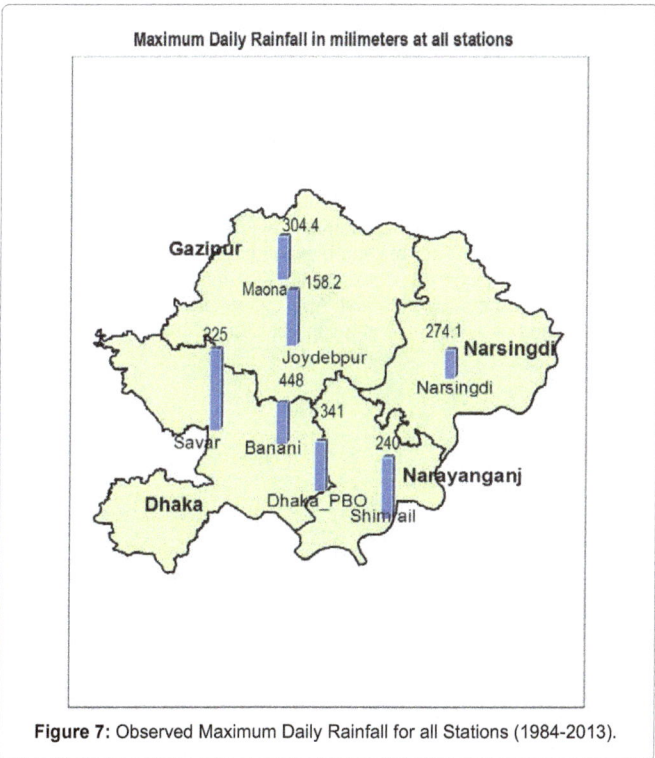

Figure 7: Observed Maximum Daily Rainfall for all Stations (1984-2013).

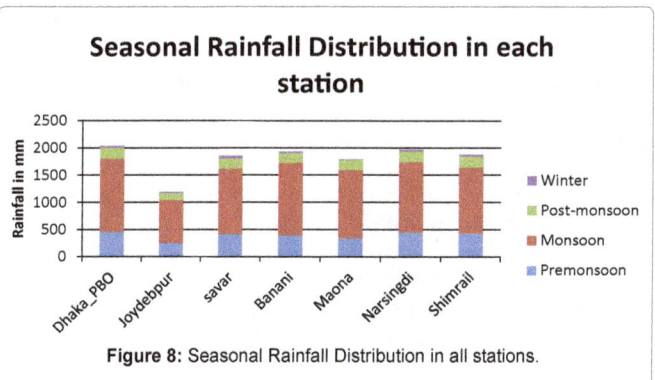

Figure 8: Seasonal Rainfall Distribution in all stations.

developed from completed set of Kriging generated data, it is indicative that the monsoon season contributes highest rainfall at all stations, followed by pre-monsoon, post-monsoon and winter respectively; the highest monsoon rainfall of 1343 mm is at Maona, the highest pre-monsoon rainfall of 465 mm is at Dhaka_PBO and the highest post-monsoon rainfall of 208 mm is at Shimrail.

References

1. Matsumoto J (1988) Synoptic features of heavy monsoon rainfall in 1987 related to the severe floods in Bangladesh. Bulletin of the Department of Geography, University of Tokyo 20: 43-56.

2. Hasan GMJ, Alam R, Islam QN, Hossain S (2012) Frequency Structure of Major Rainfall events in the North-Eastern part of Bangladesh. Journal of Engineering Science and Technology 7: 690-700.

3. Schimidli J, Frei C (2005) Trends of heavy precipitation and wet and dry spells in Switzerland during 20th century. International Journal of Climatology 25: 753-771.

4. Brifa KR, Van der Schrier G, Jones PD (2009) Wet and dry summers in Europe since 1750: evidence of increasing drought. International Journal of Climatology 29: 1894-1905.

5. Zhang Q, Xu CY, Zhang Z (2009) Observed changes of drought/wetness episodes in the Pearl River Basin, China, using the standardized precipitation index and aridity index. Theoretical and Applied Climatology 98: 89-99.

6. Isaaks EH, Srivastava, Mohan R (1989) An Introduction to Applied Geostatistics. Oxford University Press, New York 561.

7. Robinson PM (2008) Developments in the analysis of spatial data. Journal of the Japanese Society of Computational Statistics 38: 87-96.

8. Haining R (1990) Spatial Data Analysis in the Social and Environmental Sciences, Cambridge University Press, Cambridge 111-112.

9. Deutsch, Clayton V, Journel AG (1992) GSLIB: Geostatistical Software Library and User's Guide. Oxford University Press, New York 340.

10. Campozano L, Sanchez E, Aviles A, Samaniego E (2014) Evaluation of infilling methods for time series of daily precipitation and temperature: The case of Ecuadorian Andes. MASKANA 5: 99-115.

11. Boer EPJ, Beursa KM, Hartkamp AD (2001) Kriging and thin plate splines for mapping climate variables. International Journal of Applied Earth Observation and Geoinformation 3: 146-154.

12. Ahmed T, Mirjahan M (2013) Rainfall runoff analysis of eastern Dhaka city using mathematical model. 4th International Conference on Water and Flood Management 2: 113-121.

3

The Analysis of Wind Power Potential in Kerman Synoptic Stations, Iran -An Estimation Using the Weibull Density Function

Mahdi Dehghan Tezerjani*, Kamal Omidvar and Ahmad Mazidi

Department of Geography, Yazd University, Iran

Abstract

In the current research, wind energy potential at Kerman synoptic stations which contain wind data with 20 years statistic period and beyond, including Kerman synoptic station (30°.15'N, 56°.58'E, 1753.8 M high), 1951-2011; Kahnooj (27.58'N, 57°.42'E, 469.7 M high), 1989-2011; Bam (29°.06'N, 58°.21'E, 1066.9 M high), 1956-2011; Baft (29°.14'N, 56°.35'E, 2280 M high), 1986-2011; Anar (30.53'N, 55°.15'E, 1408.8 M high), 1986-2011; Shahrbabak (30°.06'N, 55°.08'E, 1834.1 M high), 1987-2011; Sirjan (29°.28'N, 55°.41'E, 1739.4 M high), 1985-2011; Miandeh-Jiroft (28°.35'N, 57°.48'E, 601 M high), 1989-2011 and Rafsanjan (30°.25'N, 55.54'E, 1580.9 M high), 1992-2011 has been studied. Wind data were obtained from the recorded data at Tehran meteorology organization. This data was recorded at an interval of 3 hrs at 10 m height above the ground level.

Using weibull distribution function, discontinuous wind data were substituted. This substitution has been applied through least square feet after computing weibull parameters C & K. Wind date were changed at 10 m height to 50 m by applying one seventh power law (50 m is the most height of turbines axis) and characteristics of both wind speed and wind power in each height has been computed.

Due to wind power density equals 292.63 and 583.29 w/m^2 at 10 and 50 m height respectively, average wind speed equals 17.42 m/s at 50 m height, wind existence equals 7438.5 h/y, probability of wind blowing in the wind existence hours with speeds between 3 to 25 m/s equals 0.95, most probability of wind speed equals 7.77 m/s, the result show that Rafsanjan synoptic station is an ideal place having high wind energy potential. Application of wind energy at this synoptic station with automatic turbines working at the speed of 3 m/s at 50 m height axis and above has been economically efficient.

Keywords: Wind energy potential; Wind turbine; Weibull distribution function; Wind power density; One seventh power law; Kerman province

Introduction

The worldwide concern and environmental awareness of air quality created a move towards pollution free energy production such as solar and wind energies. Wind is an abundant resource available in nature that can be utilized by mechanically converting wind power to electricity. Wind turbines are especially meant for this purpose.

The gradual increase in the earth population and the increasing demand for energy from natural resources have been the major causes for man to search for an appropriate substitution for the sources of energy. Having been aware of the decrease of the earth's energy resources, scientists have cautioned against the surplus use of energy. Moreover, environmental pollutions resulting from the burning of fossil fuels in the power stations have led to energy loss. This phenomenon is a threat to every living creature on the earth. Thus, the appropriate use of energy and the involvement of new sources of energy have been the focus of study for a long time. An appropriate solution to lower the impact of energy loss is the substitution of renewable forms of energies such as wind.

World winds hold almost around 2700 TW potential energy -25% of this energy occupies an environment of 100 meters above the ground level. It is noteworthy that 10% of this energy, i.e. 4 TW exceeds the capacity of the world's total water energy [1].

Although Iran has been a pioneer in the use wind energy, we have yet to witness the applications of wind energy which have been extremely limited. But the remnants of the numerous windmills all over the country support our claim that undoubtedly wind energy must have been of importance.

This project attempts to discover not only potential power of wind energy but also the feasibility of using wind power at Kerman synoptic stations in Iran.

Moreover, in the process of this study, a new formula, has been obtained that can be applied for computing wind existence hours at an area.

Material and Methods

Since wind is a vector quantity having direction and speed, it is subject to topographical and atmospheric changes. It is impossible to make an estimation of wind direction and wind speed at intervals in which wind speed is not reported. Moreover, this estimation is not error free. To reduce the impact of this error, the researcher took raw data from Tehran Meteorology Organization for a period of at least 20 years. The data reported wind direction and wind speed at Kerman synoptic stations for the consecutive years. Then wind data were converted from Knot to m/s (1 Knot=0.514 m/s).

All the graphs, tables and data in the project were analyzed

*Corresponding author: Mahdi Dehghan Tezerjani, Postal address: 6, Javaheri St., Ghods Sq., Taft, Yazd, Postal code: 8991785699, Iran
E-mail: tiam328@gmail.com

by applying Excel and SPSS 17. In order to process the data the mathematical model Weibull Probability Distribution Function is used. Probability Distribution Function is the most applicable strategy to the study and calculation of wind statistics at a specific location. After the calculation of the components of this function, some parameters relating to the calculation of wind energy can be estimated.

Theory of Analysis

There are several mathematical functions called probability density functions that can be applied to model the wind speed frequency curve. In wind power studies, Weibull and Rayleigh probability density functions are commonly used and widely adopted [2]. Herein Weibull distribution is used since the Rayleigh distribution is only a subset of it.

Weibull distribution function of wind speed

Wind power is proportionate to the cubic power of the wind speed and rotor's diameter square root power [3]. Thus, wind speed is one of the most significant factors in the optimum use of wind energy.

In the calculation of wind energy, wind speed is considered a random variable which can take every quantity in a specific distance. However, practically wind speed data recorded every 3 hours at synoptic stations. The function of which is a disconnect function. In other words, the frequency distribution should be first replaced by the connected distribution function. For this purpose Weibull probability distribution function is reliable and is the most frequently used model to describe the distribution of the wind speed [4].

Weibull distribution function is a derivative of Gamma distribution and has a higher flexibility in comparison with Rayleigh distribution. It can be defined as follows:

$$P(V) = \frac{k}{C} \cdot \left[\frac{V}{C}\right]^{K-1} \exp\left(-\left[\frac{V}{C}\right]^K\right) \tag{1}$$

where V [m/s] is the wind speed, K [-] is the Weibull shape parameter describing the dispersion data and C [m/s] is the Weibull scale parameter.

Values of the two parameters, K and C, can be calculated by using the least square fitting of the data [5] i.e.

$$Y = A + B \cdot X \tag{2}$$

Where

$$Y = \ln\left(-\ln\left(1 - P(V_i)\right)\right) \tag{3}$$

And

$$X = \ln v_i \tag{4}$$

where v_i is the mean of wind speed classes, $P(V_i)$ is the accumulative probability of the frequency of every mean speed classes. By quantities of X and Y, the values of A and B can be calculated using the following equation:

$$A = \frac{n \cdot \sum_{i=1}^{n} x_i y_i - \sum_{i=1}^{n} x_i \sum_{i=1}^{n} y_i}{n \cdot \sum_{i=1}^{n} x_i^2 - \sum_{i=1}^{n} x_i \sum_{i=1}^{n} x_i} \tag{5}$$

$$B = \frac{1}{n}\sum_{i=1}^{n} y_i - \frac{A}{n}\sum_{i=1}^{n} x_i \tag{6}$$

Here, A is the gradient of the equation of a straight line Y=AX+B, B is the width of the intersection of the line by the Y axis. In this equation, the relationship between A and B and the Weibull parameters K and C is as follows [6]:

$$K = A \quad \text{and} \quad C = \exp\left(\frac{-B}{A}\right) \tag{7}$$

A brief presentation of the observations and measurements of wind speed at Kerman synoptic stations are shown in Tables 1-9.

Considering equations (3) and (4), v_i and $P(V_i)$ are substituted with X and Y, so that A and B quantities can be calculated by liner regression equation or least square line of X and Y values. These values are shown in Tables 10-18.

After computing X and Y quantities, A and B quantities related to shape and scale parameters of Weibull function can be determined. Then, we are able to draw a line Y=AX+B which is the line nearest to points, when compared with X and Y

There is a sample of least square line for x and y which are related to Rafsanjan's synoptic station in Figure 1.

The numerical values of A=K, B and C obtained from Kerman synoptic stations are presented respectively as follows: Kerman: 1.5271, -2.6023 and 5.4963; Kahnooj: 2.4894, -5.3718 and 8.6524; Bam: 1.4619, -2.1969 and 4.494; Baft: 2.3471, -4.6212 and 7.1625; Anar: 1.9472, -3.4531 and 5.8906; Shahrbabak: 2.3925, -4.8376 and 7.553; Sirjan: 2.1545, -3.8827 and 6.0624; Miandeh-jiroft: 1.7545, -2.8968 and 5.2125; Rafsanjan:

Weibull probability function quantities (P_w) are presented in Tables 1-9. These quantities are computed using Weibull scale and Form parameters. Also, v_i numerical values, in these Tables (mean of wind speed classes) were computed through equation 1.

With the help of Weibull distribution function, the computation quantities for wind speed occurrence and factual wind speed occurrence over Kerman synoptic station has been drawn in Figures 2-10. Numerical representations of wind speed are distributed in the graphs and are connected by the continuous line drawn with the help of Weibull function. By comparing them, the wind speed disconnected quantities substituted in a curve Contiguous can be observed.

Calculation of efficacious parameters in wind energy potential estimation

As the scale and shape parameters have been calculated, two meaningful wind speeds for wind energy estimation, i.e. the most probable wind speed and the wind speed carrying maximum energy, can be simply obtained. The most probable wind speed denotes the most frequent wind speed for a given wind probability distribution and is expressed as follows [7]:

$$V_{MP} = C\left(\frac{K-1}{K}\right)^{1/K} \tag{8}$$

The wind speed carrying maximum energy represents the wind speed which carries the maximum amount of wind energy, and is expressed as follows [6]:

$$V_{MaxE} = C\left(1 + \frac{2}{K}\right)^{1/K} \tag{9}$$

The average wind speed (V) and wind speed standard deviation (σ) can be calculated through the following equations [7]:

$$V = C\Gamma\left(\frac{K+1}{K}\right) \tag{10}$$

I	Wind speed classes v[m/s]	Mean of wind speed classes v_i[m/s]	Frequency f_i	Probability $(_iv)p$	Accumulative probability $(_iv)P$	Probability in Weibull model $P_w(v_i)$
1	0.5-1.5	1	116	0.004278548	0.004278548	0.0597024
2	1.5-2.5	2	4599	0.169629684	0.173908233	0.1051664
3	2.5-3.5	3	6671	0.246053408	0.419961641	0.1333364
4	3.5-4.5	4	4830	0.178149897	0.598111537	0.1430981
5	4.5-5.5	5	3632	0.133962821	0.732074358	0.1368364
6	5.5-6.5	6	2471	0.091140454	0.823214813	0.1193128
7	6.5-7.5	7	1675	0.061780761	0.884995574	0.0960548
8	7.5-8.5	8	1209	0.0445928	0.929588374	0.0719434
9	8.5-9.5	9	326	0.012024196	0.94161257	0.0503812
10	9.5-10.5	10	724	0.026704042	0.968316613	0.0331025
11	10.5-11.5	11	103	0.003799056	0.972115668	0.0204583
12	11.5-12.5	12	343	0.012651225	0.984766893	0.0119159
13	12.5-13.5	13	96	0.003540868	0.98830776	0.0065506
14	13.5-14.5	14	125	0.004610505	0.992918265	0.003403
15	14.5-15.5	15	97	0.003577752	0.996496017	0.0016721
16	15.5-16.5	16	32	0.001180289	0.997676306	0.0007778
17	16.5-17.5	17	19	0.000700797	0.998377102	0.0003428
18	17.5-18.5	18	19	0.000700797	0.999077899	0.0001432
19	18.5-19.5	19	1	0.000037	0.999114783	0.0000567
20	19.5-20.5	20	18	0.000664	0.999778696	0.0000213
21	21.5-22.5	22	1	0.000037	0.99981558	0.0000026
22	22.5-23.5	23	2	0.000074	0.999889348	0.0000008
23	24.5-25.5	25	3	0.000111	0.999999	0.0000001
Total			27112	1		

Table 1: Arrangement of the measured three hourly time-series data in frequency distribution format for 1986-2011 and the probability density distributions calculated from the Weibull function at Anar synoptic station.

i	Wind speed classes v[m/s]	Mean of wind speed classes v_i[m/s]	Frequency f_i	Probability $(_iv)p$	accumulative probability $(_iv)P$	Probability in Weibull model $P_w(v_i)$
1	0.5-1.5	1	9	0.000326	0.000326	0.0228712
2	1.5-2.5	2	2665	0.096642	0.096968	0.0558927
3	2.5-3.5	3	5415	0.196366	0.293335	0.0891254
4	3.5-4.5	4	6664	0.241659	0.534994	0.1158754
5	4.5-5.5	5	4213	0.152778	0.687772	0.1313352
6	5.5-6.5	6	3085	0.111873	0.799645	0.1334372
7	6.5-7.5	7	959	0.034777	0.834421	0.1231778
8	7.5-8.5	8	2133	0.077350	0.911771	0.1040348
9	8.5-9.5	9	266	0.009646	0.921417	0.080687
10	9.5-10.5	10	1301	0.047179	0.968596	0.0575692
11	10.5-11.5	11	23	0.000834	0.96943	0.0378135
12	11.5-12.5	12	409	0.014832	0.984262	0.0228657
13	12.5-13.5	13	33	0.001197	0.985458	0.0127242
14	13.5-14.5	14	133	0.004823	0.990281	0.0065115
15	14.5-15.5	15	141	0.005113	0.995395	0.0030617
16	15.5-16.5	16	35	0.001269	0.996664	0.0013214
17	16.5-17.5	17	12	0.000435	0.997099	0.000523
18	17.5-18.5	18	38	0.001378	0.998477	0.0001895
19	18.5-19.5	19	1	0.000036	0.998513	0.0000628
20	19.5-20.5	20	32	0.001160	0.999674	0.000019
21	21.5-22.5	22	1	0.000036	0.99971	0.0000013
22	23.5-24.5	24	1	0.000036	0.999746	0.0000001
23	24.5-25.5	25	5	0.000181	0.999927	0
24	27.5-28.5	28	2	0.000073	0.99999	0
		Total	27576	1		

Table 2: Arrangement of the measured three hourly time-series data in frequency distribution format for 1986-2011 and the probability density distributions calculated from the Weibull function at Baft synoptic station.

i	Wind speed classes v[m/s]	Mean of wind speed classes v_i[m/s]	Frequency f_i	Probability $(_iv)p$	accumulative probability $(_iv)P$	Probability in Weibull model $P_w(v_i)$
1	0.5-1.5	1	2869	0.036216059	0.036216059	0.1453942
2	1.5-2.5	2	17502	0.220931847	0.257147906	0.1647792
3	2.5-3.5	3	12974	0.163773842	0.420921749	0.1551221
4	3.5-4.5	4	17059	0.215339754	0.636261503	0.1326244
5	4.5-5.5	5	9521	0.120185814	0.756447317	0.1061931
6	5.5-6.5	6	8499	0.107284869	0.863732185	0.0808406
7	6.5-7.5	7	3283	0.041442078	0.905174264	0.0590278
8	7.5-8.5	8	4178	0.052739873	0.957914137	0.0415805
9	8.5-9.5	9	989	0.012484379	0.970398516	0.0283729
10	9.5-10.5	10	1517	0.019149446	0.989547962	0.0188115
11	10.5-11.5	11	131	0.001653644	0.991201606	0.0121472
12	11.5-12.5	12	408	0.00515028	0.996351885	0.007654
13	12.5-13.5	13	78	0.000984612	0.997336498	0.0047134
14	13.5-14.5	14	85	0.001072975	0.998409472	0.0028405
15	14.5-15.5	15	78	0.000984612	0.999394085	0.001677
16	15.5-16.5	16	9	0.000113609	0.999507694	0.0009709
17	16.5-17.5	17	8	0.000100986	0.99960868	0.0005517
18	17.5-18.5	18	10	0.000126232	0.999734912	0.0003079
19	18.5-19.5	19	1	1.26232E-05	0.999747535	0.0001689
20	19.5-20.5	20	11	0.000138856	0.999886391	0.0000911
21	20.5-21.5	21	3	0.000038	0.999924261	0.0000484
22	23.5-24.5	24	1	0.000013	0.999936884	0.0000066
23	24.5-25.5	25	1	0.000013	0.999949507	0.0000033
24	29.5-30.5	30	1	0.000013	0.99996213	0.0000001
25	31.5-32.5	32	1	0.000013	0.999974754	0
26	34.5-35.5	35	1	0.000013	0.999987377	0
27	35.5-36.5	36	1	0.000013	0.9999999	0
	Total		79219	1		

Table 3: Arrangement of the measured three hourly time-series data in frequency distribution format for 1956-2011 and the probability density distributions calculated from the Weibull function at Bam synoptic station.

i	Wind speed classes v[m/s]	Mean of wind speed classes v_i[m/s]	Frequency f_i	Probability $(_iv)p$	accumulative probability $(_iv)P$	Probability in Weibull model $P_w(v_i)$
1	0.5-1.5	1	2	0.0000991	0.0000991	0.0115117
2	1.5-2.5	2	794	0.039336141	0.039435224	0.0316365
3	2.5-3.5	3	1459	0.072281397	0.111716621	0.0552975
4	3.5-4.5	4	2796	0.138518702	0.250235323	0.0787512
5	4.5-5.5	5	4478	0.221847907	0.47208323	0.0984773
6	5.5-6.5	6	3842	0.190339361	0.662422591	0.1115778
7	6.5-7.5	7	1619	0.080208075	0.742630666	0.1163125
8	7.5-8.5	8	2597	0.128659896	0.871290562	0.1124491
9	8.5-9.5	9	175	0.008669804	0.879960367	0.1012525
10	9.5-10.5	10	1589	0.078721823	0.95868219	0.0850939
11	10.5-11.5	11	38	0.001882586	0.960564776	0.0668046
12	11.5-12.5	12	350	0.017339609	0.977904384	0.0489959
13	12.5-13.5	13	14	0.000693584	0.978597969	0.0335557
14	13.5-14.5	14	41	0.002031211	0.98062918	0.021443
15	14.5-15.5	15	281	0.013921229	0.994550409	0.0127725
16	15.5-16.5	16	14	0.000693584	0.995243993	0.0070833
17	16.5-17.5	17	1	0.000050	0.995293535	0.0036526
18	17.5-18.5	18	20	0.000990835	0.99628437	0.001749
19	19.5-20.5	20	67	0.003319297	0.999603666	0.0003193
20	20.5-21.5	21	1	0.000050	0.999653208	0.0001214
21	21.5-22.5	22	2	0.000099	0.999752291	0.0000426
22	23.5-24.5	24	1	0.000050	0.999801833	0.0000041
23	27.5-28.5	28	1	0.000050	0.999851375	0
24	29.5-30.5	30	1	0.000050	0.999900917	0
25	37.5-38.5	38	1	0.000050	0.999950458	0
26	39.5-40.5	40	1	0.000050	0.999999	0
	Total		20185	1		

Table 4: Arrangement of the measured three hourly time-series data in frequency distribution format for 1989-2011 and the probability density distributions calculated from the Weibull function at Kahnooj synoptic station.

i	Wind speed classes v[m/s]	Mean of wind speed classes v_i[m/s]	Frequency f_i	Probability $(_iv)p$	accumulative probability $(_iv)P$	Probability in Weibull model $P_w(v_i)$
1	0.5-1.5	1	4203	0.047056584	0.047056584	0.1050749
2	1.5-2.5	2	8235	0.092198661	0.139255245	0.1317106
3	2.5-3.5	3	21770	0.243735865	0.38299111	0.1358068
4	3.5-4.5	4	9611	0.10760429	0.490595401	0.1269828
5	4.5-5.5	5	15893	0.177937258	0.668532659	0.1112482
6	5,5-6.5	6	6284	0.07035536	0.738888018	0.0927607
7	6.5-7.5	7	6155	0.068911082	0.8077991	0.0742786
8	7.5-8.5	8	5122	0.057345664	0.865144764	0.0574524
9	8.5-9.5	9	3947	0.044190421	0.909335184	0.043097
10	9.5-10.5	10	4270	0.047806713	0.957141897	0.0314463
11	10.5-11.5	11	701	0.007848362	0.96499026	0.0223699
12	11.5-12.5	12	998	0.01117356	0.976163819	0.0155425
13	12.5-13.5	13	681	0.007624443	0.983788262	0.0105627
14	13.5-14.5	14	295	0.003302806	0.987091068	0.0070302
15	14.5-15.5	15	685	0.007669227	0.994760295	0.0045872
16	15.5-16.5	16	94	0.001052419	0.995812714	0.002937
17	16.5-17.5	17	56	0.000626973	0.996439687	0.0018467
18	17.5-18.5	18	114	0.001276338	0.997716026	0.001141
19	18.5-19.5	19	23	0.000257507	0.997973533	0.0006932
20	19.5-20.5	20	89	0.00099644	0.998969972	0.0004144
21	20.5-21.5	21	47	0.00052621	0.999496182	0.0002438
22	21.5-22.5	22	8	0.0000896	0.99958575	0.0001413
23	22.5-23.5	23	12	0.000134351	0.999720101	0.0000806
24	23.5-24.5	24	5	0.000056	0.999776081	0.0000454
25	24.5-25.5	25	10	0.000112	0.99988804	0.0000252
26	25.5-26.5	26	4	0.000045	0.999932824	0.0000138
27	26.5-27.5	27	1	0.000011	0.99994402	0.0000074
28	27.5-28.5	28	2	0.000022	0.999966412	0.000004
29	29.5-30.5	30	2	0.000022	0.999988804	0.0000011
30	34.5-35.5	35	1	0.000011	0.9999999	0
		Total	89318	1		

Table 5: Arrangement of the measured three hourly time-series data in frequency distribution format for 1951-2011 and the probability density distributions calculated from the Weibull function at Kerman synoptic station.

i	Wind speed classes v[m/s]	Mean of wind speed classes v_i[m/s]	Frequency f_i	Probability $(_iv)p$	accumulative probability $(_iv)P$	Probability in Weibull model $P_w(v_i)$
1	0.5-1.5	1	6	0.000186875	0.000187	0.0188125
2	1.5-2.5	2	701	0.021833245	0.02202	0.0477558
3	2.5-3.5	3	7817	0.243467157	0.265487	0.0784585
4	3.5-4.5	4	12135	0.377954963	0.643442	0.1050525
5	4.5-5.5	5	3734	0.116298626	0.759741	0.1228575
6	5.5-6.5	6	2957	0.092098296	0.851839	0.1291674
7	6.5-7.5	7	1032	0.032142523	0.883982	0.1237958
8	7.5-8.5	8	1993	0.062073691	0.946055	0.108938
9	8.5-9.5	9	172	0.005357087	0.951412	0.0883471
10	9.5-10.5	10	834	0.025975644	0.977388	0.0661495
11	10.5-11.5	11	28	0.000872084	0.97826	0.0457582
12	11.5-12.5	12	358	0.011150216	0.98941	0.0292414
13	12.5-13.5	13	9	0.000280313	0.989691	0.0172542
14	13.5-14.5	14	143	0.004453857	0.994145	0.0093932
15	14.5-15.5	15	51	0.001588439	0.995733	0.0047133
16	15.5-16.5	16	55	0.001713022	0.997446	0.0021774
17	16.5-17.5	17	6	0.000186875	0.997633	0.000925
18	17.5-18.5	18	33	0.001027813	0.998661	0.0003609
19	18.5-19.5	19	1	0.000031	0.998692	0.0001291
20	19.5-20.5	20	30	0.000934376	0.999626	0.0000423
21	21.5-22.5	22	6	0.000186875	0.999813	0.0000035
22	23.5-24.5	24	1	0.000031	0.999844	0.0000002
23	24.5-25.5	25	2	0.000062	0.999907	0
24	27.5-28.5	28	1	0.000031	0.999938	0
25	29.5-30.5	30	1	0.000031	0.999969	0
26	34.5-35.5	35	1	0.000031	0.99999	0
		Total	32107	1		

Table 6: Arrangement of the measured three hourly time-series data in frequency distribution format for 1987-2011 and the probability density distributions calculated from the Weibull function at Shahrbabak synoptic station.

i	Wind speed classes v[m/s]	Mean of wind speed classes v_i[m/s]	Frequency f_i	Probability $(_iv)p$	accumulative probability $(_iv)P$	Probability in Weibull model $P_w(v_i)$
1	0.5-1.5	1	122	0.003856244	0.003856	0.0434674
2	1.5-2.5	2	4049	0.127983058	0.131839	0.0901226
3	2.5-3.5	3	4860	0.1536176	0.285457	0.1266384
4	3.5-4.5	4	5996	0.189524923	0.474982	0.1461868
5	4.5-5.5	5	6226	0.196794892	0.671777	0.1470099
6	5.5-6.5	6	4021	0.127098018	0.798875	0.1320683
7	6.5-7.5	7	2163	0.068369314	0.867244	0.1073469
8	7.5-8.5	8	2019	0.063817682	0.931062	0.0795051
9	8.5-9.5	9	611	0.01931283	0.950375	0.0538804
10	9.5-10.5	10	913	0.028858615	0.979233	0.0334964
11	10.5-11.5	11	85	0.002686728	0.98192	0.0191326
12	11.5-12.5	12	287	0.009071657	0.990992	0.01005
13	12.5-13.5	13	65	0.002054556	0.993046	0.0048574
14	13.5-14.5	14	86	0.002718336	0.995764	0.0021607
15	14.5-15.5	15	96	0.003034422	0.998799	0.0008847
16	15.5-16.5	16	12	0.000379303	0.999178	0.0003334
17	16.5-17.5	17	4	0.000126434	0.999305	0.0001156
18	17.5-18.5	18	12	0.000379303	0.999684	0.0000369
19	18.5-19.5	19	1	0.000032	0.999716	0.0000108
20	19.5-20.5	20	6	0.000189651	0.999905	0.0000029
21	21.5-22.5	22	2	0.000063	0.999968	0.0000002
22	23.5-24.5	24	1	0.000032	0.99999	0
		Total	31637	1		

Table 7: Arrangement of the measured three hourly time-series data in frequency distribution format for 1985-2011 and the probability density distributions calculated from the Weibull function at Sirjan synoptic station.

i	Wind speed classes v[m/s]	Mean of wind speed classes v_i[m/s]	Frequency f_i	Probability $(_iv)p$	accumulative probability $(_iv)P$	Probability in Weibull model $P_w(v_i)$
1	0.5-1.5	1	47	0.003856252	0.003856252	0.0916441
2	1.5-2.5	2	3572	0.293075156	0.296931408	0.1356201
3	2.5-3.5	3	3564	0.292418773	0.589350181	0.1518182
4	3.5-4.5	4	2460	0.201837873	0.791188054	0.1470365
5	4.5-5.5	5	1119	0.091811618	0.882999672	0.1287542
6	5.5-6.5	6	602	0.049392845	0.932392517	0.1040692
7	6.5-7.5	7	154	0.012635379	0.945027896	0.0785591
8	7.5-8.5	8	276	0.022645225	0.967673121	0.0557984
9	8.5-9.5	9	43	0.00352806	0.971201181	0.0374827
10	9.5-10.5	10	165	0.013537906	0.984739088	0.0239034
11	10.5-11.5	11	10	0.000820479	0.985559567	0.0145131
12	11.5-12.5	12	64	0.005251067	0.990810633	0.0084085
13	12.5-13.5	13	6	0.000492287	0.991302921	0.0046575
14	13.5-14.5	14	14	0.001148671	0.992451592	0.0024702
15	14.5-15.5	15	40	0.003281917	0.995733508	0.0012561
16	15.5-16.5	16	9	0.000738431	0.99647194	0.0006131
17	16.5-17.5	17	4	0.000328192	0.996800131	0.0002875
18	17.5-18.5	18	4	0.000328192	0.997128323	0.0001297
19	18.5-19.5	19	1	0.000082	0.997210371	0.0000563
20	19.5-20.5	20	23	0.001887102	0.999097473	0.0000235
21	20.5-21.5	21	2	0.000164096	0.999261569	0.0000095
22	21.5-22.5	22	3	0.000246144	0.999507713	0.0000037
23	24.5-25.5	25	6	0.000492287	0.9999999	0.0000002
		Total	12188	1		

Table 8: Arrangement of the measured three hourly time-series data in frequency distribution format for 1989-2011 and the probability density distributions calculated from the Weibull function at Miandeh-Jiroft synoptic station.

i	Wind speed classes v[m/s]	Mean of wind speed classes v_i[m/s]	Frequency f_i	Probability $(_iv)p$	accumulative probability $(_iv)P$	Probability in Weibull model $P_w(v_i)$
1	0.5-1.5	1	9	0.00022686	0.00022686	0.0157413
2	1.5-2.5	2	2582	0.065083686	0.065310546	0.0432486
3	2.5-3.5	3	5095	0.128428111	0.193738657	0.0746244
4	3.5-4.5	4	8599	0.216752369	0.410491026	0.1035553
5	4.5-5.5	5	9577	0.241404517	0.651895543	0.1243828
6	5.5-6.5	6	4775	0.120361968	0.772257512	0.1332449
7	6.5-7.5	7	2619	0.066016334	0.838273846	0.1290849
8	7.5-8.5	8	2586	0.065184513	0.903458359	0.1138468
9	8.5-9.5	9	570	0.014367816	0.917826175	0.0916788
10	9.5-10.5	10	1787	0.045044364	0.962870538	0.0674697
11	10.5-11.5	11	86	0.002167776	0.965038314	0.0453634
12	11.5-12.5	12	753	0.018980641	0.984018955	0.0278356
13	12.5-13.5	13	47	0.001184715	0.98520367	0.0155644
14	13.5-14.5	14	165	0.004159105	0.989362775	0.0079161
15	14.5-15.5	15	246	0.006200847	0.995563622	0.0036548
16	15.5-16.5	16	41	0.001033474	0.996597096	0.0015285
17	16.5-17.5	17	18	0.000453721	0.997050817	0.0005777
18	17.5-18.5	18	38	0.000957854	0.998008671	0.0001969
19	19.5-20.5	20	43	0.001083888	0.999092559	0.0000166
20	20.5-21.5	21	6	0.00015124	0.999243799	0.0000041
21	21.5-22.5	22	10	0.000252067	0.999495866	0.0000009
22	22.5-23.5	23	1	0.000025	0.999521073	0.0000002
23	23.5-24.5	24	8	0.000201654	0.999722726	0
24	24.5-25.5	25	11	0.000277274	0.99999	0
		Total	39672	1		

Table 9: Arrangement of the measured three hourly time-series data in frequency distribution format for 1992-2011 and the probability density distributions calculated from the Weibull function at Rafsanjan synoptic station.

i	$Y = \ln(-\ln(1 - P(V_i)))$	$X = \ln V_i$
1	-5.451998424	0
2	-1.655223178	0.693147181
3	-0.607591621	1.098612289
4	-0.09257517	1.386294361
5	0.275391193	1.609437912
6	0.54975009	1.791759469
7	0.771396588	1.945910149
8	0.975840664	2.079441542
9	1.044034537	2.197224577
10	1.238942994	2.302585093
11	1.275276299	2.397895273
12	1.431335631	2.48490665
13	1.492641127	2.564949357
14	1.599435321	2.63905733
15	1.732337582	2.708050201
16	1.802468091	2.772588722
17	1.859969691	2.833213344
18	1.944316865	2.890371758
19	1.950140884	2.944438979
20	2.130131358	2.995732274
21	2.151563788	3.091042453
22	2.209276046	3.135494216
23	2.779942594	3.218875825

Table 10: The numerical values of liner equations, between X and Y, for the determination of A and B, in relation to Weibull parameters K and C in Anar synoptic station.

i	$Y = \ln(-\ln(1 - P(V_i)))$	$X = \ln V_i$
1	-8.027313323	0
2	-2.28280494	0.693147181
3	-1.057859329	1.098612289
4	-0.266957784	1.386294361
5	0.151880829	1.609437912
6	0.474781312	1.791759469
7	0.586846379	1.945910149
8	0.886994051	2.079441542
9	0.933581229	2.197224577
10	1.241504457	2.302585093
11	1.249252285	2.397895273
12	1.423507429	2.48490665
13	1.442376843	2.564949357
14	1.533358734	2.63905733
15	1.682783913	2.708050201
16	1.740977015	2.772588722
17	1.76518865	2.833213344
18	1.869805029	2.890371758
19	1.873512873	2.944438979
20	2.082870223	2.995732274
21	2.09743611	3.091042453
22	2.113696938	3.17805383
23	2.254607761	3.218875825
24	2.779942594	3.33220451

Table 11: The numerical values of liner equations, between X and Y, for the determination of A and B, in relation to Weibull parameters K and C in Baft synoptic station.

i	$Y = \ln(-\ln(1 - P(V_i)))$	$X = \ln V_i$
1	-3.29986526	0
2	-1.213153752	0.693147181
3	-0.604554675	1.098612289
4	0.011256491	1.386294361
5	0.345305963	1.609437912
6	0.689707825	1.791759469
7	0.856844048	1.945910149
8	1.153114169	2.079441542
9	1.258441321	2.197224577
10	1.517532754	2.302585093
11	1.554598557	2.397895273
12	1.725182385	2.48490665
13	1.779706003	2.564949357
14	1.863101286	2.63905733
15	2.002664493	2.708050201
16	2.030305104	2.772588722
17	2.06000184	2.833213344
18	2.108447887	2.890371758
19	2.114354815	2.944438979
20	2.206376666	2.995732274
21	2.250050184	3.044522438
22	2.269083482	3.17805383
23	2.291895875	3.218875825
24	2.320558515	3.401197382
25	2.359610234	3.465735903
26	2.423028715	3.555348061
27	2.779942594	3.583518938

Table 12: The numerical values of liner equations, between X and Y, for the determination of A and B, in relation to Weibull parameters K and C in Bam synoptic station.

i	$Y = \ln(-\ln(1 - P(V_i)))$	$X = \ln V_i$
1	-9.219498309	0
2	-3.213046365	0.693147181
3	-2.133142223	1.098612289
4	-1.244809084	1.386294361
5	-0.448137813	1.609437912
6	0.082464789	1.791759469
7	0.305455532	1.945910149
8	0.717936294	2.079441542
9	0.751384632	2.197224577
10	1.158911095	2.302585093
11	1.173440145	2.397895273
12	1.338252638	2.48490665
13	1.346583584	2.564949357
14	1.372192261	2.63905733
15	1.651004846	2.708050201
16	1.676787512	2.772588722
17	1.678743455	2.833213344
18	1.721910324	2.890371758
19	2.05837794	2.995732274
20	2.075281009	3.044522438
21	2.116647862	3.091042453
22	2.143167311	3.17805383
23	2.176350754	3.33220451
24	2.221325997	3.401197382
25	2.293816262	3.63758616
26	2.779942594	3.688879454

Table 13: The numerical values of liner equations, between X and Y, for the determination of A and B, in relation to Weibull parameters K and C in Kahnooj synoptic station.

i	$Y = \ln(-\ln(1 - P(V_i)))$	$X = \ln V_i$
1	-3.032401402	0
2	-1.897404888	0.693147181
3	-0.728003987	1.098612289
4	-0.393764793	1.386294361
5	0.09914463	1.609437912
6	0.294761391	1.791759469
7	0.500298871	1.945910149
8	0.694922304	2.079441542
9	0.87571284	2.197224577
10	1.147358186	2.302585093
11	1.209595653	2.397895273
12	1.3181629	2.48490665
13	1.416343272	2.564949357
14	1.470138096	2.63905733
15	1.658511849	2.708050201
16	1.700320581	2.772588722
17	1.729512885	2.833213344
18	1.805307006	2.890371758
19	1.824784955	2.944438979
20	1.92835259	2.995732274
21	2.02726573	3.044522438
22	2.052717669	3.091042453
23	2.101824476	3.135494216
24	2.128734679	3.17805383
25	2.207985715	3.218875825
26	2.26261678	3.258096538
27	2.281414612	3.295836866
28	2.332274569	3.33220451
29	2.433609699	3.401197382
30	2.779942594	3.555348061

Table 14: The numerical values of liner equations, between X and Y, for the determination of A and B, in relation to Weibull parameters K and C in Kerman synoptic station.

i	$Y = \ln(-\ln(1 - P(V_i)))$	$X = \ln V_i$
1	-8.58497644	0
2	-3.804686255	0.693147
3	-1.175877981	1.098612
4	0.030780418	1.386294
5	0.354899419	1.609438
6	0.646818834	1.791759
7	0.767329932	1.94591
8	1.071514184	2.079442
9	1.106708841	2.197225
10	1.332175853	2.302585
11	1.342501879	2.397895
12	1.514662012	2.484907
13	1.520543523	2.564949
14	1.637128445	2.639057
15	1.696871412	2.70805
16	1.786765371	2.772589
17	1.799412777	2.833213
18	1.889434916	2.890372
19	1.89298541	2.944439
20	2.065839794	2.995732
21	2.150024635	3.091042
22	2.171039321	3.178054
23	2.227669402	3.218876
24	2.270442219	3.332205
25	2.339575374	3.401197
26	2.779942594	3.555348

Table 15: The numerical values of liner equations, between X and Y, for the determination of A and B, in relation to Weibull parameters K and C in Shahrbabak synoptic station.

i	$Y = \ln(-\ln(1 - P(V_i)))$	$X = \ln V_i$
1	-5.556130329	0
2	-1.956314973	0.693147181
3	-1.090310948	1.098612289
4	-0.43955606	1.386294361
5	0.108012043	1.609437912
6	0.472392871	1.791759469
7	0.702722588	1.945910149
8	0.983778837	2.079441542
9	1.099695602	2.197224577
10	1.354390446	2.302585093
11	1.389525039	2.397895273
12	1.549601582	2.48490665
13	1.603108938	2.564949357
14	1.698225558	2.63905733
15	1.905757044	2.708050201
16	1.960656044	2.772588722
17	1.983899355	2.833213344
18	2.086851211	2.890371758
19	2.099839338	2.944438979
20	2.226078742	2.995732274
21	2.33815324	3.091042453
22	2.779942594	3.17805383

Table 16: The numerical values of liner equations, between X and Y, for the determination of A and B, in relation to Weibull parameters K and C in Sirjan synoptic station.

i	$Y = \ln(-\ln(1 - P(V_i)))$	$X = \ln V_i$
1	-5.556128307	0
2	-1.043269862	0.693147181
3	-0.116517582	1.098612289
4	0.448729692	1.386294361
5	0.763409231	1.609437912
6	0.991040667	1.791759469
7	1.065031178	1.945910149
8	1.23310129	2.079441542
9	1.266220837	2.197224577
10	1.430899702	2.302585093
11	1.444026131	2.397895273
12	1.545370378	2.48490665
13	1.557042547	2.564949357
14	1.586459635	2.63905733
15	1.696892485	2.708050201
16	1.731125675	2.772588722
17	1.748268203	2.833213344
18	1.766930273	2.890371758
19	1.771870763	2.944438979
20	1.947382189	2.995732274
21	1.97560522	3.044522438
22	2.030310076	3.091042453
23	2.779942594	3.218875825

Table 17: The numerical values of liner equations, between X and Y, for the determination of A and B, in relation to Weibull parameters K and C in Miandeh-Jiroft synoptic station.

i	$Y = \ln(-\ln(1 - P(V_i)))$	$X = \ln V_i$
1	-8.39106291	0
2	-2.695021344	0.693147181
3	-1.53550301	1.098612289
4	-0.637778064	1.386294361
5	0.053780247	1.609437912
6	0.391731043	1.791759469
7	0.599852896	1.945910149
8	0.849202123	2.079441542
9	0.91585802	2.197224577
10	1.191903617	2.302585093
11	1.210005324	2.397895273
12	1.419814234	2.48490665
13	1.438264252	2.564949357
14	1.513674665	2.63905733
15	1.689711412	2.708050201
16	1.73750146	2.772588722
17	1.76236962	2.833213344
18	1.827601577	2.890371758
19	1.946607333	2.995732274
20	1.97230216	3.044522438
21	2.027183131	3.091042453
22	2.033916048	3.135494216
23	2.102975637	3.17805383
24	2.779942594	3.218875825

Table 18: The numerical values of liner equations, between X and Y, for the determination of A and B, in relation to Weibull parameters K and C in Rafsanjan synoptic station.

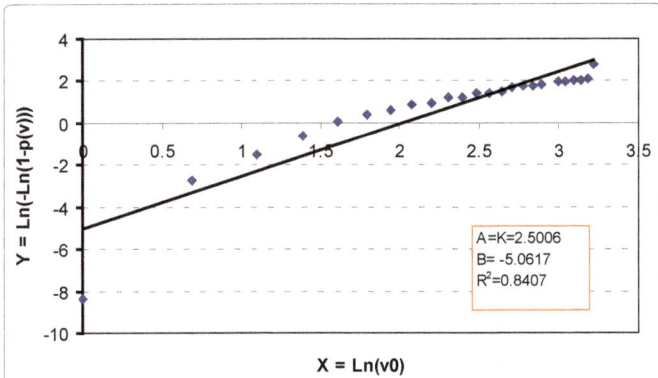

Figure 1: Least square line for Rafsanjan synoptic station, through which the parameters K and C are estimated.

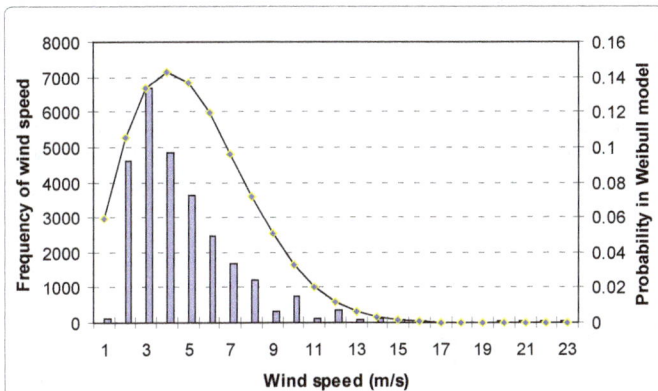

Figure 2: Wind speed frequency with fitted Weibull distribution for Anar synoptic station at 10 m height.

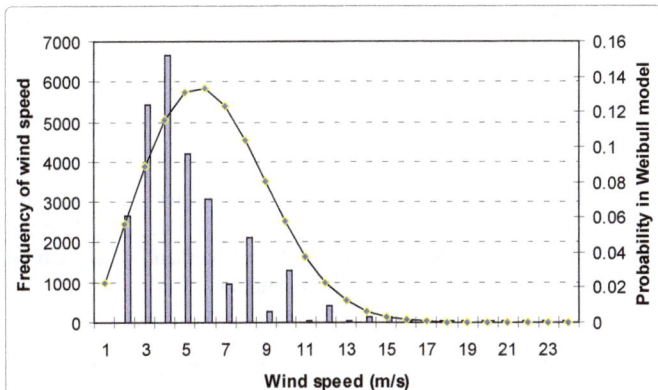

Figure 3: Wind speed frequency with fitted Weibull distribution for Baft synoptic station at 10 m height.

$$\sigma = C\sqrt{\left[\Gamma\left(\frac{K+2}{K}\right) - \Gamma^{-2}\left(\frac{K+1}{K}\right)\right]} \quad (11)$$

where Γ denotes the Gamma function.

The probability of wind speeds between v_1 and v_2 is given by [8]:

$$P\left(v_1\langle v\langle v_2\right) = \exp\left[-\left(\frac{v_1}{C}\right)^k\right] - \exp\left[-\left(\frac{v_2}{C}\right)^K\right] \quad (12)$$

Wind turbines are designed with a cut-in speed, or the wind speed

at which it begins to produce power, and a cut-out speed, or the wind speed at which the turbine will be shut down to prevent the drive train from being damaged. For most wind turbines, the range of cut-in wind speed is 3_4.5 m/s, and the cut-out speed can be as highly as 25 m/s [8].

Wind characteristics at the height of 10 m above ground level are shown in Table 19.

Wind power density

The evaluation of wind power density per unit area is of fundamental importance in assessing wind power projects. Wind power density,

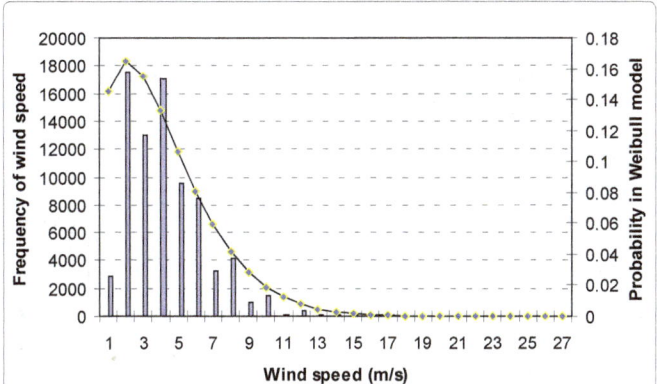

Figure 4: Wind speed frequency with fitted Weibull distribution for Bam synoptic station at 10 m height.

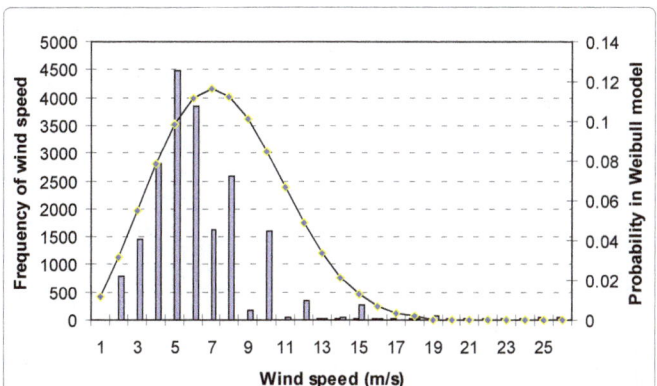

Figure 5: Wind speed frequency with fitted Weibull distribution for Kahnooj synoptic station at 10 m height.

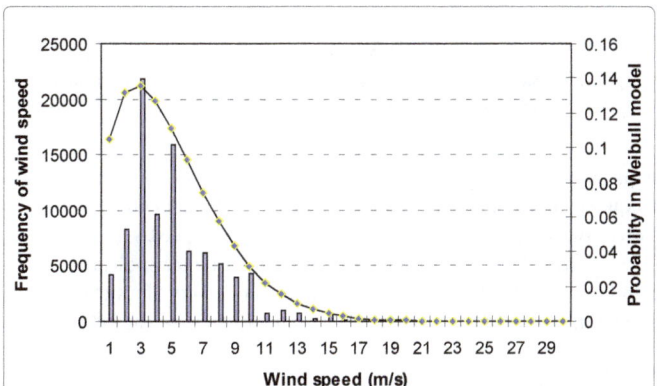

Figure 6: Wind speed frequency with fitted Weibull distribution for Kerman synoptic station at 10 m height.

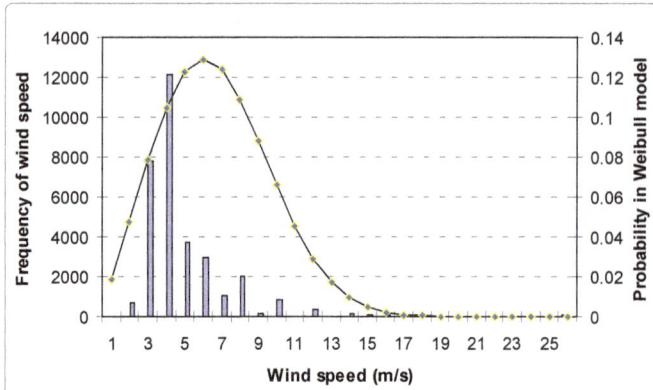

Figure 7: Wind speed frequency with fitted Weibull distribution for Shahrbabak synoptic station at 10 m height.

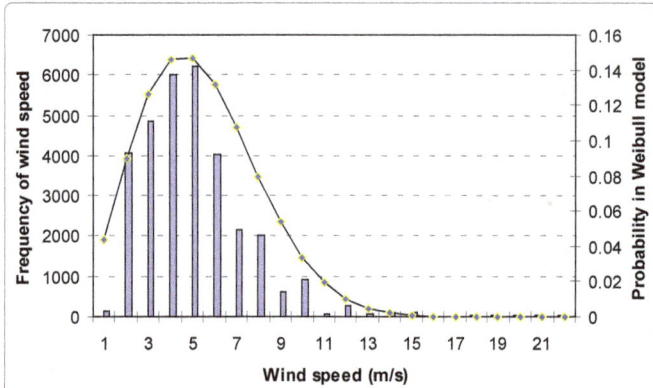

Figure 8: Wind speed frequency with fitted Weibull distribution for Sirjan synoptic station at 10 m height.

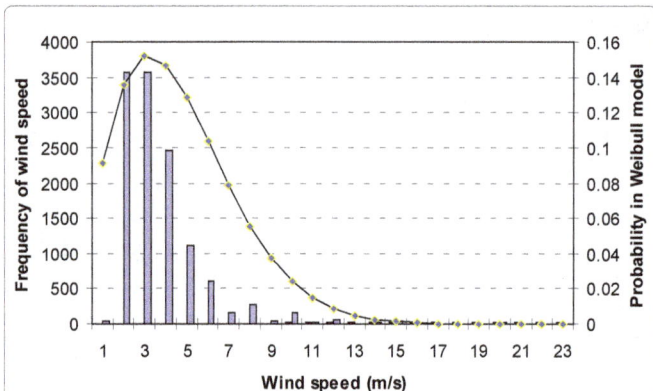

Figure 9: Wind speed frequency with fitted Weibull distribution for Miandeh-Jiroft synoptic station at 10 m height.

expressed in Watt per square meter (W/m²), takes into account the frequency distribution of the wind speed and the dependence of wind power on air density and the cube of the wind speed [9]. Therefore, wind power density is generally considered a better indicator of the wind resource than wind speed. Wind power density of a site based on a Weibull probability density function can be expressed as follows [10]:

$$\frac{P}{A} = \frac{1}{2}\rho C^3 \Gamma\left(\frac{K+3}{K}\right)$$ (13)

where ρ is the mean air density (usually taken as equal to 1.225

kg/m³ which depends on altitude, air pressure, and temperature) [11], and Γ denotes the Gamma function. Once wind power density of a site is given, the wind energy density for a desired duration (a month or a year) can be expressed as [12]:

$$\frac{E}{A} = \frac{1}{2}\rho C^3 \Gamma\left(\frac{K+3}{K}\right)T$$ (14)

where T is the time period (or duration), for example, T is 720 hr for monthly duration.

The problem of transforming Weibull parameters at the hub heights of the wind turbines can be easily solved with the compatible features of Weibull distribution. Weibull function facilitates the presentation of the wind speed distribution thereby making it possible for the researcher to transform the wind speed distribution at 10 m height to the distribution at any other height. This is done by applying the so called one seventh power law [13]:

$$\frac{C_2}{C_1} = \left(\frac{Z_2}{Z_1}\right)^{1/7}$$ (15)

where C_2 and C_1 are the Weibull scale parameters at heights Z_2 and Z_1, respectively. Even if the Weibull shape parameter, k, varies with height, the variation is small, and for the present analysis, the shape factor is assumed to be independent of the height.

Wind characteristics at the height of 50 m above ground level at studied stations are shown in Table 20.

Computing Wind Existence Hours at an Area

In the previous studies, in order to obtain total wind hour existence, wind hours in each speed class throughout a year were computed and then the cumulative quantities were achieved [14].

In the present research, a new equation is obtained which provides a simpler and easier method to estimate wind hour existence at an area. The equation is as follows (Mahdi Dehghan):

$$WE_{(h/y)} = \left(\frac{\sum f_i}{N}\right) \cdot T$$ (16)

where WE stands for Wind Existence, (h/y) is the unit of measuring the parameter, hour by year, f_i is the frequency of wind speed classes or the quantities presented in Table 1, column 4. N is the length of the statistical period under study in a year, and T is the time interval between the wind data records in hours.

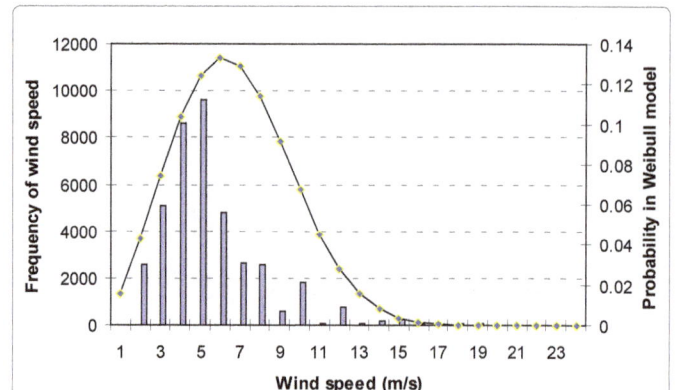

Figure 10: Wind speed frequency with fitted Weibull distribution for Rafsanjan synoptic station at 10 m height.

Station	Scale parameter C(m/s)	Shape parameter k	Mean wind power density P/A (w/m²)	Most probable wind speed V_{MP} (m/s)	Wind speed carrying maximum energy V_{MaxE} (m/s)	Mean wind speed V (m/s)
Kerman	5.4963	1.5271	196.09	2.7388	9.5090	6
Kahnooj	8.6525	2.4894	439.55	7.0393	10.9652	15.69
Bam	4.494	1.4619	116.49	2.0434	8.1047	4.81
Baft	7.1625	2.3472	259.48	5.6538	9.3132	11.74
Anar	5.8906	1.9472	171.24	4.0685	8.4676	7.62
Shahrbabak	7.553	2.3926	299.70	6.024	9.7364	12.78
Sirjan	6.0625	2.1545	168.43	4.5383	8.2227	8.79
Miandeh-jiroft	5.2126	1.7545	135.08	3.2223	8.0421	6.17
Rafsanjan	7.5694	2.5006	292.68	6.1712	9.5747	13.84

Table 19: Wind characteristics at the height of 10 m above ground level at studied stations.

Station	Scale parameter C(m/s)	Shape parameter k	Mean wind power density P/A (w/m²)	Most probable wind speed V_{MP} (m/s)	Wind speed carrying maximum energy V_{MaxE} (m/s)	Probability of wind speeds between 3 and 25 m/s	Mean wind speed V (m/s)
Kerman	6.92	1.5271	391.35	3.4467	11.97	0.76	7.55
Kahnooj	10.89	2.4894	876.33	8.8589	13.8	0.96	19.75
Bam	5.66	1.4619	232.72	2.5717	10.2	0.67	6.06
Baft	9.01	2.3472	516.51	7.1153	11.72	0.93	14.77
Anar	7.41	1.9472	340.86	5.1202	10.66	0.84	9.59
Shahrbabak	9.51	2.3926	598.24	7.5811	12.25	0.94	16.09
Sirjan	7.63	2.1545	335.77	5.7114	10.35	0.87	11.06
Miandeh-jiroft	6.56	1.7545	269.23	4.0553	10.12	0.78	7.67
Rafsanjan	9.53	2.5006	584.1	7.7665	12.05	0.95	17.42

Table 20: Wind characteristics at the height of 50 m above ground level at studied stations.

The total amount of wind hour existence at the synoptic stations under study was in Kerman 4392.69, Kahnooj 2632.81, Bam 4243.87, Baft 3181.85, Anar 3128.31, Shahrbabak 3852.84, Sirjan 3515.22, Miandeh-Jiroft 1589.74 and Rafsanjan 5950.8 hour by a year.

Conclusion

Wind power density is an essential factor in locating places suitable for the installation of wind turbines. At studied synoptic stations, wind power density in Kahnooj, Shahrbabak, Rafsenjan and Baft at heigh of 10 meters from ground level was 439.55, 299.7, 292.68 and 259.48 W/m³ respectively. Meanwhile In the classification of areas suitable for wind turbine installation based on available wind power at 10 meters from ground level, wind power density raging from 200 to 250 W/m³ is considered.

As mentioned above, most efficient wind turbines have been designed for a wind speed of 3 meters per second. Meanwhile lowest most probable monthly wind speeds for studied stations at an altitude of 50 meters (the height of the installation of most wind turbines) at all studied stations except Bam, more than 3 meters per second, can be safely observed in Kahnooj, Rafsanjan,Shahre Babak and Baft stations , respectively 8.86, 7.77 and 7.11 m.

According to equation (12), the probability of a wind speed between 3 and 25 m/s in Kahnooj, Rafsanjan, Shahrbabak, and Baft synoptic stations was respectively 96,95,94 and 93 percent of total wind hour existence that in these places has been respectively 2632.81, 5950.8, 3852.84, and 3181.85 hour by year at the height of 50 m. Therefore, the economical operations for wind turbines in mentioned stations are estimated to be about 2527, 5653, 3621 and 2959 hour by year respectively.

The mean wind speeds in all of studied stations is estimated to be more than 6 m/s. Each of these speeds can be the working speed level for wind turbines.

The difference between the most probable wind speed and the wind speed carrying maximum energy in Rafsanjan, Baft, Sirjan, Shahrbabak, Kahnooj, Anar and Miandeh-Jiroft, annually, are less than 5 which show the trivial difference between the maximum probability of wind speed and the wind speed which provides the highest amount of energy in these places.

Finally, the wind power density at 50 meters above ground level, and other mentioned features, Kahnooj, Shahre Babak and Rafsanjan synoptic stations are obtained the potential for the installation of wind turbines and extraction of wind power.

References

1. Saghafi M (2003) Renewable Energies. (2ndedn), University of Tehran Press, Tehran, Iran.

2. Patel MR (1999) Wind and solar power systems. CRC Press, Florida, USA.

3. Nasiri J (1997) Evaluation of wind power potential in Iran. (1stedn.), The ministry of energy Press, Tehran, Iran.

4. Elamouri M, Ben Amar F (2008) Wind energy potential in Tunisia. Renewable Energy 33: 758-768.

5. Vogiatzis N, Koto K, Spanomitsios S, Stoukides M (2004) Analysis of wind potential and characteristics in North Aegean- Greece. Renewable Energy 29: 1193-1208.

6. Jamil M, Parsa S, Majidi M (1995) Wind power statistics and an evaluation of wind energy density. Renewable Energy 6: 623-628.

7. Chang TJ, Wu YT, Hsu HY, Chu CR, Liao CM (2003) Assessment of wind characteristics and wind turbine characteristics in Taiwan. Renewable Energy 28: 851-871.

8. Zhou W, Yang H, Fang Z (2006) Wind power potential and characteristic analysis of the Pearl River Delta region, China. Renewable Energy 31: 739-753.

9. Al-Nassar W, Alhajraf S, Al-Enizi A, Al-Awadhi L (2005) Potential wind power generation in the State of Kuwait. Renewable Energy 30: 2149-2161.

10. Peterson EW, Hennessey JP (1977) On the use of power laws for estimates of wind power potential. J Appl Meteor 17: 390-394.

11. Karsli VM, Gecit C (2003) An investigation on wind power potential of Nurdagi-Gaziantep, Turkey. Renewable Energy 28: 823-830.

12. Jaramillo OA, Saldana R, Miranda U (2004) Wind power potential of Baja California Sur- Mexico. Renewable Energy 29: 2087-2100.

13. Amr M, Petersen H, Habali SM (1990) Assessment of wind farm economics in relation to site wind resources applied to sites in Jordan. Solar Energy 45: 167-175.

14. Dehghan Tezerjani M, Omidvar K (2010) Evaluation of wind power potential for the generation of energy at Marvast Synoptic Station-Yazd, Iran. ISESCO Science and Technology Vision 6: 30-34.

4

Prediction of Long-term Pattern and its Extreme Event Frequency of Rainfall in Dire Dawa Region, Eastern Ethiopia

Rediat Takele* and Solomon Gebretsidik

Jigjiga University, Jigjiga, Somali regional state, Ethiopia

Abstract

Rainfall is the most critical and key variable both in atmospheric and hydrological cycle. Its patterns usually have spatial and temporal variability. Its variability is assumed to be the main cause for the frequently occurring climate extreme events such as drought and flood. In this study, spectrum analysis, cross-spectral analysis as well as seasonal auto-regressive integrated moving average (SARIMA) model were used in intending to predict the pattern and its extreme event frequency of rainfall in Dire Dawa Region based on data obtained from Dire Dawa and adjacent stations: Dengego and Haramaya. The result indicated that the amount of rainfall at Dengego and Haramaya are more or less the same on average in all seasons and much higher than that of Dire Dawa during last 30 year study period. The variability of annual rainfall in Dire Dawa during the study period is a bit larger than neighboring station's rainfall (Dengego and Haramaya), indicating that climate instability is high in Dire Dawa than other nearby stations. The result also indicates that relatively there is a tendency of increasing pattern in average annual rainfall of Dire Dawa in forecasted period. In the region, the rainfall extreme event like flood predicts or inferred to be recurring at about 4.17 years. Moreover, the rainfall periodicities of Dengego and Dire Dawa are found to be more likely associated, which implies that rainfall extreme event frequency in two districts is statistical significantly associated. It is proposed that assessing the factors that cause the fluctuation in the pattern and frequency of rainfall distribution in the region is needed to be study.

Keywords: Extreme event; Climate variability; Periodicities; SARIMA; Spectral analysis

Introduction

Located within the tropics, Ethiopia has great geographical diversity with high and rugged mountainous, flat-topped plateaus, deep gorges, etc. Its altitudinal range lies between 120 m below sea level and 4600 m above sea level [1]. The differences in altitude and relief create a large variation in climate in various regions of the country [2]. Several regions receive rainfall throughout the year, but in some regions rainfall is seasonal and low making irrigation necessary [3]. Particularly, Semi-arid regions receive very small, irregular, and unreliable rainfall. Rainfall is the most critical and key variable both in atmospheric and hydrological cycle. Its patterns usually have spatial and temporal variability. These variabilities affect agricultural production, water supply, transportation, environment and urban planning and the existence of its people. Its variability assumed to be the main cause for the frequently occurring climate extreme events such as drought and flood.

Ethiopia is one of the countries whose economy is highly dependent on rain-fed agriculture and also facing recurring cycles of flood and drought. Current climate variability is imposing a significant challenge to Ethiopia in general and Dire Dawa in particular. Metrologically, Dire Dawa Administration is characterized by an arid and semi-arid climate, thus, receives low and erratic rainfall [4]. Despite the fact that, sloppy topography of the Dire Dawa administration are surrounded by the neighboring mountainous areas (Haramaya and Dengego). These are the main catchments for contributing to the disaster flood and drought event in Dire Dawa. Flood disasters are occurring more frequently in the urban area, and are having an ever more dramatic impact in terms of the costs on lives, livelihoods and environmental resources. The May 1984 catastrophic flood claimed 42 people, and property worth 10 million birr was lost. On August 2006 claimed 650 people, displaced 35,000 people, and caused direct damage of huge infrastructure estimated at 100 million Ethiopian Birr and indirect

damage of similar magnitude. On April 2010- though there was the property of worth 28 million birr had been lost. On the other hand, the area has often experienced and affected by frequent and prolonged drought claiming the lives of millions of people. The impacts of the 2004 and 2005 droughts incident posed food shortage up to 85% of the rural population [5].

Methods of prediction of rainfall extreme events have often been based on studies of physical effects of rainfall or on statistical studies of rainfall time series [2,3,6]. The study in various region of African shows that rainfall has Periodic tendencies [7,8]. A study made by Mersha and Seifu [9,10] on rainfall cyclicity over selected stations in Ethiopia shows that there appears to be periodic tendency in the annual rainfall series, particularly, Gode, Dire Dawa, Jigjiga, Negelle, and Debre Zeit station. Haile [11] found that drought occur at every 6-8 years in the semi-arid regions of Ethiopia. Tsegay [12] on his study the occurrences of drought and frequencies of rainfall suggest that drought occur every 3-5 years in Eastern Ethiopia. On the other side, the study by Yilma and Alamerew et al. [2,3] reach the conclusion of rainfall extreme event recurs in Addis Ababa region between 10 to 11 years.

Awareness about the characteristics of the rainfall over an area such as the source, quantity, variability, distribution and the frequency of rainfall is essential for efficient control and management of water resources, the implication in utilization and associated problems.

---

*Corresponding author: Rediat Takele, Jigjiga University, Jigjiga, Somali regional state, Ethiopia, E-mail: redtakele@gmail.com

The impacts of climate variability and climate change are evident on almost all socio-economic sectors. The output of this study can play an important role in the design of hydrological structures in the area of study and help the policy makers in improve their decisions by taking into consideration the available and future water resources.

Materials and Methods

Data and variable of the study

A time series of monthly 30 years rainfall data in mm for the period January, 1984 to January, 2014 collected by the National Meteorological Agency of Ethiopia were used in the study. The data were collected from the synoptic stations of Dire Dawa and adjacent station: Dengego and Haramaya for further investigation.

Methodology

The development of climatology as a science has given rise to growing statistical applications on climatic information. Conducting investigations using standard statistical methodologies is an essential step in the development of climatology [13]. For instance, time series analysis is used in order to evaluate the temporal and spatial behavior of rainfall [14]. In this study a univariate Box-Jenkins Methods, in particular, Seasonal Autoregressive Integrated Moving Average (SARIMA) methods and spectral analysis and cross-spectral analysis are employed [15].

Seasonal autoregressive integrated moving average (SARIMA) modeling is one of the most widely implemented methods for analyzing univariate time series data that exhibits a seasonal variation such as temperature and rainfall, which have strong components corresponding to seasons. Hence, the natural variability of many physical, biological, and economic processes tends to match with seasonal fluctuations [16]. The seasonal autoregressive integrated moving average model, denoted by SARIMA $(p,d,q)*(P,D,Q)_s$ as in Shumay and Stoffer [17] and is given by:

$$\Phi_P(B^s)\,\phi(B)\nabla_S^D x_t = \alpha + \Theta_Q(B^s)\theta(B)\,w_t \qquad (1)$$

where w_t is the usual Gaussian white noise process. The ordinary autoregressive and moving average components are represented by polynomials $\phi(B)$ and θ (B) of orders p and q, respectively and the seasonal autoregressive and moving average components by Φ_P (B^s) and Θ_Q(B^s) of orders P and Q. The ordinary and seasonal difference components can be written as: $\nabla^d = (1-B)^d$ and $\nabla_S^D = (1-B^S)^D$

Test for stationary assumption

Before developing a Box-Jenkins modeling process, it is important to check whether the data under study meets basic assumptions such as stationary. A time series is said to be stationary if there is no systematic change in mean (no trend), if there is no systematic change in variance and if periodic variations have been removed. Many testing procedures for testing stationary are employed in the literature.

Time plot: This procedure may reveals seasonality, trends either in the mean level or the variance of the series, long- term cycles, and so on. If any such patterns are present, then these are signs of non-stationary.

The correlagram test: test based on the plot its sample autocorrelation function and sample partial autocorrelation function. Both function used to reveal important information regarding given series stationary. Wei [18] states that if the sample ACF decays very slowly, which indicates that series is not stationary.

The augmented dickey-fuller (ADF) test: This test first poses the null hypothesis that the given time series has a unit root, which means that the time series is non-stationary. That is:

H$_0$: ϕ=1 (has a unit root) Vs H$_1$: ϕ< 1 (has root outside the unit circle)

To use the Augmented Dickey-Fuller test the following test equation will be used:

$$\nabla x_t = \theta + \beta t + \varphi x_{t-1} + \theta_1 \nabla x_{t-1} + ... + \theta_p \nabla x_{t-p} + w_t \qquad (2)$$

Where: ∇x_t is the first differenced value of series (x_t), w_t is the error term, ∇x_{t-j} is the jth lagged first differenced of values of x_t, $\theta_o, \beta, \Phi = \phi - 1, \theta_1, \theta_2,...,\theta_p$ are parameters to be estimated

Variance comparisons: The characteristics of the sample variances associated with different orders of differencing can provide a useful means of deciding the appropriate order of differencing [19] cited also in Alamarew et al. [3]. The sample variances will decrease until a stationary sequence has been found.

Building seasonal -ARIMA models

To identify a perfect ARIMA model for a particular time series data, Box and Jenkins [16] proposed a methodology that consists of four phases: i) Model identification; ii) parameters Estimation; iii) Diagnostic checking for the identified model and iv) Forecasting.

Model identification: The purpose of the identification stage is to determine the differencing required achieving stationarity and also the order of both the seasonal and the non- seasonal AR and MA operators for the residual series. The autocorrelations function (ACF) and the partial autocorrelation functions (PACF) are the two most useful tools in any attempt at time series model identification. To identify the order of model parameters, we examine the ACF and PACF based on the theoretical pattern using Tables 1 and 2 as summarized in Shumway and Stoffer [17].

Parameter estimation: After choosing the most appropriate order of the model, the next step is to estimate the model parameters ($\phi_1,...,\phi_p, \Phi_1,...\Phi_P, \theta_1,...,\theta_q, \Theta_1,...,\Theta_Q, \delta_k^2$) by using several estimation procedures like maximum likelihood methods and least square methods.

In time series analysis, there may be several adequate models that can be used to represent a given data set, and hence, information criteria are used for model comparison. The AIC and SBC is a mathematical selection criterion of model building by select the model that gives the minimum of the AIC and SBC defined by Shumway and Stoffer [17]:

	AR(P)	MA(Q)	ARMA(P,Q)
ACF	Tails off	Cuts off after lags Q	Tails off
PACF	Cuts off after lags P	Tails off	Tails off

Source: Shumway and Stoffer, [17].

Table 1: Behavior of the ACF and PACF for ARMA Models.

	AR(P)s	MA(Q)s	ARMA(P,Q)$_s$
ACF	Tails off at lags Ks, K=1, 2 ,…	Cuts off after lags Qs	Tails off at lags Ks
PACF	Cuts off after lags Ps	Tails off at lags ks	Tails off at lag Ps

Source: Shumway and Stoffer, [17].

Table 2: Behavior of the ACF and PACF for Pure SARMA Models.

$$AIC = \log \hat{\delta}_k + \frac{n+2k}{n} \text{ and } SBC = \log \hat{\delta}_k^2 + \frac{k \log n}{n} \qquad (3)$$

Where: $\hat{\delta}_k^2 = \frac{SSE_k}{n}$ denotes the maximum likelihood estimator for the error variance and kis the order of seasonal and non-seasonal autoregressive and moving average parameters to be estimated in the model, that is, according to Wei [18], k=p + q + P+ Q + 1 and n is the number of observations in the series.

Diagnostic checking: After fitting a provisional time series model, we can assess its adequacy in various ways. Most diagnostic tests deal with the residual assumptions in order to determine whether the residuals from fitted model are independent, have a constant variance, and are normally distributed. Several diagnostic statistics and plots of the residuals can be used to examine the goodness of fit of the tentative model to the time series data.

The first approaches that can be used to evaluate the adequacy of a model are the plot of the errors over time, which can be written Shumway and Stoffer [17]:

$$w_t = \left(x_t - x_t^{t-1} \right) / \sqrt{p_t^{t-1}} \qquad (4)$$

Where $x_t - x_t^{t-1}$ is the one-step-ahead prediction of (x_t) based on the fitted model and $\sqrt{p_t^{t-1}}$ is the estimated one-step-ahead error variance. If visual inspections of the errors reveal that they are randomly distributed over time, then we have a good model.

The autocorrelations function (ACF) of the series can also be used to examine whether the residual of the fitted model is white noise or not. Under the null hypothesis that residual follows a white noise process, roughly 95% of the autocorrelation coefficient should fall within the range $\pm 1.96/\sqrt{T}$ [20].

Ljung-Box Q (LBQ) test: It can be employed to check independence of residual instead of visual inspection of the sample autocorrelations. This test uses the null hypothesis that

Ho: $\rho_1 = \rho_2 = ... = \rho_k = 0$ and thetest statistic is calculated as Ljung and Box [21].

$$Q(r) = n'(n'+2) \sum_{j=1}^{k} \frac{\mu_j^2}{n-j} \qquad (5)$$

Where: $n' = (n-d)$, n is the number of observations in the original time series, μ_j^2 is the sample autocorrelation of the residuals at lag j and d is the degree of non- seasonal differencing used to transform the original time series values into stationary time series. If a model is correctly specified, residuals should be uncorrelated and Q(r) should be small (p- value should be large).

Runs test: The runs test can be used to decide if a data set is from a random process. The first step in the runs test is to compute the sequential differences $(Y_i - Y_{i-1})$. If $Y_i > Y_{i-1}$a 1 (one) is assigned for an observation and a 0 (zero) otherwise. let T be the number of observations, T_a be the number above the mean, T_b be the number below the mean and R be the observed number of runs. Then the test statistic for this test is given as Cromwell et al. [22]:

$$Z_{N=} \frac{R-E(R)}{\sqrt{V(R)}} \approx N(0,1) \text{ where: } \begin{array}{l} E(R) = T + 2T_a T_b / T \\ V(R) = \dfrac{2T_a T_b (2T_a T_b - T)}{2T(T-1)} \end{array} \qquad (6)$$

The test of series randomness is rejected if the calculated Z_N value exceeds the selected critical value obtained from the standard normal distribution table.

Turning point test: A turning point means when the series changes from increasing to decreasing or vice versa. That is, $X_{t-1} < X_t > X_{t+1}$ or $X_{t-1} > X_t < X_{t+1}$. Let T=the number of turning points in an n- period series. The hypothesis of residual whitenoise should be rejected if the absolute value of N_T(eqn.(7)) $> N_{T(1-\alpha/2)}$, where $N_{T(1-\alpha/2)}$ is the (1- α /2) quartile of standard normal distribution Cromwell et al. [22]:

$$N_T = \frac{|T - \mu_T|}{\sqrt{Var(T)}} \approx N(0,1) \text{ Where: } \begin{array}{l} \mu_T = (2/3)(n-2) \\ Var(T) = (16n - 29)/90 \end{array} \qquad (7)$$

Test for normality of the residuals: The Shapiro-Wilk Test, histogram and Q-Q plot of a data set give information related to normality. The null hypothesis for any test of normality is that the data are normally distributed. The Shapiro-Wilk statistic "W", which used as a base for the test of normality of residual in Shapiro-Wilks test, is given as [23]:

$$W = \frac{(\sum_{i=1}^{n} a_{i,n} x_i)^2}{\sum_{i=1}^{n} (x_i - \bar{x})^2} \qquad (8)$$

Procedure for checking of homosecdasticity

Goldfeld-Quandt test: If there is a change in variance (heteroscedasticity) of residuals, a transformation is necessary for the data. For this test:

If test statistic: $F_{cal} = S_2^2 / S_1^2$ is smaller than the critical value $(F_\alpha(n_1 - 2, n_2 - 2))$, the residuals are assumed to be homoscedastic.

Breusch-Pagan test: This involves applying ordinary least-square (OLS) to:

$\dfrac{\hat{\varepsilon}_i^2}{\hat{\sigma}^2} = \gamma_0 + \gamma_1 X_{1i} + \gamma_2 X_{2i} + ... + \gamma_K X_{Ki} + u_i$ and calculating the regression sum of squares (RSS). The test statistic is: $\chi_{cal}^2 = \dfrac{RSS}{2}$. Reject the null hypothesis of homoscedasticity: if: $\chi_{cal}^2 > \chi_\alpha^2(K)$.

If the constant variance and normality assumptions are not true, they are often reasonably well satisfied when the observations are transformed by a Box-Cox transformation [17]:

$$x_t^{(\lambda)} = \begin{cases} \dfrac{x_t^\lambda - 1}{\lambda}, \lambda \neq 0 \\ \ln x_t, \lambda = 0 \end{cases} \text{Choose the value of } \lambda \text{ that maximizes:}$$

$$l(\lambda) = -\frac{n}{2} \text{h} \left[\frac{1}{n} \sum_{j=1}^{n} (x_j^{(\lambda)} - \bar{x}^{(\lambda)})^2 \right] + (\lambda - 1) \sum_{j=1}^{n} \text{h} \, x_j \qquad (9)$$

If the selected model is inadequate, the three-step model building process is typically repeated several times until a satisfactory model is finally obtained. The final selected model can then be used for prediction purposes [18].

Forecasting: The last step in time series modeling is forecasting.

There are two kinds of forecasts: sample period forecasts and post-sample period forecasts. The former are used to develop confidence in the model and the latter to generate genuine desired forecasts. In forecasting, the goal is to predict future values of a time series, x_{t+m}, m =1, 2,... based on the data collected to the present, $x=\{x_t,\ x_{t-1},...,\ x_1\}$. Then explicit forms of the model for the observation x_{t+m} generated by the ARIMA process may be expressed as follows (Box and Jenkins) [16]:

$$x_{t+m} = \phi_1 x_{t+m-1} + ... + \phi_{p+d} x_{t+m-p-d} - \theta_1 w_{t+m-1} - ... - \theta_q w_{t+m-q} \quad (10)$$

Forecasting accuracy measures

Once forecasts are made they can be evaluated if the actual values of the series to be forecasted are observed. To assess the precision of the forecasts, prediction interval can be written as [17]:

$$x^t_{t+m} + C_{\frac{\alpha}{2}} \sqrt{P^t_{t+m}}\ ,\ \text{Where}\ C_{\alpha/2}\ \text{is chosen to get the desired}$$

degree of confidence. Where: Mean square prediction error,

$$P^t_{t+m} = E(x_{t+m} - \tilde{x}_{t+m})^2 = \delta_w^2 \sum_{j=0}^{m-1} \psi_j^2$$

There are also some measurements of the accuracy of forecasts, which evaluate the accuracy by comparing with observed value are Mean Square Error (MSE); Mean Absolute Error (MAE); Mean Absolute Percentage Error (MAPE); Theil's inequality coefficient (U-Statistics). The scaling of Uis such that it will always lie between 0 and 1. If U=0, there is a perfect fit; if U=1 the predictive performance is as bad as it possibly could be. Moreover, in evaluating the performance of the forecasting models, the lower the RMSE, MAE, MAPE, the better the forecasting accuracy [24].

Spectral analysis: Quite often, hydrologic phenomenon depicts cyclic and stochastic processes. The Spectral analysis based on periodogram method has dominated hydrology for years because of underlying periodicities in hydrologic processes [25]. The main objective of spectral analysis is to estimate and study the spectrum. It is therefore concerned with estimating the unknown spectrum of the process from the data and quantifying the relative importance of different frequency bands to the variance of the process [26]. The spectrum of a time series is the distribution of variance of the series as a function of frequency. The area under the curve in spectrum is proportional to the variance. Hence, we can be estimating the power spectrum by the raw periodogram method as [17]:

$$\bar{f}(\omega) = n[R(\omega)]^2 = n|d(\omega)|^2 \quad (11)$$

$$\text{Where:}\ |d(\omega)|^2 = \frac{1}{n}\left[\left(\sum_{t=1}^n (x_t - \bar{x})\cos(2\pi\omega t)\right)^2 + \left(\sum_{t=1}^n (x_t - \bar{x})\sin(2\pi\omega t)\right)^2\right]$$

and the frequencies (ω_j) are related to the size (n) by: $\omega_j = j/n$, $1<j< n/2$

By considering the fact that the periodogram estimates (eqn.11) are independent and identically approximated is χ^2-distributed with v-degree of freedom, the approximate $100(1-^\alpha)\%$ confidence interval for the spectrum:

$$\frac{2I(\omega_j)}{x_v^2(1-\alpha/2)} \le f(\omega) \le \frac{2I(\omega_j)}{x_v^2(\alpha/2)} \quad (12)$$

The raw periodogram is a wildly fluctuating estimate of the

spectrum with high variance. For a stable estimate, the periodogram must be smoothed. Bloomfield [27] recommends the Daniell window as a smoothing filter for generating an estimated spectrum from the periodogram. The smoothed periodogram is defined as [17]:

$$\hat{f}(\omega) = \sum_{k=-m}^m h_k I(\omega_j + k/n)\ ;\ \text{Where the weights}\ h_k > 0\ \text{satisfy,}$$

$$\sum_{k=-m}^m h_k = 1 \quad (13)$$

The result in eqn. (13) can be rearranged to obtain an approximate $100(1-\alpha)\%$ confidence interval for the best estimator of spectrum, $f(\omega)$ of the form

$$\frac{2L_h \hat{f}(\omega)}{x_{2Lh}^2(1-\alpha/2)} \le f(\omega) \le \frac{2L_h \hat{f}(\omega)}{x_{2Lh}^2(\alpha/2)} \quad (14)$$

Cross- spectral Analysis: Cross-spectrum analysis examines the relationship between a pair of series. The main objective of cross-spectrum analysis is to determine the existence of correlation between two variables and to simultaneously separate each of two time series in to its harmonic component. The principal tool that measures the strength of relation between two series in cross-spectral analysis is squared coherence function defined as [17]:

$$\rho_{xy}^2(\omega) = \frac{|f_{xy}(\omega)|^2}{f_{xx}(\omega)f_{yy}(\omega)},\ -\frac{1}{2} \le \omega \le \frac{1}{2} \quad (15)$$

Then spectral Matrix of a Bivariate Process is given by:

$$f(\omega) = \begin{pmatrix} f_{xx}(\omega) & f_{xy}(\omega) \\ f_{yx}(\omega) & f_{yy}(\omega) \end{pmatrix} \quad (16)$$

An estimate of the above spectral matrix is similar to the eqn. (13) defined in power spectra. Then an estimate of the squared coherence between two series, x_t and y_t is given by:

$$\hat{\rho}_{xy}^2 = \frac{|\hat{f}_{xy}(\omega)|^2}{\hat{f}_{xx}(\omega)\hat{f}_{yy}(\omega)} \quad (17)$$

At a particular significance level (α), the approximate critical value that must be exceeded for the original squared coherence to be able to reject $\rho_{xy}^2 = 0$ at a specified frequency is [17]:

$$C_\alpha = \frac{F_{2,2L-2}(\alpha)}{L-1+F_{2,2L-2}(\alpha)} \quad (18)$$

Result and Discussions

The statistical software package used for most of the analysis is by the help of R-15.10.1 and for some test and graphs; however, SAS-9.2 was used.

Descriptive analysis

Here it can be described rainfall of Dire Dawa, Haramaya and Dengego station and mainly focused on analyzing monthly as well as annually rainfall data of Dire Dawa station recorded by the National

Meteorological Agency of Ethiopia (NMAE). From descriptive statistics result, it can be observed that the mean annual rainfall of Dire Dawa, Dengego and Haramaya are 611 mm, 774 mm and 772 mm, respectively. In Dire Dawa, the minimum annual rainfall of 228.3 mm was recorded during the year 1984 and the maximum of 719.7 mm in the year 1997. The amount of rainfall at Dengego and Haramaya are more or less the same on average in all seasons, and is much higher than that of Dire Dawa over the study period. NMA [28] documented that a series with coefficient of variation(CV) less than 0.20 can be considered as less variable, between 0.20 and 0.30 is moderately variable, and greater than 0.30 is highly variable. The monthly rainfall data in all stations show that there is high variability since their CV is greater than 0.30. However, relatively the monthly variability of rainfall of Dire Dawa and Dengego are similar. When we see the overall variability, the high variability occurs during months of dry seasons while the low variability is observed during the rainy season in all stations. The variability of annual rainfall in Dire Dawa during the last 30 year period is a bit larger than neighboring stations. This may indicate that climate instability is high in Dire Dawa than other stations.

Testing stationary assumption

Visual Inspection: Regardless of which technique is used, the first step in any time series analysis is to plot the observed values against time. The pattern of the time series plot in Figure 1 does not show any systematic upward and downward change about the mean. This indicates that the series is non-seasonally stationary. Moreover, the sample autocorrelation function (SACF) plot in Figure 2, the presence of seasonality behavior and seasonally non-stationarity of the mean monthly rainfall series is clear. Because it shows strong seasonal wave pattern at the multiple of seasonal intervals(s=12) and declining slowly while non-seasonal lags are relatively decaying quite rapidly. This can be interpreted as a 6-month seasonal pattern that cycles between summer when there is little to no rainfall, and winter when rain is at its peak.

Therefore, from the actual series plot and sample autocorrelation plot (Figures 1 and 2) we observed that our series has seasonal variation, and seasonal differencing is needed to make it stationary.

Unit root test: This test first impose that the mean monthly rainfall

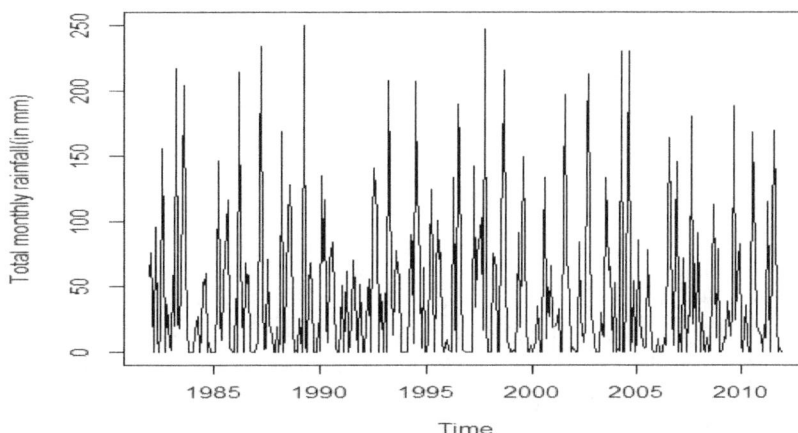

Figure 1: Time plot for total monthly rainfall in mm of Dire Dawa.

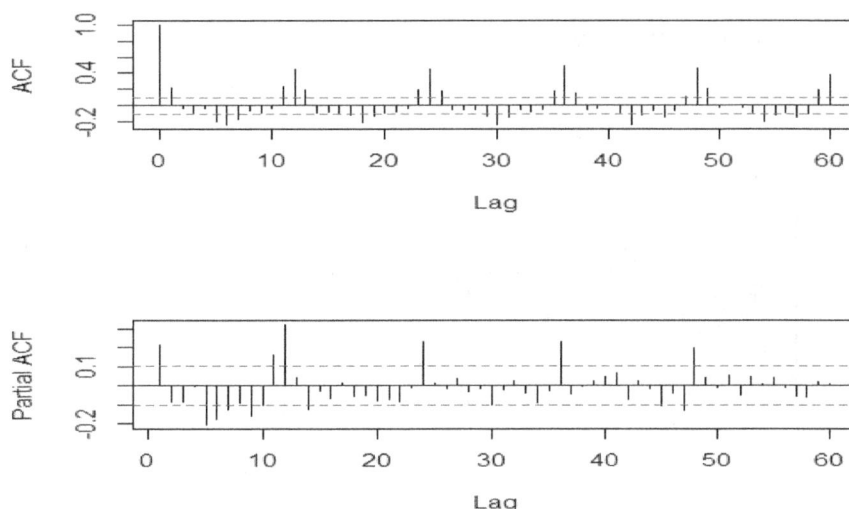

Figure 2: Sample ACF and PACF plots for the original rainfall series of Dire Dawa.

series has a unit root, versus the series is stationary. We test the null hypothesis using the Augmented Dickey-Fuller (ADF) testto know whether our series are stationary or not at level or after differencing. From Figure 1, we observe that the rainfall series of Dire Dawa doesn't have a trend and potentially slow- turn around zero. Therefore, we can use the test equation given in equation (2). For our series, maximum lag lengths (P) of the ADF test to remove serial correlation from the residuals of the regressions were selected as 12. The results of ADF test for the mean monthly rainfall series of Dire Dawa at level (without differencing), after first regular and seasonal differencing shows that, the ADF test statistic for the original and first regular differencing monthly rainfall series are greater than the critical values at 1%, 5% and 10% significance levels. This implies that, the null hypothesis, which has a unit root, for the series should not be rejected at all significance levels. This figure further confirms that original series as well as series obtained after first regular differencing are not stationary. However, as also suggested by visual analysis, after first seasonal differencing, the computed ADF test statistic are smaller than the critical values at 1%, 5% and 10% significant levels. This leads to the rejection of the test that there is a unit-root problem.

Variance Comparison: Increase in the differencing order tends to increase the variance indicating over differencing [29]. In this study the following results were obtained:

$Var(\nabla x_t)$ =5127.7, $Var(x_t)$=3252.7, $Var(\nabla_{12}x_t)$ =2704.76, and $Var(\nabla_{12}^2 x_t)$ =7981.4 values. From these it is clear that: $Var(\nabla x_t)$

> $Var(x_t)$ *and* Var $(\nabla_{12}^2 x_t)$ >$Var(x_t)$ >$Var(\nabla_{12}x_t)$. These results again suggest that non-seasonal first differencing (∇x_t) has been over-differenced and hence the original series is non-seasonally stationary. The first seasonal differencing would rather be important (Figure 3).

Consequently, all tests for stationarity seem to agree and suggest that the first-seasonal differencing of the series make stationarity around a constant mean, which is approximately zero and calculated its standard deviation of 52.05 mm (Figure 4).

Moreover, to determine whether stationarity has been achieved, either by trend removal or by differencing, one may examine the autocorrelation function (ACF) and partial autocorrelation function (PACF) of series [30]. The sequence corresponding to a stationary process should converge quite rapidly to zero as the value of the lag increases. From this point of view, Sample autocorrelation and partial autocorrelation plot shown in Figure 5 are in support of monthly rainfall series stationary after having first seasonal difference.

To test whether residual from fitted model come from normally distributed series, we use histogram, QQ-plot of the residual and Shapro-wilks test. The histogram and QQ-plot of the residual shown in Figure 6a, show skewed rather than normally distribution. Shipiro-Wilks test also confirms this fact since it results in P-value of 7.03e-09, which is smaller than 0.05. This implies that the residuals from the fitted models do not come from a normal distribution. This therefore, suggests some kind of transformation for our monthly rainfall series

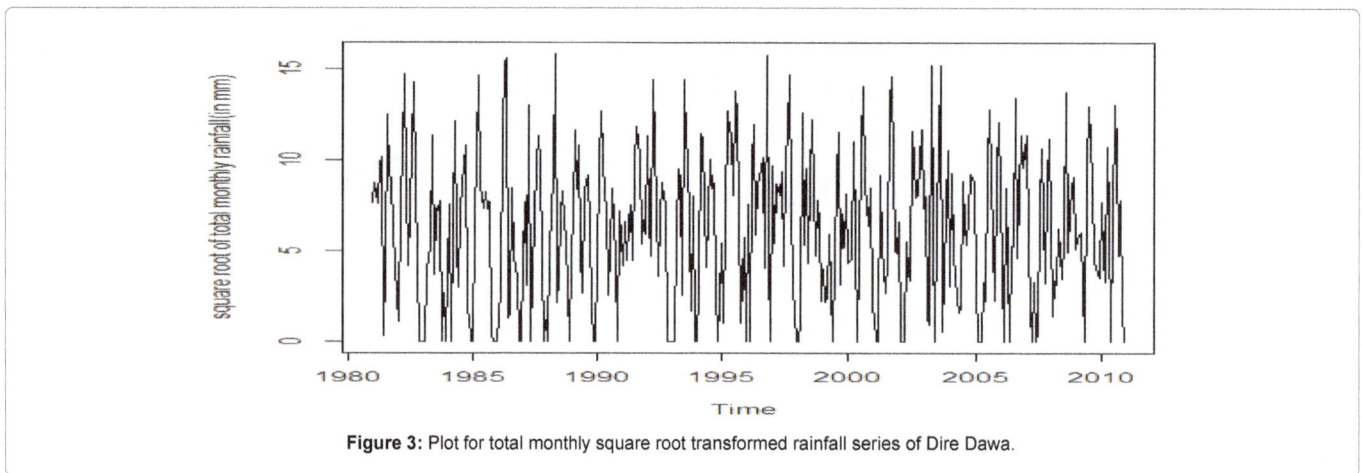

Figure 3: Plot for total monthly square root transformed rainfall series of Dire Dawa.

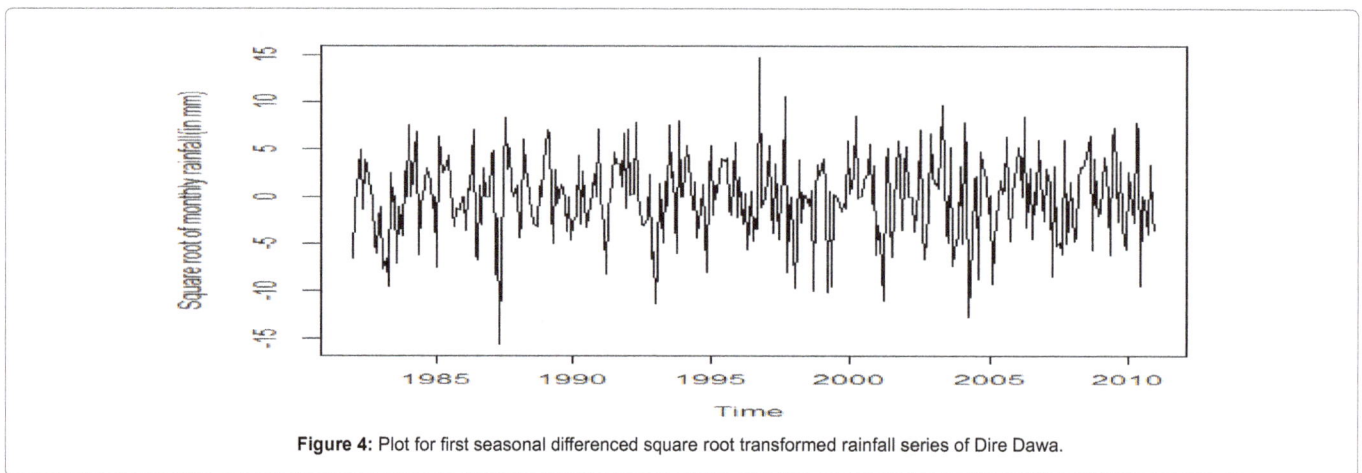

Figure 4: Plot for first seasonal differenced square root transformed rainfall series of Dire Dawa.

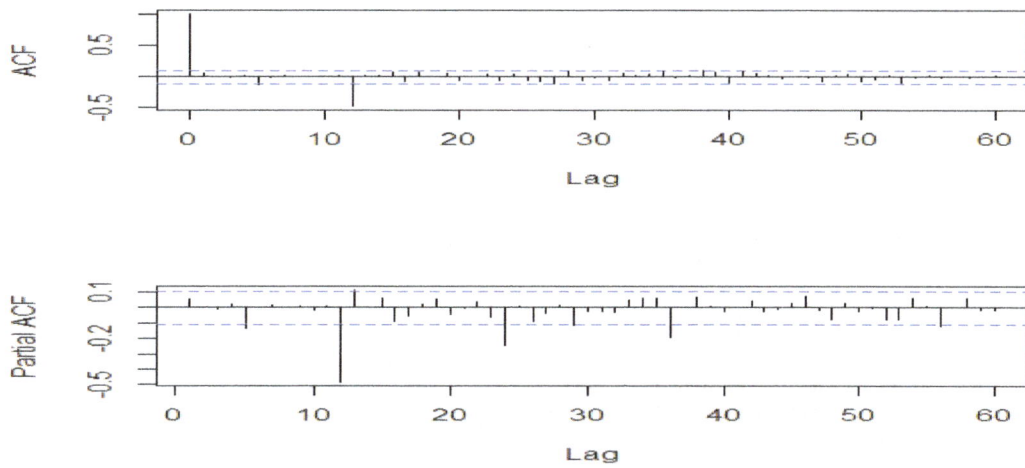

Figure 5: Sample ACF and PACF for first seasonal differenced transformed rainfall of Dire Dawa.

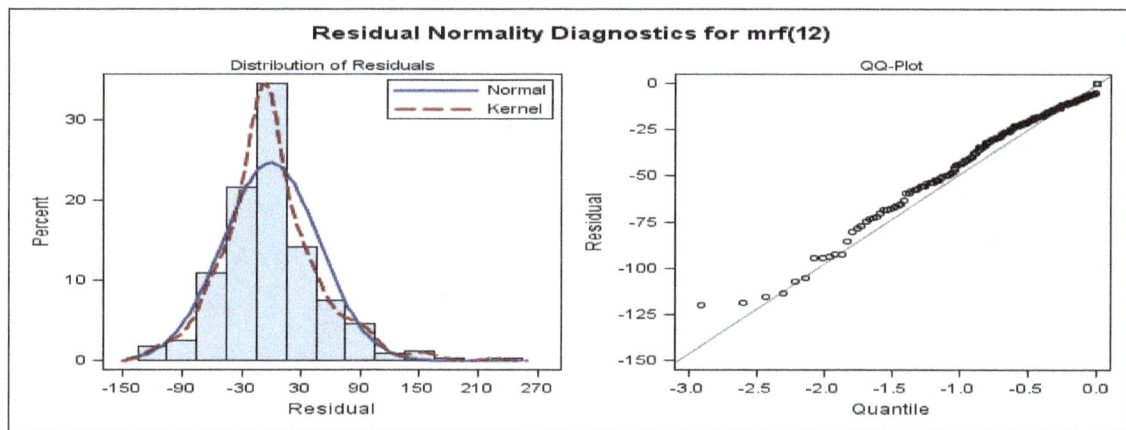

(a)

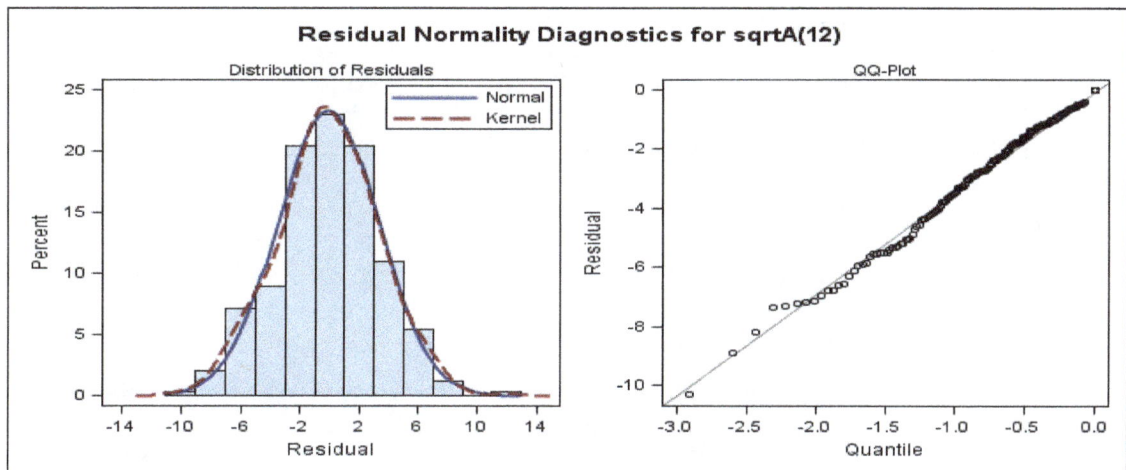

(b)

Figure 6: Normality Diagnostics plot for: (a) untransformed series model, (b) square root transformed series.

to achieve normality. By applying Box-Cox transformation defined in eqn.(9) to the monthly rainfall series, the normality assumptions of residual will be achieved using optimum value $\lambda = 0.5$, which maximizes the likelihood function with various iteration by the help of SAS 9.2-software. Figure 6b reveals that after this transformation the problem of non-normality seems dealt with. The Shapiro–Wilk test result based on test statistic defined on eqn.(8) (p-value=0.397) is also in support of the normality of transformed series.

Since normality is not fulfilled for the original series, three stages model building is performed with help of Sample ACF and PACF plot for the square root transformed rainfall data of Dire Dawa.

Model building for monthly rainfall of dire dawa

Since monthly rainfall data has seasonality variation, the univariate Seasonal Authoregressive Integrated Moving Average methodology is used to model rainfall series of Dire Dawa region.

Model identification: Once the degree of differencing has been determined, the autoregressive and moving-average orders are selected by examining the sample autocorrelations and sample partial autocorrelations function. Thus, it considered the SACF and SPACF shown in Figure 5 above and Tables 1 and 2 [17] as a guide and preliminary values of p, q, P and Q are chosen. Because we are dealing with estimates, it will not always be clear whether the sample ACF or PACF is tailing off or cutting off. Also, two models that are seemingly different can actually be very similar. With this in mind, we should not worry about being so precise at this stage of the model fitting.

At this stage, a few preliminary values of p, q, P and Q should be at hand, and we can start estimating the parameters. First, concentrating on the seasonal lags, the characteristics of the ACF and PACF of our transformed series in Figure 2 tend to show a strong peak at h=12 in the autocorrelation function, combined with peaks at h=12, 24, 36 in the partial autocorrelation function. Hence it appears that either: the ACF is cutting off after lag 12 and the PACF is tailing off in the seasonal lags, or the ACF and PACF are both tailing off in the seasonal lags. Consequently, either (i) a seasonal moving average of order Q=1, or (ii) due to the fact that both the ACF and PACF may be tailing off at the seasonal lags, perhaps both components, P=1 and Q=1, are needed. To identify the between-season model, we focus the lags h=1, 2...11 and identify order based on Table 1. First, we set the ACF to be tailing-off and the PACF to be cut-off after lag 5, we identify p=5 and q=0. Also it is possible to think of the PACF to be tailing-off and the ACF to cut-off after lag 5, leading to identify P=0 and Q=5. Fitting the models suggested by these observations, it can be obtained:

$$\text{SARIMA } (0, 0, 5) \times (0, 1, 1)_{12}; \text{SARIMA } (5, 0, 0) \times (0, 1, 1)_{12}$$

$$\text{SARIMA } (0, 0, 5) \times (1, 1, 1)_{12}; \text{SARIMA } (5, 0, 0) \times (1, 1, 1)_{12}$$

Parameter estimation: Here the maximum likelihood estimation methods for transformed total monthly rainfall are used to estimate the model parameters and the results are summarized in Table 3, which displays the list of the parameters for each temporally entertained model. For each model parameter, the table presents the estimated value, standard error, t- value, AIC, SBC and variance $(\hat{\delta}_k^2)$ for the estimate. As indicated by Shumway et al. [17], parameters must differ

Model	Parameter	Estimate	Std.Error	T-ratio	P-value	Criteria
(i)	μ	-0.002	0.039	-0.05	0.959	AIC=1863.206
	θ_5	0.165	0.053	3.09	0.002	SBC=1874.763
	Θ_{12}	0.775	0.037	20.88	<.0001	$\delta_k^2 = 11.81$
(ii)	μ	-0.002	0.039	-0.05	0.959	AIC=1860.17
	ϕ_5	-0.181	0.053	-3.41	0.0006	SBC=1870.73
	Θ_{12}	0.777	0.037	20.99	<.0001	$\delta_k^2 = 10.71$
(iii)	μ	-0.002	0.039	-0.05	0.959	AIC=1865.19
	Θ_{12}	0.165	0.053	3.09	0.002	
	Θ_{12}	0.772	0.037	20.88	<.0001	SBC=1879.61
	Θ_{12}	-0.007	0.071	0.09	0.925	$\delta_k^2 = 12.19$
(iv)	μ	-0.002	0.039	-0.05	0.959	AIC=1864.13
	ϕ_5	-0.181	0.053	-3.41	0.0006	
	Θ_{12}	0.775	0.037	20.88	<.0001	SBC=1878.54
	Θ_{12}	-0.011	0.071	-0.16	0.877	$\delta_k^2 = 11.85$

Table 3: Summary of model Parameter Estimates and model selection measures.

significantly from zero and all significant parameters must be included in the model. The result indicated that the seasonal and non-seasonal moving averages as well as non-seasonal autoregressive parameters are all significant since their p-values is smaller than 0.05 and should be retained in the model. However, the constant (μ), non-seasonal autoregressive parameters $(\phi_1, \phi_2, \phi_3, \phi_4)$ and seasonal autoregressive parameters (ϕ_2) in all selected models are insignificant; hence it should be omitted from the model. Moreover, the correlation of the parameter estimates is also used to assess the extent to which collinearity may have influenced the results. If two parameter estimates are very highly correlated, you might consider dropping one of them from the model. However, in our case, correlation between seasonal moving average and non-seasonal autoregressive parameter is -0.079, which indicate that less correlation is observed between parameter estimates.

Finally, the AIC, SBC and variance of the estimate $(\hat{\delta}_k^2)$ in Table 3 are computed according to eqn. (3) and used to compare selected models fit best to the monthly rainfall series. From this, SARIMA (5, 0, 0)*(0, 1, 1)$_{12}$ model has lower AIC, SBC and variance of estimate than other model, so it fits the transformed monthly rainfall series of Dire Dawa better than other, and can be used for further analysis.

Diagnostic checking: If the model fits the rainfall data well, the residuals of the fitted model are random or white noise [31].

Inspection of the time plot of the standardized residuals for square root transformed rainfall series in Figure 7, which can be obtain by help of eqn.(4) above, shows no clear patterns (trend or seasonality behavior).

Under the test that residual follows a white noise process, roughly 95% of the residual autocorrelations should fall within the range of $\pm 1.96 / \sqrt{360}$. It is clear, as shown in Figure 7 that there is no pattern in the residuals autocorrelation plot for the selected model, which means there is no autocorrelation coefficient which lies outside the two standard errors significantly for the fitted models. The Ljlung Q-statistic again for each group of six lags are computed using Eq. (5) and the results are summarized in Table 4 below and Figure 6, indicate that the p-value is greater than 0.05 for all lags, which implies that the white noise hypothesis is not rejected. Moreover, two alternative methods, namely runs and turning point tests, are applied to check the independence assumption of residuals of the fitted model. The fitted model was found to be consistent with the independence assumption for both methods since their test statistic is smaller than the critical values. Thus, those results also closely follow with the result plot of the standardized residuals in Figure 7 and Table 4.

The results related to the normality of residuals using Shapiro–Wilk tests indicate the residuals of the fitted models is normally distributed since p-value is greater than 0.05. In addition to these tests, Figure 6b shows the histograms and QQ-plot of the residuals. As expected,

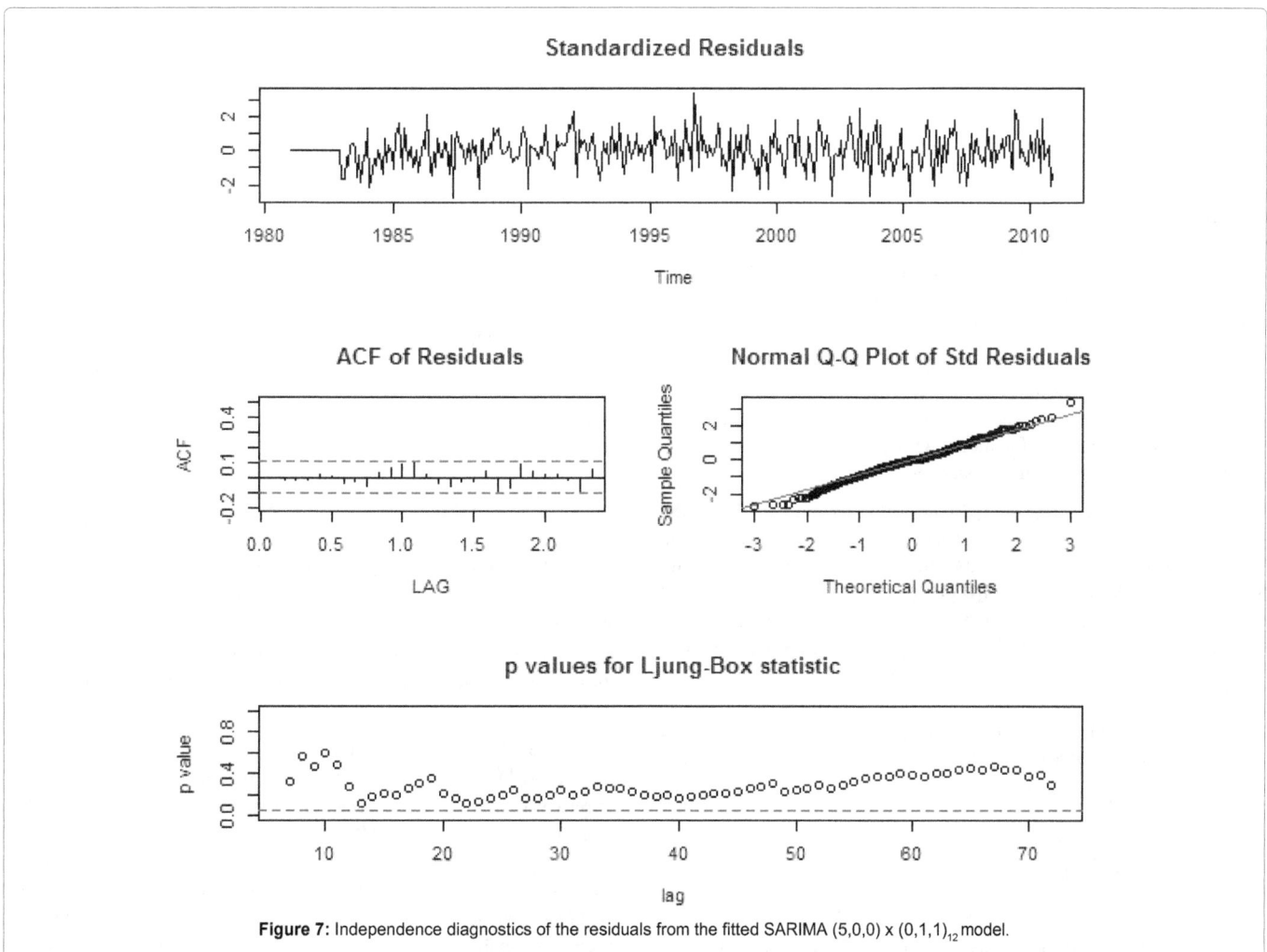

Figure 7: Independence diagnostics of the residuals from the fitted SARIMA (5,0,0) x (0,1,1)$_{12}$ model.

To Lag	Chi- Square	DF	P-Value	Autocorrelations					
6	7.60	4	0.1072	0.03	0.033	-0.021	-0.017	0.008	-0.032
12	12.24	10	0.2695	-0.064	-0.034	-0.023	0.040	0.073	-0.006
18	17.67	16	0.3435	0.095	0.024	-0.007	-0.063	-0.025	-0.025
24	22.56	22	0.4267	0.014	-0.074	-0.033	0.069	0.030	0.024
30	27.21	28	0.5069	-0.021	-0.036	-0.084	0.042	-0.038	-0.012
36	32.48	34	0.5420	-0.062	-0.008	0.016	0.070	0.066	0.012
42	39.07	40	0.5120	0.055	0.079	0.043	-0.055	0.042	0.030
48	40.61	46	0.6970	0.019	-0.024	-0.027	-0.007	-0.042	0.018
54	45.06	52	0.7411	0.059	-0.039	0.009	-0.006	-0.073	-0.020
60	47.08	58	0.8469	-0.000	-0.019	-0.029	-0.055	-0.023	-0.001
66	50.60	64	0.8885	0.026	0.012	0.065	0.029	0.036	-0.032
72	57.42	70	0.8594	-0.019	0.051	-0.044	-0.078	-0.013	-0.067

Table 4: White noises check of SARIMA (5, 0, 0) × (0, 1, 1)12 model using Ljung-Box.

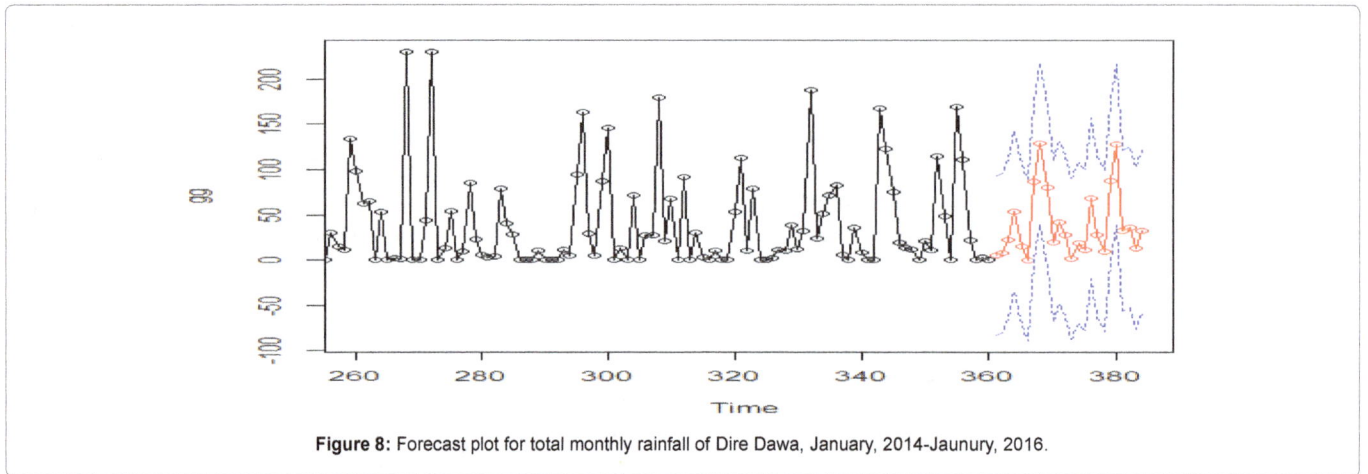

Figure 8: Forecast plot for total monthly rainfall of Dire Dawa, January, 2014-Jaunury, 2016.

the curves significantly reflect a normal distribution. Moreover, all calculated values are found to be smaller than the respective critical values for Goldfeld-Quandt (G-Q) Test and Breusch and Pagan(B-P) test, in the test of homoscedasticity of the residual, which indicate that the residuals has constant variance.

Therefore, from above all result it can be concluded that the hypothesis that the residuals are white noise cannot be rejected indicating that the fitted model is adequate. That is, SARIMA (5,0,0) × $(0,1,1)_{12}$ model is adequate for modeling the square root transformed monthly rainfall series of Dire Dawa region.

Forecasting: Since the model diagnostic tests show that all the parameter estimates are significant and the residual series is white noise, the estimation and diagnostic checking stages of the modeling process are complete. We can now proceed to forecasting the rainfall series with fitted SARIMA (5, 0, 0)*(0, 1, 1)$_{12}$ model. According to eqn. (1) above, the SARIMA (5, 0, 0) × (0, 1, 1)$_{12}$ model can be written as:

$$(1 - \phi_5 B^5)(1 - B^{12})y_t = (1 + \theta_{12} B^{12})e_t \qquad (19)$$

This equation (19) can also be multiplied out and rewritten in a form that is used in forecasting as:

$$y_t = y_{t-12} + \phi_5(y_{t-5} - y_{t-17}) + e + \theta_{12}e_{t-12} \qquad (20)$$

Where $\beta^5 y_t = y_{t-5}$ and $(1 - \beta^{12})y_t = y_t - y_{t-12}$. This equation further re-expressed as:

$$y_{t+m} = y_{t+m-12} + \phi_5(y_{t+m-5} - y_{t+m-17}) + e_{t+m} + \theta_{12}e_{t+m-12} \qquad (21)$$

After substituting the estimated parameter values in Eq. (21) above,

we obtain the following equation:

$$\hat{y}_{t+m} = \hat{y}_{t+m-12} - 0.181(\hat{y}_{t+m-5} - \hat{y}_{t+m-17}) \\ + \hat{e}_{t+m} + 0.777\hat{e}_{t+m-12} \qquad (22)$$

Forecasts are to be made at the origin, t=January, 2014 for lead times m= 1, ..., 24 for the total monthly rainfall series in the coming 24-month. Then the one-step ahead forecast at the origin, t=Jan, 2014, which give us forecast of actual series for the month of January, 2012 are given by:

$$\hat{x}_{Jan,2014+m} = \hat{x}_{Jan,2011+m-12} - 0.033(x_{Jan,2014+m-5} - \hat{x}_{Jan,2014+m-17}) \\ + \hat{e}_{Jan,2014+m} + 0.604\hat{e}_{Jan,2014+m-12} \qquad (23)$$

By the help of eqn.(23), it can be obtain 24 month(2-years)-steps a head forecast and prediction interval for total monthly rainfall series of Dire Dawa, as presented in Figure 8. Results reveal that there is a tendency of relatively increasing in average pattern of monthly and annual rainfall over the forecast period from January 2014 to January, 2016.

Forecasting accuracy evaluation: If the fitted SARIMA (5, 0, 0)*(0, 1, 1)$_{12}$ model has to perform well in forecasting, the forecast error will be relatively small. The accuracy of forecasts is usually measured using root mean square error (RMSE), mean absolute error (MAE), Mean absolute percentage error (MAPE) and Theil's inequality coefficient (Theil-U). The result shows that the Mean Absolute Percentage Error (MAPE) turn out to be 3.56%, which is relatively less than 4% and Theil's inequality coefficient (U-statistic) turn out to be 0.018, which is relatively close to zero. Besides this result, the bias and variance proportion are also very small, which are 0.047 and 0.001, respectively. Thus, measures indicate that the forecasting inaccuracy is low.

To assess the out-of-sample forecasting ability of the model it is advisable to retain some observations at the end of the sample period. Therefore, with the hold-out last monthly rainfall values from Jan, 2013 to Jan, 2014 for evaluating forecasting accuracy and model validation, The RMSE, MAE, MAPE values are less than 5% and Theil-U statistics is close to zero, which indicates the difference between the actual value and the predicted value is very small. That is, the predictive powers of the models are better and suitable for m-step ahead forecasting.

Spectral analysis result

Power spectral Analysis is applied for annual and monthly rainfall series of Dire Dawa region. The first step in estimation of the spectrum by the periodogram method is subtraction of the sample mean and removing any obvious trend. Figure 9 show that raw periodogram of annual rainfall of Dire Dawa, in each point of it represents the variance of the series contributed by a frequency range centered at the point. Since raw periodogram is a wildly fluctuating estimate of the spectrum with high variance, the periodogram must be smoothed to ensure stable estimate. Proper amount of smoothing is somewhat subjective and depends on the characteristics of the data, generally, it is a good idea to try several bandwidths that seem to be compatible with the general overall shape of the spectrum, as suggested by the periodogram [17]. Considering the tradeoff in smoothness, stability and resolution in selecting widths of modified Daniel filters, which leads to span of length, c (1,1) and 15 as a suggested value for smoothing the annual and monthly rainfall of Dire Dawa station respectively (Figure 10). The result is displayed in Figure 2a-2c, which provide a sensible compromise between the noisy version, shown in Figure 1a-1c, and a more heavily

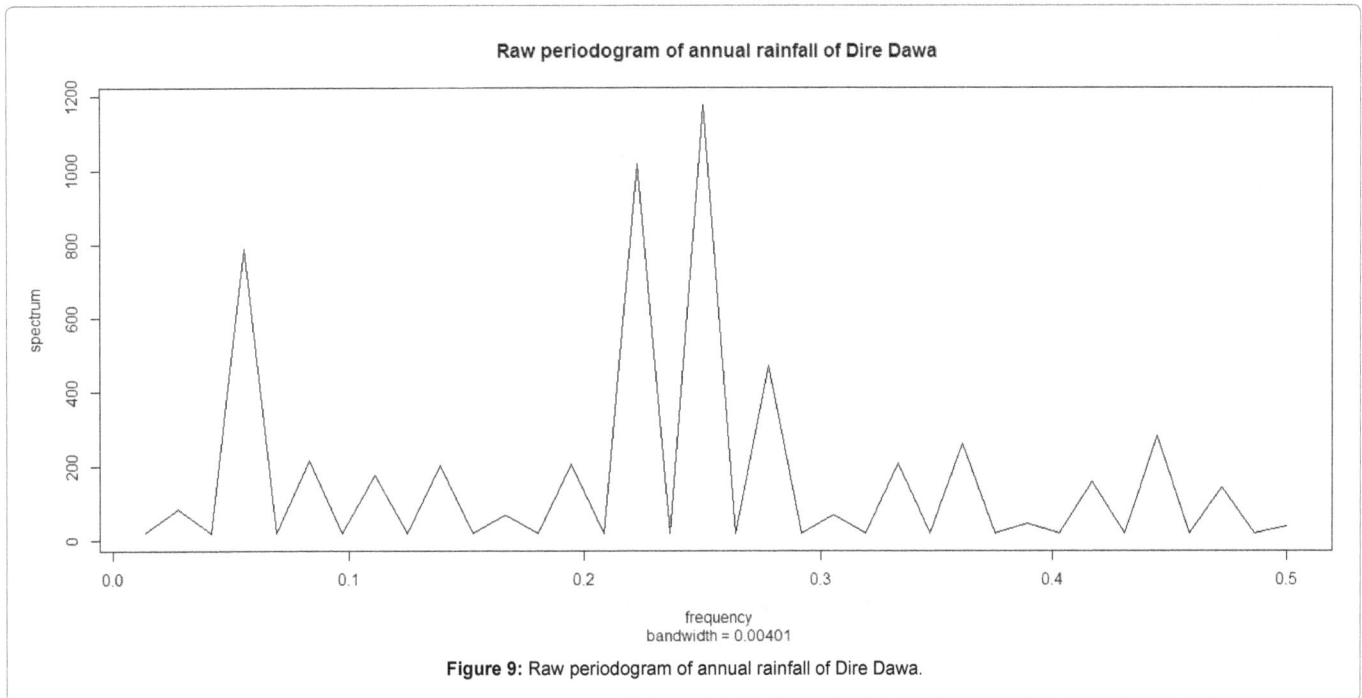

Figure 9: Raw periodogram of annual rainfall of Dire Dawa.

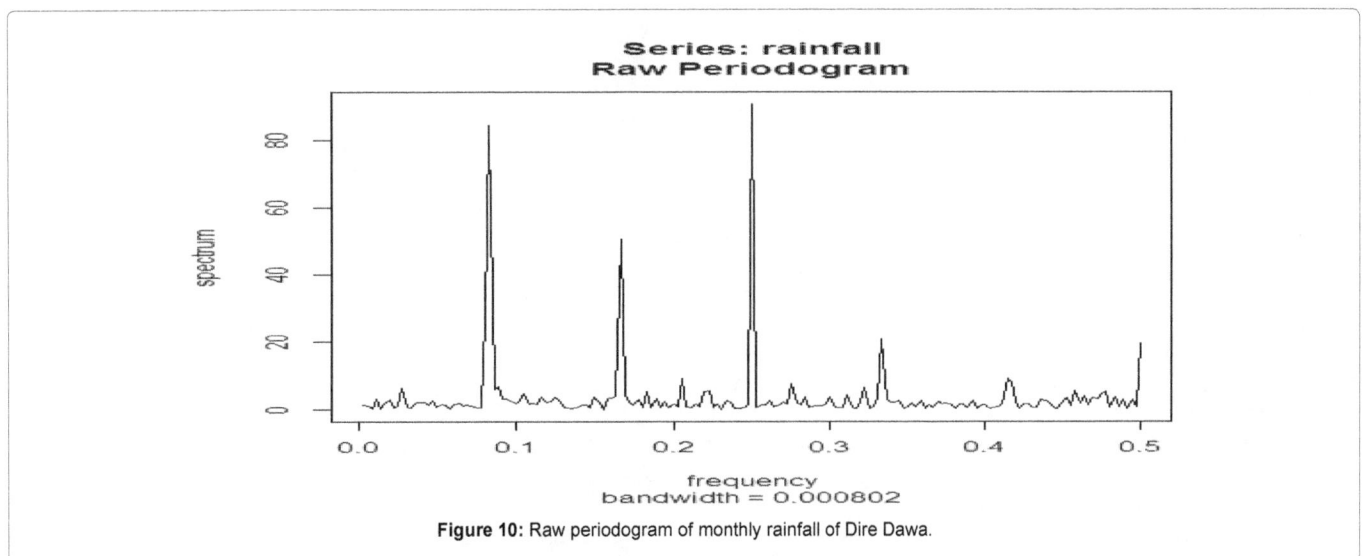

Figure 10: Raw periodogram of monthly rainfall of Dire Dawa.

smoothed spectrum, which might lose some of unnecessary peaks in estimation of spectrum. The width of the center mark on the 95% confidence interval indicator indicates the band width [32] cited also in Alemrew et al. [3]. The bandwidth, is equal to 0.1 cycles per year and 0.0417 cycles per month for the spectral estimator. This bandwidth means we are assuming a relatively constant spectrum over about 20% and 8.3% for monthly and annual rainfall with entire frequency interval (0, 0.5), respectively. By using degrees of freedom (d_f) of 6 and 30 for annual and monthly rainfall respectively, it can constructed approximate 100(1-α) %confidence intervals (eqn.(14)) for spectral density for the frequency bands identified as having the maximum power, as shown in Table 5 above. If the lower confidence limit for the spectral value is greater than the baseline level at some predetermined level of significance, we may claim that frequency value as a statistically significant peak [17]. As we observed from Table 5, an approximate 95% confidence interval for the spectrum $f_s(0.37)$ is [39.34, 255.75], which is again too wide to be of much use, but we do notice that the lower value 39.34 is higher than any other periodogram ordinate, so it is safe to say that this value is statistically significant. Similarly, an approximate 95% confidence interval for the spectrum f_s (0.083) is [6.17, 17.25].

The smoothed spectra shown in Figure 11 and summarized in Table 3 provide a sensible compromise between the noisy version, shown in Figure 9, and a more heavily smoothed spectrum, which might lose some of the peaks. As displayed in the Figure, the peak with an average period of 4.17 years, contributing the highest percent to total variance

of the annual rainfall series of Dire Dawa, which corresponds to the existence of "strong" peak at a frequency band centered at 0.24 cycles per year. This indicates that the periodicity of 4.17 years the most dominant in the annual rainfall of Dire Dawa. This can be attributed to the El Nino phenomenon. By looking for the other large spectral ordinate, it can be identified peaks of 16.67, 2.86 and 2.27 year periodicities.

Moreover, the monthly rainfall data are also used and their spectrum will exhibit narrow and sharp peaks at seasonal frequencies, as shown in Figure 12. For such type of data, commonly occurred periodic oscillation: Semi-Annual Oscillation (SAO), Annual Oscillation (AO), Quasi-Biennial Oscillation (QBO) and El-Nino/Southern Oscillation (ENSO) are approximately obtained by 4-7 months, 10-14 months, 22-32 months, and 40-66 months, respectively [33]. The peak with an average period of 12 month contributes a higher percent to total variance of the monthly rainfall series. This indicates that the most dominant peak observed in monthly rainfall of Dire Dawa is annual Oscillation (AO). There is also a subsidiary peak corresponding to a periodicity of 4 month and 6 month with associated frequency band centered around 0.25 and 0.167 cycles per month, respectively. These can also be attributed to the Semi-Annual Oscillation (SAO).

Using spectral analysis there have been observed prominent periodicities of 1.8-2.6 years [34], 2.7-3.3 years [35] and 2.5-8 years [36]. The study result by Perterbough [37] various regions of African rainfall indicate that there is no apparent statistically significant spectral peak in most cases. However, opposite of this result obtained by Nicholson [38], who discovered a statistically significant spectral

Series	Frequency	Period	Power (variance)	Lower limit	Upper limit
Annual rainfall	0.37	2.5 years	62.3	39.34	255.75
	0.2	5 years	38.2	16.76	156.74
Monthly rainfall	0.083	12 month	9.66	6.17	17.25
	0.167	6 month	5.51	3.52	9.85
	0.250	4 month	7.61	4.86	13.59

Table 5: The Smoothed Spectra of the annual and monthly rainfall Series of Dire Dawa.

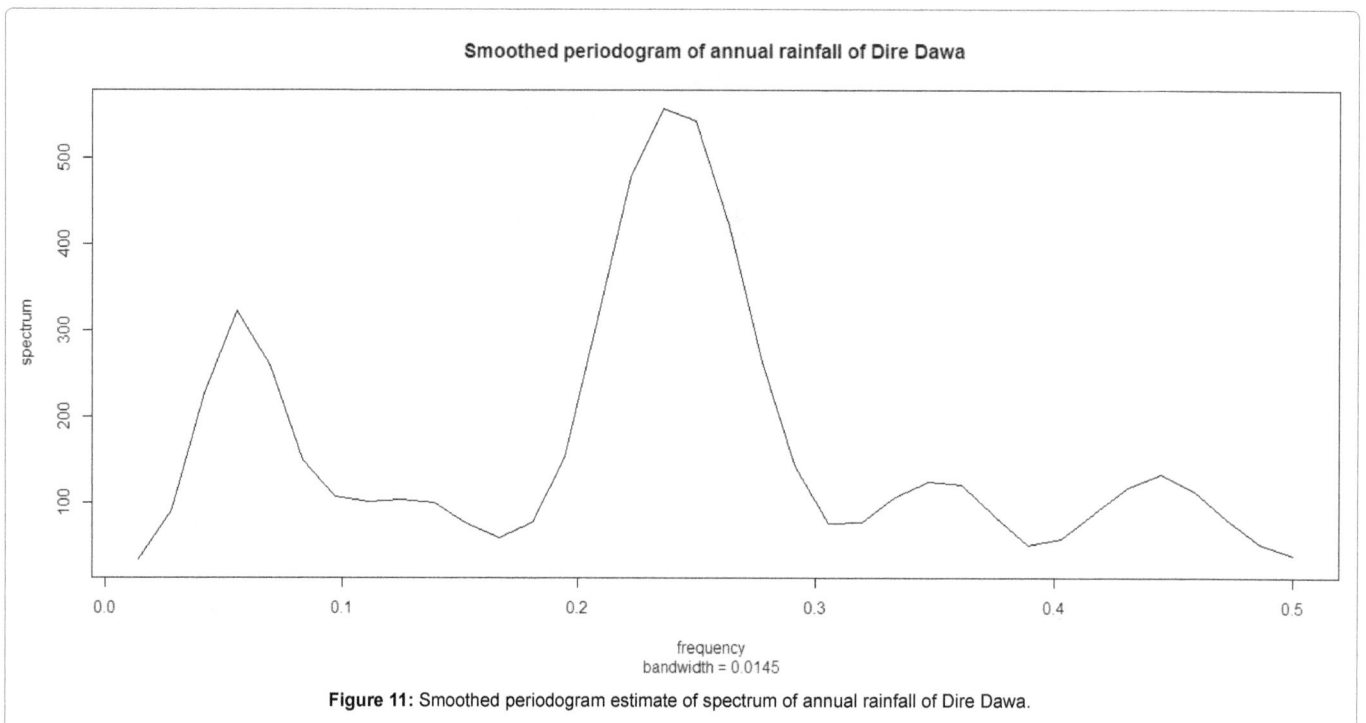

Figure 11: Smoothed periodogram estimate of spectrum of annual rainfall of Dire Dawa.

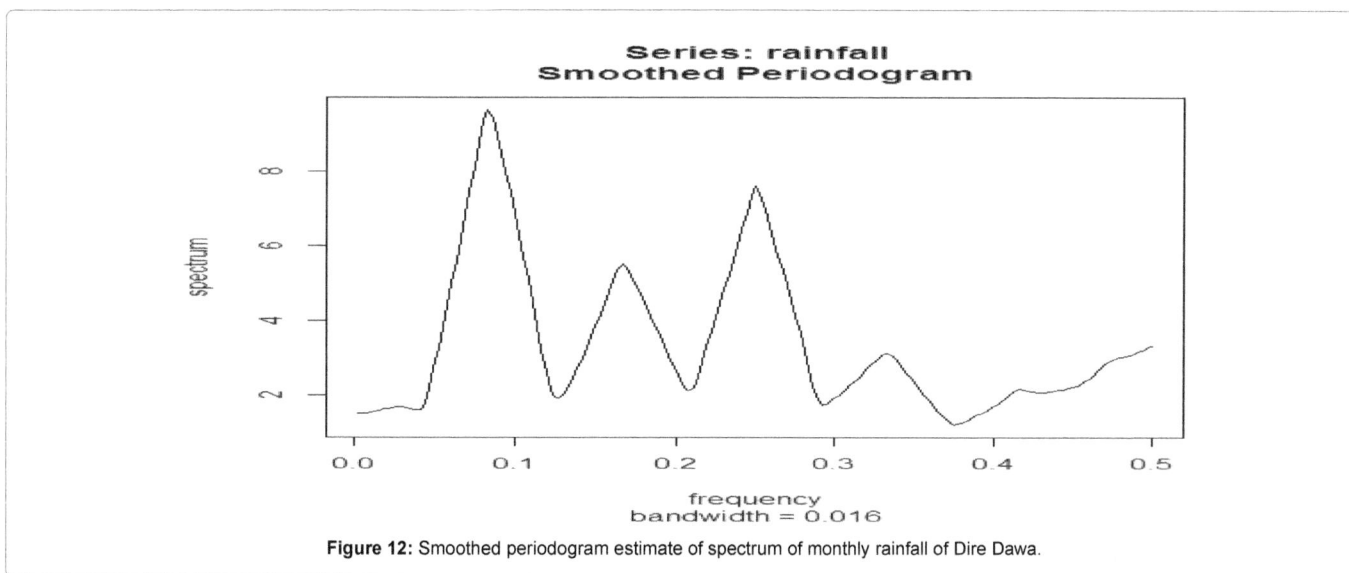

Figure 12: Smoothed periodogram estimate of spectrum of monthly rainfall of Dire Dawa.

peaks of approximately 2.3, 2.8, 3.6 and 5-6 years is observed in different regions of African rainfall. Moreover, Perterbough [37] study and our result opposed each other. By conducted a detailed power spectrum analysis of African rainfall series, Nicholson et al. [39] found that quasi-periodicities clustered in four bands at 2.2–2.4, 2.6–2.8, 3.3–3.8 and 5.0-6.3 years, common throughout equatorial and southern Africa. Ayoade [26] detected periodicities of 2-6 years while Adejuwon et al. [40] detected prominent periodicities of between 2 and 8 years cycles and less prominent periodicities of 16, 18 and 21 years cycles both in study of precipitation pattern of Southern Nigeria. Alamerew et al. [3] using spectral analysis method found that the dominant cycle with periodicity of 11.24 year for the annual rainfall of Addis Ababa, but they didn't test the statistical significance of spectral peak.

Many study shows that Ethiopian rainfall strongly related to the ENSO phenomena via rainfall extreme event like drought and flood [4,11,41]. The ocean–atmosphere interaction is believed to be a dominant factor in the weather processes in Eastern Ethiopia and contributes the largest share of the total annual rainfall the region. With this inter-action, many studies were used for estimating possible future rainfall frequency of extreme events of the area. Yilma [2] found that a dominant cycle for the annual rainfall of Addis Ababa is 10.22 years. Alemrew et al. [3] inferred that drought recurs in Addis Ababa between 10 to 11 years. Haile [11] also found that drought occur every 6-8 years in the semi-arid regions of Ethiopia. The study by Tsegay [12] based on the frequencies of rainfall deviation from the average suggest that drought occur every three to five years in Eastern Ethiopia. The possibility of using prominent peaks in the spectrum to predict the long-range behavior of the rainfall is attractive [42]. Oduro et al. [6] predict that a rainfall extremes event in Ghana recurs every 5.6 years. Stringer [43] also estimated that the climatic events recur every 2 to 2.5 year in Africa.

Cross-spectra analysis result

The cross-spectrum analysis is used to provide a means of determining the contributions of fluctuations in various frequency bands to covariance quantities such as the fluctuations in rainfall series at different nearby stations.

Figure 13 shows the squared coherence between rainfall series of Dire Dawa station with Haramaya and Dengego stations with L_h =15

and 30 degree of freedom and $F_{2,30}(0.05)$=3.34 at 5% significance level. Hence, the hypothesis of no coherence is rejected for the values of estimated square coherence (eqn.17) that exceed $C_{0.05}$=0.16 (eqn.18). As it can be observed in the same Figure that, at lower seasonal frequency, the Dire Dawa rainfall characteristics are significantly associated with both Haramaya and Dengego rainfall, which mean there is strongly coherent. However, the degree of relatedness or coherence is higher in Dengego than Haramaya rainfall to the rainfall of Dire Dawa.

Conclusion of the Study

In this study, 30 years annual and monthly rainfall records were analyzed, using data from Dire Dawa region and neighboring area in the Eastern Ethiopia, mainly to study the rainfall pattern and its extreme event frequency of Dire Dawa. Based on the overall results of the research, the following conclusions could be drawn:

➢ A time series model for monthly rainfall series of Dire Dawa was adjusted, processed, diagnostically checked and lastly SARIMA(5,0,0)*(0,1,1)$_{12}$ model is established and adequately be used to forecast 2 years monthly rainfall values. Further results reveal that there is a tendency of relatively increasing in average pattern of monthly and annual rainfall over the forecast period from January 2014 to January, 2016.

➢ Rainfall extreme event like flood predict or inferred to be recurring at about 4.17 years in Dire Dawa region.

➢ The annual rainfall of Dire Dawa region found to dominate with El Nino phenomenon. On other hand, Annual oscillations dominate monthly rainfall of the area.

➢ Rainfall extreme event like flood and drought highly association between Dire Dawa and Dengego area.

According to the result in this study, the pattern and frequency in rainfall extreme event is increasing. So, every effort should be made to minimize such risks as deaths, loss of property and damage to infrastructure, associated with rainfall extreme event like drought and flood, crop failure by pre-planning and make decision against adverse effects of it on the region. Moreover, more study need to assess the factors that cause the change in the pattern of rainfall distribution in the region.

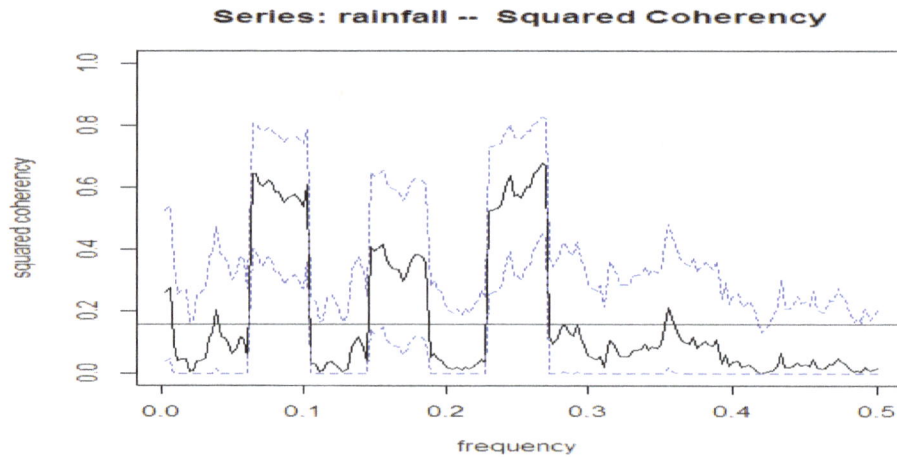

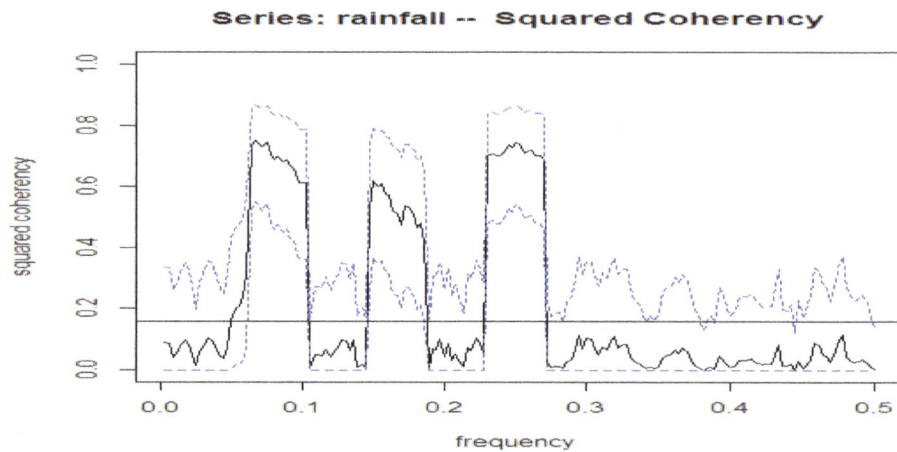

Figure 13: Squared coherency between rainfall series of: (a) Dire Dawa and Haramaya station; (b) Dire Dawa and Dengego stations.

Acknowledgment

I am indebted to Dr. Solomon Harrar, University of Kentucky for his comments and suggestion and am also grateful to the National meteorological Agency (NMA) of Ethiopia for providing the data for study.

References

1. Admasu G (1989) Regional flood frequency Analysis. Technical report, Royal institute of technology stockholm, Sweden 46-84.

2. Yilma S, Demaree GR, Delleur JW (1994) Sunspot numbers as a possible indicator of annual rainfall at Addis Ababa, Ethiopia. International Journal of Climatology 14: 911-923.

3. Alamerew B, Eshetu W (2009) Assessment of local climate in Addis Ababa. Journal of Ethiopian Statistical Association 18: 46-84.

4. Bekele F (1997) Ethiopian use of ENSO information in its seasonal forecast. Internet Journal for African Studies (IJAS).

5. Dire Dawa Environmental Protection Authority (DDAEPA) (2011) Dire Dawa administration program of adaptation to climate change.

6. Oduro-AfriyieK, Adukpo DC (2006) Spectral characteristics of the annual mean rainfall series in Ghana. West Africa Journal of Applied Ecology 9: 15-18.

7. Wood CA, Loveth PR (1974) Rainfall, drought and the solar cycles. Nature 251: 594-596.

8. Dyer TGJ, Tyson PH (1977) Estimating above and below-normal rainfall periods over South Africa 1972-2000. Journal of Applied Meteorology 16: 145-147.

9. Mersha E (2002) Determination of rainfall cyclicity over selected location in Ethiopia. Ethiopian J Water Science and Technology.

10. Seifu A (2004) Rainfall variation and its effect on crop production in Ethiopia. Addis Ababa University, unpublished MSC thesis, Ethiopia.

11. Haile T (1988) Causes and characters of drought in Ethiopia. Ethiopian Journal of Agricultural Sciences 10: 85-97.

12. Tsegay W (1998) El Nino and drought early warning in Ethiopia. In Using Science Against Famine: Food Security Early Warning System and El Nino, Internet Journal of African Studies.

13. Polyak I (1996) Computational statistics in climatology, Oxford University Press, Oxford.

14. Bewket W, Conway D (2007) A note on the temporal and spatial variability of rainfall in the drought-prone Amhara region of Ethiopia. International Journal of Climatology 27: 1467-1477.

15. Gibbons RD (1994) Statistical methods for groundwater monitoring. John Wiley & Sons, New York.

16. Box GEP, Jenkins GM (1976) Time series analysis: forecasting and control, Holden day.

17. Shumway RH, Stoffer DS (2010) Time series analysis and its applications with R Examples. Springer (3rdedn).

18. Wei WWS (1990) Time series analysis: univariate and multivariate methods. Addison-Wesley Publishing Company, Inc, New York, USA, p. 478

19. Mills T (1999) The econometric modeling of financial time series. Cambridge University Press, Cambridge.

20. Lehmann A, Rode M (2001) Long-term behaviour and cross-correlation water quality analysis of the River Elbe, Germany. Water Res 35: 2153-2160.

21. Ljung GM, Box GEP (1978) On a measure of lack of fit in time series models. Biometrika 65: 297-303.

22. Cromwell JB, Labys WC, Terraza M (1994) Univariate tests for time series models. A Sage Publications, London, 96: 07-99.

23. Shapiro SS, Wilk MB (1965) An analysis of variance test for normality (complete samples). Biometrika 52: 591-611.

24. Makridakis S, Wheelwright SC, Hyndman RJ (1998) Forecasting methods and application. John Wiley & Sons, New York.

25. Yevjevich V (1972) Stochastic processes in hydrology, Water Resources Publications, Fort Collins.

26. Ayoade JO (1973) Annual rainfall trends and periodicities in Nigeria. Nigeria Geographical Journal 16: 167-176.

27. Bloomfield P (2000) Fourier analysis of time series: an Introduction. (2ndedn) John Wiley & Sons, Inc., New York.

28. NMA (1996) Climate & agro climate resources of Ethiopia. NMSA, Meteorological Research Report Series, Addis Ababa.

29. Hamilton J (1994) Time series analysis. Princeton University Press, New Jersey.

30. Janacek G, Swift L (1993) Time series forecasting, Simulation, Application, Ellis Horwood, New York, USA.

31. Chatfield C (1991) The Analysis of time series - an introduction (4thedn.), Chapman and Hall: London.

32. Venables W, Ripley B (1999) Modern applied statistics with S-PLUS. Springer-Verlag, New York.

33. Sinta BS, Hariadi TE (2003) The spectrum analysis of rainfall in Indonesia. Indonesian Journal of Physics 14.

34. Klaus B (1977) Spatial distribution and periodicities of mean annual precipitation south of the Sahara. Archives for Meteorology Geophysics and Bioclimatology Series 26: 17-27.

35. Ogallo L (1979) Rainfall variability in Africa. Monthly weather review 107: 1133-1139.

36. Farmer G, Wigley TML (1985) Climatic trends for tropical Africa. A Research Report for the Overseas Development Administration, England.

37. Peterbough TL (1983) A regional comparison of seasonal African rainfall anomalies. In Proceedings of the 2nd International Meeting in Statistical Climatology. Institute national de meteorologiae Geofisca: Lisbon, Portugal 8.8.1-8.5.8.

38. Nicholson SE (1985) African rainfall fluctuations from 1850 to present: spatial coherence, periodic behavior and long-term trends. In 3rd Conference on Climate Variations and Symposium on Contemporary Climate 1850-2100, Los Angeles, American Metrological Society Extended Abstracts, 62-63.

39. Nicholson SE, Entekhabi D (1986) The quasi-periodic behaviour of rainfall variability in Africa and its relationship to the Southern Oscillation. Journal of Climate and Applied Meteorology 34: 311-348.

40. Adejuwon JO, Balogun EE, Adejuwon SA (1990) On the annual and seasonal patterns of rainfall fluctuations in Sub-Sahara West Africa. International Journal of Climatology 10: 839-848.

41. NMA (2007) National Adaptation Program of Action of Ethiopia (NAPA). Final draft report. National Meteorological Agency, Addis Ababa, Ethiopia.

42. Zangvil A (1979) Temporal fluctuations of seasonal precipitation in Jerusalem. The Institute for Desert Research, Ben-Gurion University, Israel, 31: 413-420.

43. Stringer ET (1972) Techniques of climatology. WH Freeman & Co., San Francisco, p. 539

5

SMHI-RCA Model Captures the Spatial and Temporal Variability in Precipitation Anomalies over East Africa

Asmelash T. Reda*

Mekelle University, Mekelle Institute of Technology,Mekelle Tigray Ethiopia, Ethiopia

Abstract

This work assesses the performance of the Sweden's Meteorological and Hydrological Institute-the Rossby Centre Regional Atmospheric Climate Model, SMHI-RCA, in reproducing precipitation variability over Ethiopia. The simulated datasets, which are generated in the frame work of Coordinate Regional climate Downscaling Experiment (CORDEX) project, are validated by Global Precipitation Climatology Project (GPCP). Comparison of means, correlation coefficients, value of chi-square, bias test and test of significance showed that there is a good agreement between SMHI-RCA and GPCP rainfall at each grid point. Rotated Principal Components (RPCs) and the associated spectra for both datasets showed that the improvement of rainfall simulations is remarkable over different parts of the country. Classification using Hierarchical Clustering Analysis (HCLA) reasonably agrees with the reduction of data using Principal Component Analysis (PCA). Small scaled condensed homogeneous groups are identified from SMHI-RCA and GPCP datasets; several of them being shared by both. Zones of rainfall maxima for each cluster are primarily associated with the migration of Inter Tropical Convergence Zone (ITCZ); even though, elevation differences induce rainfall peaks to have a phase shift at local scale.

Keywords: Regional climate model; Rainfall; SMHI-RCA

Background

Ethiopia lies in the northeast part of Africa, north of the equator, covering a total area of 1,221,900 km². It's topography is composed of massive highlands along with complex mountains and dissected plateaus divided by Great Rift Valley running from northeast to southwest [1,2]. Elevation varies from a height at below sea level in the northeastern part of the country to higher than 3500 m above sea level in the northern highlands. Climate of Ethiopia is a typical of equatorial regions, but topography complicates its pattern and character [3-5]. It creates diverse microclimates ranging from hot deserts over the lowlands to cool, very wet over highlands [3,5,6].

Ethiopia's rainfall is highly variable both in amount and distribution across regions and seasons. There are regions that experience three seasons (bimodal type-1) with two rainfall peaks (where one peak is more prominent than the other) while some regions have four seasons with two similar rainfall peaks (bimodal type-2). There are still some regions which have two seasons with single rainfall peak (monomodal rainfall type) [7]. On the other hand, some areas have rainfall for 10 consecutive months; others receive rainfall for just a few months; while still others are characterized by three distinct rainfall seasons [2,3,7,8]. Mean annual rainfall distribution over the country is characterized by large spatial variations which range from about 2000 mm over some areas in the southwest to less than 250 mm over the Afar and Ogaden low lands [7].

The seasonal and annual rainfall variations are results of the macro-scale pressure systems and monsoon flows which are related to the changes in the pressure systems [7,9,10]. The most important weather systems that cause rain over Ethiopia include Sub-Tropical Jet (STJ), Inter Tropical Convergence Zone (ITCZ), Red Sea Convergence Zone (RSCZ), Tropical Easterly Jet (TEJ) and Somali Jet [7]. The spatial variation of the rainfall is, thus influenced by the changes in the intensity, position and direction of movement of these rain-producing systems over the country [11-13]. However, the fine scale spatial distribution of rainfall in Ethiopia is significantly influenced by topography. For example the detail spatial and temporal variability of rainfall along the Rift Valley show many abrupt changes and not well known yet [3,4].

So far, three identified seasons exist in Ethiopia. The first is the main rainy season from June to September, the second is the dry season from October to December/January, and the third is the small rainy season from February/March to May, known locally as Kiremt, Bega and Belg respectively. A brief description of the mechanisms for rainfall formation for each season is discussed below.

During Bega, most of the country is generally dry except the south and southeast of Ethiopia that receives its second important seasonal rainfall in this period. In this season, the country predominantly falls under the influence of warm and cool northeasterly winds. These dry air masses originate either from the Saharan anticyclone or from the ridge of high pressure extending into Arabia from the large high over central Asia (Siberia). Occasionally the northeasterly winds are interrupted when migratory low pressure systems originating in the Mediterranean area move southwards and interact with the tropical systems resulting into unseasonal rains over central and northern Ethiopia [7]. The development of the RSCZ also produces rains over northeastern coastal areas [14].

During Belg (the small rainy season), which is from March to May, coincides with the domination of the Arabian high as it moves towards the north Arabian Sea and the development of thermal low (cyclone) over the south Sudan. Winds from the Gulf of Aden and the Indian Ocean highs that are drawn towards this centre blow across central and southern Ethiopia [7,15]. These easterly and southeasterly moist winds produce the main rains in southern and southeastern Ethiopia and the Belg rains to the east-central part of the northwestern highlands.

*****Corresponding author:** Asmelash T. Reda, Mekelle University, Mekelle Institute of Technology, Mekelle Tigray Ethiopia
E-mail: fasika20@gmail.com

Sometimes when the low-level westerly trough penetrates along the Rift Valley, the rainfall activity could linger for some days. The relationship between Ethiopian rainfall during Belg and the tropical cyclones over the southwest Indian Ocean indicates that low/high frequency of the cyclones resulted in excess/deficit rainfall [16]. There is also a meridional arm of the ITCZ due to the difference in heat capacity of the land surface and the Indian Ocean. This produces rainfall during February/March over southwest of Ethiopia [17]. The formation of intense and frequent tropical disturbances over the southeast Indian Ocean occurs simultaneously with Belg and Kiremt rainfall deficiency in Ethiopia [10].

During Kiremt the air flow is dominated by a zone of convergence in low-pressure systems accompanied by the oscillatory ITCZ extending from West Africa through the north of Ethiopia towards India. Major rain-producing systems during Kiremt are: the northward migration of the ITCZ; development and persistence of the Arabian and the Sudan thermal lows along 20°N latitude; development of quasi-permanent high-pressure systems over the south Atlantic and south Indian Oceans; development of the tropical easterly jet and its persistence and the generation of the low-level Somali jet which enhances low-level southwesterly flow. It is to be noted that Kiremt rainfall covers most of the country with the exception of the south and southeast of Ethiopia [3,7,8,13,17,18].

Data and Methodology

The SMHI-RCA simulated data, which are found from CORDEX downscaling experiment, are implemented for Ethiopia for the period 1996-2008. The model output, which is driven by ERA-interim runs, is validated by GPCP dataset [19,20]. Both datasets have horizontal resolution of 50 km. For analysis, the monthly averaged precipitation

data and its anomalies are used for the period 1996 to 2008. Validation and inter-comparison are performed using comparison of means, mean error, correlation coefficients, bias test and test of significance. Model validation is begun by comparing model's mean precipitation with observations. Mean error calculates measure of dissimilarity. To see the model be consistent with GPCP in time at each grid point, we have utilized correlation, bais and T- tests. Clustering and classification of precipitation are done based on RPCA, spectral density and HCLA. RPCA is applied to determine dominant precipitation variability and uniform rainfall regions. Computation of periodicities and pseudo periodicities of RPCs are done using power spectral density. Finally, HCLA is used for further regrouping of precipitation in more localized scale [21].

Results and Discussion

Comparison of RCA and GPCP datasets

Mean: Figure 1 shows contour plots of mean precipitation over Ethiopia for the period October 1996 to December 2008. It reveals that RCA simulated rainfall is in a good agreement with GPCP observational rainfall at each grid point. Both datasets depict mean precipitation greater than 2.0 mm day^{-1} over the highlands but show relatively less rainfall lower than 2.0 mm day^{-1} over the rest part.

More sensitive measure of discrepancy can be obtained by subtracting GPCP from RCA at each grid point. Figure 2a shows distribution of differences of means. Model precipitation matches GPCP with some mixed positive and negative biases oscillating around zero in several parts of the country. Positive biases, with higher magnitude in the high altitudes, are seen over western part while along the Rift Valley, RCA appeared to be somehow underestimated by 1.0 mm day^{-1}.

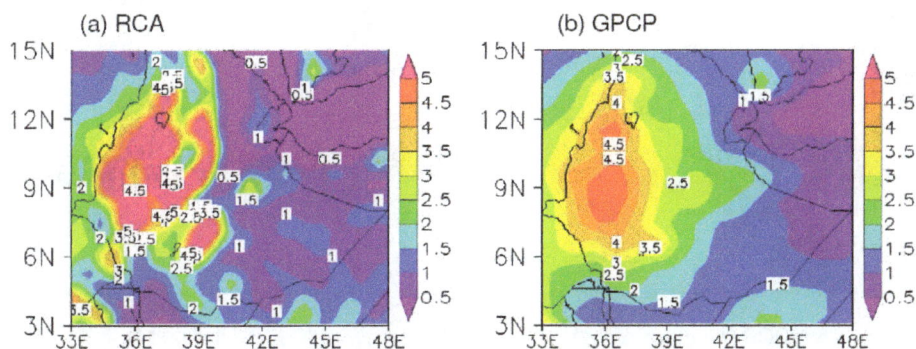

Figure 1: Mean precipitation in mm day^{-1}.

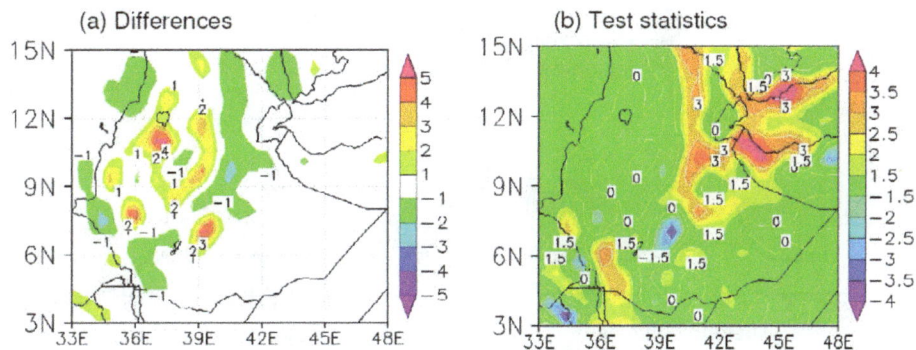

Figure 2: (a) Differences between RCA and GPCP. (b) Value of test statistics setting H_0: $\mu_{RCA}=\mu_{GPCP}$.

These observed differences might be due to lack of observational gauge dataset in these regions and/or due to efficient dynamic topographic adaptation behavior of RCA model over area where there is no gauge/less gauge density measurement within the grid box available.

It is important to see how the means estimate agree or disagree at each grid point. We have utilized the two-tailed T-test to calculate the frequency of agreement/disagreement between the two datasets at each grid point. To handle the temporal dependence at each location, effective number of degrees of freedom is computed from lag-1 autocorrelation coefficient as described by Janowiak and Wilks et al. [21,22]. This method is often reasonable approximations for representing the effective sample size. Critical values of Z for two tailed test are taken from Zwiers et al. [23]. Setting the null hypothesis 'mean of both datasets are equal at each grid point', Z value shows (Figure 2b), the differences are not significant at the level of 99% confidence interval (i.e. Z<±4) over the entire region.

Bias test: Figure 3a shows model-simulated versus GPCP average precipitation datasets over Ethiopia from 1996 to 2008. RCA precipitation is in line with GPCP having apparent low partiality at two extreme points. RCA is somehow overestimated over the area that receives highest precipitation, but to some extent underestimated over the area that experiences lowest precipitation in spite of the existence of less number of disregarded outliers with large values. Evaluation of the model through difference versus grid points over the region (Figure 3b) supports this idea. Differences approximately equally distributed above and below the zero value lying mostly between ±2.0 mm day^{-1}.

Distribution of errors caused by differences between RCA and GPCP means are displayed in Figure 4. It is approximately Gaussian shaped having mean and variance respectively -0.264 and 0.820 delimited by maximum values ± 2.5 (Figure 4a). This shows 95% and 67% of all values of the differences lie in the interval of -0.264 ± 1.812 and -0.264 ± 0.906 respectively. Figure 4b strengthens this idea. Out of 980 grid points' values which are distributed in both directions from the mean, about 850 (90%) lie in the interval ± 1.5. Inside the interval ±0.5 mm, the number of positive difference in rainfall is identical to negative differences. However, outside this limit, negative differences are roughly twice as large as positive differences.

Correlation

For the time series assessment, Spearman's rank and Pearson correlation coefficients are calculated for RCA and GPCP monthly mean precipitation values at each grid point for the period 1996-2008. Spearman correlation (Figure 5a) shows RCA and GPCP are highly related by monotonic function with positive correlation coefficient (r>0.7) corresponding to simultaneous trend swing of them. Correlation is greater than 0.9 over all part of the region; however, in the northeast coast, near border areas and over the Red Sea, correlation is less but still higher than 0.7.

The better way to examine how the two datasets agree linearly is to look at Pearson correlation coefficient that requires linear relationship of variables. It is less sensitive than Spearman and only gave good value in the area where GPCP and RCA are related by linear function. From Figure 5b, high correlation and high linear relationship exist over the moorland located above the Rift Valley. Lowlands experience less precipitation and less correlation coefficient but still higher than 0.5. Pearson correlation coefficient fluctuates highly along the Rift valley which runs from northeast to southwest across Ethiopia. This can arise

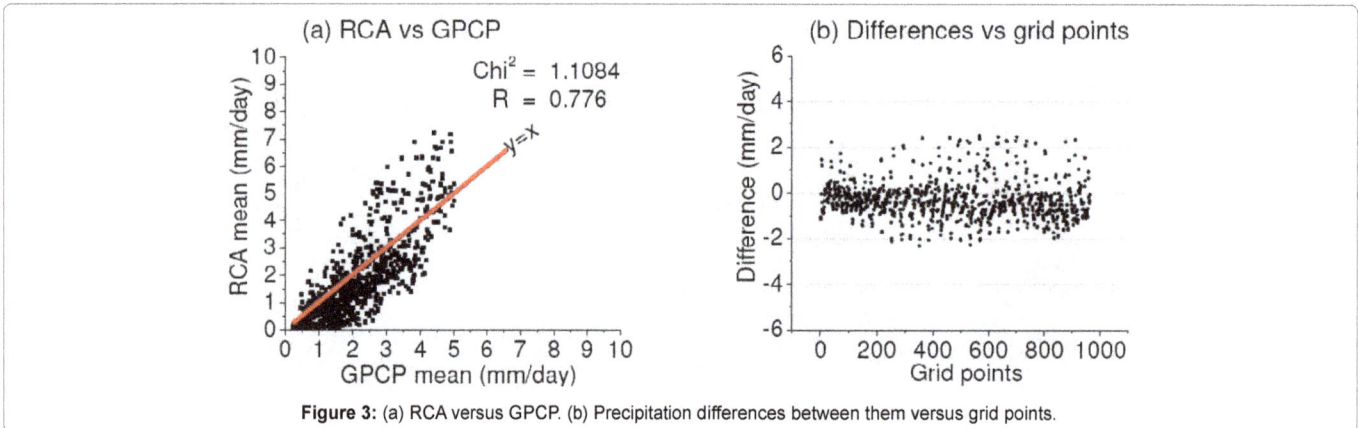

Figure 3: (a) RCA versus GPCP. (b) Precipitation differences between them versus grid points.

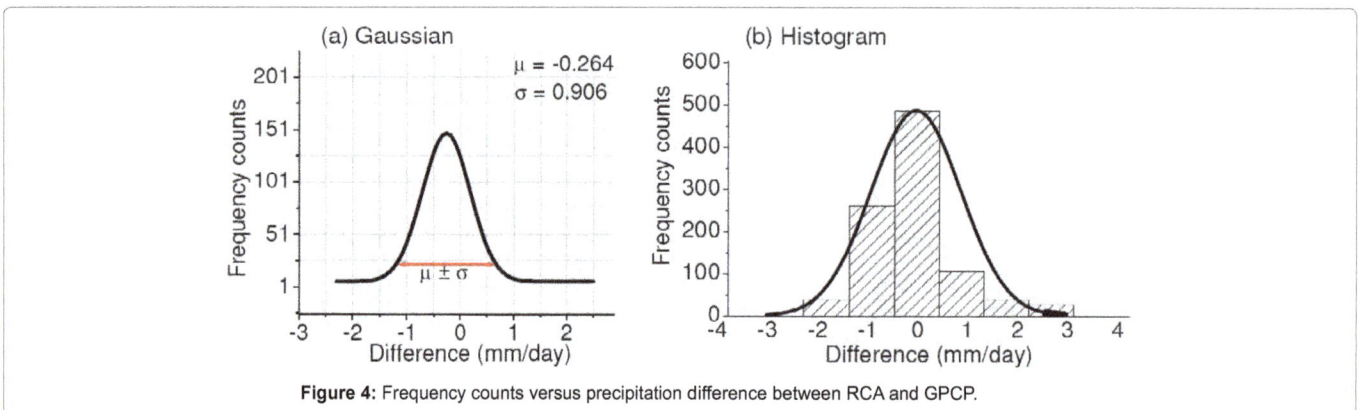

Figure 4: Frequency counts versus precipitation difference between RCA and GPCP.

while the model adapts the complex orography during simulation. An area of low correlation (<0.5) is observed over the sea where gauge data are absent.

Test of significance has made for Pearson and Spearman correlation coefficients (Figure 6). It is calculated using a two-tailed T-test under the assumption 'there is no relationship between GPCP and RCA data (r=0)'. The effective number of degrees of freedom is calculated from first-order auto regressions following the method used by Zwiers et al. [23]. It is observed that Z lies above 7 for Spearman and above 5 for Pearson. Hence, the null hypothesis of 'no relationship in the paired population' totally rejected. There is significant relation between the two data in the whole domain.

Rotated empirical orthogonal functions, rotated principal components and spectral densities

To identify and compare dominant modes of rainfall variability, RPCA based on VARIMAX rotation method is used. It is applied to the monthly mean precipitation RCA and GPCP datasets. The leading four spatial patterns of REOFs and the corresponding scatter plots of RPC time series are displayed in Figures 7 and 8 respectively. Spectral

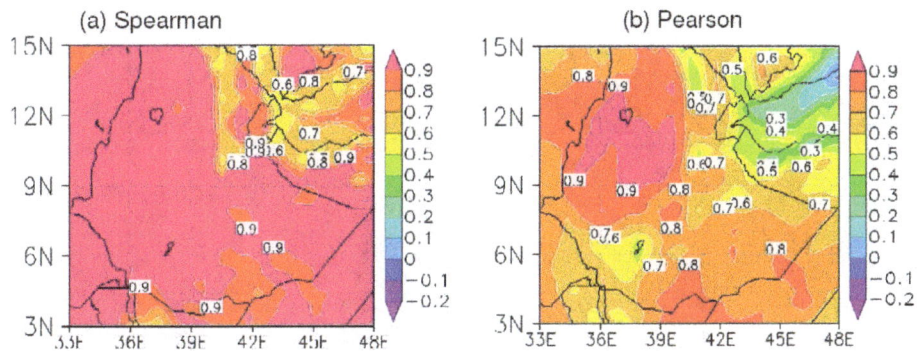

Figure 5: Correlation coefficient between RCA and GPCP.

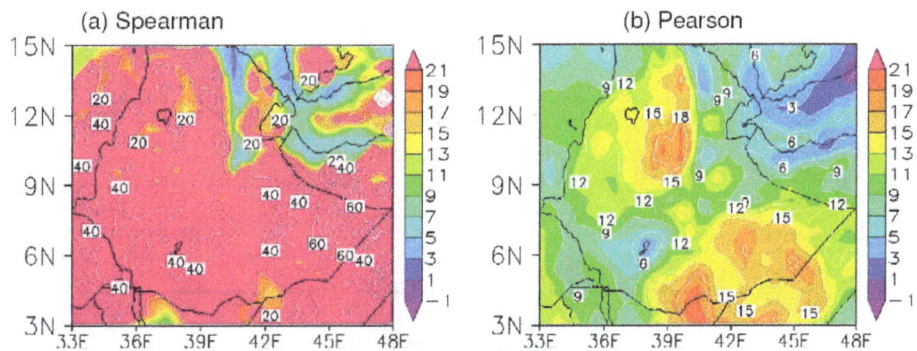

Figure 6: Value of test statistics setting H_0: r=0.

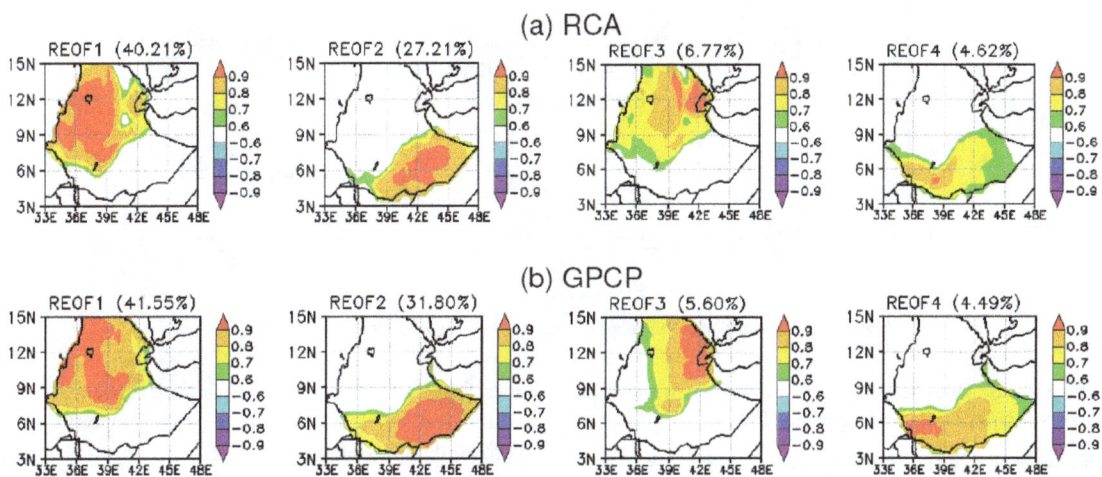

Figure 7: Correlation based leading four REOFs.

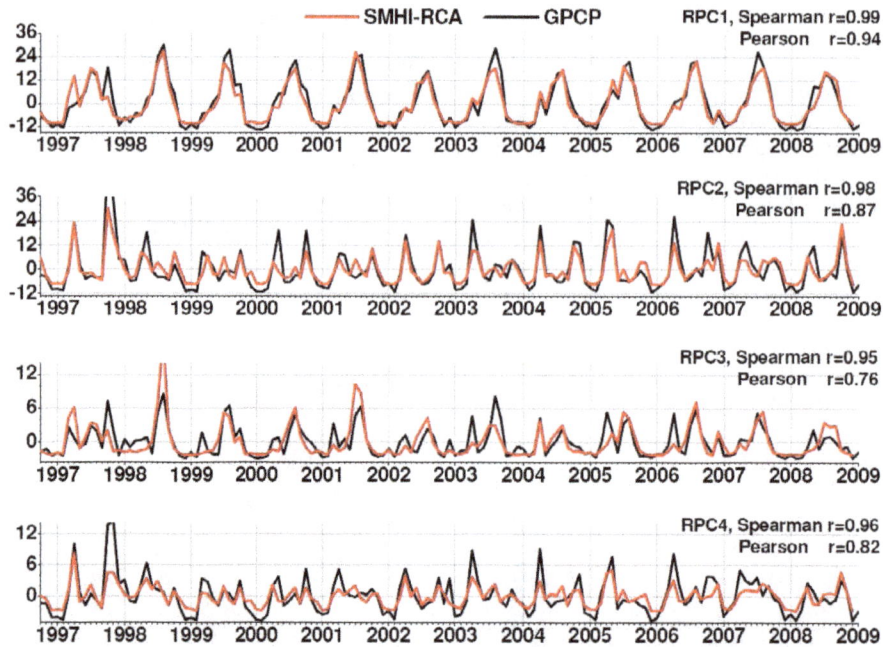

Figure 8: Scaled leading four RPCs and correlation values between RCA's and GPCP's RPCs.

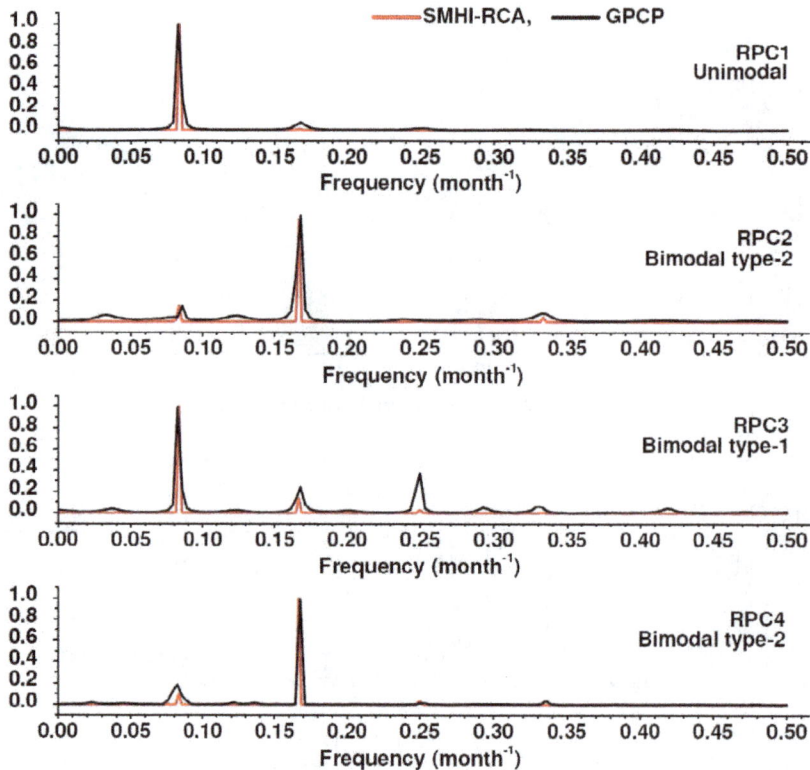

Figure 9: Normalized spectral density of the four leading RPCs for RCA and GPCP.

density associated with the autoregressive (AR) model was utilized to identify pseudo periodicities of the leading RPCs (Figure 9).

The results show the first four REOFs/RPCs account for 78.81% and 84.4% (respectively for RCA and GPCP) fraction of the total variance. They depict spatial and temporal pattern of homogeneous precipitation zones. Spectra of the RPCs exhibit dominant peaks at frequency f=0.167 month^{-1} (period T=6 months) and f=0.83 month^{-1} (T=12 months). Based on the spectral peaks, rainfall can be divided into unimodal (one season) and bimodal (two seasons) types. Bimodal is further splited up to bimodel type-1 and bimodal type-2.

REOF1 captures the well-known precipitation pattern exists in the north Ethiopia and have better correlated with northwestern and central highlands. This mode accounts for the largest amount of variance (40.21% and 41.55% for RCA and GPCP respectively). RPC1 time series, which is associated with REOF1, shows inter-seasonal variability. It has one dominant broad width associated with the long rainy season (JJAS) attaining maximum near July. Spectrum of RPC1 has strong signal at 12 months that corresponds to unimodal rainfall type (Kiremt season) [7].

REOF2 explains 27.21% and 31.8% (respectively for RCA and GPCP) fractions of the total variability. It involves rainfall anomalies over the southern Ethiopia and is strongly correlated with southeastern portion. Its spectrum has strong signal at 6 months and hence explains two seasons of roughly equal length and amplitude (bimodal type-2) peaking in October and April months. It has two distinct rainy periods (SON and MAM/Belg), separated by well-marked dry periods (DJF and JJA). Forward (northward movement) and retreat (southward movement) of ITCZ give rise to a bimodal rainfall pattern [3].

The spatial patterns of REOF3 appear to capture the east and northeastern precipitation pattern. It explains 6.77% and 5.6% (respectively for RCA and GPCP) of the total variance. Spectrum of RPC3 exhibits two unequal peaks that stand for bimodal type-1 mode. The two stronger peaks below one year suggest occurrence of two unequal precipitation patterns or existence of long (Kiremt) and short (Belg) rain seasons.

REOF4 explain only about 4.62% and 4.49% of the total variance for RCA and GPCP respectively. RPC2 and RPC4 signals, that have a periodicity of around 6 months, capture similar zone though RPC4 is more correlated with southwest and has better peak value in April than in October. The other unmentioned principal components capture the rest random variations departing from overall regional value.

These methods confirm the previous classification made by NMSA (1996). Monomodal rainfall pattern in the northern Ethiopia (June-September (Kiremt)) and bimodal rainfall pattern in southern Ethiopia (March-May (Belg) and September-November (SON). During Kiremt the air flow is dominated by a zone-of convergence in low-pressure systems accompanied by the oscillatory ITCZ extending from West Africa through the north of Ethiopia towards India. There is convergence between the air stream of African southwest monsoons diverted from the south Atlantic southeast trades and the Indian southwest monsoon on the Ethiopian highlands, especially on the western, central and eastern high grounds, resulting in heavy rainfall over the region [7]. Major rain-producing systems during Kiremt are: the northward migration of the ITCZ; development and persistence of the Arabian and the Sudan thermal lows; development of quasi-permanent high-pressure systems over the south Atlantic and south Indian Oceans; development of the tropical easterly jet and its persistence and the generation of the low-level Somali jet which enhances low-level southwesterly flow [17]. The Belg season coincides with the domination of the Arabian high as it moves towards the north Arabian Sea. It is the short and long rainy season for eastern and southern Ethiopia respectively [7,15]. Major systems during the Belg are: the domination of the Arabian high as it moves towards the north Arabian Sea and the development of thermal low (cyclone) over the south Sudan. Winds from the Gulf of Aden and the Indian Ocean highs that are drawn towards this centre blow across central and southern Ethiopia [1,7,15]. These easterly and southeasterly moist winds produce the main rains in southern and southeastern Ethiopia and the Belg rains to the east and central part of the northwestern highlands.

The value on the right top of each layer in Figure 8 describes Spearman and Pearson correlation coefficients between RCA and GPCP rainfall signals. For RPC1, both RCA and GPCP data are highly linearly correlated; whereas for RPC2, RPC3 and RPC4; RCA and GPCP are better nonlinearly correlated than linear.

Clusters

As the climate is rather complex, another broad classification of rainfall is found important to identify diffused homogeneous precipitation zones. Two criteria are used to divide the country into similar rainfall zones: identical peaks and minimum-distance clustering method. These criteria are applied to RCA and GPCP precipitation datasets. First, similar rainfall peaks above grid mean have been computed and categorized to get initial classes. Then clustering using simple-linkage criterion is applied for further regrouping within each class. Finally, clustered precipitation zones are assigned to the same color. Figure 10 shows homogeneous precipitation series lying within groups. Peaks above grid mean each cluster/zone possesses are displayed on top of the legend. Uppermost dominant color is assigned to the corresponding cluster name. Asterisk show significant peaks which might not be ignored.

In this way ten small scale distinct homogeneous zones are identified, nine being shared by both. Naming of zones is done in line with the previous works labeled as Zone- A_1 – A_2, Zone- B_1 – B_3 Zone- C_1 – C_4, Zone-D and Zone-E. Except zone-C and zone-E all have maxima that fall in the JJAS (Kiremt). Out of these, zone-A has additional peak in March/April associated with Belg. Zone-C and Zone-E (southern and southeastern parts) has two distinct maxima (in April/May and October/November). Terrain induced peak precipitation is observed for zone-D. Higher altitude has earlier climax month than lower altitude.

Peaks of rainfall primary follow forward and retreat of ITCZ when the low-pressure trough follows the sun's apparent movement in the northern hemisphere. The exact position of the ITCZ changes over the course of the year, migrating across the equator from its northern most position over northern Ethiopia in July and August to its southern most position over southern Kenya in January and February. However, the great terrain diversity in the country have induced rainfall pattern to have wide variations at local scale.

As soon as the ITCZ migrates northwards and enters south Ethiopia, zones start attain climax rainfall. The discussion below is synchronized with the movement of ITCZ. Zone-C and zone-E have rainfall crest in April/May. Zone-B_3 and zone-B_2 have maxima in June and July respectively. Zone-B_1 (north of zone-B_2) and zone-A (the lowland east of zone-B_2) have climax rainfall in August. Zone-E and zone-C have second maxima in October and October/November respectively connected to retreat of ITCZ back to the south. For zone-D (Arisi mountain), topography shows a considerable part in bringing rainfall. Higher altitude experiences earlier and higher monthly mean rainfall. In general, this and previous works have showed the intra-seasonal precipitation variation over Ethiopia is the result of large scale changes in macro-scale pressure systems and monsoon flows [9,10]. However, the fine scale spatial distribution of precipitation is significantly influenced by topography [4,7].

Zone-B_3 represents west highlands and experiences the longest rainy season of all zones. Major weather systems accompanied by high elevation play a considerable part in bringing rainfall throughout most of the year [17]. Winds carrying moisture fluxes from different directions forced up to these high grounds easily exhausted and cool to

Precipitation maxima above grid mean

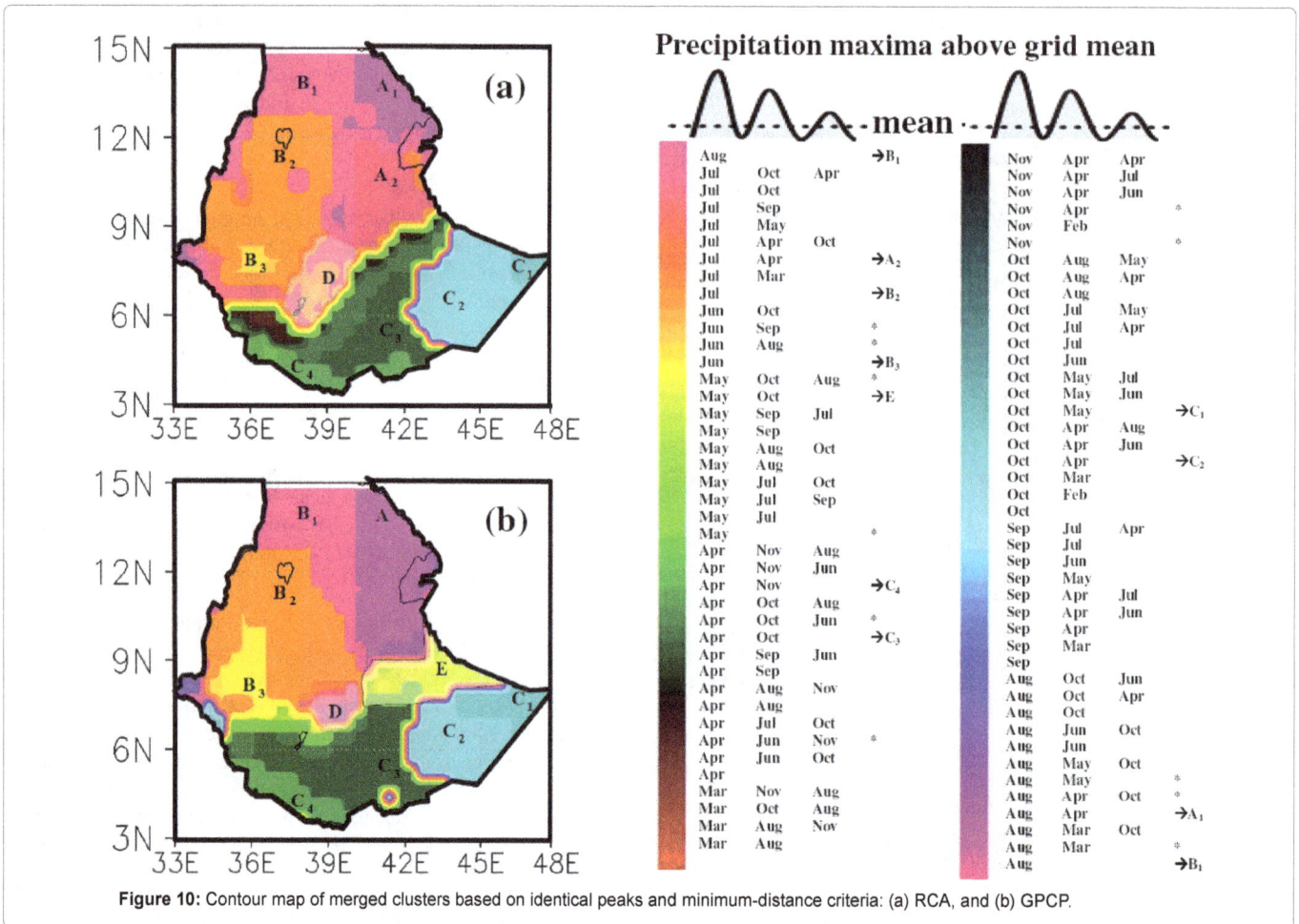

Figure 10: Contour map of merged clusters based on identical peaks and minimum-distance criteria: (a) RCA, and (b) GPCP.

give orographic rain [2]. Dominant peaks for zone-B_3 fluctuate between June and August. Zone-B_2 stands for the highlands found northwest of the Rift Valley and experiences heavy rainfall in Kiremt having climax in July. Elevation still affects this region. Zone-B_1 describes northern part of Ethiopia and has one dominant peak in August. The northeast lowland is captured by zone-A. It has two peaks in July and April (for Zone-A_2) or in August and March (for Zone-A_1) associated with Kiremt and Belg seasons (i.e. bimodal type-1). Zone-C represents bimodal type-2 with similar rainfall durations in MAM and SON having two dominant peaks in April/May and October/November. It describes precipitation series occurred over south and southeast. Well distinct four sub-groups (C_1-C_4) are found as a result of position of ITCZ, differences in elevation and wind velocity blowing over it. Zone-C_1 has highest peaks in October and May; zone-C_2 has peaks in October and April. Magnitudes of the two peaks are reversed for zone-C_3. Peaking in April than in October is because of the differences in the amount of moisture flux carried into and released over these zones by two winds. In April wind (with relatively higher moisture) from south Indian ocean crossing the flat lands of eastern Kenya [24,25]. Ellen et al. [2] starts to dominate the easterly/southeasterly wind associated with the Arabian Sea High Camberlin et al. [15] and brings more rainfall to zone-C_3 and zone-C_4. The southeastern part (zone-C_1 and zone-C_2) do not experience significant rainfall as the west part (zone-C_3 and zone-C_4) owing to: the development of low-level Somali Jet (also called East African low-level jet); simultaneous transition of low-level trade from southeast (ascending wind) to southwest (descending wind);

consequently weakening of rainfall activity over these zones [26,27]. In JJAS, the Jet intensifies and enhances divergence and dryness over zone-C [24,25]. Clear distinction is observed aligned with 42°E due to splitting/deflecting of wind near Marsabit Mountain towards the two low pressure systems. One across southwestern Ethiopia that persist as Turkana Jet towards south Sudan and the other across southeastern Ethiopia, Indian Ocean to India [3,28,29]. In October month the Somali Jet starts to weaken and the associated wind having little/no deflection accompanied by enhanced orography effect and convective activity dominates over zone-C_1 and zone-C_2. On the contrary swelling of Turkana Jet weakens the convective activity over zone-C_3 and zone-C_4 [29]. Zone-C_4 (which is relatively lowland) has highest peak in April as zone-C_3 with the second peak shifted to November. On the other hand, rainfall over zone-E has bimodal type-2 feature with climaxes in May and October although RCA merges it with Zone-C. Navel of Ethiopia, which is identified by zone-D, is highly influenced by local elevation and hence by orographic rain. It has combined features just as zone-A and zone-B in both datasets. The pockets of matchless rainfall activity observed in both RCA and GPCP might be caused by the influence of local elevation on wind velocity and hence orographic rainfall [30]. In general, this and previous works confirm the large scale spatial variation of rainfall is influenced by the changes in intensity, position and direction of the rain-producing systems [11,12]. However, the small scale spatial distribution of rainfall in Ethiopia is significantly influenced by topography [4,7].

The results of hierarchical CLA more or less agree with the

reduction of data using PCA. Zone-B is captured more by REOF1 to some extent by REOF3. It represents the distribution of precipitation over western, central and northern Ethiopia. Zone-A is more described by REOF3 and to some extent by REOF1. Zone-C is captured by both REOF2 and REOF4; likewise, the southeast and southwest parts are well explained by REOF2 and REOF4 respectively. Zone-D and Zone-E have intermediate characters and described by all REOFs.

Conclusion

This study examines the performance of the SMHI-RCA focusing on seasonal variations in precipitation over Ethiopia based on GPCP datasets. The basic statistical results showed that the model has good performance in simulating the precipitation variability and the spatial distribution of rain bands. However, it has been observed that RCA exhibits a little tendency to overestimate the highest rainfall over the highlands but underestimate the lowest rainfall over the eastern lowlands. RPCs and the associated spectra for both datasets demonstrate that the improvement of rainfall simulations is remarkable over the analysis domain. The leading four RPCs computed from PCA explain 37.55%, 26.84%, 10.30%, 3.16% and 41.55%, 31.8%, 5.6%, 4.49% for RCA and GPCP respectively. Spectra of the leading RPCs show dominant peaks at period of 6 months and 12 months that corresponds to unimodal (RPC1), quasi-bimodal (RPC3) and bimodal (RPC2 and RPC4) rainfall types. Classification using hierarchical clustering analysis (CLA) agrees with the reduction of data by PCA. Rainfall peaks at each zone primary follows the intensity, position and direction of the rain-producing systems; however, at the finer scale is significantly influenced by topography.

Acknowledgements

The authors would like to acknowledge for the data providers: KNMI Climate Explorer which is the primary web site for GPCP and other high resolution gridded observational datasets and Sweden's Meteorological and Hydrological Institute for making available RCA data on CORDEX data archive.

References

1. Yilma S, Zanke U (2004) Recent changes in Rainfall and Rainy Days in Ethiopia. International Journal of Climatology 24: 973-983.
2. Ellen V, Asgeir S (2011) Moisture transport into the Ethiopian highlands. International Journal of Climatology 10: 249-263.
3. Nicholson SE (1996) A review of climate dynamics and climate variability in eastern Africa. In: Johnson TC, Odada E (eds) The limnology, climatology and palaeoclimatology of the East African Lakes. Gordon and Breach, Amsterdam, pp. 25-56.
4. Slingo J, Spencer H, Hoskins BJ, Berrisford P, Black E (2005) The meteorology of the western Indian Ocean and the influence of the East African Highlands Philosophical Transactions of the Royal Society 363: 25-42.
5. Dinku T, Ceccato P, Grover-Kopec E, Lemma M, Connor SJ, et al. (2007) Validation of satellite rainfall products over East Africa's complex topography. International Journal of Remote Sensing 28: 1503-1526.
6. Segele ZT, Lamb PJ, Leslie LM (2009) Large-scale atmospheric circulation and global sea surface temperature associations with Horn of Africa June-September rainfall. International Journal of Climatology 29: 1075-1100.
7. National Meteorology Service Agency (1996) Climatic and agroclimatic resources of Ethiopia. National Meteorological Services Agency of Ethiopia. Meteorological Research Report Series 1: 1-137.
8. Korecha D, Barnston AG (2007) Predictability of June-September Rainfall in Ethiopia. Monthly Weather Review 135: 628-650.
9. Haile T (1987) A case study of seasonal forecasting in Ethiopia. WMO Regional Association I, Geneva pp 53-76.
10. Bekele F (1997) Ethiopian use of ENSO information in its seasonal forecasts. Internet J Afr 2: 2.
11. Tadesse T (1994) The influence of the Arabian Sea storms/depressions over the Ethiopian weather. Proc. Int. Conf. On Monsoon Variability and Prediction, WCRP-84 and WMO Tech. Doc. 619, World Meteorological Organization, Geneva, pp. 228-236.
12. Camberlin P (1995) June-September rainfall in northeastern Africa and atmospheric signals over the tropics: a zonal perspective. International Journal of Climatology 15: 773-783.
13. Camberlin P (1997) Rainfall anomalies in the source region of the Nile and their connection with the Indian summer monsoon. J Climate 10: 1380-1392.
14. Pedgley DE (1966) The Red Sea Convergence Zone. Weather 21: 394-406.
15. Camberlin P, Philippon N (2002) The East African March-May Rainy season: Associated Atmospheric dynamics and predictability over the 1968-1997 periods. J Climate 15: 1002-1019.
16. Shanko D, Camberlin P (1998) The Effect of the Southwest Indian Ocean Tropical Cyclones on Ethiopian Drought. International Journal of Climatology 18: 1373-1388.
17. Kassahun B (1987) Weather systems over Ethiopia. Proc. First Tech. Conf. on Meteorological Research in Eastern and Southern Africa, Nairobi, Kenya, UCAR: 53-57.
18. Segele ZT, Lamb PJ (2005) Characterization and variability of Kiremt rainy season over Ethiopia. Meteorology and Atmospheric Physics 89: 153-180.
19. Adler RF, Huffman GJ, Chang A, Ferraro R, Xie P, et al. (2003) The Version 2 Global Precipitation Climatology Project (GPCP) Monthly Precipitation Analysis (1979-Present). J Hydrometeor 4: 1147-1167.
20. Huffman GJ, Adler RF, Morrissey M, Bolvin DT, Curtis S, et al. (2001) Global Precipitation at One-Degree Daily Resolution from Multi-Satellite Observations. J Hydrometeor 2: 36-50.
21. Wilks DS (2006) Statistical Methods in the Atmospheric Sciences. (2ndedn) Academic, Amsterdam.
22. Janowiak JE, Gruber A, Kondragunta CR, Livezey RE, Huffman GJ (1998) A comparison of the NCEP–NCAR reanalysis precipitation and the GPCP rain gauge–satellite combined dataset with observational error considerations. J Climate 11: 2960-2979.
23. Zwiers FW and Von Storch H (1995) Taking Serial Correlation into Account in Tests of the Means. Journal of climate 8: 336-350.
24. Findlater J (1969) A major low-level air current near the Indian Ocean during the northern summer. Quarterly Journal of the Royal Meteorological Society 95: 362-380.
25. Findlater J (1977) Observational aspects of the low-level cross equatorial jet stream of the western Indian Ocean. Pure and Applied Geophysics 115: 1251-1262.
26. Flohn H (1965) Studies on the Meteorology of Tropical Africa. Bonner Meteorologische Abhandlungen 5: 57.
27. Flohn H (1987) Rainfall teleconnections in northern and northeastern Africa. Theoretical and Applied Climatology 38: 191-197.
28. Kinuthia JH, Asnani GC (1982) A Newly Found Jet in North Kenya (Turkana Channel). Monthly Weather Review 110: 1722-1728.
29. Kinuthia JH (1992) Horizontal and Vertical Structure of the Lake Turkana Jet. Journal of Applied Meteorology 31: 1248-1274.
30. Barbro J, Deliang C (2003) The influence of wind and topography on precipitation distribution in Sweden: statistical analysis and modelling. International Journal of Climatology 23: 1523-1535.

6

Reproductive Structure of Invading Fish, *Oreochromis niloticus* (Linnaeus, 1757) in Respect of Climate from the Yamuna River, India

Amitabh Chandra Dwivedi[1*], **Priyanka Mayank**[1] **and Sheeba Imran**[2]

[1]*Regional Centre, ICAR-Central Inland Fisheries Research Institute, 24 Panna Lal Road, Allahabad 211002, India*

[2]*Department of Biological Sciences, SHIATS, Allahabad, India*

*Corresponding author:** Amitabh Chandra Dwivedi, Regional Centre, ICAR-Central Inland Fisheries Research Institute, 24 Panna Lal Road Allahabad 211002, India
E-mail: saajjjan@rediffmail.com

Abstract

Fish is a rich source of protein, vitamins and minerals. New knowledge on the role of omega-3 fatty acids in human physiology and their high contents in fish has added a new dimension to their importance in health and nutrition. *Oreochromis niloticus* (Nile tilapia) is one of the fish species of great economic importance in the globe. Reproductive biology of *O. niloticus* in the Yamuna was studied. Samples of *O. niloticus* were collected monthly during August 2011 to July 2012. The fishes breed in months of March to June and September to November. Sex ratio of male and female fishes was equal in 43.1-46.0 cm size group. Male proportion was higher in all size groups except 43.1-46.0 cm size group. In 10.1-13.0 cm, 25.1-28.0 cm and 28.1-31.0 cm size groups' female proportion was half as compared to male fishes. Observed difference was not significant in all size groups, except 19.1-22.0 cm size group. Chi-square values were recorded maximum in 19.1-22.0 cm size group with 4.56 and minimum in 31.1-34.0 cm size group with 0.02. Higher proportion of male was observed in the stock, sex ratio of male and female was 1:0.78 and Chi-square value was 7.94 and the difference was significant. The fecundity and breeding frequency of *O. niloticus* indicated that the climate condition of the Yamuna river most suitable for *O. niloticus*.

Keywords: Fecundity; Sex ratio; Yamuna River; Climate; *Oreochromis niloticus*

Introduction

The large network of inland water masses will continue to provide great potential for economic capture fishery which consequently will compete well with fast growing fish culture practices. The freshwater inland water bodies fall into five major categories, the Ganga, the Brahmaputra, the Indus system, the East and the West coast of India. These water bodies have certain characteristics of their own with respect to their ecology, climatic conditions and fish populations of commercial food fishes.

Oreochromis niloticus (Nile tilapia) is an African native fish species. This species is naturally distributed in the Nile River as well as most parts of African rivers, reservoirs and lakes. It is exotic fish species for India. It is now transplanted to many other countries of the globe especially the tropical and subtropical parts of the world. It has become the second most commonly consumed farmed fish after carps [1]. It is commercially exploited in the Yamuna and Ganga River India [2-4]. Natural fisheries resources are dense by its stock in central India [5,6]. Natural fisheries are worked on a catch and return basis [7-9]. It is one of the most popular fish species for human consumption and profitable purposes by majority of poor community and fishermen [10]. *O. niloticus* is a deep-bodied fish with cycloid scales [11]. Culture of *O. niloticus* is increasing significantly in African and several Southeast Asian countries due to its fast growth rate under high stocking rates and seed availability in culture ponds [12,13]. It grows to a maximum length of 62 cm with weight 3.65 kg (at an estimated 9 years of age) [14]. Its survival and growth are superior in polluted water [15,16]. Heavy metal pollution is harmful for gonadal development [17].

Gonad condition (development) of fishes are total responsible for the propagation of the stock.

Breeding is synonymous to fish broadcast which simply means multiplication. Most riverine stocks are composed of multiple distinct breeding populations. Breeding is important for wild sock of fishes in natural water bodies. The sex ratio is very important to the reproduction of a population [18]. Information on the breeding and fecundity of *O. niloticus* can provide basic knowledge for the proper management of the resource. However, such knowledge is not recently available for this species in the Yamuna and this has hindered proper management of the fishery. Knowledge of sex ratio is considered essential in the management of the fisheries as it enables to follow the movement of the sexes in relation to the season [19]. The sex structure is very important to the sustenance of a population, and consequently there are mechanisms for adjusting it to changes in the food supply. The last is itself dependent on the population density, so that the sex ratio naturally reflects the density [20].

Studies on the various aspects of reproductive biology of *O. niloticus* have been carried out in the different parts of world. Considerable literature is available on the fecundity [21-26] spawning [24,27-29]. Sex ratio has been studied [22,24,29,30-34]. But nothing is known about *O. niloticus* with respect of reproductive biology from the Yamuna River, India. A large number of studies assess the main biological parameters and population dynamics of the *O. niloticus* for fisheries but lock in case of reproductive parameters. The present study was carried out to investigate sex ratio, sex structure, breeding season and fecundity of *O. niloticus* from the Yamuna River at Allahabad, India. Record and assessment of the present research work is necessary to formulate informed decisions about restoration and management of the fishery and rivers.

Materials and Method

Climate

The climate of the Gangetic plain is characterised by dry and wet periods. The dry period is longer than the wet period. The maximum air temperature is about 45°C to 46°C at Allahabad during May and June. The major wet period (monsoon) extends from mid-June to September. Though rainfall occurs in winter, it is negligible to be categorised as a wet period. The dry period therefore extends from October to mid-June, for nearly about 8 months. Occasional showers of local rainfall may occur in dry period following disturbances.

Methodology

During the course of study 516 specimens of *O. niloticus* (290 males and 226 females) were considered for the estimation of sex ratio and sex structure but in case of fecundity we 23 samples were collected. The female sex (fish) was determined by microscopic examination of the ovary as they show sexual dimorphism only in the breeding season. The numbers of fish samples were segregated on the basis of their sex (male and female) to determine the percentage composition of each sex in different age groups. This helped to understand the distribution of sexes in different age groups. Their ratio (M:F) was computed for each age group. The significant deviation from the hypothetical 1:1 sex ratio was tested by Chi-square [35] at 5% significance level.

Twenty three ripe female fishes were used for fecundity estimations in present study. To obtain representative samples of the whole gonads, 3 g eggs were taken from the posterior, middle and anterior of both lobes of the ovary. The numbers of ripe eggs were counted. The total number of ripe eggs in the ovary was estimated by multiplying the number of ripe eggs in the sample by the ratio of the ovary weight to the sample weight.

Sokl and Rohlf [35] have given the following formula especially for two classes calculation of chi-square.

$$Chi - Square(\chi^2) = \sum \frac{O_i^2}{E_i} - n$$

Where

O_i=Observed frequency,

E_i=Expected frequency and

n=Number of total sample

Result

Breeding seasons

The environment condition (especially temperature) of the Yamuna River at Allahabad is slightly warm which is very favourable for *O. niloticus*. The water current velocity of the Yamuna River is very poor. Both parameters are very helpful for the stability of *O. niloticus* in the Yamuna River at Allahabad. The frequency of temporal variation between ripe males and females was similar. The smallest spent female fish was recorded with 17.8 cm total length. The fishes breed in months of March to June and September to November. A lot of spent fishes were recorded in this period. Current study indicated that the *O. niloticus* breeds two times in per year. The first breeding season was reported long with 4 months and second was short with 3 months. *O.*

niloticus were caught at various stages of gonad development and reproduction in almost all months. However, their frequency varied with the month of capture. The seasonal pattern of gonad development for both sexes was almost similar.

Fecundity

A total of 23 ripe female were used for fecundity estimation. Their total length ranged from 14.6 to 40.5 cm. In the current findings, *O. niloticus* were found to be mature starting below 14.6 cm total length, taking individual fish. The number of eggs per individual ranged from 410 to 4008 (Table 1). The number of ripe eggs increased with ovarian weight. Fecundity was linearly related to total length (Figure 1). Fecundity was also linearly related to weight of gonad (Figure 2). The fecundity of *O. niloticus* indicated that the climate condition of the Yamuna river most suitable for *O. niloticus*. The number of eggs closely related to size of fishes compared to the gonad weight.

SN	Total length (cm)	Gonad weight (g)	No. of eggs
1	14.6	3.41	410
2	16.8	7.89	768
3	16.9	8.73	790
4	17.8	11.65	1286
5	17.6	10.49	1016
6	18.5	10.46	987
7	20.0	12.46	1310
8	21.1	14.21	1613
9	21.9	14.00	1706
10	23.0	13.19	1920
11	25.5	16.41	1985
12	25.6	12.20	2165
13	28.2	16.49	2065
14	28.6	19.58	2428
15	30.0	16.89	2760
16	30.6	20.49	2562
17	32.5	21.68	2941
18	33.5	22.01	2946
19	35.6	26.45	3462
20	36.0	29.65	3865
21	38.0	29.46	3659
22	39.0	27.49	3669
23	40.5	31.62	4008

Table 1: Fecundity of *Oreochromis niloticus* from the Yamuna River, India.

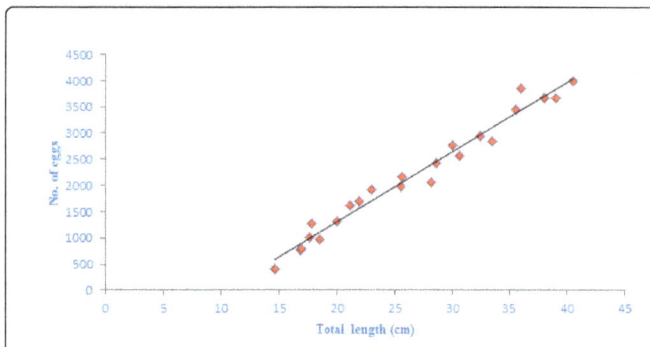

Figure 1: Relationship between fecundity and total length of *O. niloticus* in the Yamuna River.

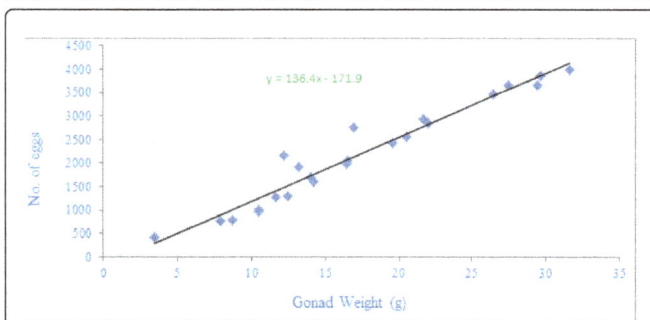

Figure 2: Relationship between fecundity and gonad weight of *O. niloticus* in the Yamuna River.

According to size, sex ratio of male and female fishes was equal in 43.1-46.0 cm size group. Male ratio was higher in all size groups except 43.1-46.0 cm size group (Table 2). The oral incubation behaviour (female, eggs and larvae) is responsible for the higher male proportion of *O. niloticus* in the Yamuna River. After spawning in a nest made by a male, the young fry or eggs are carried in the mouth of the mother for a period of 12 days. During oral incubation behaviour period female fish's movement are limited. In 10.1-13.0 cm, 25.1-28.0 cm and 28.1-31.0 cm size groups' female proportion was half as compared to male fishes. Observed difference was not significant in all size groups, except 19.1-22.0 cm size group. Chi-square values were recorded maximum in 19.1-22.0 cm size group with 4.56 and minimum in 31.1-34.0 cm size group with 0.02. Higher proportion of male was observed in the stock, sex ratio of male and female was 1:0.78 and Chi-square value was 7.94 and the difference was significant.

Size Classes (cm)	Male	Female	Sex ratio	Chi-square	Significance
10.1-13.0	10	5	01:00.5	0.72	NS
13.1-16.0	33	23	01:00.8	0.98	NS
16.1-19.0	39	35	01:00.8	1.76	NS
19.1-22.0	43	39	01:00.7	4.56	S
22.1-25.0	34	31	01:00.8	0.38	NS
25.1-28.0	39	25	01:00.5	1	NS
28.1-31.0	42	33	01:00.5	0.34	NS
31.1-34.0	15	14	01:00.9	0.02	NS
34.1-37.0	23	13	01:00.6	2.78	NS
37.1-40.0	7	5	01:00.7	0.34	NS
40.1-43.0	4	2	01:00.5	0.66	NS
43.1-46.0	1	1	01:01.0	-	-
Total	290	226	01:00.8	7.94	S

Table 2: Sex ratio of *Oreochromis niloticus* from the Yamuna River at Allahabad.

Sex structure of male and female was equal in 43.1-46.0 cm size group. Male proportion was higher than the female in all size groups. Sex structure of male was observed maximum in 10.1-13.0 cm size group (66.67%) while female was observed in 43.1-46.0 cm size group with 50% (Figure 3).

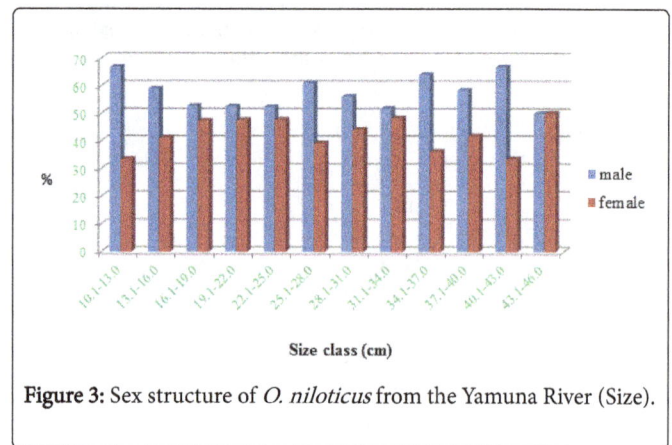

Figure 3: Sex structure of *O. niloticus* from the Yamuna River (Size).

Discussion

The fecundity of *O. niloticus* varied from 410 to 4008 eggs/fish from the Yamuna river, India. The high fecundity of *O. niloticus* in the Yamuna River at Allahabad could probably be a result of combination of different causes such as climatic condition, habitat structure, food abundance, water quality and quality of food. [28] Was observed fecundity between 241 to 709 eggs/fish in 125-209 mm size of fishes. Shalloof and Salama [36] reported more or less similar fecundity in same fish from the Abu-zabal Lake, Egypt. Lower fecundity was recorded from the lake Beseka, Ethiopia [24].

The largest size of *O. niloticus* indicated that the well stable in the Yamuna river. The size composition is good indicator for estimation of stock health in water bodies [37,38]. Male fishes of *O. niloticus* were dominated in the lower stretch of the Yamuna River at Allahabad, India. Theoretically, the expected ratio of males to females should be 1:1 [39]. The sex ratio of most fish species in the wild tends to be 1:1, but deviations can occur and seasonal variations are common [7,40]. The sex ratio is influenced by several factors, including mortality, longevity and growth rate, these in turn lead to differences in the catch rate [41]. Generally, Tilapia species in temperate areas have very short breeding season and limited to some months only, but in tropical countries their breeding seasons are extended, and in most cases throughout the year [28]. Fecundity of *O. niloticus* depends mainly on the body condition of the fish [42].

Novaes [31] was recorded male *O. niloticus* comprised 56.1% and females 43.9% of the catches, with a sex ratio (M:F) of 1.3:1 from Barra Bontia reservoir, Brazil. Gomez-Marquez et al. [43] stated that the female *O. niloticus* (51.8%) was more than male (48.2%) from a tropical shallow lake in Mexico. Pena-Mendoza [44] reported that the 624 fishes of *O. niloticus* caught with 272 females (43.59%) and 352 males (56.41%) and sex ratio was 1:1.29 (females: males) in Emiliano Zapata dam Morelos, Mexico. Sex ratio is helpful in understanding the recruitment of fishes in population. The sex ratio in the spawning population and in the various age and size groups varies with the species, reflecting the relationship of that species to its environment. The sex structure is also adaptive to the food supply, which thereby influences the reproductive rate and the variability of the offspring [2,4].

Garcia-Lizarraga [45] stated that, in the Aguamilapa reservoir (Mexico), dominance of males *Cichlasoma beani* over females in most months. Ramos-Cruz [46] found that male *O. aureus* accounted for 72% of the population in the Benito Juarez reservoir while Biotecs [47] reported that males of *O. aureus* shared for 67% of the population in the Temazcal reservoir. Inuwa [25] reported that the sex ratio of *O. niloticus* was 1:1 in Jakara dam, Kano, Nigeria. Sex ratio differs from one population to another of the same species, and may vary from year to year in the same population [20]. The unequal sex ratio for mature fish reflects different survival rates for males and females [48]. Fecundity of fishes is associated to the growth and climatic condition [49].

Conclusion

It may be concluded that the fecundity and breeding frequency of *O. niloticus* indicated that the climatic condition of the Yamuna River is most suitable. The fecundity of *O. niloticus* in the Yamuna River is higher compared to other places of world. *O. niloticus* is powerfully invaded in the Yamuna River, India.

References

1. Kevin F (2008) Tilapia products quality and new product forms for international markets Arizona, USA.

2. Mayank P, Dwivedi AC (2015) Biology of *Cirrhinus mrigala* and *Oreochromis niloticus*. LAP LAMBERT Academic Publishing GmbH & Co. KG, Dudweiler Landstr. 99, 66123 Saarbrucken, Germany188.

3. Mayank P, Dwivedi AC (2015) Role of exotic carp, *Cyprinus carpio* and *Oreochromis niloticus* from the lower stretch of the Yamuna river. Advances in biosciences and Technology 93-97.

4. Pathak RK, Gopesh A, Dwivedi AC (2015) Invasion potential and biology of *Cyprinus carpio* (Common carp). LAP LAMBERT Academic Publishing GmbH & Co. KG, Dudweiler Landstr. 99, 66123 Saarbrucken, Germany.

5. Pathak RK, Gopesh A, Dwivedi AC (2011) Alien fish species, *Cyprinus carpio var. communis* (common carp) as a powerful invader in the Yamuna river at Allahabad, India. National Academy of Science Letter 34: 367-373.

6. Dwivedi AC, Mishra AS, Mayank P, Tiwari A (2016) Persistence and structure of the fish assemblage from the Ganga river (Kanpur to Varanasi section), India. Journal of Geography and Natural Disasters 6: 159.

7. Dwivedi AC, Nautiyal P (2010) Population dynamics of important fishes in the Vindhyan region, India. LAP LAMBERT Academic Publishing GmbH & Co. KG, Dudweiler Landstr. 99, 66123 Saarbrucken, Germany 220.

8. Dwivedi AC, Nautiyal P (2012) Stock assessment of fish species, *Labeo rohita, Tor tor and Labeo calbasu* in the rivers of Vindhyan region, India. Journal of Environmental Biology 33: 261-264.

9. Imran S, Jha DN, Thakur S, Dwivedi AC (2015) Age structure of *Labeo calbasu* (Hamilton 1822) from the river Yamuna. Journal of the Inland Fisheries Society of India 47: 81-85.

10. Zambrano L, Martinez-Meyer E, Menezes N, Peterson AT (2006). Invasive potential of common carp (*Cyprinus carpio*) and Nile tilapia *Oreochromis niloticus* in American freshwater systems. Canadian Journal of Fisheries and Aquatic Sciences 63: 1903-1910.

11. Picker MD, Griffiths CL (2011) Alien and invasive animals-a South African perspective. Randomhouse/Struik, Cape Town, South Africa 240.

12. Sukadi F (2001) Tilapia culture in Indonesia. INFOFISH Int 4: 37-39.

13. Mayank P, Dwivedi AC (2015) Population structure of alien fish species, *Oreochromis niloticus* (Linnaeus 1758) from lower stretch of the Yamuna river, India. Journal of the Kalash Science 3: 35-40.

14. FAO (2012) Cultured aquatic species information programme. *Oreochromis niloticus*. Cultured aquatic species information programme. FAO Fisheries and Aquaculture Department.

15. Tiwari A, Dwivedi AC (2014) Assessment of heavy metals bioaccumulation in alien fish species Cyprinus carpio from the Gomti river, India. European Journal of Experimental Biology 4: 112-117.

16. Dwivedi AC, Tiwari A, Mayank P (2015) Seasonal determination of heavy metals in muscle, gill and liver tissues of Nile tilapia, *Oreochromis niloticus* (Linnaeus, 1758) from the tributary of the Ganga River, India. Zoology and Ecology 25: 166-171.

17. Tiwari A, Dwivedi AC, Shukla DN, Mayank P (2014) Assessment of heavy metals in different organ of *Oreochromis niloticus* from the Gomti river at Sultanpur, India. Journal of the Kalash Science 2: 47-52.

18. Dwivedi AC, Nautiyal P, Joshi KD (2011) Sex ratio and structure of certain cyprinids of Vindhyan region in Central India. Journal of the Inland Fisheries Society of India 43: 77-82.

19. Somvanshi VS (1980) Study on some aspects of spawning biology of a hill stream fish *Garra mullya* (Sykes). Proceeding of the Indian National Science Academy 46: 105-113.

20. Nikolskii GV (1980) Theory of fish population dynamics as the biological background for rational exploitation and management of fishery resources. Otto Koeltz Science Publishers Kooenigstein, W. Germany 1-323.

21. Babiker MM, Ibrahim H (1979) Studies on the biology of reproduction in cichlid Tilapia nilotica, gonadal maturation and fecundity. Journal of Fish Biology 5: 437-447.

22. Admassu D (1994) Maturity, fecundity, brood size and sex ratio of Tilapia (*Oreochromis niloticus*) in lake Awassa. SINET: Ethiopian Journal of Sciences 17: 53-96.

23. Balirwa JS (1998) Lake Victoria wetlands and the ecology of Nile tilapia *Oreochromis niloticus*. A Balkema Rotterdam 83-88.

24. Hirpo LA (2013) Reproductive biology of *Oreochromis niloticus* in Lake Beseka, Ethiopia. Journal of Cell and Animal Biology 7: 116-120.

25. Inuwa B (2013) Fecundity and length relationship of fish species collected from jakara dam, kano, Nigeria. Journal of Biological Sciences and Bioconservation 5: 143-153.

26. Singh AK, Verma P, Srivastava SC, Tripathi M (2014) Invasion, biology and impact of feral population of Nile tilapia (*Oreochromis niloticus Linnaeus* 1757) in the Ganga river (India). Asia Pacific Journal of Research 1: 151-163.

27. Bhujel RC (2000) A review of strategies for the management of Nile tilapia (*Oreochromis niloticus*) broodfish in seed production systems, especially hapa-based system. Aquaculture 181: 37-59.

28. Gómez-Márquez JL, Peña-Mendoza B, Salgado-Ugarte IH, Guzmán-Arroyo M (2003). Reproductive aspects of *Oreochromis niloticus* (Perciformes: Cichlidae) at Coatetelco lake, Morelos, Mexico. Revista De Biologia Tropical 51: 221-228.

29. Offem BO, Akegbejo-Samsons Y, Omoniyi IT (2007) Biological assessment of *Oreochromis niloticus* (Pisces: Cichlidae; Linne, 1958) in a tropical floodplain River. African Journal of Biotechnology 6: 1966-1971.

30. Lemma AH (2012) Breeding seasons and condition factor of *Oreochromis niloticus* (Pisces: Cichlidae) in Lake Babogaya, Ethiopia. International Journal of Agriculture Sciences 2: 116-120.

31. Novaes JLC, Carvalho ED (2012) Reproductive, food dynamics and exploitation level of *Oreochromis niloticus* (Perciformes: Cichlidae) from artisanal fisheries in Barra reservoir, Brazil. Revista De Biologia Tropical 60: 721-734.

32. Berihun A, Dejenie T (2012) Population dynamics and condition factor of *Oreochromis niloticus L.* in two tropical small Dams, Tigray (Norther Ethiopia). Journal of Agricultural Science and Technology B2: 1062-1072.

33. El-Zaeem SY, Salam GM (2013) Production of genetically male tilapia through interspecific hybridization between *Oreochromis niloticus* and *O aureus.* Journal of Fisheries Science 12: 802-812.

34. Mortuza MG, Al-Misned FA (2013) Length-weight relationships, condition factor and sex ratio of Nile Tilapia, *Oreochromis niloticus* in Wadi hanifah, Riyadh, Saudi Arabia. World J Zool 8: 106-109.

35. Sokal RR, Rohlf FJ (1973) Introduction of Biostatistics. W. H. Freeman and Company San Francisco Toppan Company, Limited, Tokyo Japan 1-368.

36. Shalloof KA, Salama HMM (2008) Investigations on Some Aspects of Reproductive Biology in *Oreochromis niloticus* (Linnaeus, 1757) Inhabited Abu-zabal Lake, Egypt. Global Veterinaria 2: 351-359.

37. Tripathi S, Gopesh A, Joshi KD, Dwivedi AC (2015) Size composition, exploitation pattern, sex ratio and sex structure of *Eutropiichthys vacha* (Hamilton, 1822) from the middle stretch of the river Ganga at Allahabad, India. Advances in biosciences and Technology 116-120.

38. Imran S, Thakur S, Jha DN, Dwivedi AC (2015) Size composition and exploitation pattern of *Labeo calbasu* (Hamilton 1822) from the lower stretch of the Yamuna river. Asian Journal of Bio Sciences 10: 171-173.

39. Holick J, Hensel K, Nieslanik J, Skacel J (1988) The Eurasian Huchen, Hucho hucho largest Salmon of the World.

40. Helfman EJ, Collette BB, Facey DE, Bowen BW (2007) The diversity of fish: biology, evolution and ecology. Wiley-Blackwell, Oxford, England.

41. King RP, Etim L (2004) Reproduction, growth, mortality and yield of Tilapia mariae Boulenger 1899 (Cichlidae) in a Nigerian rainforest wetland stream. Journal of Applied Ichthyology 20: 502-510.

42. Teferi Y, Admassu D, Mengistou S (2001) Breeding season, maturation and fecundity of *Oreochromis niloticus L.* (pisces: Cichlidae) in lake Chamo, Ethiopia. Ethiopian Journal of Science 24: 255-264.

43. Gomez-Marquez JL, Pena-Mendoza B, Salgado-Ugarte IH, Arredondo-Figueroa JL (2008). Age and growth of the tilapia Oreochromis niloticus (Perciformes: Cichlidae) from a tropical shallow lake in Mexico. Revista De Biologia Tropical 56: 875-884.

44. Pena-Mendoza B, Gomez-Marquez JL, Salgado-Ugarte IH, Mamirez-Noguera D (2005) Reproductive biology of *Oreochromis niloticus* (Perciformes: Cichlidae) at Emiliano Zapata dam, Morelos, Mexico. Revista De Biologia Tropical 53: 515-522.

45. Garcia-Lizarraga MA, Soto-Franco FE, de JM, Velazco-Arce JR, Velazquez-Abunader JI, et al. (2011) Population structure and reproductive behavior of Sinaloa cichlid Cichlasoma beani (Jordan, 1889) in a tropical reservoir. Neotropical Ichthyology 9: 593-599.

46. Ramos-Cruz S (1995) Reproduccion Y crecimiento de la mojarra tilapia (*Oreochromis aureus*) en la presa Benito Juarez, Oaxaca, Mexico en 1993. INP-SEMARNAP Ciencia Pesquera 11: 54-61.

47. Biotecs S (1990) Determinacion del potencial acuicola de los embalses epicontinentales mayors de 10000 hectareas Y nivel de aprovechamiento. Presa Miguela Aleman Temazcal. Informe Final Pesca Mexico City.

48. Leonardos I, Sinis A (1999) Population, age and sex structure of *Aphanius fasciatus Nardo*, 1827 (Pisces: Cyprinodontidae) in the Mesolongi and Etolikon lagoons (W. Greece). Fisheries Research 40: 227-235.

49. Mayank P, Tyagi RK, Dwivedi AC (2015) Studies on age, growth and age composition of commercially important fish species, *Cirrhinus mrigala* (Hamilton, 1822) from the tributary of the Ganga river, India. European Journal of Experimental Biology 5: 16-21.

7

Using the WRF Regional Climate Model to Simulate Future Summertime Wind Speed Changes over the Arabian Peninsula

Hussain Alsarraf[1]* and Matthew Van Den Broeke[2]

[1]Kuwait Meteorology Department, Kuwait
[2]Department of Earth and Atmospheric Sciences, University of Nebraska-Lincoln, USA

Abstract

The normal surface pressure distribution in the Middle East includes high pressure over the eastern Mediterranean Sea and low pressure over the southeastern Arabian Peninsula. The resulting west-east pressure gradient leads to summertime northerly or northwesterly shamal winds across the Arabian Peninsula, which typically result in many days per month with substantial lofted dust, leading to considerable human health and transportation impacts. It would be helpful to understand how the regional pressure gradient may change in the future, as the strength of this gradient exerts predominant control over the strength of the shamal wind. One factor possibly leading to changes in the strength of the pressure gradient is climate variability. We have simulated the regional climate under a present-day scenario (2006-2010) and a mid-century scenario (2056-2060) using the Weather Research and Forecasting (WRF) model. Our results indicate a weakening of the regional pressure gradient by mid-century, resulting in lower average wind speeds and fewer days conducive to dust storms across the Arabian Peninsula.

Keywords: Downscaling; Regional climate modeling; Sea-level pressure; Arabian peninsula; Shamal

Introduction

Variability of the Indian monsoon strongly affects the atmospheric circulation of the Middle East. Changes in sea-surface temperature (SST), soil moisture, sea ice, and Himalayan snow cover can lead to interannual variability of the Indian monsoon. It is important to understand the behavior of the Indian monsoon and its relationship with global climate during the summertime to be able to elaborate on how changes may affect regions influenced by the monsoon regime [1].

There is a link between global climate variability and Indian monsoon variability, and the forces that drive their variability can result from both external and internal forcings. One contributor to Indian summer monsoon precipitation variability is SST variability across the equatorial Pacific. For example, warm SST anomalies in the eastern equatorial Pacific and cool anomalies in the western equatorial Pacific result in the positive phase of the El Niño-Southern Oscillation (ENSO), which is one of the external forces that influence Indian monsoon precipitation [2]. The positive phase of ENSO weakens or reverses the Walker circulation across the equatorial Pacific, resulting in sinking air and decreased precipitation over Southeast Asia [3]. SSTs in the Indian Ocean can weaken or strength the subsidence, and cold SSTs over the Arabian Sea can lead to a decrease of monsoon precipitation [2]. Observations taken for longer periods of time suggest that Pacific Ocean SSTs are more strongly related to the Indian monsoon than Indian Ocean SSTs [2].

The Indian monsoon plays a key role in controlling the summer circulation over Asia and the Middle East. The reversal in the land-ocean temperature contrast during summertime drives the wind from the Arabian Sea toward the Indian subcontinent. Southeasterly wind over the southern Indian Ocean crosses the equator and is deflected due to the earth's rotation to become the southwesterly Somali Jet over the Arabian Sea to India. The onset of these winds coincides with the development of low pressure over India, resulting in a pressure gradient between the Indian subcontinent and Indian Ocean. This pressure gradient steers southerly flow into India to mature the summer monsoon, ultimately producing huge amounts of precipitation over India [1].

The Indian monsoon governs the northerly shamal wind in summertime over the Arabian Peninsula. The geographic location of the Zagros Mountains in western Iran forces a lee trough between the Persian Gulf and Iran. The pressure gradient across the Arabian Peninsula intensifies between this lee trough and high pressure over the Mediterranean Sea, which normally drives strong northerly (shamal) winds [4]. Intensification of the shamal wind in summer may cause frequent episodes of dust. Such dust storms can influence the earth's radiation budget, as the size and composition of dust aerosols can change the amount of shortwave radiation reaching the ground, affecting surface temperature [5]. A study by Francis et al. suggests that recent loss of sea ice may have consequences for atmospheric circulation on a hemispheric level. Sokolova et al. showed via a modeling study an association between low sea ice cover and the negative phase of the Arctic Oscillation (AO), indicating weaker zonal wind, although regional details are complicated by tracks of individual storm systems and atmospheric long-wave/low-frequency dynamic processes.

Observations indicate a reduction in the mean sea level pressure (MSLP) difference from east to west along the equatorial Pacific during the 20th century [6]. Observations of the pattern and strength of MSLP biases over the tropical Indo-Pacific from 1861-1992 concur with model-simulated changes when the model is forced with anthropogenic changes in radiative forcing [7].

Gillett et al. used four global climate models to investigate how these models explain observed changes in MSLP over the last 50 years. They found that that both polar regions have experienced MSLP decreases over the last 50 years, and that the North Atlantic Ocean,

*****Corresponding author:** Hussain Alsarraf, Kuwait Meteorology Department, Kuwait. E-mail: minnesotta@yahoo.com

Europe, North Africa, India, and other tropical to mid-latitude regions have experienced a MSLP increase. Global climate models run both with and without the effects of greenhouse gases show a qualitatively similar pattern of global pressure redistribution. In the Arabian Peninsula, changes in the MSLP distribution would produce changes in wind speed, affecting magnitude and frequency of future dust storms.

In summertime, a thermal low is present over Pakistan and Afghanistan, extending west across southern Iran and the Persian Gulf (Figure 1). Along with the semi-permanent high pressure zone to the west, the resulting pressure gradient enhances southward airflow across the Arabian Peninsula similar to that described in Aurelius [4,8]. The intense pressure gradient may be enhanced by a dry cold frontal passage, allowing especially strong northerly and northwesterly winds.

Surface winds vary diurnally. Strong daytime solar heating causes intensification of low pressure over the southeastern Arabian Peninsula. The thermal low remains nearly stationary during the day, and weakens at night due to surface cooling. The daytime deepening of the thermal low leads to a steeper pressure gradient, causing surface wind speed to increase. Strong surface heating and friction leads to development of a deep well-mixed layer during the daytime. When such a well-mixed layer is created over dry regions, it tends to become relatively dry and characterized by a high lapse rate [9].

Any change in the strength of the high to the west or low to the east/southeast may change the intensity of the pressure gradient, which may modify wind speed. Wind speed is a critical atmospheric variable for the Arabian Peninsula, since strong shamal winds in summertime may loft substantial quantities of dust with severe impacts on transportation and human health.

Changes in mean wind speed have been investigated globally and for different regions in the past, present, and future. Based on the results of Jiang et al. and observational data from the last 50 years, biases of the yearly and seasonal mean wind speeds over the globe have been calculated for the Northern Hemisphere. Decreasing trends in surface wind speed over land were found for middle and low latitudes, with increasing trends over high latitudes and some oceanic regions.

Reductions in surface wind speed by 5% to 15% have been found in Central Asia, Eastern and Southern Asia, Europe, and North America in the last 30 years [10]. Several studies over the past 50 years in East Asia have found that the winter and summer meridional circulation indices in this region had experienced decreasing trends, whereas the zonal circulation indices had shown increasing trends. Vautard et al. associated the reduction of Northern Hemisphere wind speed with increased land-surface roughness. Another study by Bichet et al. [5] using simulated wind speed observations and anomalies averaged over 30 years indicates a wind speed decrease by 0.3% over the Northern Hemisphere. In this study, the Northern Hemisphere was divided into domains, where the Arabian Peninsula was located in the south Asia domain. The south Asian domain wind speed declined by 0.24%. A decrease of wind speed globally and regionally in the Arabian Peninsula would decrease the potential for future dust storms.

Precipitation may also play a role in increasing or decreasing dust storm potential over the Arabian Peninsula. For instance, different precipitation characteristics of 1997, 2000, and 2008 were associated with differing numbers of days with summertime dust events in Kuwait. In past years, a slight increase in precipitation and decrease in wind speed has been observed in Kuwait. For the future, an increase of precipitation and decrease of wind speed been simulated over Kuwait by Al-Zawad [11], indicating decreasing dust events in the future.

If future global warming reduces the temperature gradient between the poles and equator, mid-latitude winds should also be reduced. Such a reduction of the temperature gradient may change the location of the jet stream, which may modify tracks of mid-latitude cyclones. Recent evaluation of a past climate reanalysis determined that "jet streams have risen in altitude and moved poleward in both hemispheres" [12]. A poleward shift of the jet stream may result in fewer mid-latitude cyclones able to influence the Arabian Peninsula, leading to fewer days favorable for significant dust storms. Such a poleward shift in mid-latitude cyclones has been noted in prior studies [13-16].

The mean summertime wind speed at Kuwait International Airport (KIA) decreased by 12% from 1973 to 2012, averaging 32.4 km hr^{-1} from 1973-1983 but only 28.5 km hr^{-1} from 2002-2012. This decrease in wind speed may be driven partially by a decrease in the pressure gradient across the Arabian Peninsula, which may be partially attributable to global climate change. Given these observed changes, there is a need to investigate how wind speed, and therefore dust storm potential, may change regionally in the future under a warning climate scenario. We utilize a regional climate model to simulate the mid-century (2056-2060) summertime MSLP distribution across the Arabian Peninsula to answer the following questions:

(1) Will the pressure gradient weaken or strengthen across the Arabian Peninsula in the future?

(2) How will summertime mean surface wind speed be affected by changes to the MSLP distribution?

By determining whether wind speed may be expected to continue deceasing under a future climate scenario, for the first time we will be able to infer how dust storm potential may change by mid-century across the Arabian Peninsula.

Experimental Design and Model Validation

WRF simulations and physics options

Simulations in this study were completed using the Weather Research and Forecasting (WRF) model, version 3.3. The Advanced Research WRF (ARW) dynamic core was utilized, which is non-

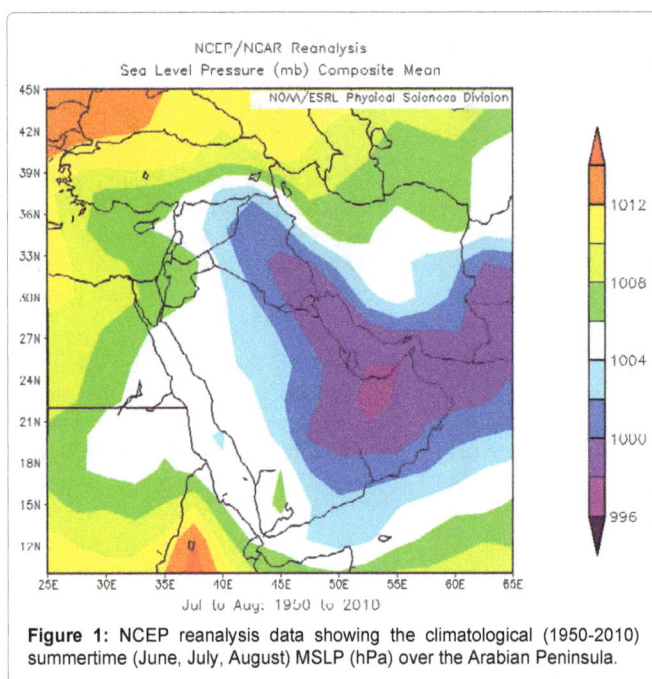

Figure 1: NCEP reanalysis data showing the climatological (1950-2010) summertime (June, July, August) MSLP (hPa) over the Arabian Peninsula.

hydrostatic and fully compressible [17]. The Eta (Ferrier) microphysics scheme was specified-this is the microphysics scheme used in the North American Mesoscale (NAM) model, and has demonstrated validity in broad-scale simulations of mid-latitude precipitation microphysics. For shortwave radiation, the Dudhia simple downward calculation was selected [18], and for longwave radiation the Rapid Radiative Transfer Model (RRTM) scheme was used [19]. These radiation schemes are valid for cloudy or clear-sky conditions. Monin-Obukhov similarity theory was used for the surface layer, and for the land surface the Noah Land Surface Model was used by Ek [20], which includes information about the dominant vegetation and allows soil moisture to influence the overlying atmosphere. The Yonsei University scheme was used for the planetary boundary layer [21,22]. The Kain-Fritsch scheme was used for the cumulus parameterization [23]. More details on the physics options can be found in Table 1. WRF simulations were completed for a domain centered on the Arabian Peninsula for 1997, 2000, and 2008, three years with different precipitation regimes across the region. Choosing three years with different climate history provides a more reliable test of the climate model output, and gives a better understanding of the model simulation under different climate regimes. WRF simulations with dynamic downscaling were forced by reanalysis using the National Centers for Environment Prediction (NCEP)/ National Center for Atmospheric Research (NCAR) Reanalysis Project output [24] for 1997, 2000, and 2008. These simulations are hereafter referred to as WRF-NCEP. Three WRF simulations were run: the first for a domain with 48-km horizontal resolution (D01), the second for a domain with 12-km horizontal resolution (D02), and the third for a domain with 4-km horizontal resolution (D03).

Model validation

A preliminary verification was done to compare the output of the WRF simulations with observational data from Kuwait for the years 1997, 2000, and 2008. The variables that were compared with observational data include maximum temperature, MSLP, precipitation, and wind speed. These data were compared with observations in each downscaling nested domain.

In 1997, total precipitation at KIA (215.3 mm) was well above average. In 2000, the precipitation (80.27 mm) was slightly lower than

average, while in 2008 total precipitation (50.9 mm) was less than half the annual average. Number of dust days generally followed the precipitation, with 41 days recorded in 1997, 38 days in 2000, and 56 days in 2008. High annual precipitation in 1997 was associated with few dust storms, while very low precipitation in 2008 was associated with a high number of dust storms. Precipitation was low in 2000, but so was the number of dust storm days.

When more precipitation falls during winter, a higher wind speed is required to loft dust particles. In a normal precipitation year, an average wind speed of 35-40 km hr^{-1} can cause blowing dust in summer, but in a high-precipitation year an average wind speed of at least 45 km hr^{-1} is required. The friction and surface roughness normally play a key role in identifying the speed of dust formation in the Arabian Peninsula and desert areas. As wind speed increases in the desert, the chance of dust events become greater and lower visibility is probable.

Downscaling results were first validated by comparing model output with observational data. Model output MSLP at (29°N, 48°E) was compared with observations from KIA (also 29°N, 48°E). A summary of the validation is presented in Table 2, which shows averages and standard deviations for June, July, and August of 1997, 2000, and 2008. The averages of the KIA observations and the WRF-NCEP 12 km and 4 km domains are very close, whereas the average of the 48 km output contains much higher pressure values. The root-mean-square error (RMSE) and the mean absolute error (MAE) calculations show the smallest magnitude of error at 4 km grid resolution (Table 2). A summary of the validation for wind speed in the 4-km WRF-NCEP run is shown in Table 3. The model tends to give slightly lower daily average wind speed values than the KIA observations, with similar standard deviation. Root-mean-square-error (RMSE) and mean-absolute-error (MAE) also indicate a relatively small magnitude of error at 4 km grid resolution (Table 3). Thus, we conclude that the results of D03 are the most representative, and show future wind speed results from D03 through the remainder of this paper.

Model Results

GCM forcing

In this part of the study, the Community Climate System Model

Variable	Value	Description
mp_physics	4	
sst_update	1	allow sea surface temperature update
ra_lw_physics	1	(longwave radiation) RRTM scheme. Spectral scheme K-distribution Look-up table fit to accurate calculations. Interacts with clouds Ozone/CO2 from climatology
ra_sw_physics	1	(short wave radiation) MM5 shortwave (Dudhia) Simple downward calculation. Clear-sky scattering. Water vapor absorption. Cloud albedo and absorption
radt	20	minutes between radiation physics calls
sf_sfclay_physics	1	(surface layer) Monin-Obukhov similarity theory. Taken from standard relations used in MM5 MRF PBL. Provides exchange coefficients to surface (land) scheme. Should be used with bl_pbl_physics=1 or 99.
sf_surface_physics	2	(land surface) Noah Land Surface Model: Unified NCEP/NCAR/AFWA scheme with soil temperature and moisture in four layers, fractional snow cover and frozen soil physics
bl_pbl_physics	1	(Planetary Boundary layer) Yonsei University scheme: Non-local-K scheme with explicit entrainment layer and parabolic K profile in unstable mixed layer.
bldt	0	minutes between boundary-layer physics calls. 0 = call every time step
cu_physics	1	(cumulus Parameterization) Kain-Fritsch scheme: Deep and shallow convection sub-grid scheme using a mass flux approach with downdrafts and CAPE removal time scale
cudt	5	minutes between cumulus physics calls
icloud	1	cloud effect to the optical depth in radiation
num_soil_layers	4	number of soil level or layer in WPS output

Table 1: Parameterizations used in WRF simulations for this study.

MSLP	1997				2000				2008			
	Obs	4 km	12 km	48 km	Obs	4 km	12 km	48 km	Obs	4 km	12 km	48 km
AVE	994.9	999.4	998.8	1011.3	993.6	997.5	997.3	1009.9	993.6	997.5	995.8	1007.8
STDEV	3.12	3.64	3.24	4.02	2.42	2.16	2.36	2.83	2.49	2.33	2.71	3.59
MAE		4.10	5.10	6.30		3.81	4.13	6.27		2.88	4.20	4.20
RMSE		4.67	5.66	6.68		4.35	4.69	6.64		3.51	4.85	4.69

Table 2: Daily average summertime MSLP values (June, July, and August) simulated by WRF- NNRP for 48 km-, 12 km-, and 4 km-resolution domains at (29°N, 48°E) compared with Kuwait International Airport (KIA; 29°N, 48°E) observations for 1997, 2000, and 2008. Standard deviation (STDEV), mean absolute error (MAE), and root mean square error (RMSE) are also shown for simulation results vs. observations.

Wind	1997		2000		2008	
	Obs	4 km	Obs	4 km	Obs	4 km
AVE	5.58	4.66	4.35	4.85	5.38	4.67
STDEV	2.13	2.94	1.93	2.56	2.25	3.47
MAE		2.30		2.30		2.82
RMSE		2.89		2.88		3.45

Table 3: Daily average summertime wind speed (June, July, and August) simulated by WRF- NNRP for 4-km resolution domain at (29°N, 48°E) compared with KIA observations (29°N, 48°E) for 1997, 2000, and 2008. Standard deviation (STDEV), mean absolute error (MAE), and root mean square error (RMSE) are also shown for simulation results vs. observations.

CCSM4 under the RCP8.5 scenario was used to provide lateral forcing for WRF runs for two five-year time periods, fifty years apart, to allow any climate change signal to be revealed. These runs are hereafter designated WRF-CCSM4. Two five-year periods were simulated: 2006-2010 (present-day climate) and 2056-2060 (mid-century climate). The second five-year period was chosen because going 50 years out helps assure the long-term signal will be clearly distinguished from shorter-term climate variability. The WRF runs were initialized at 0000 UTC on 27 April and ended on 3 October of each year, covering a total of 143 summer months (May-September). WRF model output was archived every three hours.

The left column of (Figure 2) shows the difference between the present-day (2006-2010) and mid-century (2056-2060) MSLP for May through September simulated by WRF-CCSM4. The right column of (Figure 2) shows the same MSLP difference simulated by WRF at 48-km horizontal resolution (D01) forced by CCSM4. The CCSM4 simulates MSLP globally but does not sufficiently represent regional and local MSLP differences. The WRF runs, on the other hand, simulate effects that can change MSLP distributions on a synoptic or regional scale, such as coastlines.

WRF model results

Mean sea level pressure: The maps produced from WRF-CCSM4 runs show the difference in average MSLP values from May to September (Figures 3-5) between the present (2006-2010) and mid-century (2056-2060). In May, MSLP increases over the Arabian Peninsula by up to 1.0 hPa, whereas west of the Arabian Peninsula it decreases up to 0.5 hPa (Figure 3a).

The month of June is normally dominated by a strong pressure gradient, and is characterized by strong shamal winds and dust storms. The gradient between low pressure to the southeast and the high pressure normally centered over the Mediterranean Sea is normally strong (Figure 1). The WRF-CCSM4 simulated differences in pressure for June indicate an increase in MSLP of 0.5 to 1.0 hPa in the northern and central Arabian Peninsula, while in southern portions of the Arabian Peninsula an increase of 1.5 to 2.0 hPa is indicated. In Iran, an increase of 1.5 to 2.0 hPa is shown whereas pressure remains nearly the same or slightly decreases in the Mediterranean. An increase of up to 3.0 hPa is indicated over the Himalayan Plateau, a large change between present and future climate conditions (Figure 3b).

July is also usually influenced by a strong pressure gradient across the Arabian Peninsula (Figure 1). The WRF-CCSM4 simulation results indicate an increase in MSLP of up to 1.0 hPa across most of the Arabian Peninsula, with greater increases south of the Peninsula and to the northeast over Iran, where an increase of up to 3.0 hPa is predicted. The western Arabian Peninsula is predicted to experience a MSLP decrease of approximately 0.5 hPa (Figure 3c).

The WRF-CCSM4 results for August show a pressure decrease of up to 1.5 hPa in the Arabian Peninsula. A decrease in MSLP of 1 to 2 hPa throughout most of the northern part of the Middle East, the Mediterranean region, North Africa, and Western Europe is predicted, with a decrease of up to 2.5 hPa in south Asia (Figure 3d). In September, the WRF-CCSM4 results indicate a pressure increase of up to 1.5 hPa in the Arabian Peninsula, and an increase of up to 2 hPa in Iran and India. North Africa and the Mediterranean Sea show a MSLP decrease from 1 to 2.5 hPa (Figure 3e).

The downscaling 12 km horizontal resolution domain simulation (D02; Figure 4) indicates results similar to D01, with an increase in May of up to 1.0 hPa over the Arabian Peninsula. In June, the WRF simulation shows an increase of 0.5 to 1.5 hPa in the northern and central Arabian Peninsula, but the southern region near the Indian Ocean experiences an increase of 1.5 to 2.0 hPa. The Zagros Mountains in Iran show even more substantial pressure increases (Figure 4b). D02 in June shows that the thermal low and extension of the Indian monsoon is weakened, leading to a lessening of the pressure gradient across the region. In July, D02 indicates a MSLP increase by midcentury of up to 1.0 hPa in the Arabian Peninsula (Figure 4c). In August, D02 shows similar results to D01, with a MSLP decrease by midcentury from 1 to 1.5 hPa in the northeast Arabian Peninsula. Pressure in the central and southern Arabian Peninsula decreases by a lesser amount. The east Mediterranean Sea shows a pressure decrease up to 2.0 hPa, indicating that the region's semi-permanent high pressure area is weaker (Figure 4d). In September, D02 shows an increase in MSLP across the Arabian Peninsula, with a decrease over the Mediterranean Sea (Figure 4e).

The third downscaling domain (4 km grid resolution, D03; Figure 5) in May (Figure 5a) indicates a pressure increase of 0.5 to 1.0 hPa in Kuwait, and from 0.5 to 1.5 hPa in eastern Saudi Arabia, Qatar, and the United Arab Emirates. In June (Figure 5b), pressure increases from 1.0 to 2.0 hPa across much of the same region. In July (Figure 5c), pressure

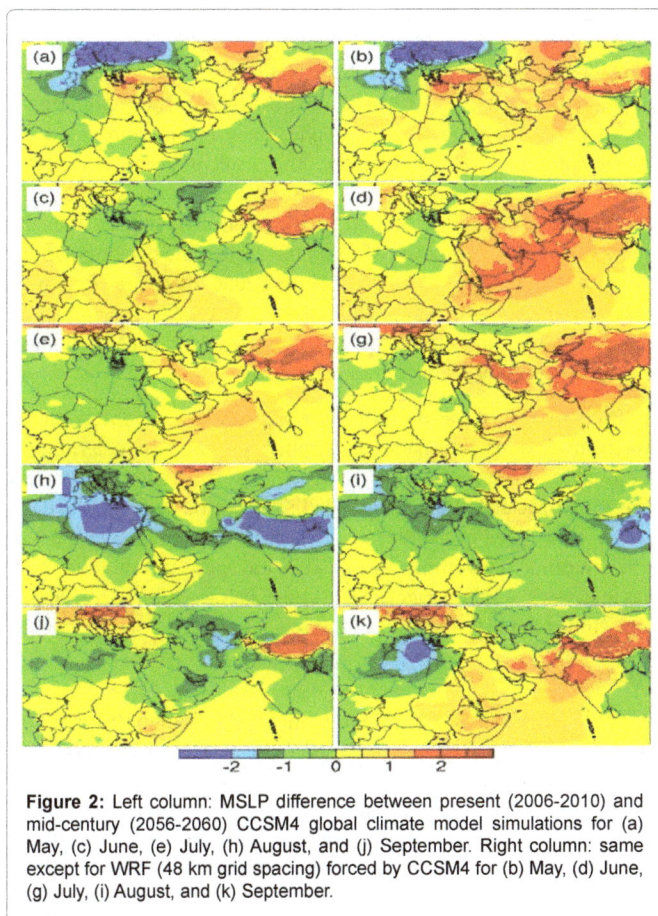

Figure 2: Left column: MSLP difference between present (2006-2010) and mid-century (2056-2060) CCSM4 global climate model simulations for (a) May, (c) June, (e) July, (h) August, and (j) September. Right column: same except for WRF (48 km grid spacing) forced by CCSM4 for (b) May, (d) June, (g) July, (i) August, and (k) September.

Arabian Peninsula, and the region becomes more favorable to the development of thermal lows. The WRF simulations shows a decrease of MSLP in the Arabian Peninsula in August, which indicates that there will be a shorter period of strong pressure gradients that will

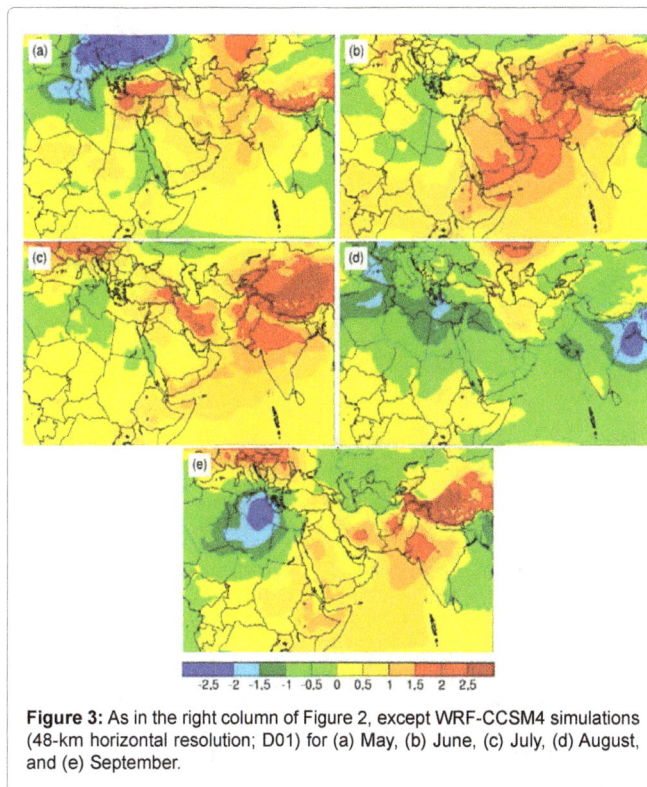

Figure 3: As in the right column of Figure 2, except WRF-CCSM4 simulations (48-km horizontal resolution; D01) for (a) May, (b) June, (c) July, (d) August, and (e) September.

increases across most of the same region, but less than in June. In August D03 shows a pressure decrease from 1.0 to 2.0 hPa region-wide (Figure 5d). By September, D03 indicates a pressure increase from 1.5 to 2.0 hPa over most of the region (Figure 5e). The three simulations, while qualitatively similar, yield slightly different quantitative results since grid resolution varied. Given that results from the finer-resolution domain (D03) most closely matched observations at KIA (Table 2), this simulation likely provides the best illustration of how the pressure field may vary across the Arabian Peninsula in the future.

The main motivation for simulating the MSLP using WRF is to examine the future changes in pressure gradients. The three downscaling simulations tend to provide similar results in terms of the changes in MSLP strength, and the indication of the weakening of the pressure gradient that develops in summertime in the Arabian Peninsula. The circulation patterns in summertime are controlled by the summer Indian monsoon leading to northerly winds in summer; the steeper the pressure gradient the stronger the northerly to northwesterly wind. The thermal low that develops in the southeastern Arabian Peninsula steepens the pressure gradient over the Arabian Peninsula as was seen in the case study. The WRF simulations for summertime show that the southeastern region where the thermal low normally develops over southeastern Saudi Arabia, United Arab Emirates (UAE), or northeast Oman will weaken in the future. The weakening of this feature will lead to a flattening of the regional pressure gradient. The WRF simulations for MSLP indicate a weakening pressure gradient, which results in fewer days with moderate to strong northerly to northwesterly wind in summertime. In late August, the pressure gradient begins to weaken, leading to the development of a more mesoscale circulation over the

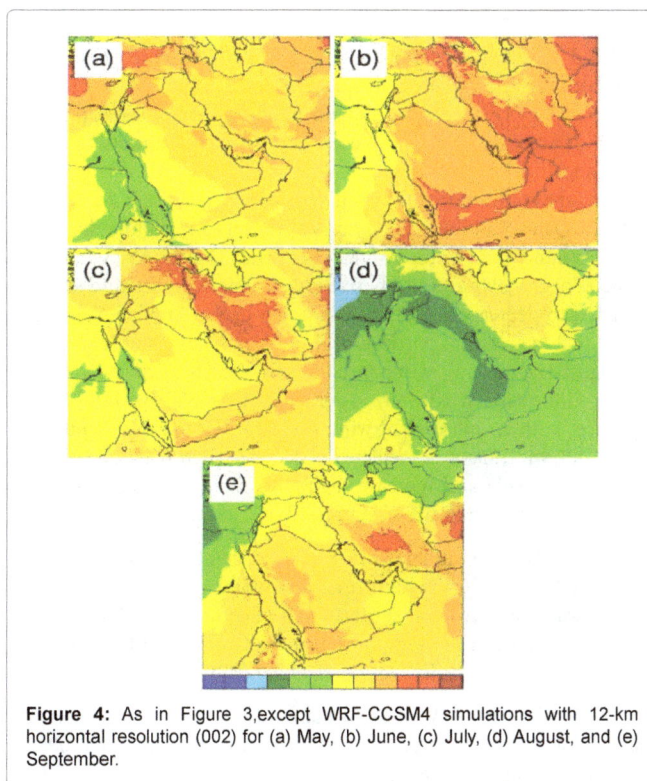

Figure 4: As in Figure 3, except WRF-CCSM4 simulations with 12-km horizontal resolution (002) for (a) May, (b) June, (c) July, (d) August, and (e) September.

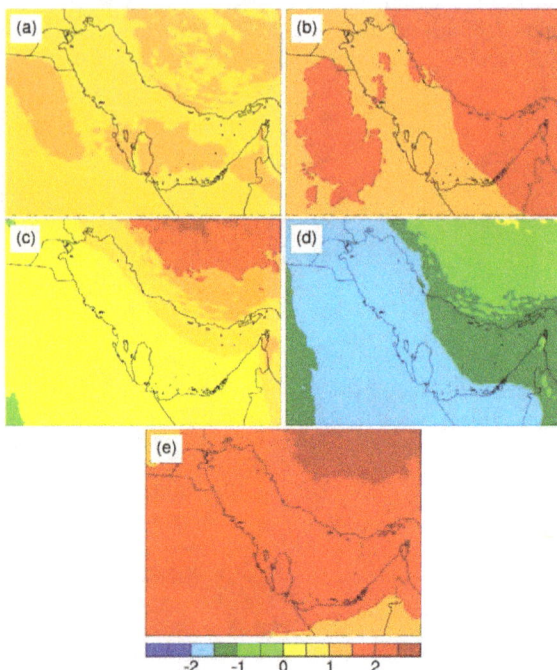

Figure 5: As in Figure 3, except WRF-CCSM4 simulations with 4-km horizontal resolution (D03) for (a) May, (b) June, (c) July, (d) August, and (e) September.

weaken very early in August. The weak synoptic forcing will lead to humid air advecting inland from the Persian Gulf into the peninsula, and more days with light to moderate easterly and southeasterly winds rather than strong shamal. This indicates that in the future, August will experience fewer dust storms and more humid days, and an increase in number of days with a mesoscale circulation such as a land/sea breeze in June, July, and August.

Wind speed: For the 4-km domain (D03), difference in wind speed between the present and mid-century was also plotted (Figure 6). During May (Figure 6a), a slight increase in wind speed (up to 1 m s⁻¹) is indicated over Kuwait, eastern Saudi Arabia, Bahrain, southern Qatar, and the UAE, with a slight decrease in wind speed (up to 1 m s⁻¹) along the south shore of Kuwait, eastern shore of Saudi Arabia, and northern Qatar. In June, mid-century wind speed is slightly decreased over most land areas of the Arabian Peninsula (Figure 6b), with slight increases over Qatar and southern Saudi Arabia. In July, wind speed is predicted to decrease slightly across much of the Arabian Peninsula, and up to 2 m s⁻¹ across portions of Kuwait, Iraq, and Iran (Figure 6c). By August (Figure 6d), wind speed is predicted to decrease over the Persian Gulf, eastern Kuwait, eastern Saudi Arabia, Bahrain, Qatar, and UAE, with an increase of wind speed farther west. In September (Figure 6e), wind speed is predicted to decrease slightly in the northern Arabian Peninsula and increase slightly in the southeastern Arabian Peninsula. These results are summarized in Table 4.

Daily average wind speed in Kuwait: The daily mean of surface wind speed (m s⁻¹) for the high-resolution domain (D03) of the WRF-CCSM4 was computed. The motivation is to compare the daily average wind speed between present and future in Kuwait, and to examine if there may be days in the future with stronger daily average wind speeds.

The results of the WRF-RCM daily mean wind speed prediction was arranged from highest to lowest values in the present (2006 to 2010). These five years were then averaged to produce a graph of daily

averaged wind speed arranged from higher to lower values. A similar calculation was performed for the future (2056-2060) to compare the slopes of both graphs. The graph in Figure 7 represents daily wind speed in red for the present and blue for the future. The graph implies that there are more days with a higher daily mean wind speed for the present compared to the future. The result indicates that in the future (2056-2060) fewer days will be experienced with strong winds compared to the present in Kuwait. This conclusion is supported by the results of a Wilcoxon-Mann-Whitney test. When the two populations of daily mean wind speed (2006-2010 and 2056-2060) are compared, p=0.0209, indicating a high likelihood that the two mean wind speed populations are indeed different, e.g., the mean wind speed will likely become lower under a future climate scenario.

Threshold wind speed for dust days: Thirty years of observational data from Kuwait International Airport was used to determine a threshold for dust events (blowing dust, dust storms, and sand storms) from an average daily 10-m wind speed. We investigated values of daily average wind speed of ≥ 4.5 m s⁻¹, ≥ 5.0 m s⁻¹, ≥ 5.5 m s⁻¹, and ≥ 6.0 m s⁻¹ to determine the optimal mean daily wind speed threshold for dust days in Kuwait. A percentage of 88.8% dust days were found for a mean daily wind speed threshold of ≥ 5.0 m s⁻¹.

Discussion

The main motivation for simulating MSLP using WRF-CCSM4 is to examine future changes in the pressure gradient and wind speed across the Arabian Peninsula, which allows an inference about possible future regional dust storm frequency changes. The three downscaling simulations for MSLP produce qualitatively similar results, and indicate weakening of the summertime pressure gradient across the Arabian Peninsula. The regional circulation pattern in summer is controlled by the summer Indian monsoon, with a steeper pressure gradient

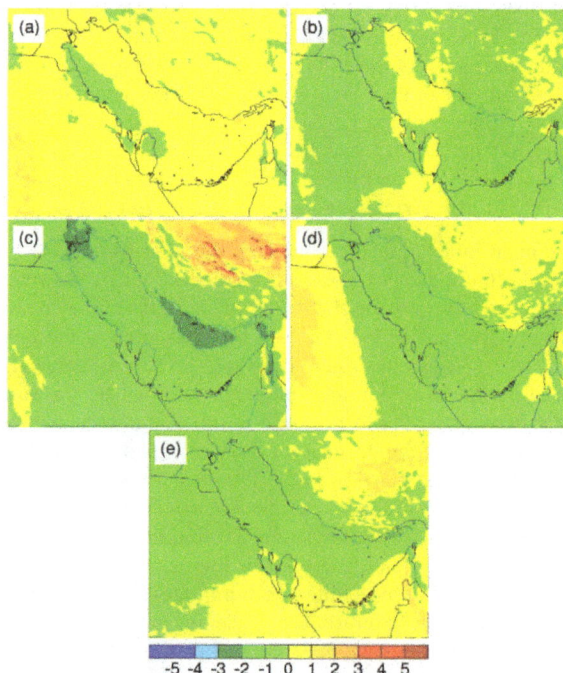

Figure 6: As in Figure 5, except wind speed difference (m s⁻¹) between present (2006-2010) and mid-century (2056-2060) for (a) May, (b) June, (c) July, (d) August, and (e) September.

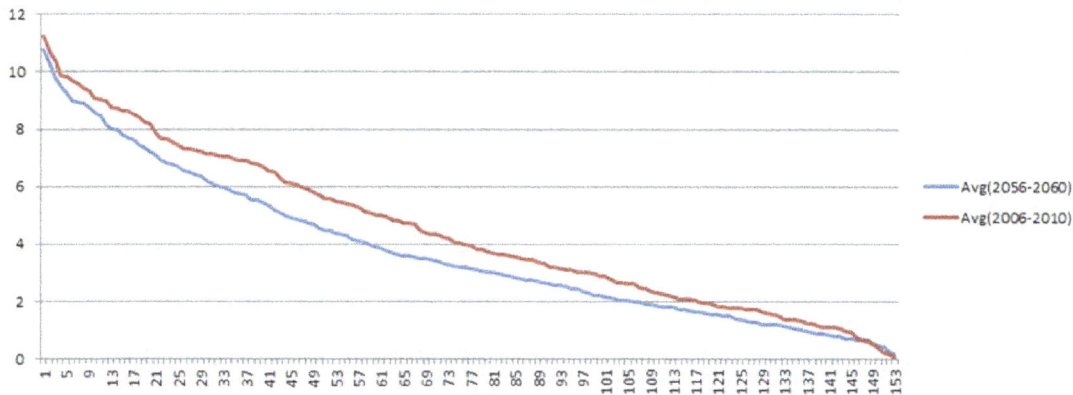

Figure 7: Ranked daily average wind speed for present (2006-2010) averaging the five years in red, and the ranked daily average wind speed for the future (2056-2060) averaging the five years in blue. Individual years are ranked separately prior to averaging over the 5-year simulation.

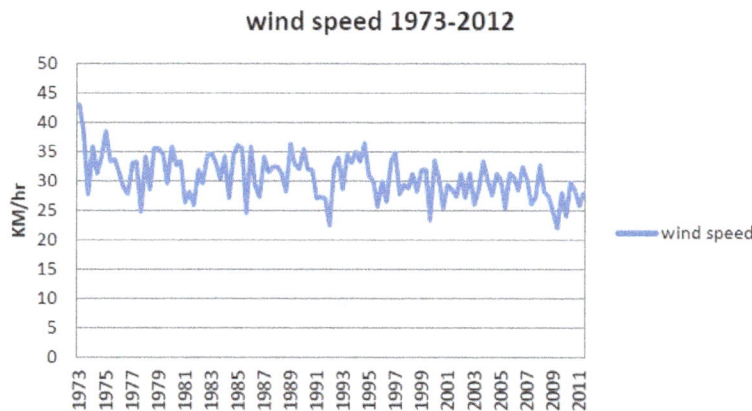

Figure 8: Summer average wind speed at Kuwait International Airport 1973-2012 in Km/hr.

leading to stronger northerly to northwesterly wind. Prior studies have indicated little change in intensity of the Indian monsoon circulation by the end of the 21st century [25]. A thermal low that develops in the southeastern Arabian Peninsula steepens the pressure gradient in this region (Figure 1). WRF-CCSM4 simulations for summertime show that this thermal low may weaken in the future, leading to a flattening of the pressure gradient. In addition, the high pressure area over the Mediterranean Sea is also anticipated to weaken. A weakening pressure gradient would result, allowing fewer summertime days with moderate to strong northerly to northwesterly wind. By late August the strong pressure gradient begins to weaken, making the region more conducive to the development of thermal lows. The WRF-CCSM4 simulations show a MSLP decrease in the Arabian Peninsula during August, which indicates that there may be a shorter period dominated by a strong pressure gradient, which will weaken very early in August. The weakened synoptic forcing will lead to moisture advection inland from the Arabian Gulf into the Arabian Peninsula, and more days with light to moderate easterly and southeasterly winds rather than strong shamal. The WRF-CCSM4 simulations of future wind speed indicates a decreasing wind speed in the eastern portion of the Arabian Peninsula in June, July, August, and September. This indicates that, in the future, the region will experience a decrease in the number of summertime days with strong pressure gradients and dust storms, and an increase in the number of days with mesoscale circulations such as the land-sea breeze. Confidence in this result is strengthened by a Wilcoxon-Mann-

Whitney test, which indicated ~98% certainty that the mean wind speed distributions are different between the present and mid-century.

WRF dynamical downscaling runs were investigated for three domains to investigate if there will more or fewer days with winds exceeding a mean daily 10-m threshold of 5.0 m s^{-1} in the future, comparing the present (2006-2010) with the future (2056-2060). The results of D03 indicate 40.8 days on average of daily mean wind speed thresholds $\geq$ 5.0 m s^{-1} in the present-day, with the results for mid-century showing 27 days. These results indicate less future days with a wind speed exceeding the threshold for dust, consistent with the finding of a lower mean future wind speed. The KIA summer mean wind speed observation from 1973-2012 indicates a declining summertime wind speed (Figure 8).These observation from the past 40 years show a similar trend to what the D03 high resolution model predicts in the future.

These results provide further support for prior research which indicates decreasing future wind speed over southern Europe [11,26,27]. Decreasing future wind speed is consistent with a continuation of the trend of decreasing wind speed over land globally and in the Arabian Peninsula as observed over recent decades [5]. The reduction in future wind speed seen in our simulations occurred without accounting for any vegetation changes, which may also be locally important through roughness length considerations [27]. Changing precipitation distributions through the year, leading to changes in soil moisture and the ability of dust to be lofted, should be investigated in more detail. Such changes are closely tied to the mean tracks of mid-latitude cyclones, which may shift poleward [12]. If less mid-latitude weather disturbances influence the Arabian Peninsula, precipitation may decline and wind speed may decrease, which have opposite implications for the number of dust storm-favorable days in the region.

Conclusion

The WRF-CCSM4 results show similar MSLP patterns between the present and mid-century, but they indicate a change in the magnitude of the surface pressure. The WRF-CCSM4 indicates an increase in pressure east of the Arabian Peninsula where low pressure forms in the summer. This low pressure normally forms east of the region due to the Indian monsoon or the thermal low pressure created due to solar heating. The summertime Mediterranean high pressure area is expected to experience a future decrease in intensity. The Mediterranean high pressure west of the Arabian Peninsula and the low pressure east of the region usually result in northerly to northeasterly flow across the region. The southeast thermal low that develops usually steepens the pressure gradient across the Arabian Peninsula. This gradient will subsequently weaken by mid-century, resulting in decreasing summertime wind speed and a possible reduction in the number of days favorable for dust events. This change will alter the amount of the aerosol in the atmosphere and will cause regional changes in absorption and scattering of long wave and shortwave radiation. Potential next steps in this research area include using high-resolution WRF climate modeling to examine such changes in summertime environmental physics over the region by mid-century.

References

1. Boos WR, Emanuel KA (2009) Annual intensification of the Somali jet in a quasi-equilibrium framework: Observational composites. Quarterly Journal of the Royal Meteorological Society 135: 319-335.

2. Ihara C, Kushnir Y, Cane MA (2008) July droughts over Homogeneous Indian monsoon region and Indian Ocean dipole during El Niño events. International Journal of Climatology 28: 1799-1805.

3. Fischer AS, Terray P, Guilyardi E, Gualdi S, Delecluse P (2005) Two independent triggers for the Indian Ocean Dipole/zonal mode in a coupled GCM. Journal of Climate 18: 3428-3449.

4. AlSarraf H, Van Den Broeke M (2015) Using high-resolution WRF model simulations to investigate the relationship between mesoscale circulations and aerosol transport over Kuwait. Journal of Climatology and Weather Forecasting 3: 3-126.

5. Bichet A, Wild M, Folini D, Schär C (2012) Causes for decadal variations of wind speed over land: Sensitivity. Geophysical Research Letters 39: 11701.

6. Karnauskas KB, Seager R, Kaplan A, Kushnir Y, Cane MA (2009) Observed strengthening of the zonal sea surface temperature gradient across the equatorial Pacific Ocean. J. Climate 22: 4316-4321.

7. DiNezio PN, Vecchi GA, Clement AC (2013) Detectability of changes in the Walker Circulation in response to global warming. J. Climate 26: 4038-4048.

8. Aurelius L, Buttgereit V, Cammelli S, Zanina M (2008) The impact of Shamal winds on tall building design in the Gulf. Dubai Building Government of Dubai.

9. Salmond JA, McKendry IC (2005) A review of turbulence in the very stable nocturnal boundary layer and its implications for air quality. Progress In Physical Geography 29: 171-188.

10. Zhao Z, Luo Y, Jiang Y (2011) Is global strong wind declining? Adv. Climate Change Res 2:225-228.

11. Al-Zahrani M (2008) Impacts of climate change on water resources in Saudi Arabia. Arabian Journal for Science and Engineering 38: 1959-1971.

12. Koch P, Weirnli H, Davies HC (2006) An event-based jet-stream climatology and typology. Int. J. Climatology 26: 283-301.

13. Bengtsson L, Hodges K, Froude L (2005) Global observations and forecast skill. Tellus A Dynamic Meteorology and Oceanography 57: 515-527.

14. Solomon S, Qin D, Manning M, Chen Z, Marquis M, et.al, (2007) The Physical Science Basis. Contribution of Working Group I to the Fourth Assessment Report of the IPCC. Cambridge University Press, Cambridge, United Kingdom and New York, USA.

15. Mesquita MDS, Nils GK, Asgeir S, David EA (2008) Climatological properties of summertime extra-tropical storm tracks in the Northern Hemisphere. Tellus 60: 557-569.

16. Chu C, Yang X, Ren X, Zhou T (2013) Response of Northern Hemispheric storm tracks to Indian-western Pacific Ocean warming in atmospheric general circulation models. Climate Dynamics 40: 1057-1070.

17. Skamarock W, Klemp J, Dudhia J, Gill D, Barker D, et.al, (2008) A description of the Advanced Research WRF version 3. NCAR Tech. Note NCAR/TN-4751STR 113.

18. Dudhia J (1989) Numerical study of convection observed during the Winter Monsoon Experiment using a mesoscale two-dimensional model. J. Atmos. Sci 46: 3077-3107.

19. Mlawer EJ, Taubman SJ, Brown PD, Iacono MJ, Clough SA (1997) Radiative transfer for inhomogeneous atmospheres: RRTM, a validated correlated-k model for the longwave. J. Geophys. Res. Atmos 102: 16663–16682.

20. Ek MB, Mitchell KE, Lin Y, Rogers E, Grunmann P, et.al, (2003) Implementation of Noah land surface model advances in the National Centers for Environmental Prediction operational mesoscale Eta model. J. Geophys. Res. Atmos 108.

21. Hong S, Noh Y, Dudhia J (2006) A new vertical diffusion package with an explicit treatment of entrainment processes. Mon. Weather Review 134: 2318-2341.

22. Hong SY (2010) A new stable boundary-layer mixing scheme and its impact on the simulated East Asian summer monsoon. Quart. J. Roy. Meteor. Soc 136: 1481-1496.

23. Kain JS (2004) The Kain-Fritsch convective parameterization: An update. J. Appl. Meteor 43: 170-181.

24. Kistler R, William C, Suranjana S, Glenn W (2001) The NCEP-NCAR 50-year reanalysis: Monthly means CD-ROM and documentation. Bull. Amer. Meteor Soc 82: 247-267.

25. Kumar KK, Kamala K, Rajagopalan B, Hoerling MP, Eischeid JK, et.al, (2011) The once and future pulse of Indian monsoonal climate. Climate Dynamics 36: 2159-2170.

26. Jiang Y, Luo Y, Zhao Z (2009) Review of research on wind resources changes in China and in the world. Review on Science & Technology 27: 96-104.

27. Tobin I, Vautard R, Balog I, Bréon F, Jerez S, et al. (2015) Assessing climate change impacts on European wind energy from ENSEMBLES high-resolution climate projections. Climatic Change 128: 99-112.

8

A Simplified Approach for Stormwater Drainage Networks Sizing

Luigi Cimorelli[1], Luca Cozzolino[2], Andrea D'Aniello[2], Francesco Morlando[2] and Domenico Pianese[2]

[1]Department of Civil, Architectural and Environmental Engineering, University of Naples Federico II, Via Claudio n.21 – 80125, Napoli, Italy
[2]Department of Engineering, Parthenope University of Naples - Centro Direzionale di Napoli, 80142 Napoli, Italy

Abstract

In this work, a modification and a generalization of the "Italian-Storage method" (ISM), a renowned method for sizing rainstorm drainage systems, are proposed and applied. The approach adopts a steady non-linear rainfall-runoff transformation probabilistic model within a "variational" or "maximizing" procedure. In this model, the transformation of rainfall in "excess (or effective) rainfall" is accomplished by means of the Rational Method, while the travel time is neglected. The flow within each link of the system, induced by a rectangular shaped discharge hydrograph for any rainfall duration, is assumed to be unsteady but locally uniform (Kinematic modeling). Combining the hydrological and hydraulic sub-models, and considering different couples of rainfall heights-durations, consistent with the local IDF curve, a "critical duration" is evaluated. Once the critical duration has been obtained, a maximum of maxima instantaneous flow discharges is evaluated for such duration, and subsequently, with such a discharge each link of the storm water drainage system can be designed.

Keywords: Italian-storage method; Sewer/rural drainage networks

Introduction

In the last century, several methods have been proposed in the technical literature in order to allow the sizing and verification of storm water drainage networks, given a probabilistic framework. In particular, starting from several pioneering works [1-6], a few approaches, based on the adoption of optimization procedures, have been proposed for the optimal sizing of drainage networks (see, among others, [7-11]). Independently on the specific procedure adopted for the optimization process (e.g., linear or quadratic programming, heuristic algorithms, etc.), the considerable number of evaluations of network performances required in optimization problems imply the adoption of a tool for the fast evaluation of the network hydraulic features (e.g., discharges and flow depths values). As a consequence, very often simplified approaches are preferred to the most accurate, and meanwhile slow, ones [9].

Starting from these considerations, in this work a slight modification and generalization of a method for the sizing of rainstorm drainage networks, well-known in Europe as the "Italian-Storage method" (ISM), is proposed and then applied to some case studies. This method was firstly proposed [12] for sewer systems and then extended [13] for drainage networks. Subsequently the approach was developed in order to give the possibility to straightly and quickly carry out the design of both sewer/rural drainage networks [14,15]. In particular, one of these approaches is nowadays widely adopted in Italy by technicians in the design of drainage systems, especially given the noticeable simplicity in software implementation [14]. Moreover, it allows the estimation of the hydraulic features of the whole design network in a very limited time, allowing its adoption in optimization procedures where it is usually needed the iterative modification of the network physical characteristics (e.g., longitudinal slopes, shapes and sizes of pipes, roughness parameters, upstream/downstream elevations or crown/bed elevations, etc.).

Given that this method has, 'in nuce', the whole characteristics of whatever semi-distributed, probabilistically based, hydrologic model, it has been already adopted, within several papers of these authors, together with an 'extremal', or 'variational', procedure [16-18] for choosing the 'critical value' for the rainstorm duration, in order to allow the comprehension of the whole procedure.

Fundamentals

Generally speaking, if lateral inflow/outflows are left aside, the *de Saint-Venant*'s unsteady flow equations could be written as:

$$\frac{\partial \Omega}{\partial t} + \frac{\partial Q}{\partial x} = 0 \tag{1}$$

$$\frac{1}{g}\frac{\partial u}{\partial t} + \frac{V}{g}\frac{\partial u}{\partial x} + \frac{\partial h}{\partial x} + (i - j) = 0 \tag{2}$$

where $i=sin\beta$ and β=angle between the bed and the horizontal (Figure 1).

If the kinematic wave approximation could be considered, equation (2) becomes $i=j$ (i.e., locally and instantaneously uniform flow), and then a resistence formula, such as the one proposed by Chezy and shown hereinafter, could be applied:

$$U = k_c \cdot \sqrt{R \cdot i} \Rightarrow Q = k_c \cdot \omega \cdot \sqrt{R \cdot i} \cong \mu \cdot \omega^\alpha \tag{3}$$

where the coefficient μ and the exponent α could be considered constant during the filling/emptying process (with the value of the constants to be defined properly), and calibrated by an iterative procedure.

Applying the equation (1) to a discrete reach of length L (which could be assumed, for instance, equal to the pipe length), the governing equations become:

$$\frac{d}{dt}\left(\frac{w}{L}\right) + \frac{[q_{out} - q_{in}]}{L} = 0 \quad \text{or} \quad q_{in} - q_{out} = \frac{dw(t)}{dt} \tag{4}$$

and

*Corresponding author: Francesco Morlando, Department of Civil, Architectural and Environmental Engineering, University of Naples Federico II, Via Claudio n.21 – 80125, Napoli, Italy, E-mail: francesco.morlando@unina.it

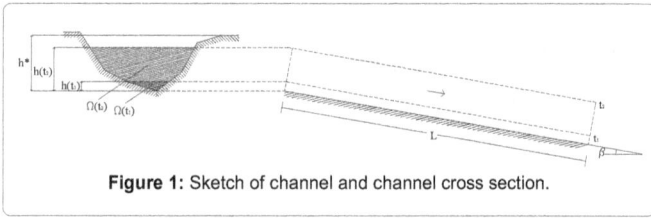

Figure 1: Sketch of channel and channel cross section.

$$q_{out} = \mu \cdot \omega^{\alpha} \qquad \text{or} \qquad w = k \cdot q_{out}^{\nu} \qquad (5)$$

In the previous equations, $w = L\omega$ is the volume of water stored in the reach of length L when the discharge flowing out from the channel at time t is q_{out} and the area of cross section is ω; $\nu = 1/\alpha$, $k = (W^*/Q^{*\nu})$ $W^* = L\Omega^*$ is the maximum volume of water that can be stored in the reach when the discharge flowing out from the reach itself is the maximum flow discharge capacity Q^*, and Ω^* is the corresponding channel cross section.

By differentiating the first of equations (7), one obtains:

$$d\omega = \frac{\Omega^*}{\alpha Q^{*1/\alpha}} \cdot q_{out}^{1/\alpha - 1} dq_{out} \quad \text{or} \quad dw = \frac{W^*}{\alpha Q^{*1/\alpha}} q_{out}^{\frac{1}{\alpha}-1} dq_{out} \qquad (6)$$

Substitution of Equation (8) into the Equation (6) gives:

$$(q_{in} - q_{out}) = \frac{W^*}{\alpha Q^{*1/\alpha}} \cdot q_{out}^{1/\alpha - 1} \frac{dq_{out}}{dt} \qquad (7)$$

and then:

$$dt = \frac{W^*}{\alpha Q^{*1/\alpha}} \cdot \frac{q_{out}^{1/\alpha - 1}}{q_{in} - q_{out}} dq_{out} \qquad (8)$$

or

$$t_2 - t_1 = \frac{W}{\alpha Q^{*1/\alpha}} \cdot \int_{(q_{out})_1}^{(q_{out})_2} \frac{q_{out}^{1/\alpha - 1}}{(q_{in} - q_{out})} dq_{out} \qquad (9)$$

If, at the beginning of the rainstorm ($t_1 = 0$,) the discharge flowing out from the reach is $(q_{out})_1 = 0$ and, at time T_f, the maximum permissible discharge flowing out from the reach is $(q_{out})_2 = Q$, the Equation (11) becomes:

$$T_f = \frac{W^*}{\alpha Q^{*1/\alpha}} \cdot \int_0^Q \frac{q_{out}^{1/\alpha - 1}}{(q_{in} - q_{out})} dq_{out} \qquad (10)$$

where T_f, namely filling time, is the time needed to reach the maximum permissible discharge Q.

If q_{in} = constant in time and the "attenuation ratios" $z = q_{out}/q_{in}$ and $Z^* = Q^*/q_{in}$ are considered, the time T_f could be evaluated as:

$$T_f = \frac{W^*}{q_{in}} \cdot \frac{1}{\alpha Z^{*1/\alpha}} \int_0^z \frac{z^{1/\alpha - 1}}{1 - z} dz = \frac{W}{q_{in}} \cdot \Phi_{\alpha}(Z^*) \qquad (11)$$

where

$$\Phi_{\alpha}(Z^*, Z) = \frac{1}{\alpha Z^{*1/\alpha}} \int_0^z \frac{z^{1/\alpha - 1}}{1 - z} dz = \frac{1}{\alpha Z^{*1/\alpha}} \int_0^z z^{1/\alpha - 1}(1 + z + z^2 + ... + ...) dz \qquad (12)$$

Equation (13) describes the hydraulic response; to an input discharge q_{in} constant in time, of a reach where the volume $W^* = L \cdot \Omega^*$ can be stored.

In order to evaluate q_{in}, a hydrologic approach could be used. In the approach proposed by Supino and Chow [14,15] the evaluation of discharge q_{in} is attained considering:

1) Storms characterized by a return period T, constant rainfall intensity $i_{d,T}$ in the time interval $t[0,d]$ and constant rainfall intensity 0 for $t > d$;

2) As the input values at most upstream cross section of channels considered in the calculations, the runoff discharges drained from the whole area A subtended by the most downstream cross section of the channel reach;

3) The infiltration processes, by using the runoff coefficient $c = A_{imp}/A$, being A_{imp} and A, respectively, the impervious and total drainage area of the basin;

4) Initially neglecting lag phenomenon due to the formation and subsequent arrival of the surface runoff to the upstream section of the reach considered in the analysis, for which is $q_{in} = c\, i_d A$;

An Intensity-Duration-Frequency relationship having the simple structure

$$i_{d,T} = (a \cdot d^{n-1}) \cdot k_T \Leftrightarrow i_{d,T} = (a \cdot k_T) \cdot d^{n-1} = a_T \cdot d^{n-1} \qquad (13)$$

where a and n have to be evaluated by using regional and/or local rainfall data (partial series of the annual maximum rainfall depth in the duration d).

If the storm duration d was higher than the channel filling time T_f ($d > T_f$), the *Surface Runoff* $SR = q_{in}d$ would be higher than the storage availability W^*, and, then, the maximum value of the flow depth h^*_{max}, corresponding to the maximum available flow area Ω^*, would be exceeded (causing, for instance, a pressurized flow instead of a free surface flow).

On the other hand, if $d < T_f$ also $SR < W^*$. As a direct consequence the maximum cross-sectional flow area, Ω, and the maximum flow depth, h_{max}, both attained during the filling process, should follow the following relations: $\Omega < \Omega^*$, $h_{max} < h^*_{max}$ and $(Q_{out})_{max} < (Q^*_{out})_{max}$. In this case, the sizes and/or slope considered should be higher than those strictly needed, and, as a consequence, the construction costs would grow.

Given the previous framework, it is evident that, in order to avoid both under- and over- sizing of cross-sections/slopes/conveyance capability, it would be preferable to consider the following conditions:

$$d = T_f, \ \Omega = \Omega^*, \ h_{max} = h^*_{max}, \ Q_{out} = Q^*_{out} \ \text{and} \ Z = \frac{Q_{out}}{q_{in}} = \frac{Q^*_{out}}{q_{in}} = Z^*$$

As a consequence, it would result:

$$\Phi_{\alpha}(Z^*, Z) = \Phi_{\alpha}(Z) = 1 + \frac{Z}{1 + \alpha} + \frac{Z^2}{1 + 2\alpha} + ... + \frac{Z^j}{1 + j\alpha} + ... \qquad (14)$$

By substituting equation (16) in equation (13), together with the condition $d = T_f$, one would obtain:

$$Q_{out} = \frac{(c\, a\, A)^{1/n}}{W^{1/n - 1}} \cdot Z \cdot \left[\Phi_{\alpha}(Z)\right]^{\frac{n-1}{n}} \qquad (15)$$

or

$$Q_{out} = \left[\frac{\mu^{\frac{1}{\alpha}\left(\frac{1}{n} - 1\right)} (c\, a\, A)^{\frac{1}{n}}}{L^{\frac{n-11}{n}}} \cdot Z \cdot \left[\Phi_{\alpha}(Z)\right]^{\frac{n-1}{n}} \right]^{\left[\frac{\alpha \cdot n}{1 - n \cdot (1 - \alpha)}\right]} \qquad (16)$$

If the '*udometric coefficient*' (i.e., the contribution of the basin unitary area to the formation of peak discharge), $u = Q_{out}/A = Q/A$, and the specific storage (i.e., volume of water stored in the reach of length L per basin unitary area), $w = W/A$, are introduced, equation (17) becomes:

$$u = \frac{(ca)^{1/n}}{w^{1/n-1}} \cdot Z \cdot \left[\Phi_\alpha(Z)\right]^{\frac{n-1}{n}} \tag{17}$$

For a fixed value of Z, is it possible to evaluate:

- by equation (17'), the values of Q_{out} ;

- by equation (18), the corresponding value of u;

- by the following equation

$$d = \left(\frac{u}{Zca}\right)^{\frac{1}{n-1}} \tag{18}$$

the corresponding rainstorm duration.

Then, by changing the Z value (i.e.: by varying the storm duration d and, subsequently, inflow discharge q_{in}), the evaluation of the maximum discharge value, $Q_{max} = \max_Z\left[Q_{out}(Z)\right]$, becomes possible. This is the main reason why in the technical literature the procedure above described is known as '*variational*' (or '*extremal*') approach. In particular, the duration d_{crit} for which $Q_{out} = Q_{max}$ is given by:

$$d_{crit} = \left[\frac{Q_{max}}{Z_{crit} c a A}\right]^{\frac{1}{n-1}} \tag{19}$$

and is usually defined as '*critical storm duration*'.

As a consequence, the maximum $u_{max} = \max_Z\left[u(Z)\right] = \max_d\left[u(d)\right] = u(d_{crit})$ could be evaluated by:

a) differentiating equation (18) with respect to Z;

b) putting the above obtained derivative equal to zero, in order to evaluate Z_{crit} (and, then, the critical rainstorm duration d_{crit});

c) Substituting this value of Z within equation (18).

Following these steps, it is possible to show that, at "critical" conditions, it needs:

$$(1 - Z_{crit}) \cdot \Phi_\alpha(Z_{crit}) = \frac{1-n}{1+n\cdot(\alpha-1)} \tag{20}$$

Using the Equation (21) for each couple of n and α values, it is possible to evaluate, by a trial-and-error approach, corresponding value of Z_{crit} and, then, corresponding value of the product $Z_{crit} \cdot \left[\Phi_\alpha(Z)_{crit}\right]^{\frac{n-1}{n}}$, both present in equation (18).

A few values of Z_{crit} and $Z_{crit} \cdot \left[\Phi_\alpha(Z)_{crit}\right]^{\frac{n-1}{n}}$ are summarized, for fixed α and n values, in Table 1.

It is worth noticing that the values of the product $Z_{crit} \cdot \left[\Phi_\alpha(Z)_{crit}\right]^{\frac{n-1}{n}}$ can be approximated by using different interpolation formulas. Puppini and Supino [13,15] proposed the following expression:

$$Z_{crit} \cdot \left[\Phi_\alpha(Z)_{crit}\right]^{\frac{n-1}{n}} \cong (\lambda_1\alpha + \lambda_2) \cdot n \tag{21}$$

where $\lambda_1=0.259$ and $\lambda_2=0.518$ (with maximum errors about 5%) for [13] formulation and $\lambda_1=0.221$ and $\lambda_2=0.574$ for [15] one, with a maximum error lower than 3%.

Other and better approximation can also be obtained if the following relation is considered [15]:

$$Z_{crit} \cdot \left[\Phi_\alpha(Z)_{crit}\right]^{\frac{n-1}{n}} \cong (43,3 \cdot \alpha + 40,7) \cdot n - (47,2 \cdot \alpha + 68,5) \cdot n^2 \tag{22}$$

where the variables involved are expressed considering the following units of measurement: u [l/(s·ha)]; a [m/day] and w [m].

In order to evaluate the maximum peak discharges, flow depths and volumes stored in the reach of length L, the following first order approximation relation can be adopted, starting from equation (21), (with errors ranging within the interval $[-7.80\%, +13.44\%]$, (Figure 2):

n	Z_{crit}					$Z_{crit} \cdot \left[\Phi_\alpha(Z)_{crit}\right]^{\frac{n-1}{n}}$				
	$\alpha = 1.00$	$\alpha = 1.25$	$\alpha = 1.50$	$\alpha = 1.75$	$\alpha = 2.00$	$\alpha = 1.00$	$\alpha = 1.25$	$\alpha = 1.50$	$\alpha = 1.75$	$\alpha = 2.00$
0.100	0.187	0.206	0.224	0.241	0.258	0.075	0.083	0.091	0.099	0.107
0.150	0.271	0.296	0.319	0.340	0.360	0.113	0.125	0.137	0.148	0.159
0.200	0.350	0.378	0.404	0.427	0.449	0.153	0.168	0.182	0.196	0.209
0.250	0.423	0.453	0.480	0.504	0.526	0.193	0.211	0.228	0.244	0.259
0.300	0.491	0.522	0.549	0.573	0.594	0.233	0.254	0.273	0.291	0.307
0.350	0.554	0.584	0.610	0.633	0.653	0.275	0.298	0.318	0.338	0.355
0.400	0.612	0.641	0.666	0.687	0.705	0.318	0.342	0.364	0.384	0.403
0.450	0.666	0.693	0.715	0.734	0.751	0.362	0.387	0.410	0.431	0.449
0.500	0.715	0.740	0.760	0.777	0.791	0.407	0.433	0.456	0.477	0.496
0.550	0.761	0.782	0.800	0.814	0.827	0.454	0.480	0.503	0.524	0.542
0.600	0.802	0.820	0.835	0.848	0.858	0.502	0.528	0.551	0.570	0.588
0.650	0.839	0.855	0.867	0.877	0.886	0.552	0.577	0.599	0.618	0.635
0.700	0.873	0.886	0.895	0.904	0.910	0.604	0.628	0.648	0.666	0.681
0.750	0.903	0.913	0.920	0.927	0.932	0.658	0.680	0.699	0.715	0.729
0.800	0.930	0.937	0.942	0.947	0.950	0.715	0.735	0.751	0.765	0.777

Table 1: A few values of Z_{crit} and $Z_{crit} \cdot \left[\Phi_\alpha(Z)_{crit}\right]^{\frac{n-1}{n}}$ for fixed α and n values.

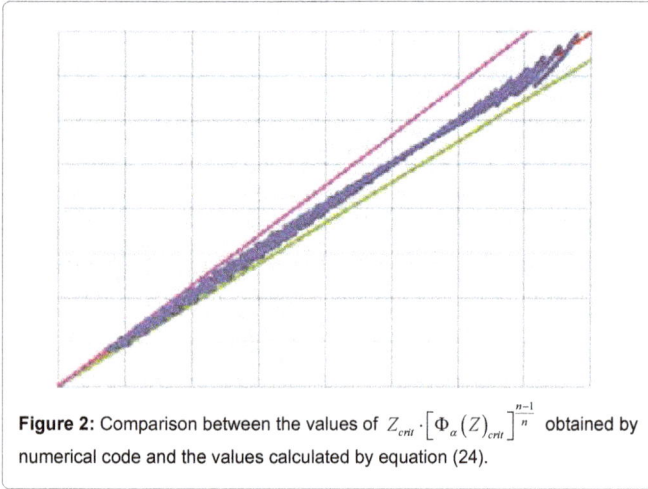

Figure 2: Comparison between the values of $Z_{crit} \cdot \left[\Phi_\alpha (Z)_{crit} \right]^{\frac{n-1}{n}}$ obtained by numerical code and the values calculated by equation (24).

$$u = \left[(\lambda_1 \alpha + \lambda_2) \right] \cdot n \cdot \frac{(ca)^{1/n}}{w^{1/n-1}} \quad (23)$$

with $n \in [0.10, 0.80]$, $\alpha \in [1,2]$, $\lambda_1 = 0.140$; $\lambda_2 = 0.709$.

A second order approximation is reported hereinafter, (with errors ranging within the interval $[-2.66\%, +1.26\%]$, (Figure 3):

$$u = \left[(\lambda_3 \alpha + \lambda_4) + (\lambda_5 \alpha + \lambda_6) \cdot n \right] \cdot n \cdot \frac{(ca)^{1/n}}{w^{1/n-1}} \quad (24)$$

with: $\lambda_3 = 0.347$; $\lambda_4 = 0.364$; $\lambda_5 = -0.340$; $\lambda_6 = 0.566$.

A third order of approximation has the following structure, (errors ranging within the interval $[-0.44\%, +1.07\%]$, (Figure 4):

$$u = \left[(\lambda_7 \alpha + \lambda_8) + (\lambda_9 \alpha + \lambda_{10}) \cdot n + (\lambda_{11} \alpha + \lambda_{12}) \cdot n^2 \right] \cdot n \cdot \frac{(ca)^{1/n}}{w^{1/n-1}} \quad (25)$$

with: $\lambda_7 = 0.358$; $\lambda_8 = 0.389$; $\lambda_9 = -0.384$; $\lambda_{10} = 0.464$; $\lambda_{11} = 0.041$; $\lambda_{12} = 0.094$.

Finally, a fourth order approximation leads to the following form (errors ranging within the interval $[-0.38\%, +0.82\%]$, see (Figure 5):

$$u = \left[(\lambda_{13} \alpha + \lambda_{14}) + (\lambda_{15} \alpha + \lambda_{16}) \cdot n + (\lambda_{17} \alpha + \lambda_{18}) \cdot n^2 + (\lambda_{19} \alpha + \lambda_{20}) \cdot n^3 \right] \cdot n \cdot \frac{(ca)^{1/n}}{w^{1/n-1}} \quad (26)$$

with: $\lambda_{13} = 0.368$; $\lambda_{14} = 0.364$; $\lambda_{15} = -0.456$; $\lambda_{16} = 0.646$; $\lambda_{17} = 0.193$; $\lambda_{18} = -0.291$; $\lambda_{19} = 0.099$; $\lambda_{20} = 0.250$.

Summarizing:

a) the errors obtained by each approximation have a decreasing trend as the order of approximation and formal complexity increase (Figure 6);

b) the *udometric coefficient u* can be expressed as it follows:

$$u = f(\alpha, n) \cdot n \cdot \frac{(ca)^{1/n}}{w^{1/n-1}} \quad (27)$$

where $f(\alpha, n)$ is a simple polynomial, function of α and n.

Furthermore, also Z_{crit} can be expressed by means of different interpolation formulas. Indeed, the Z_{crit} values obtained by equation (21) can be approximated by the following:

$$Z_{crit} \cong \left(k_1 \cdot n^3 + k_2 \cdot n^2 + k_3 \cdot n + k_4 \right) \cdot \alpha + \left(k_5 \cdot n^2 + k_6 \cdot n + k_7 \right) \quad (28)$$

where $k_1 = 0.839$; $k_2 = -1.434$; $k_3 = 0.609$; $k_4 = 0.026$; $k_5 = -0.600$; $k_6 = 1.700$; $k_7 = -0.058$; (with errors ranging within the interval $[-0.04\%, +0.02\%]$ (Figure 7).

Equation (28) can be easily applied in order to evaluate the main flow characteristics (i.e: maxima of discharge, velocity and flow depth) of the drainage network source channels.

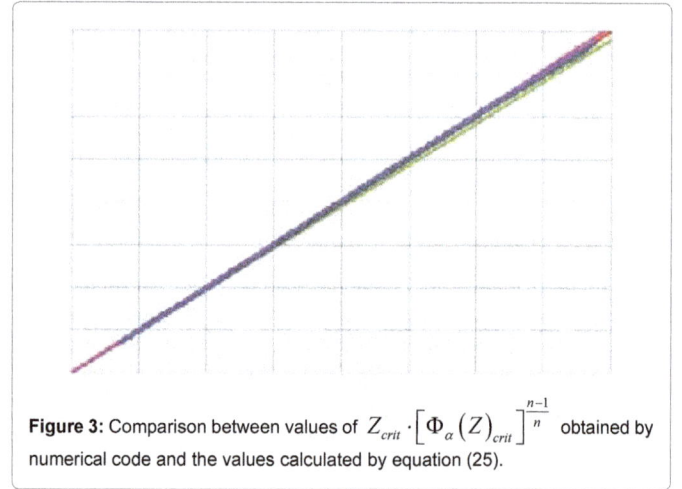

Figure 3: Comparison between values of $Z_{crit} \cdot \left[\Phi_\alpha (Z)_{crit} \right]^{\frac{n-1}{n}}$ obtained by numerical code and the values calculated by equation (25).

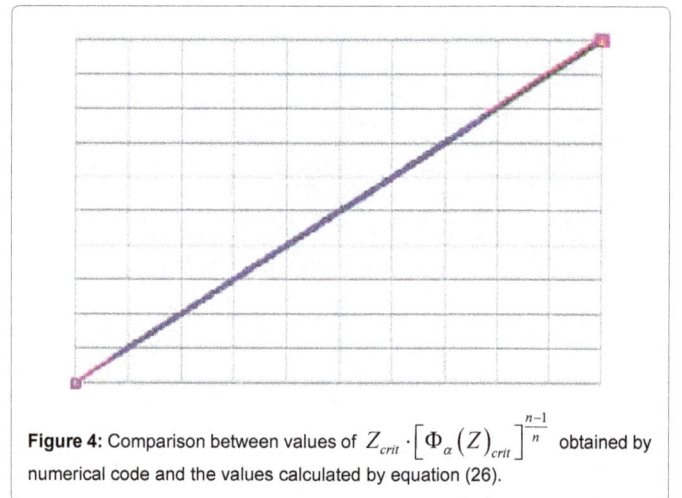

Figure 4: Comparison between values of $Z_{crit} \cdot \left[\Phi_\alpha (Z)_{crit} \right]^{\frac{n-1}{n}}$ obtained by numerical code and the values calculated by equation (26).

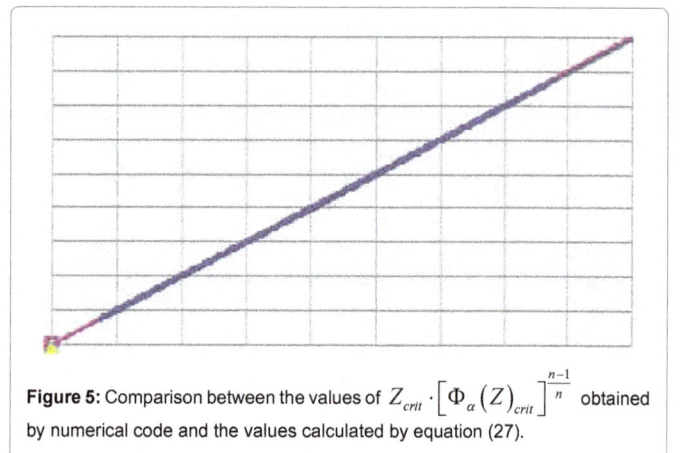

Figure 5: Comparison between the values of $Z_{crit} \cdot \left[\Phi_\alpha (Z)_{crit} \right]^{\frac{n-1}{n}}$ obtained by numerical code and the values calculated by equation (27).

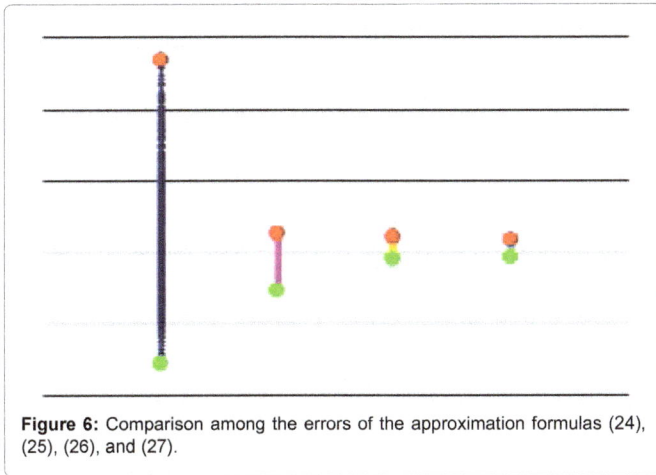

Figure 6: Comparison among the errors of the approximation formulas (24), (25), (26), and (27).

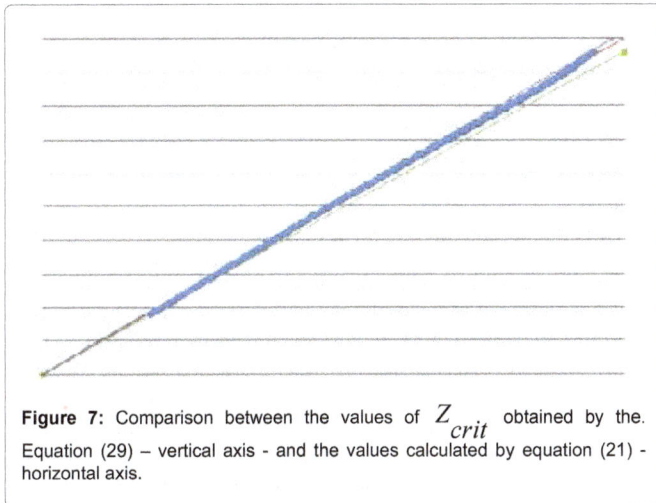

Figure 7: Comparison between the values of Z_{crit} obtained by the Equation (29) – vertical axis - and the values calculated by equation (21) - horizontal axis.

Once the hydraulic (i.e.: longitudinal slope; cross section; roughness parameter), the hydrological (i.e.: the drainage area of the catchment; the surface runoff coefficient), the climatic (i.e.: the coefficient a and the exponent n of the IDF curve) and the design (i.e.: the return period T) characteristics of the reach have been defined, the proposed design procedure follows the steps described hereinafter:

1) a first guess value of α' of the exponent α in equation (28) has to be assumed (e.g.: if $\alpha'=\alpha=1$, the routing is initially carried out by using the *linear reservoir* conceptual model);

2) a guess value u' of u is considered;

3) a first guess maximum peak discharge value Q'_{out} is evaluated (i.e.: $Q'_{out} = u' \cdot A$);

4) by means of the chosen steady and uniform roughness formula, the first guess values of the maximum peak flow depth h'_{max}, the maximum peak flow cross section area Ω'_{out}, the maximum water volume stored in the reach $W' = \Omega'_{out} L$, and the maximum specific stored volume $w' = W'/A$ are evaluated;

5) equation (28) is applied, and a new value of u (let say u'') is evaluated;

6) then, if

$$\left| \frac{u'' - u'}{\left(\frac{u'' + u'}{2} \right)} \right| \leq \varepsilon_u \qquad (29)$$

with $\varepsilon_u \to 0$ (for instance, $\varepsilon_u = 0,0001$), the iteration process is stopped; otherwise

7) a second guess value of u ($u = u'''$) is considered, and the procedure is repeated from stage 1) to 6) until the relation (30) is satisfied;

8) the value of h_{max} evaluated during the first iteration process is then subdivided in N steps (for instance: $N=100$), obtaining $h_1 = 1 \times h_{max}/N$; $h_2 = 2 \times h_{max}/N$; ...; $h_N = h_{max}$, and $\Omega_1 = \Omega_1(h_1)$; $\Omega_2 = \Omega_2(h_2)$; ...; $\Omega = \Omega(h_{max})$;

9) by means of the roughness formula $q = g(h)$, the discharges $q_1 = g(h_1)$, $q_2 = g(h_2)$, ..., $Q_{max} = g(h_{max})$ are evaluated;

10) then, the values of couples (Q_i, Ω_i) are interpolated by using the expression $q = \mu \cdot \Omega^\alpha$, thus obtaining a new guess value for α ($\alpha = \alpha''$);

11) the iterative procedure is repeated again from step 1) to 11);

12) then, if

$$\left| \frac{\alpha'' - \alpha'}{\left(\frac{\alpha'' + \alpha'}{2} \right)} \right| \leq \varepsilon_\alpha \qquad (30)$$

with $\to$ (for instance $\varepsilon_\alpha = 0,0001$), the second iterative stage is stopped; otherwise

13) a second guess value of α ($\alpha = \alpha''$) is considered, and the procedure is repeated from stage 1) to 12) until convergence is reached, equation (31).

Usually, after few iterations both the conditions (30) and (31) are satisfied. Thanks to modern computers, the whole procedure described usually takes less than (0.1 - 0.2) s [*]1.

Extension of the italian storage method to the whole channel network

In order to apply the approach illustrated above to the whole drainage network, the approach is modified introducing in the equation a new parameter w_o, obtained by splitting the volume $w = W/A$ in two parts, namely W_0 and W_r, for which $w = w_0 + w_r$:

- the first term, $w_r = W_r/A_r$, where W_r represents the maximum water volume actually stored within the reach considered in the calculation, and A_r is the area of the whole basin drained from the most downstream cross-section of the reach r;

- the second term, $w_0 = W_0/A_r$, in which W_0 represents the whole volume of water stored either in the reaches upstream the reach considered in the calculations or in other channels hydraulically linked to the reaches of the main network or at the surface of basins constituting the catchment whose ending drain is just the reach considered in the calculations. In particular:

---

1 Please note that at the beginning of the computations, it is not possible to assign a correct guess value of u. As a matter of fact, because of the iteration processes obtain the solution oscillating around the final (true) values of u and α, it is possible that: *i)* by using a first value u' too low, the second iteration could give a u'' value too big, for which the second guess Q''_{out} value could be higher than the flow capability of the channel (and, then, the flow could overcome one or both the banks if the channel is open, or the flow could became 'pressurized flow' if the channel is closed); *ii)* on the opposite, by using a first value u' too high, the second iteration could give a u'' value too low, for which the second guess Q''_{out} value could be lower than the flow capability of the channel (the channel itself becomes oversized). As a consequence, it is suggested to perform initial computation of a 'trial' value of u'. This value could be evaluated by reminding that the final (true) value of u is independent with respect to: the sizes, the roughness parameter/s, the longitudinal slope and the shape of the channel. Thus, a good' initial value u' could be obtained by considering, at beginning of the computations, a 'virtual' cross section, characterized by a very large size, and by carrying out the whole iteration process for this 'virtual channel'.

$$W_0 = \sum_{up=1}^{N_{up}^r} \left[W\right]_{up}^r + \left(\sum_{up=1}^{N_{up}^r} \left[W_0'\right]_{up}^r + \sum_{up=1}^{N_{up}^r} \left[W_0''\right]_{up}^r \right) + \left(\left[W_0'\right]_r + \left[W_0''\right]_r \right) \quad (31)$$

where:

$\left[W\right]_{up}^r$ is the volume of water stored within the link up of the network, located upstream of the reach r considered in the calculations and, then, which does not flow into the reach r itself;

$\left[W_0'\right]_{up}^r$ is the volume of water stored on the surface of the basin directly drained from the reach up of the network and, then, which does not flow into the reach up itself;

$\left[W_0''\right]_{up}^r$ is the volume of water stored within the whole reaches/tanks directly linked to the reach up of the network, located upstream of the reach r considered, but which were not explicitly considered in the calculations ("ghost reaches"); also this volume does not flow into the reach r;

$\left[W_0'\right]_r$ and $\left[W_0''\right]_r$ are the same of $\left[W_0'\right]_{up}^r$ and $\left[W_0''\right]_{up}^r$, but they apply to the reach r under consideration;

A_{up}^r is the area of the basin directly drained from the up^{th} reach existing upstream the reach r;

A_r is the area of the basin directly drained from the reach r;

$A = \sum_{up=1}^{N_{up}^r} A_{up}^r + A_r$ is the whole area drained from the most downstream cross section of the reach r considered in the calculation;

N_{up}^r is the number of reaches located upstream the link under consideration.

As a consequence, the relationships (24), (25), (26), (27) and (28) become, respectively:

$$u = \left[\left(\lambda_1 \alpha + \lambda_2\right)\right] \cdot n \cdot \frac{(c a)^{1/n}}{\left(w_o + w_r\right)^{1/n - 1}} \quad (32)$$

$$u = \left[\left(\lambda_3 \alpha + \lambda_4\right) + \left(\lambda_5 \alpha + \lambda_6\right) \cdot n\right] \cdot n \cdot \frac{(c a)^{1/n}}{\left(w_o + w_r\right)^{1/n - 1}} \quad (33)$$

$$u = \left[\left(\lambda_7 \alpha + \lambda_8\right) + \left(\lambda_9 \alpha + \lambda_{10}\right) \cdot n + \left(\lambda_{11} \alpha + \lambda_{12}\right) \cdot n^2\right] \cdot n \cdot \frac{(c a)^{1/n}}{\left(w_0 + w_r\right)^{1/n - 1}} \quad (34)$$

$$u = \left[\left(\lambda_{13} \alpha + \lambda_{14}\right) + \left(\lambda_{15} \alpha + \lambda_{16}\right) \cdot n + \left(\lambda_{17} \alpha + \lambda_{18}\right) \cdot n^2 + \left(\lambda_{19} \alpha + \lambda_{20}\right) \cdot n^3\right] \cdot n \cdot \frac{(c a)}{(w \quad w)} \quad (35)$$

$$u = f\left(\alpha, n\right) \cdot n \cdot \frac{(c a)^{1/n}}{\left(w_0 + w_r\right)^{1/n - 1}} \quad (36)$$

In order to better understand the meaning of W_0, it is possible to observe, preliminarily, that to arrive to the relationships (33), (34), (35), (36) and (37), it needs to introduce in the equation (6) the sum $\left[W_0 + w_r\left(t\right)\right]$ instead of $w(t)$, preserving W_0 = constant in time during the filling/empting processes afflicting the reach r under consideration.

Usually, the values of $w_{0,r} = \dfrac{\left[W_0'\right]_{up}^r + \left[W_0''\right]_{up}^r}{\left[A\right]_{up}^r}$ are considered parameters of the model (though variable sub-basin by sub-basin) whereas the values of $\left[W\right]_{up}^r$ are taken to be equal to the values of W_r already evaluated for all the reaches hydraulically upstream with respect to that considered in the calculations.

The last hypothesis (usually defined '*synchronism hypothesis*', because it implies that the peak time for all the reach of the networks has to be the same) has to be considered together with the second (usually defined '*independence hypothesis*'), which allows, by using equation (37), the performance of a reach upstream another (and also, the performance of a reach upstream a confluence) without evaluating the eventual influence of the filling/empting phenomena which are simultaneously developing in the downstream reaches.

These hypotheses, though seeming (together with that related to not consider the lag times) the worst present in the model, could, in certain cases, balance one each other, because the backing up prorogued by

Branch	Length	Long. Slope	Subcatch. Area	c	Cross Section Bottom Width	Cross Section Lateral slope	Manning Coeff.	Cross Section Max. Height
n.	m	m/m	m²	m²/m²	m	m/m	m^{-1/3}s	m
1	220	0.001	14.9	0.2	3	1	0.025	1
2	530	0.0015	33.8	0.2	3.2	1	0.025	1
3	450	0.001	20.4	0.3	3.5	1	0.025	1.2
4	300	0.002	19.7	0.3	3.2	1	0.025	1
5	250	0.002	8	0.3	3.1	1	0.025	1.1
6	350	0.0008	12.1	0.3	3	1	0.025	1.5
7	450	0.0012	11.5	0.22	3.8	1	0.025	2
8	310	0.0016	12.1	0.25	3.2	1	0.025	1.2
9	260	0.002	19.4	0.25	3.1	1	0.025	1.2
10	260	0.0014	12	0.22	3.8	1	0.025	1.5
11	510	0.0009	19.5	0.2	4	1	0.025	2.2
12	350	0.0022	24	0.25	2.7	1	0.025	1.2
13	380	0.0017	24.5	0.3	2.5	1	0.025	1
14	150	0.0015	13	0.22	2.3	1	0.025	1
15	130	0.0018	19.6	0.22	2.5	1	0.025	1.2
16	150	0.002	17.4	0.2	2.5	1	0.025	1
17	180	0.0021	13.5	0.2	2.6	1	0.025	1.1

Table 2: Geometrical characteristics of the network.

filling phenomena occurring in the downstream reaches could enhance the flow depths in the upstream reaches at peak discharge flows.

Application to a case study

In this subsection the ISM is applied to a case study. The case study consists of a 17 branch rural drainage network with each branch having a trapezoidal cross section. The network geometrical characteristics are summarized in Table 2, while a schematic representation of the network is reported in Figure 8.

The simulation with the ISM has been carried out by considering $W_{0,i} = 40$ m for each sub-catchment while the following intensity duration frequency (IDF) curve has been considered:

$$i_{d,T} = \begin{cases} 0.0003674417526 \; d^{-0.4}, & 0 < d < 3600 \, s \\ 0.0042862673310 \; d^{-0.7}, & d \geq 3600 \, s \end{cases} \qquad (37)$$

For Comparisons purpose, the same network has been simulated with numerical solution of the Kinematic Wave Model (KWM) within a variational approach (see, for instance, [7]) using equation 38 as IDF curve and evaluating the rainstorm runoff with the runoff coefficient method (see point 4 at section 2)

Results of the simulation are reported in Table 3 in terms of both peak discharge Q_{max} and maximum flow depth H_{max} obtained with ISM and KWM.

By inspection of Table 3 it is evident that the ISM model can be successfully employed in the design of rainstorm drainage networks.

Conclusion

In this work a slight modification and a generalization of a well-known method for sizing rainstorm drainage networks, very common in European contexts, the "*Italian-Storage method*" (ISM), are proposed and applied. The application to a case study has shown that the ISM is equivalent to the numerical solution of the Kinematic Wave Model, used within a variational approach in order to find the maximum discharges and flow depths flowing through the branch of a drainage

Branch	ISM		KWM	
	Qmax [m³/s]	Hmax [m]	Qmax [m³/s]	Hmax [m]
1	0.843	0.8	0.91	0.836
2	0.567	0.613	0.573	0.616
3	1.869	1.164	1.872	1.165
4	0.64	0.657	0.586	0.624
5	1.462	1.077	1.391	1.047
6	2.478	1.469	2.178	1.368
7	4.478	1.821	4.48	1.826
8	0.954	0.829	0.961	0.833
9	0.512	0.548	0.497	0.538
10	1.628	1.029	1.65	1.037
11	6.443	2.169	6.487	2.175
12	0.594	0.692	0.548	0.66
13	0.767	0.834	0.675	0.776
14	0.304	0.512	0.317	0.525
15	0.475	0.634	0.397	0.571
16	0.357	0.537	0.342	0.523
17	0.267	0.442	0.316	0.489

Table 3: Result of the simulation in terms of peak discharge Q_{max} and maximum flow depth H_{max}.

network. For this reason, the ISM can be satisfactory employed in the design of both urban and rural rainstorm drainage network.

References

1. Deininger RA (1970) Systems Analysis for Water Supply and Pollution Control. In Natural Resource Systems Models in Decision Making 45-65.

2. Wilson AJ, Britch AL, Templeman AB (1974) The Optimal Design of Drainage Systems. Engineering Optimization 1: 111-123.

3. Mays LW, Yen BC (1975) Optimal Cost Design of Branched Sewer Systems. Water Resources Research 11: 37-47.

4. Tang WH, Mays LW, Yen BC (1975) Optimal Risk-Based Design of Storm Sewer Networks. ASCE Journal of the Environmental Engineering Division 101: 381-398.

5. Mays LW, Wenzel HG (1976) Optimal Design of Multi-Level Branching Sewer Systems. Water Resources Research 12: 913-917.

6. Mays LW, Wenzel HG, Liebman JC (1976) Model for Layout and Design of Sewer Systems. ASCE Journal of the Water Resources Planning and Management Division 102: 385-405.

7. Cimorelli L, Cozzolino L, Covelli C, Mucherino C, Palumbo A, et al. (2013) Optimal design of rural drainage networks. Journal of Irrigation and Drainage Engineering 139: 137-144.

8. Palumbo A, Cimorelli L, Covelli C, Cozzolino L, Mucherino C, et al. (2014) Optimal design of urban drainage networks. Civil Engineering and Environmental Systems 31: 79-96.

9. Cimorelli L, Cozzolino L, Covelli C, Della Morte R, Pianese D (2014) Enhancing the efficiency of the automatic design of rural drainage networks. Journal of Irrigation and Drainage Engineering 140: 04014015

10. Cozzolino L, Cimorelli L, Covelli C, Mucherino C, Pianese D (2015) An innovative approach for drainage network sizing. Water (Switzerland) 7: 546-567.

11. Cimorelli L, Morlando F, Cozzolino L, Covelli C, Della Morte R, et al. (2015) Optimal positioning and sizing of detention tanks within urban drainage networks. Journal of Irrigation and Drainage Engineering

12. Fantoli G (1904) Final Report of Committee for the analysis of the sewer network of Milano City (in Italian) Milano, Italy.

13. Puppini F (1923) Design of Rural Drainage Channels (in Italian). Original title: "Il calcolo dei canali di Bonifica"). Monitore Tecnico.

14. Supino G (1947) A few observations on the using of "volume di invaso" method to evaluate the specific flood contribution given by unit drained area (in Italian) 227-231.

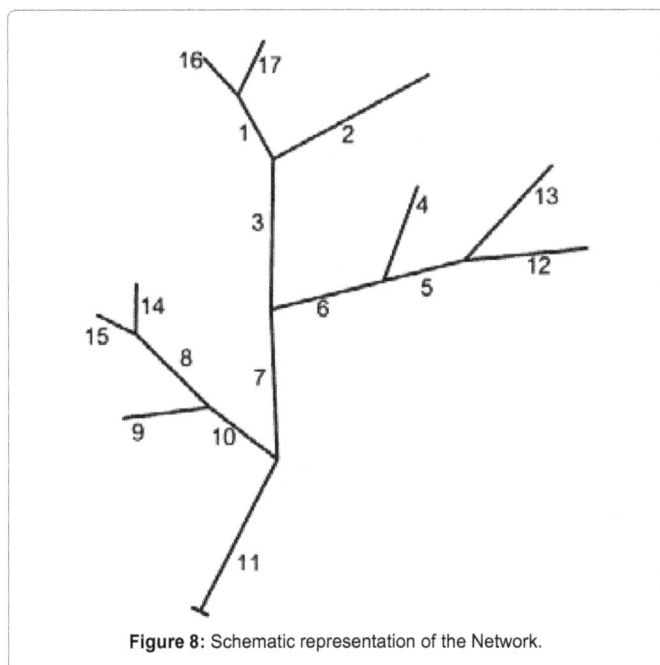

Figure 8: Schematic representation of the Network.

15. Chow VT (1962) Hydrologic design of culverts. ASCE Journal of the Hydraulics Division 88: 39-55.

16. Della Morte R, Iavarone V, Pianese D (2001) Evaluation of maximum flow depths and discharges in drainage networks by a simplified model coupled to a variational approach. Proceedings of 29th IAHR Congress - Theme D2, Beijing, China.

17. Denisov VM, Pak AV (2009) A Methodology of Determination of the Maximum Water Discharges and Volume of Rain Floods in Small Catchments. Russian Meteorology and Hydrology 34: 818-827.

18. Ponce VM, Li RM, Simons DB (1978) Applicability of kinematic and diffusion models. Journal of Hydraulic Division 104: 353-360.

9

Forecasting Climate Change Pattern for Bolero Agriculture Extension Planning Area in Malawi

Wales Singini*, Mavuto Tembo and Chikondi Banda

Mzuzu University, Private Bag 201 Mzuzu 2, Malawi

Abstract

Bolero community in Malawi, like most rural communities in Sub-Saharan Africa, is very vulnerable to climate variability and change because of its reliance on local biological diversity, ecosystem services, cultural and religious landscapes as source of sustenance and well-being. For this reason, a study was conducted to forecast climate change pattern for Bolero Agriculture Extension Planning Area based on temperature data from 1982 to 2013 in order to inform the policy makers and community on the future prospects of climate change and its effects. The data was collected by Malawi Government Department of Meteorology and Climate Change at Bolero weather station using fixed temperature recording thermometer. The study used Univariate Autoregressive Integrated Moving Average to model and forecast temperature variability. Based on ARIMA and its components autocorrelation and partial autocorrelation functions, Normalized Bayesian Information Criterion, Box-Ljung Q statistics and residuals estimated, ARIMA (1,1,3) was selected for the maximum temperature data which helped in explaining the temperature time series and forecasting the future values. From the forecast available from the fitted ARIMA model, it is concluded that forecasted maximum temperature will increase by 1.6°C from 27.7°C in 1982 to 29.3°C in 2030. The temperature increase suggests that climate change could continue to negatively impact on agricultural livelihood options in Bolero community and this call for increased adaptive capacity for the community.

Keywords: Forecasting; Climate change; Adaptation; stochastic

Introduction

There is a widely-held agreement now amongst scientists and even policy makers across the entire globe that climate change is real and will continue to impact on future adaptation strategies [1-6]. For most people in developing areas like Bolero, climate change has just added an extra layer of burden on top of the already existing socio-economic challenges such as high levels of poverty and inequality, poor agriculture production leading to chronic and acute food shortages and undeveloped markets. For instance, by merely altering the productivity of natural resources, climate change exerts far reaching implications on the people in Bolero who usually depend on the climate sensitive natural resources for their livelihoods [3]. No wonder that the majority of people in the area, and developing nations in general look severely exposed to the impacts of climate variability and change [7-9]. By forecasting future temperature values for Bolero, the research in effect enriched the body of knowledge about climate variability. Greater understanding about the recent past to present (1982 to 2013) and future (forecasted) temperature variability was important to both practitioners and policy makers in establishing the extent of vulnerability of the Bolero Community to climate change and setting strategies for adaptation.

The main impact of climate change is predicted to be an increase in global temperature beyond acceptable levels over the most land surfaces. The temperature records for a period of over the recent 100 years denote a warming of surface temperature, with the most obvious increases observed between 1983 to 2012 period [10]. As far as temperature is concerned, it is expected that the temperature will continue to rise. However, choices that people make now and in the years to come will determine by how much the earth's average temperature will rise. Higher temperatures have effects on droughts, changing rainfall patterns and availability of surface water whose consequences range from less food supply to general fewer water supplies. Research has shown that although the average global trend shows an increase in temperature, there are localized places that have

not become warmer yet. Research has further proven that changes in temperature are easier to project because temperature in contrast to precipitation is a large-scale variable [10,11].

Bolero community, like most rural communities in Sub-Saharan Africa, is very vulnerable to climate variability and change because of its reliance on local biological diversity, ecosystem services, cultural and religious landscapes as source of sustenance and well-being. The degree of vulnerability to climate change in the area is in actual fact dependent on the magnitude of climate variability and the socio-economic characteristics of households. Although, general knowledge about climate change exists in Bolero, Malawi, the extent of temperature variability remains a gap of knowledge in the area. There is also little knowledge about how the Bolero communities modify their livelihoods, traditional practices and religious values in order to adapt to climate change [12] suggested that the understanding of how communities modify their livelihoods, traditional practices, religious values, and production and consumption practices in amidst climate change is prerequisite in designing and implementation of any adaptation strategies particularly at a local level. Therefore, the study was conducted to forecast climate change pattern for Bolero Agriculture Extension Planning Area based on temperature data from 1982 to 2013 in order to inform the policy makers and community on the future prospects of climate change and its effects. The study hypothesis is that maximum and minimum temperatures of Bolero have not increased

*****Corresponding author:** Wales Singini, Mzuzu University, Private Bag 201 Mzuzu 2, Malawi, E-mail: walessingini@gmail.com

over time to influence climate change and variability. The results of the study will inform policy makers, stakeholders and community to design appropriate adaptation measures to the effects of climate change.

Materials and Methods

The research was conducted in Bolero Extension Planning Area (EPA) in Rumphi district located in the northern region of Malawi. The EPA has 12 functional sections with 58,550 people living in 112 villages. The area has 11,710 farm families holding a mean land size of 2.7 hectares per household of 5 persons. The area presented itself suitable for this nature of study mainly because of its vulnerability to climate variability and change besides being deeply rich in both culture and modern religion. The area is characterized by droughts and erratic rains resulting in crop failure and perpetual food shortages. Average annual rainfall range from 300mm in bad seasons to far less than 800mm in good seasons. With the majority of the inhabitants being rain fed agrarians, the area is more vulnerable to climate variability and change.

Secondary data on maximum temperature were obtained from the meteorological section of the Planning and Crops Departments at the Rumphi District Agriculture Development Office. The temperature data covered a period of 31 years from 1982 to 2013. The data was collected by the Department of Meteorology and Climate Change at Bolero weather station using fixed temperature recording thermometer. The data were used to fit the ARIMA model to forecast the future temperature patterns.

As the aim of the study was to model and forecast climate change pattern for Bolero Agriculture Extension Planning Area, various forecasting techniques were considered for use. ARIMA model introduced by [13] was frequently used for modeling and forecasting the pattern of climate change in Bolero. Among the methods based on univariate techniques, the ARIMA models by [13] stand out because of their wide range of application Singini et al.used univariate ARIMA model to model and forecast small *Haplochromine* fish species production in Malaŵi. Singini W [14] also used univariate ARIMA to model and forecast *Oreochromis* fish species production in Malawi. Tsitsika [15] used univariate and multivariate ARIMA models to model and forecast the monthly pelagic production of fish species in the Mediterranean Sea during 1990-2005. Jai Sankar et al. [16] used ARIMA model to model and forecast milk production in Tamilnadu during 1978-2008. Kannan et al. [17] also used stochastic modeling for cattle production and forecast the yearly production of cattle in the Tamilnadu state during 1970-2010. Jai Sankar et al. [18] used a stochastic model approach to model and forecast fish product export in Tamilnadu during 1969-2008.

ARIMA modelling and forecasting involved four steps: Identification, estimation, diagnostic checking and forecasting. To check for stationarity of the catch data, graphical analysis method was used. Model identification involved examining plots of the sample autocorrelograms and partial autocorrelograms and inferring from patterns observed in these functions the correct form of ARMA model to select. Gujarati [19] pointed out that when the PACF has a cutoff at p while the ACF tails off it gives an autoregressive of order p (AR (p)). If the ACF has a cutoff at q while the PACF tapers off, it gives a moving-average of order q (MA (q)). However, when both ACF and PACF tail off, it suggests the use of the autoregressive moving-average of order p and q (ARMA (p, q)).

Autoregressive process (AR) of order p

$$Y_t = \varphi_1 Y_{t-1} + \varphi_2 Y_{t-2} + ..., \varphi_t Y_{t-p}$$

Moving Average process (MA) of order q

$$Y_t = \phi_1 \in_{t-1} + \phi_2 \in_{t-2} + ..., \phi_t \in_{t-p}$$

Autoregressive Moving Average (ARMA) of order (p, q)

$$Y_t = \varphi_1 Y_{t-1} + \varphi_2 Y_{t-2} + ..., \varphi_t Y_{t-p}$$
$$+ \phi_1 \in_{t-1} + \phi_2 \in_{t-2} + ..., \phi_t \in_{t-p}$$

The general form of ARIMA Model of order (p, d, q)

$$Y_t + \sum_{i=1}^{p} \varphi_i Y_{i-1} + \sum_{j-i}^{q} \phi_j Y_{j-i} + \in_t$$

Where Y_t is the observation at φ and $\emptyset$ are coefficients and ε is an error term.

Model fitting consisted of finding the best possible estimates for the parameters of the tentatively identified models. In this stage, maximum likelihood estimation (MLE) method was considered to estimate the parameters. (MLE) method for estimation of ARIMA was applied in SPSS version 16.0. MLE runs an algorithm several times, using as the starting point the solution obtained in the previous iteration/run. Basically SPSS maximizes the value of a function by choosing the set of coefficient estimates that would maximize the function. Each time, it uses the estimates obtained in the previous iteration/run. In model diagnostics, various diagnostics such as the method of autocorrelation of the residuals and the Ljung-Box-Pierce statistic were used to check the adequacy of the identified models. If the model was found to be inappropriate, the process was returned back to model identification and cycle through the steps until, ideally, an acceptable model was found.

Plots of autocorrelation and partial autocorrelation of the residuals were used to identify misspecification. For evaluating the adequacy of ARMA and ARIMA processes, various statistics like Correlogram of the residuals; Normalized Bayesian Information Criterion (BIC), R-square, Stationary R-square, Root Mean Square Error (RMSE), Mean Absolute Percentage Error (MAPE), Maximum Absolute Percentage Error (MaxAPE), Mean Absolute Error (MAE) and Maximum Absolute Error (MaxAE) were used.

Results and Discussion

Data stationarity testing

Data stationarity was tested by means of sequence charts. Figure 1 shows that the time series for temperature data were not stationary. The non-stationarity was explained by the unstable means which increased and decreased (sharp ups and downs) at certain points throughout the 1982 to 2013 period. According to Pierce et al. [20] and Georgakarakos et al. [21] the use of ARIMA models requires that the time series is stationary. Therefore, the non-stationarity in mean was corrected through first differencing of temperature data and the newly constructed Y_t for the data set was then reexamined for stationarity (Figure 1).

Model identification

Since, Y_t was stationary in mean after first differencing (d = 1), the autocorrelations (ACF and PACF) of various orders of Y_t were computed and are presented in Figures 2 and 3 in order to identify the values of p and q. Gutíerrez-Estrada et al. [22]. Indicated that a good autoregressive model of order p (AR (p)) has to be stationary, and a

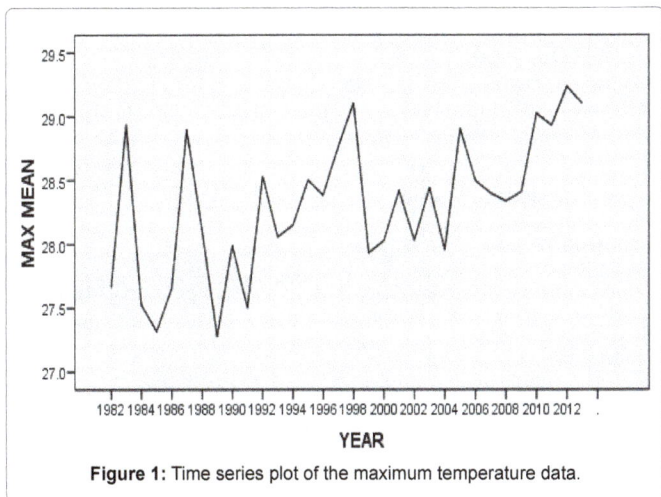

Figure 1: Time series plot of the maximum temperature data.

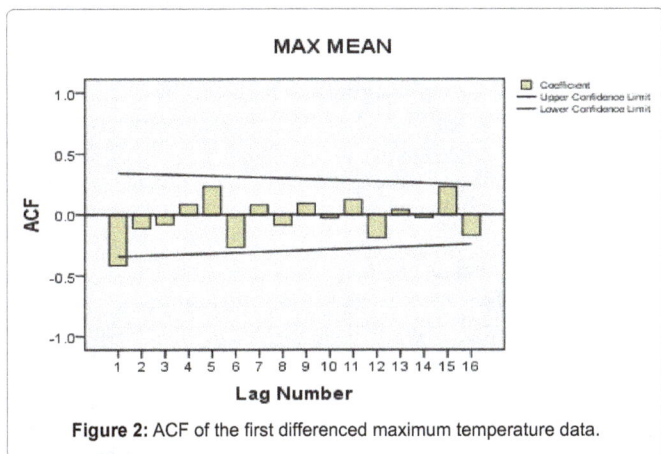

Figure 2: ACF of the first differenced maximum temperature data.

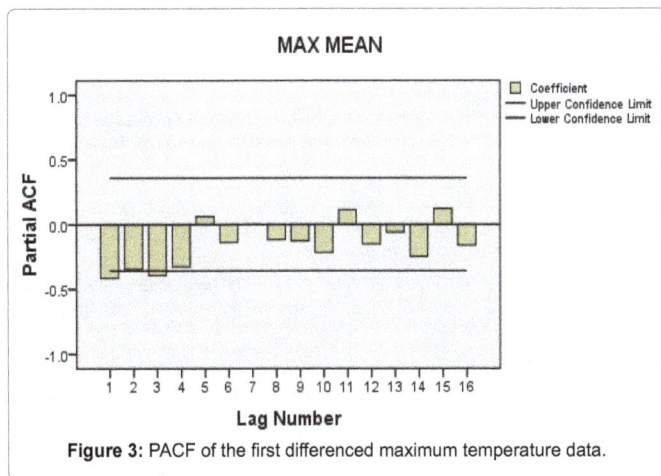

Figure 3: PACF of the first differenced maximum temperature data.

good moving average model of order q (MA (q)) has to be invertible. Invertibility and stationarity will give a constant mean, variance, and covariance (Figures 2 and 3).

The tentative ARIMA models are discussed with values differenced once (d=1) and the model which had the minimum Normalized BIC was selected. The various ARIMA models and the corresponding Normalized BIC values are given in Table 1. The value of the Normalized BIC of the selected ARIMA model was -0.544 (Table 1).

Model estimation

Model parameters estimated are presented in Table 2. Having obtained some suggested models the best possible estimates for the parameters were found by considering the final estimates of parameters and the model selection criteria. Estrada et al.[22] indicated that quality of the coefficients has to meet the following requirement; it must be statistically significant for each coefficient of the estimated model. Czerwinski et al. [23] Indicated that the Normalized BIC test reveals that the model with the least Normalized BIC is better in terms of forecasting performance than the one with a large Normalized BIC. The most suitable model for temperature forecasting is ARIMA (1, 1, 3), as this model had statistically significant coefficient, the lowest Normalized BIC, good R^2 and model fit statistics (MaxAPE and MAPE) (Table 2).

Diagnostic checking

In trying to verify the model, the residuals of the model were checked if they contained any systematic pattern for possible removal to improve the selected ARIMA model. This was done by examination of the autocorrelations and partial autocorrelations of the residuals of various orders. Abrahart et al. [24] Argued that for the model to be acceptable, the residuals should be independent from each other and constant in mean and variance over time. For this cause, various autocorrelations up to 9 lags were computed. The plots of the ACF and PACF residuals as indicated in Figure 4 shows that the sample autocorrelation coefficients of the residuals were low and lay within the limit of (-0.5 and +0.5) implying that none of the autocorrelations is significantly different from zero and any reasonable level. This proves that the chosen ARIMA model (1, 1, 3) is appropriate model for forecasting the temperature data for Bolero Area (Figure 4).

Hence the fitted ARIMA model for minimum temperature data is:

$$Y_t = -1.723 - 0.782Y_{t-1} + 0.105Y_{t-2}$$
$$+0.997Y_{t-3} - 0.107Y_{t-4} + 5.346 \in_t$$

Forecasting

Table 3 and Figure 5 show the actual and forecasted value of maximum temperature with 95% confidence limit. In order to ascertain

Model	SR-squared	R-squared	RMSE	MaxAPE	MAE	MaxAE	MAPE	NBIC
1,1,1	0.390	0.125	0.537	4.351	0.393	1.259	1.389	-0.800
1,1,2	0.388	0.123	0.548	4.352	0.398	1.260	1.405	-0.649
1,1,3	0.421	0.170	0.544	4.350	0.393	1.259	1.386	-0.544

Table 1: Model selection criterion for the maximum temperature data for Bolero.

Model	Model type	Coefficient	SE-Coefficient	T-value	P-value
1,1,1	Constant	-1.726	5.258	-0.328	0.745
	AR1	0.008	0.223	0.038	0.970
	MA1	0.994	2.87	0.346	0.732
1,1,2	Constant	-1.726	5.259	-0.328	0.745
	AR1	0.125	6.874	0.018	0.986
	MA1	1.155	5.614	0.206	0.839
	MA2	-0.159	6.535	-0.024	0.981
1,1,3	Constant	-1.723	5.346	-0.322	0.750
	AR1	-0.782	0.343	-2.281	0.031
	MA1	0.105	37.390	0.003	0.998
	MA2	0.997	40.677	0.025	0.981
	MA3	-0.107	3.853	-0.028	0.978

Table 2: Final estimates of the maximum temperature ARIMA models.

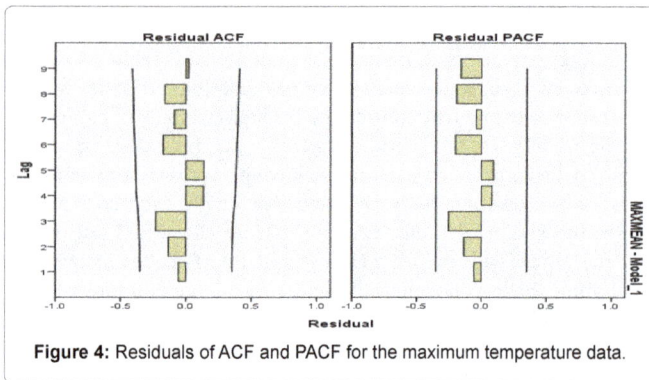

Figure 4: Residuals of ACF and PACF for the maximum temperature data.

Year	Forecasted maximum temperature	95% LCL	95% UCL
2015	27.7	26.3	29.1
2016	28.4	27.1	29.7
2017	27.9	26.8	29.1
2018	28.1	26.9	29.2
2019	27.8	26.7	28.9
2020	28.2	27.1	29.3
2021	27.9	26.8	28.9
2022	28.2	27.1	29.2
2023	28.0	26.9	29.0
2024	28.1	27.0	29.1
2025	28.1	27.1	29.2
2026	28.0	27.0	29.1
2027	28.2	27.2	29.3
2028	28.2	27.1	29.2
2029	28.3	27.3	29.3
2030	28.3	27.3	29.3

Table 3: Forecast of maximum temperature for Bolero EPA.

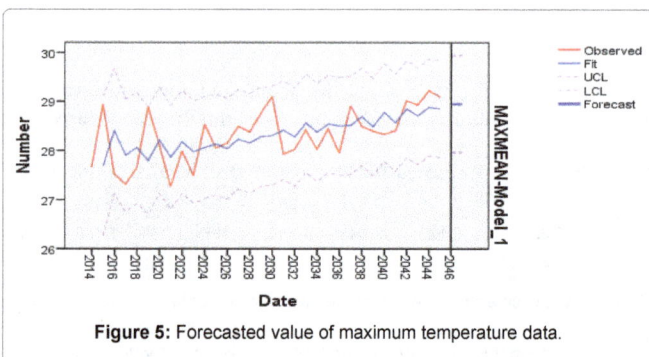

Figure 5: Forecasted value of maximum temperature data.

the forecasting ability of the fitted ARIMA model, the measures of the sample period forecasts' accuracy were computed. The results of the computation show that this measure had low forecasting inaccuracy. Czerwinski et al. [23] stated that a good model should show low forecasting error. This study found that the magnitude of the difference between the forecasted and actual values were low for the selected model. The noise residuals were combinations of both positive and negative errors which shows that, the model is not forecasting too low on the average or too high on the average. Having this positive, the model has outperformed as far as the forecasting power of the model is concerned. The temperature increase suggests that climate change could continue to negatively impact on agricultural livelihood options in Bolero community and this call for increased adaptive capacity for the

community. With the magnitude of future climate change as forecasted in this study the less privileged and marginalized social groups would continue to remain vulnerable and unless deliberate efforts are put in place to help the community to adapt to climate change effects. Higher temperatures have effects on droughts, changing rainfall patterns and availability of surface water whose consequences range from less food supply to general fewer water supplies in Bolero (Table 3 and Figure 5).

Conclusion

The most appropriate ARIMA model for temperature forecasting was found to be ARIMA (1,1,3). The fitted ARIMA model shows that temperature will increase by 1.6 C from 27.7°C in 1982 to 29.3°C in 2030. The results of the modelling show that the measure had low forecasting inaccuracy. This study found that the magnitude of the difference between the forecasted and actual values were low for the selected model. The noise residuals were combinations of both positive and negative errors which shows that, the model is not forecasting too low on the average or too high on the average. Having this positive, the model has outperformed as far as the forecasting power of the model is concerned. The temperature increase suggests that climate change could continue to negatively impact on agricultural livelihood options in Bolero community and this call for increased adaptive capacity for the community. With the magnitude of future climate change as forecasted in this study the less privileged and marginalized social groups would continue to remain vulnerable and unless deliberate efforts are put in place to help the community to adapt to climate change effects. Higher temperatures have effects on droughts, changing rainfall patterns and availability of surface water whose consequences range from less food supply to general fewer water supplies in Bolero.

Acknowledgements

The success of the work undertaken in this study was made possible due to support and assistance of many people and institutions. We owe a special thanks to the Irish Aid and Higher Education Authority of Ireland including the University of Maynooth team for the financial and material support. We would like to thank staff from District Agriculture Development Office whose kindness and availability were precious.

References

1. Bryan E, Deressa TT, Gbetibouo GA, Ringler C (2009) Adaptation to climate change in Ethopia and South Africa: Options and constraints. Environmental Science and Policy 12: 413-426.

2. Pareek A, Trivedi PC (2010) Cultural values and indigenous knowledge of climate change and disaster prediction in Rajasthan. Indian Journal of Traditional Knowledge 10: 183-189.

3. Sofoluwe NA, Tijani AA, Baruwa OI (2011) Farmers' perception and adaptation to climate change in Osun State, Nigeria. African Journal of Agricultural Research 20: 4789-4794.

4. Swai OW, Mbwambo JS, Magayane FT (2012) Gender and perception on climate change in Bahi and Kondoa Districts, Dodoma Region, Tanzania. Journal of African Studies and Development 4: 218-231.

5. Bagamba F, Bashaasha B, Claessens L, Antle J (2012) Assessing Climate Change Impacts and Adaptation Strategies for Smallholder Agricultural Systems in Uganda. African Crop Science Journal 20: 303-316.

6. Etwire PM, Al-Hassan RM, Kuwornu JKM, Osei-Owusu Y (2013) Smallholder farmers' adoption of technologies for adaptation to climate change in Northern Ghana. Journal of Agricultural Extension and Rural Development 5: 121-129.

7. IPCC (2007) Climate change 2007: Synthesis report, contribution of working groups I, II, and III to the fourth assessment of the Intergovernmental Panel on Climate Change.

8. Marshall NA, Park SE, Adger WN, Brown K, Howden SM (2012) Transformational capacity and the influence of place and identity. Environmental Research Letters 7: 034022.

9. Goulden MC, Adger WN, Allison EH, Conway D (2012) Limits to Resilience from livelihood diversification and social capital in lake social ecological systems. Annals of Association of American Geographers 103: 906-924.

10. Macchi M, Oviedo G, Gotheil S, Cross K, Boedhihartono A et.al, (2008) Indigenous and Traditional Peoples and Climate Change, Issue Paper.

11. Frame D (2007) Indigenous Peoples and Climate Change Models: Environmental Change Institute, University of Oxford.

12. Kpadonou RAB, Adegbola PY, Tovignan SD (2012) Local Knowledge and Adaptation to Climate Change in Oueme, African Crop Science Journal 20: 181-192.

13. Box GEP, GM Jenkins (1970) Time Series Analysis-Forecasting and Control. Holden-Day Inc., San Francisco.

14. Singini W, Kaunda E, Kasulo V, Jere W (2013) Modeling and forecasting Oreochromis (Chambo) production in Malawi- A stochastic model approach. Recent Research in Science and Technology 9: 1-6.

15. Tsitsika EV, Maravelias CD, Haralabous J (2007) Modeling and forecasting pelagic fish production using univariate and multivariate ARIMA models. Fisheries Science 73: 979-988.

16. Jai Sankar, Prabakaran R (2012) Forecasting milk production in Tamilnadu. International Multidisciplinary Research Journal 2: 10-15.

17. Jai Sankar, Prabakaran R, Senthamarai KK (2010) Stochastic Modeling for Cattle Production Forecasting. Journal of Modern Mathematics and Statistics 4: 53-57.

18. Jai Sankar (2011) Stochastic Modeling Approach for forecasting fish product export in Tamilnadu. Journal of Recent Research in Science and Technology 3: 104-108.

19. Gujarati DN (2004) Basic Econometrics. (4thedtn) The McGraw−Hill Companies.

20. Pierce GJ, Boyle PR (2003) Empirical modelling of interannual trends in abundance of squid (Loligo forbesi) in Scottish waters. Fisheries Research 59: 305-326.

21. Georgakarakos S, Haralabous J, Valavanis V, Koutsoubas D (2002) Loliginid and ommastrephid stock prediction in Greek waters using time series analysis techniques. Bull Mar.Sci 71: 269-288.

22. Guti errez- Estrada JC, Pedro-SE, L'opez-Luque R, Pulido-Calvo I (2004) Comparison between traditional methods and artificial neutral networks for ammonia concentration forecasting in an eel (Anguilla Anguilla L.) intensive rearing system. Aquat. Eng 31: 183-203.

23. Czerwinski IA, Gutierrez-Estrada JC, Hernando JA (2007) Short-term forecasting of halibut CPUE: Linear and non-linear univariate approaches. Fisheries Research 86: 120-128.

24. Abrahart RJ, See L (2000) Comparing neural network and autoregressive moving average techniques for the provision of continuous river flow forecasts in two contrasting catchments. Hydrological Process 14: 2157-2172.

10

Detection of Recent Changes in Climate Using Meteorological Data from South-eastern Bangladesh

Farzana Raihan[1]*, Guangqi Li[1] and Sandy P. Harrison[1,2]

[1]Department of Biological Sciences, Macquarie University, North Ryde, NSW 2109, Australia
[2]School of Archaeology, Geography & Environmental Sciences, Reading University, Whiteknights, Reading, UK

Abstract

Analysis of meteorological records from four stations (Chittagong, Cox's Bazar, Rangamati, Sitakunda) in south-eastern Bangladesh show coherent changes in climate over the past three decades. Mean maximum daily temperatures have increased between 1980 and 2013 by *ca.* 0.4 to 0.6°C per decade, with changes of comparable magnitude in individual seasons. The increase in mean maximum daily temperature is associated with decreased cloud cover and wind speed, particularly in the pre- and post-monsoon seasons. During these two seasons, the correlation between changes in maximum temperature and clouds is between -0.5 and -0.7; the correlation with wind speed is weaker although similar values are obtained in some seasons. Changes in mean daily minimum (and hence mean) temperature differ between the northern and southern part of the basin: northern stations show a decrease in mean daily minimum temperature during the post-monsoon season of between 0.2 and 0.5°C per decade while southern stations show an increase of *ca.* 0.1 to 0.4°C per decade during the pre-monsoon and monsoon seasons. In contrast to the significant changes in temperature, there is no trend in mean or total precipitation at any station. However, there is a significant increase in the number of rain days at the northern sites during the monsoon season, with an increase per decade of 3 days in Sitakunda and 7 days at Rangamati. These climate changes could have a significant impact on the hydrology of the Halda Basin, which supplies water to Chittagong and is the major pisciculture centre in Bangladesh.

Keywords: Monsoon; Rain days; Maximum daily temperature; Recent climate trends; Climate change detection

Introduction

Global average temperature has increased by 0.85°C (0.6-1.05°C) during the post-industrial period (1880-2012) with an increase over the most recent decade (2003-2012) of *ca.* 0.78°C (0.72-0.85°C) [1,2] and it is extremely likely that more than half of the observed increased in the latter part of the 20th century is anthropogenic [3]. The increase in global temperature is reflected at a regional scale across much of Asia [4-8]. Changes in mean temperature are accompanied by an increase in minimum temperature and in the frequency of extreme temperatures and heatwaves in many regions, including Asia [2,7]. Increases in global temperature are expected to lead to increases in precipitation [9]. Although increases in precipitation over mid-latitude land areas have been detected during the latter part of the 20th century [2], it is more difficult to detect changes in monsoon regions where high-quality records are shorter and the impact of short-term climate variability is large. Nevertheless, several studies have suggested that monsoon precipitation is increasing in parts of Asia [10,11].

Bangladesh is a predominantly agricultural country, and hence likely to be very sensitive to changes in climate. It has been predicted that Bangladesh could lose up to 17% of its land area and 30% of its food production as a result of the impacts of anticipated climate changes by 2050, while Hertel et al. [12] have suggested that Bangladesh will experience a 15% net increase in poverty by 2030 as a result of climate change. From this perspective, the Halda River Basin in south-eastern Bangladesh, is particularly important because it is the major spawning ground for Indian carp and sustains pisciculture in Bangladesh [13]. Fish constitute *ca.* 63% of the daily protein consumption in Bangladesh and pisciculture is responsible for 5-6% of the nation's gross domestic product [13].

There have been several studies on the detection of climate changes in Bangladesh [14-26] but the focus has largely been on individual climate variables and there has been little attempt to determine how changes in e.g. mean temperature are related to or influence other aspects of the regional climate. Furthermore, these studies have focused on the country as a whole rather than on areas that are most important from an agricultural point of view. In this study, we examine meteorological records from south-eastern Bangladesh to determine whether there are coherent trends in climate over recent decades, taking the opportunity to extend the analyses to cover the interval to 2013. In addition to individual temperature and precipitation trends that have been the focus of previous analyses of the climate of Bangladesh, we also explore trends in related climate variables (humidity, cloud cover, wind speed) in order to explain the mechanisms underpinning the changes in temperature and precipitation.

Study Area and Methods

South-eastern Bangladesh has a typical tropical monsoon climate, with most of the rainfall occurring during the monsoon season from June to September. The seasonal cycle of temperature is mediated by the monsoon, with lowest temperatures in January (*ca.* 16-20°C) but highest temperatures (*ca.* 38-40°C) in April just before the onset of the monsoon.

Daily data on maximum and minimum daytime temperature,

---

*Corresponding author: Farzana Raihan, Department of Biological Sciences, Macquarie University, North Ryde, NSW 2109, Australia
E-mail: farzana.raihan@students.mq.edu.au

precipitation, relative humidity, cloud cover and wind speed are available from four meteorological stations in south-eastern Bangladesh: Chittagong, Sitakunda, Rangamati and Cox's Bazar (Figure 1 and Table 1). Mean daily temperature was derived as the average of maximum and minimum daytime temperature, and diurnal temperature range as the difference between these variables. Sunshine data was only available for 3 of the stations, and thus is not used in our analyses. The records cover the period of 1980 to 2013, except in the case of Chittagong. The Chittagong station was moved in 2003. To avoid problems of inhomogeneity, analyses were performed on data from the interval 1980 to 2002 only for this station. The climate of south-eastern Bangladesh has three distinct seasons: the pre-monsoon season is from March through May; the monsoon season is June through September; and the post-monsoon season is October through February. Monthly, seasonal and annual averages were calculated for all the climate variables. No attempt was made to infill missing values of the daily observations, and the averages were only calculated for those months, seasons or years for which there were no missing values. The number of observations used therefore varies between variables and stations (Table 1).

Linear regression between climate index and year number was used to determine whether there were trends in the observations [27,28]. The significance of the trends was determined using a t-test, with a 95% cut-off for significance (i.e. the probability of accepting the null hypothesis of no change is <0.05). A 95% cut-off, rather than a 99% cut-off, was used because of the comparatively short length of the records examined. The r^2 value provides a measure of the goodness-of-fit of the relationships, and is lower when there is a large scatter around the regression line. It will always be close to zero when there is no significant trend. The slope coefficients from the regressions were

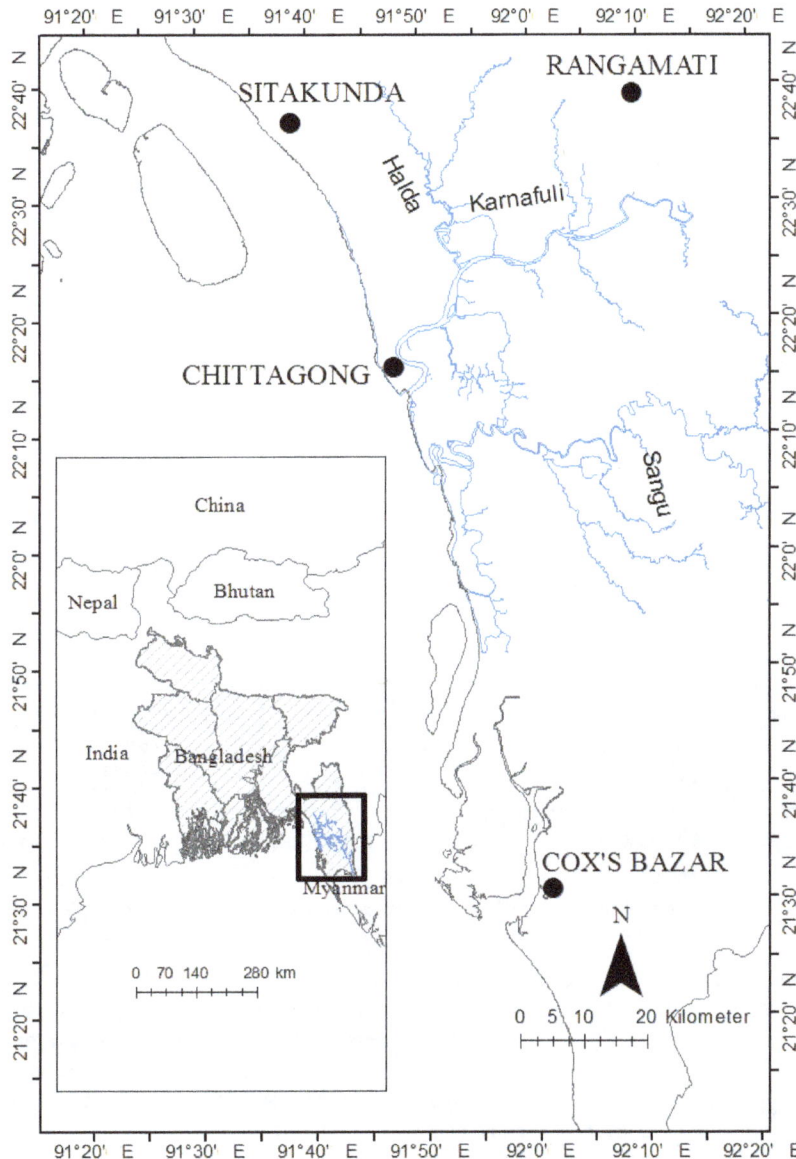

Figure 1: The location of the four meteorological stations used in this study. The inset map shows the location of the Halda Basin (in box) within Bangladesh, and the regional context of Bangladesh.

Station	Lat (°N)	Long (°E)	Elevation (m)	Period covered	Maximum temperature (°C)			Minimum temperature (°C)			Precipitation (mm)			Cloud (tenths)			Wind speed (km per hr)			Humidity (%)		
					Pre	Mon	Post	Pre	Mon	Post	Pre	Mon	Post	Pre	Mon	Post	Pre	Mon	Post	Pre	Mon	Post
Cox's Bazar	21.45	91.97	2	1980-2013	83	106	136	84	122	154	86	127	108	99	131	164	99	132	165	99	131	164
Chittagong	22.22	91.80	6	1980-2002	55	79	102	60	82	100	58	86	82	65	88	109	66	88	110	65	88	109
Rangamati	22.63	92.15	69	1980-2013	87	125	161	98	133	164	90	130	113	98	132	163	99	132	165	98	132	163
Sitakunda	22.63	91.70	7	1980-2013	98	125	162	94	126	161	94	128	107	99	132	165	99	132	165	102	136	170

Table 1: Information about the meteorological stations. Latitude (Lat) and longitude (Long) are given in decimal degrees, elevation (Elev) in meters. Averages were only calculated for periods (months, seasons, years) when there were no missing daily observations, and therefore the number of intervals used in these calculations varies by site and by variable. The number of monthly observations within the pre-monsoon (Pre), monsoon (Mon) and post-monsoon (Post) seasons for each site and variable is given.

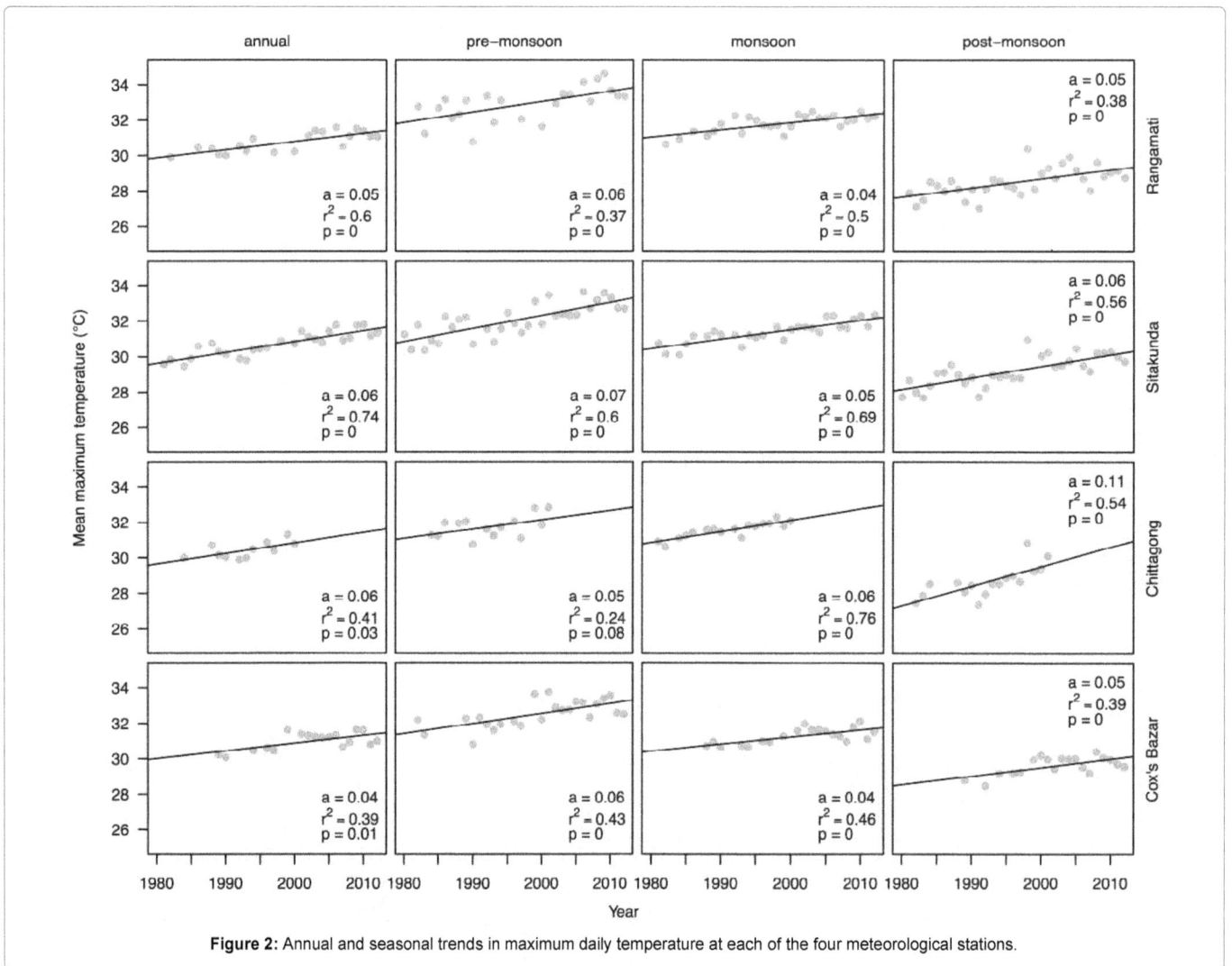

Figure 2: Annual and seasonal trends in maximum daily temperature at each of the four meteorological stations.

then used to estimate the magnitude of the change, over the length of the record, expressed as change per decade.

Results

All of the stations show a significant increase in maximum daily temperature (Figure 2) in the annual mean over the interval of observations. This positive trend is present in all seasons (and most individual months) at Cox's Bazar, Rangamati and Sitakunda (Tables 2 and 3). A positive trend (i.e. an increase in maximum daily temperature

over the interval with observations) is also seen in every season at Chittagong, but is only significant for the monsoon and post-monsoon seasons, and for the month of April during the pre-monsoon season. The lack of significance in the pre-monsoon season overall probably reflects the shortness of the length of record (22 years) for this station. The increase in maximum daily temperature is less marked in the monsoon season than in either the pre- or post-monsoon seasons, with values in the range of 0.4 to 0.6°C per decade compared to a range of 0.5 to 0.7°C per decade in the pre-monsoon and 0.5 to 1.1°C per decade in

Station	Variables	Code	Annual	Pre-monsoon	Monsoon	Post-monsoon
Cox's Bazar	Mean daily temperature (°C)	Tmean	**0.02**	**0.02**	**0.01**	0.01
	Maximum daily temperature (°C)	Tmax	**0.04**	**0.06**	**0.04**	**0.05**
	Minimum daily temperature (°C)	Tmin	**0.02**	**0.04**	0.01	0.01
	Diurnal temperature range (°C)	DTR	**0.03**	**0.03**	**0.05**	**0.03**
	Maximum daily precipitation (mm)	Pmax	**2.41**	**2.32**	**2.25**	0.32
	Mean daily precipitation on rain days (mm)	Pmean	0.08	**0.40**	0.04	0.02
	Total precipitation (mm)	Ptot	12.41	**8.13**	6.90	-1.88
	Number of rain days	Pwet	0.03	0.02	0.10	-0.09
	Cloud cover (tenths)	Cloud	-0.01	-0.01	0.00	-0.01
	Humidity (%)	Hum	-0.02	-0.05	**-0.04**	0.00
	Wind speed (km per hr)	Wind	**-0.04**	**-0.05**	-0.03	**-0.04**
Chittagong	Mean daily temperature (°C)	Tmean	**0.03**	0.02	**0.03**	**0.04**
	Maximum daily temperature (°C)	Tmax	**0.06**	0.05	**0.06**	**0.11**
	Minimum daily temperature (°C)	Tmin	0.02	0.02	0.04	0.02
	Diurnal temperature range (°C)	DTR	**0.03**	0.03	0.01	**0.05**
	Maximum daily precipitation (mm)	Pmax	**-7.39**	2.04	**-7.46**	2.33
	Mean daily precipitation on rain days (mm)	Pmean	0.10	0.38	0.01	0.15
	Total precipitation (mm)	Ptot	7.31	3.37	0.81	4.31
	Number of rain days	Pwet	-0.10	-0.31	-0.08	0.19
	Cloud cover (tenths)	Cloud	0.02	0.00	0.01	0.02
	Humidity (%)	Hum	**0.12**	0.07	0.02	**0.18**
	Wind speed (km per hr)	Wind	**-0.11**	**-0.18**	**-0.12**	**-0.06**
Rangamati	Mean daily temperature (°C)	Tmean	-0.01	0.00	0.00	-0.03
	Maximum daily temperature (°C)	Tmax	**0.05**	**0.06**	**0.04**	**0.05**
	Minimum daily temperature (°C)	Tmin	**-0.02**	0.00	0.00	**-0.05**
	Diurnal temperature range (°C)	DTR	**0.08**	**0.06**	**0.05**	**0.10**
	Maximum daily precipitation (mm)	Pmax	-0.32	-0.32	0.31	0.30
	Mean daily precipitation on rain days (mm)	Pmean	-0.08	-0.18	-0.07	-0.01
	Total precipitation (mm)	Ptot	6.84	-1.72	9.52	1.23
	Number of rain days	Pwet	**0.90**	0.14	**0.72**	0.13
	Cloud cover (tenths)	Cloud	**-0.02**	-0.01	0.00	**-0.03**
	Humidity (%)	Hum	**0.07**	0.02	0.03	**0.13**
	Wind speed (km per hr)	Wind	-0.02	-0.01	**-0.03**	0.00
Sitakunda	Mean daily temperature (°C)	Tmean	**-0.03**	-0.02	-0.01	**-0.06**
	Maximum daily temperature (°C)	Tmax	**0.06**	**0.07**	**0.05**	**0.06**
	Minimum daily temperature (°C)	Tmin	0.00	-0.01	0.00	**-0.02**
	Diurnal temperature range (°C)	DTR	**0.07**	**0.08**	**0.05**	**0.09**
	Maximum daily precipitation (mm)	Pmax	-0.32	-0.24	1.26	-1.15
	Mean daily precipitation on rain days (mm)	Pmean	-0.02	-0.04	0.20	-0.29
	Total precipitation (mm)	Ptot	9.20	-0.41	**23.30**	-4.62
	Number of rain days	Pwet	0.40	0.07	**0.29**	0.02
	Cloud cover (tenths)	Cloud	**-0.01**	-0.02	-0.01	**-0.02**
	Humidity (%)	Hum	**0.27**	**0.18**	**0.17**	**0.39**
	Wind speed (km per hr)	Wind	**-0.05**	**-0.06**	**-0.05**	**-0.04**

Table 2: Observed trends in annual and seasonal climate at individual meteorological stations. The values given are the slope coefficients from the linear regression; values in bold are significant (i.e. show a trend) at the 95% confidence level.

the post-monsoon interval (absolute values per decade calculated from slope coefficients given in Table 2).

Although the recent change in maximum daily temperature is coherent at all four stations, the trends in minimum daily temperature differ between the southern (Cox's Bazar, Chittagong) and northern (Sitakunda, Rangamati) stations (Figure 3). There is a significant increase in minimum daily temperature at Cox's Bazar, both in annual average and in the pre-monsoon season. Minimum daily temperatures

also increase during the monsoon and post-monsoon seasons, but the trend is not statistically significant although the trend for some individual months (August in the monsoon, October in the post-monsoon season) within each season is significant. The increase in both maximum and minimum daily temperature results in an overall increase in mean daily temperature at Cox's Bazar, which is significant in annual, and pre-monsoon and post-monsoon season, and positive though not significant in the monsoon season (Figure 4). Increases in

Station	Variables	Code	Jan	Feb	Mar	Apr	May	Jun	Jul	Aug	Sep	Oct	Nov	Dec
Cox's Bazar	Mean daily temperature (°C)	Tmean	0.00	0.02	0.02	0.03	0.00	**0.03**	0.01	0.01	0.00	0.01	0.01	0.02
	Maximum daily temperature (°C)	Tmax	**0.04**	**0.08**	**0.08**	**0.05**	**0.04**	**0.06**	0.03	**0.05**	**0.04**	**0.04**	**0.05**	**0.04**
	Minimum daily temperature (°C)	Tmin	0.02	0.03	**0.05**	0.03	0.01	0.02	0.01	**0.01**	0.01	**0.02**	0.01	0.04
	Diurnal temperature range (°C)	DTR	0.01	**0.05**	0.02	**0.03**	**0.03**	**0.05**	**0.04**	**0.05**	**0.04**	0.02	0.03	0.01
	Maximum daily precipitation (mm)	Pmax	-0.10	-0.17	0.18	0.18	**2.16**	1.80	1.25	1.22	0.06	**1.62**	-1.44	-0.34
	Mean daily precipitation on rain days (mm)	Pmean	-0.02	0.37	0.08	-0.05	**0.49**	0.08	0.21	0.15	0.00	0.27	**-0.72**	-0.10
	Total precipitation (mm)	Ptot	-0.08	-0.30	-0.32	-1.20	**8.37**	-0.60	4.42	5.78	2.40	3.61	-2.46	-0.22
	Number of rain days	Pwet	-0.01	-0.05	-0.04	-0.06	0.11	-0.06	-0.03	0.09	**0.12**	0.10	-0.02	-0.02
	Cloud cover (tenths)	Cloud	-0.01	-0.02	-0.01	-0.02	0.01	-0.01	0.00	0.01	0.01	0.01	-0.02	0.00
	Humidity (%)	Hum	0.04	-0.07	-0.05	-0.07	-0.02	**-0.12**	-0.05	-0.03	0.03	0.07	0.02	0.06
	Wind speed (km/hr)	Wind	**-0.05**	**-0.03**	**-0.05**	**-0.05**	**-0.05**	-0.04	**-0.05**	-0.03	-0.02	-0.03	**-0.06**	**-0.06**
Chittagong	Mean daily temperature (°C)	Tmean	-0.01	0.04	0.04	0.05	0.01	**0.04**	0.03	0.03	0.02	**0.04**	0.05	0.04
	Maximum daily temperature (°C)	Tmax	0.04	0.08	0.08	**0.11**	0.01	**0.07**	**0.05**	0.03	**0.05**	**0.07**	**0.08**	**0.10**
	Minimum daily temperature (°C)	Tmin	-0.04	0.03	0.00	0.04	0.04	0.04	0.03	**0.04**	0.02	0.04	0.06	-0.01
	Diurnal temperature range (°C)	DTR	**0.07**	0.04	0.06	0.03	-0.01	0.01	0.02	0.00	0.02	0.01	0.03	**0.07**
	Maximum daily precipitation (mm)	Pmax	0.18	0.83	-0.25	-0.74	2.10	-3.91	-4.79	-1.89	-0.45	3.06	0.49	0.15
	Mean daily precipitation on rain days (mm)	Pmean	0.30	0.68	0.00	-0.06	0.72	-0.23	-0.15	-0.05	-0.08	0.37	-0.51	0.60
	Total precipitation (mm)	Ptot	0.22	1.44	-0.22	-4.86	8.44	-2.30	-7.97	-1.16	0.56	4.35	0.91	0.47
	Number of rain days	Pwet	0.00	0.01	-0.09	-0.24	0.03	-0.02	-0.14	-0.06	0.12	0.24	0.08	-0.02
	Cloud cover (tenths)	Cloud	0.01	0.02	-0.02	**-0.04**	**0.06**	-0.01	0.01	0.01	0.02	**0.07**	0.03	0.00
	Humidity (%)	Hum	**0.27**	**0.29**	0.09	0.01	0.12	-0.01	-0.01	0.03	0.06	0.13	**0.32**	**0.20**
	Wind speed (km/hr)	Wind	**-0.11**	**-0.14**	**-0.11**	**-0.31**	-0.12	**-0.13**	**-0.15**	**-0.14**	-0.06	-0.04	-0.06	-0.01
Rangamati	Mean daily temperature (°C)	Tmean	**-0.05**	-0.01	0.00	0.01	0.00	0.00	0.01	0.00	0.00	-0.01	**-0.03**	**-0.05**
	Maximum daily temperature (°C)	Tmax	**0.04**	**0.08**	**0.09**	**0.07**	0.04	**0.04**	**0.05**	**0.05**	**0.05**	**0.04**	**0.05**	**0.05**
	Minimum daily temperature (°C)	Tmin	**-0.08**	**-0.06**	-0.03	0.00	0.00	0.01	0.00	0.00	-0.01	-0.02	**-0.06**	**-0.07**
	Diurnal temperature range (°C)	DTR	**0.12**	**0.15**	**0.10**	**0.06**	**0.03**	**0.04**	**0.04**	**0.05**	**0.06**	**0.06**	**0.10**	**0.11**
	Maximum daily precipitation (mm)	Pmax	0.10	-0.75	-0.05	-0.69	-0.06	1.76	0.21	**-2.45**	0.60	0.48	-0.59	-0.18
	Mean daily precipitation on rain days (mm))	Pmean	0.01	-0.50	-0.24	-0.20	-0.18	0.23	-0.22	-0.27	-0.11	0.02	-0.15	-0.35
	Total precipitation (mm)	Ptot	0.13	-1.26	-0.39	-1.84	0.62	6.42	-1.85	-2.30	2.23	2.78	-0.99	-0.05
	Number of rain days	Pwet	0.01	-0.07	0.01	-0.02	0.15	0.09	**0.14**	**0.18**	**0.20**	**0.16**	0.00	0.01
	Cloud cover (tenths)	Cloud	-0.02	**-0.03**	-0.01	**-0.03**	0.01	0.00	0.00	0.00	0.00	0.00	**-0.06**	**-0.04**
	Humidity (%)	Hum	**0.17**	-0.07	-0.09	0.02	0.06	0.04	0.00	0.04	**0.06**	**0.15**	**0.18**	**0.19**
	Wind speed (km/hr)	Wind	0.02	0.01	0.00	-0.03	-0.01	-0.01	-0.03	**-0.06**	-0.03	-0.03	-0.02	0.00
Sitakunda	Mean daily temperature (°C)	Tmean	**-0.08**	**-0.05**	-0.02	-0.01	-0.01	-0.01	-0.01	-0.01	**-0.02**	**-0.03**	**-0.05**	**-0.07**
	Maximum daily temperature (°C)	Tmax	**0.05**	**0.09**	**0.09**	**0.07**	**0.06**	**0.05**	**0.06**	**0.05**	**0.05**	**0.07**	**0.06**	**0.05**
	Minimum daily temperature (°C)	Tmin	**-0.05**	-0.03	-0.02	-0.02	0.00	0.00	0.01	0.00	0.00	0.00	-0.01	-0.02
	Diurnal temperature range (°C)	DTR	**0.10**	**0.14**	**0.10**	**0.09**	**0.05**	**0.05**	**0.05**	**0.05**	**0.06**	**0.06**	**0.08**	**0.07**
	Maximum daily precipitation (mm)	Pmax	0.10	-0.18	-0.70	-0.64	0.37	1.08	1.05	1.79	1.06	-0.11	-0.90	-0.23
	Mean daily precipitation on rain days (mm)	Pmean	0.01	0.06	-0.53	-0.14	0.06	0.22	0.06	0.12	-0.08	-0.24	-0.71	-0.39
	Total precipitation (mm)	Ptot	0.14	-0.25	-1.60	-2.32	3.52	4.25	2.45	4.73	2.88	-0.72	-1.64	-0.20
	Number of rain days	Pwet	0.01	-0.04	-0.02	-0.05	0.13	0.01	0.05	0.08	**0.17**	0.10	0.00	0.00
	Cloud cover (tenths)	Cloud	-0.01	-0.02	-0.02	**-0.04**	-0.01	-0.01	-0.01	0.00	0.00	0.00	-0.03	-0.02
	Humidity (%)	Hum	**0.47**	**0.28**	0.17	**0.18**	**0.19**	**0.18**	**0.12**	**0.16**	**0.21**	**0.31**	**0.42**	**0.52**
	Wind speed (km/hr)	Wind	-0.04	-0.02	-0.04	**-0.09**	**-0.05**	**-0.05**	**-0.06**	**-0.06**	**-0.05**	-0.03	**-0.05**	-0.04

Table 3: Observed trends in monthly climate at individual meteorological stations. The values given are the slope coefficients from the linear regression; values in bold are significant (i.e. show a trend) at the 95% confidence level.

minimum daily temperature are also recorded at Chittagong; although a positive trend is seen in all seasons, the trend is only significant during the monsoon season (Figure 3). Again, the increase in both maximum and minimum daily temperature results in an overall increase in mean daily temperature at the Chittagong station, in all seasons although the trend is only significant for the monsoon and post-monsoon seasons (Figure 4).

In contrast to the two southern stations, both Rangamati and Sitakunda show no significant changes in minimum daily temperature during the pre-monsoon and monsoon seasons (Figure 3). However,

they both show a significant decrease in minimum daily temperature during the post-monsoon season. This change is larger than the increase in maximum daily temperature, and as a result the mean daily temperature during the post-monsoon season is reduced by 0.3°C per decade at Rangamati and 0.6°C per decade at Sitakunda (in contrast to the increase in mean daily temperature of 0.1°C per decade at Cox's Bazar and 0.4°C per decade at Chittagong).

There is a significant increase in diurnal temperature range, annually and in all seasons in Cox's Bazar, Rangamati and Sitakunda (Figure 5). There is also an increase in diurnal temperature range at

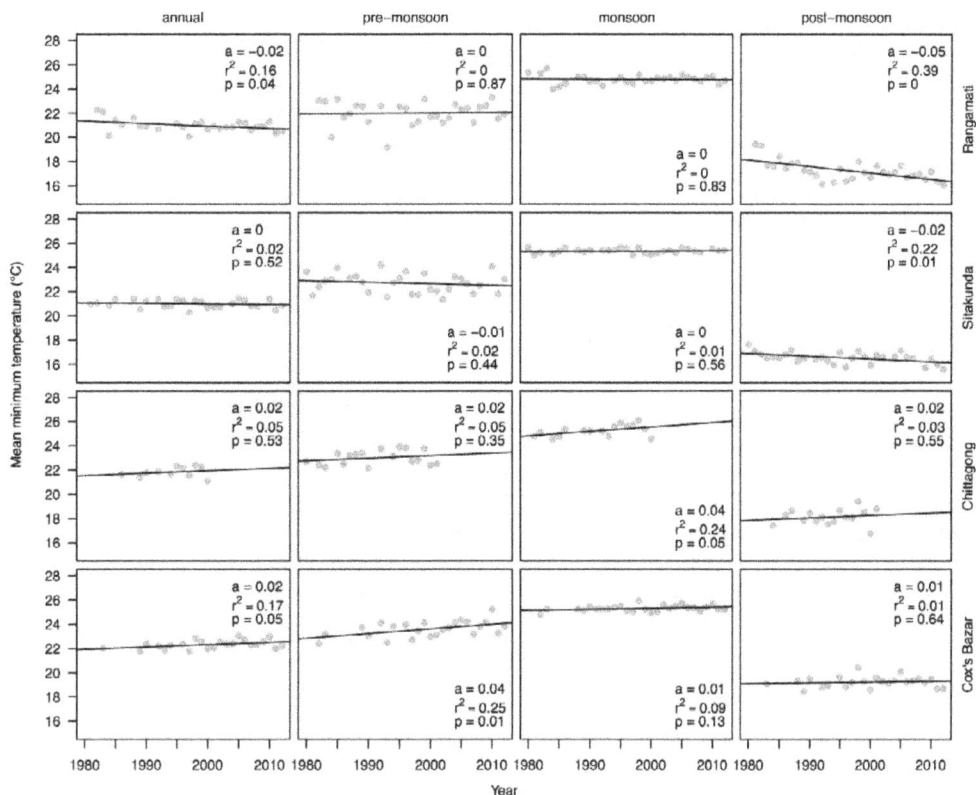

Figure 3: Annual and seasonal trends in minimum daily temperature at each of the four meteorological stations.

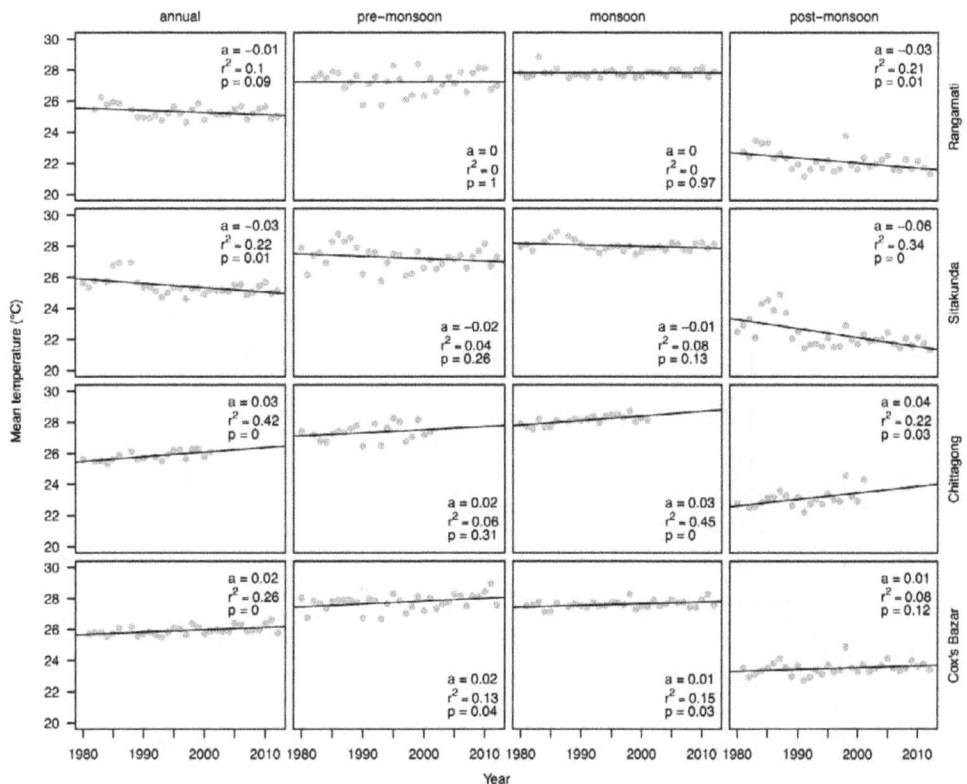

Figure 4: Annual and seasonal trends in mean daily temperature at each of the four meteorological stations.

Chittagong, but the trend is only significant in the post-monsoon season. The increase in the diurnal temperature range in the northern sites (Sitakunda, Rangamati) reflects the fact that minimum daily temperature is decreasing while maximum daily temperature is increasing. However, the increase in the diurnal temperature range in the southern sites (Cox's Bazar, Chittagong) reflects the fact that the increase in maximum daily temperature is larger than the increase in minimum daily temperature. Thus, the coherent response in diurnal temperature range across the region results from two different causes.

The recent trends in temperature are relatively coherent, but changes in precipitation are less coherent. There are no discernible trends in precipitation in any season at Chittagong, which may reflect the shortness of this record. However, there are changes in precipitation during the monsoon season at Sitakunda and Rangamati (Figure 6). At Sitakunda, there is a significant increase in total precipitation during the monsoon season, which is associated with a significant increase in the number of rain days (Figure 6). Maximum daily precipitation during the monsoon season also shows an increase, but this is not significant (p=0.26) and there is no discernible trend in mean precipitation on rain days. There is also a significant increase in the number of rain days during the monsoon season at Rangamati,

although this is not accompanied by significant trends in other precipitation characteristics. Although total precipitation increases at Cox's Bazar during the monsoon season, the trend is not significant. However, there is a significant increase in mean daily precipitation on rain days during the pre-monsoon season, largely driven by a significant increase (in mean and maximum precipitation on rain days and in total precipitation) during the month of May. The change in total precipitation during the monsoon season at Sitakunda is 233 mm per decade and the change in the pre-monsoon season at Cox's Bazar is 81 mm per decade. As might be expected, the increases in precipitation at Rangamati and Cox's Bazar are strongly correlated with increases in relative humidity (Table 4). However, the relationship between relative humidity and precipitation is also positive at the other stations, even when the change in precipitation is not significant. Overall, the emerging pattern is towards increased monsoon rainfall, although the way this is expressed and the exact timing differ between stations.

There are no discernible or consistent trends in cloud cover during the monsoon season at any station, despite the apparently significant changes in either total precipitation or the number of rain days. However, there are decreases in cloud cover during the pre- and post-monsoon seasons. The decrease in cloud cover at Sitakunda

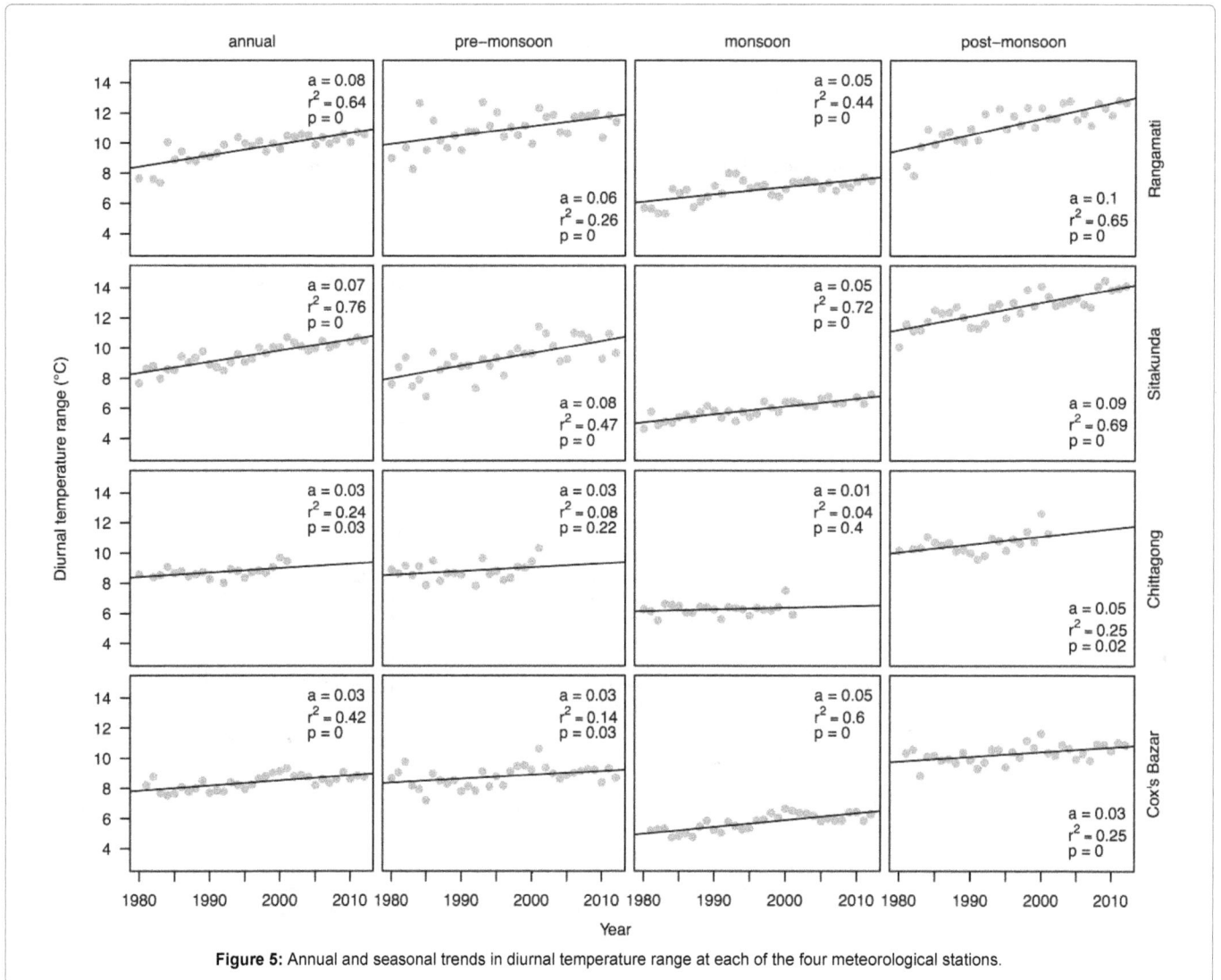

Figure 5: Annual and seasonal trends in diurnal temperature range at each of the four meteorological stations.

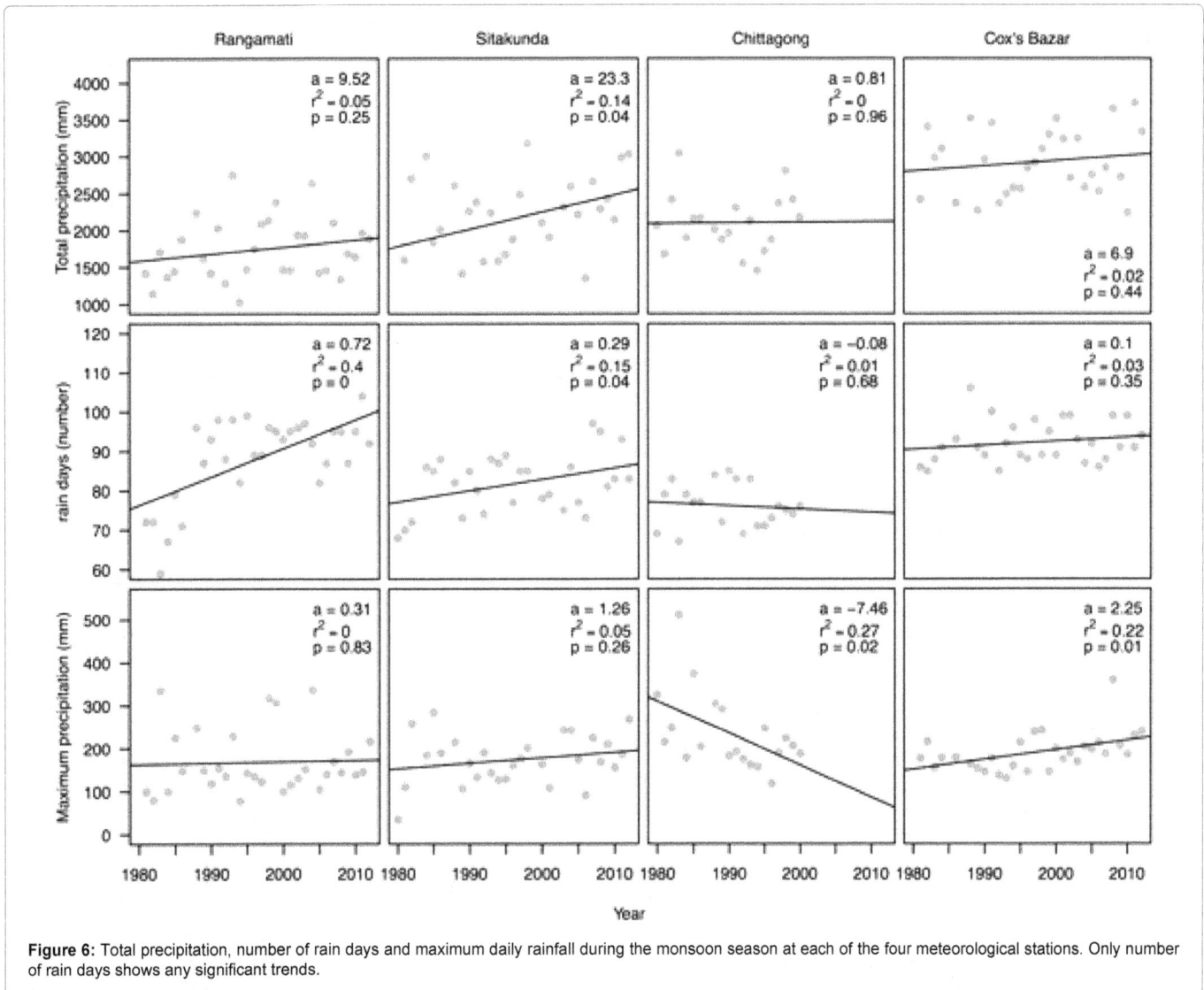

Figure 6: Total precipitation, number of rain days and maximum daily rainfall during the monsoon season at each of the four meteorological stations. Only number of rain days shows any significant trends.

is significant for both seasons (Table 2). There is a decrease in cloud cover in both seasons at Rangamati, but the decrease is only significant during the post-monsoon season. Cloud cover also appears to decrease during both seasons at Cox's Bazar, although the changes are not statistically significant. The reduction in cloud cover is significantly correlated (Table 4) with an increase in maximum daily temperature (although not with minimum or mean daily temperature) at all three sites, suggesting that changes in cloud cover contribute to the strong observed increase in maximum daily temperature.

The analyses indicate a general decrease in wind speed across this region. At Sitakunda and Chittagong, the decrease in wind speed is significant in all three seasons (Table 2). Although wind speeds decrease in all three seasons at Cox's Bazar, the trends are only significant in the pre- and post-monsoon seasons. On the other hand, there is no significant change in wind speed during these two seasons at Rangamati, but a strong decrease (by 0.3 km/hr per decade) during the monsoon season. The change in wind speed is significantly negatively correlated with changes in maximum temperatures (Table 4) suggesting that decreased wind speed is potentially contributing to the observed increases in maximum daily temperatures.

Discussion and Conclusions

There is a significant increase in maximum daily temperature in south-eastern Bangladesh over the past three decades. This trend is seen throughout the year, although the largest changes occur outside the monsoon season. The trends in minimum daily temperature and in mean temperature are different between the northern and southern parts of the region, with a decrease in both minimum and mean daily temperature in the north (Sitakunda, Rangamati) and an increase in both in the south (Cox's Bazar, Chittagong). The diurnal temperature range has increased at all stations, but this reflects the fact that maximum and minimum daily temperatures show opposite tendencies in the northern stations while in the southern stations the increase is due to larger increases in maximum than in minimum daily temperatures. Our results are broadly consistent with earlier analyses of temperature changes in Bangladesh [20,25]. Islam [20] found an overall increase of maximum daily temperature in south-eastern Bangladesh between 1948-2007 and also identified the contrast between northern and southern stations in the region in terms of the trend in minimum temperature. The reconstructed sign and magnitude of the annual change in these variables are comparable for Cox's Bazar (0.26°C versus

Station	Season	Variables	Diurnal temperature range (°C)	Maximum daily precipitation (mm)	Mean daily precipitation on rain days (mm)	Total precipitation (mm)	Number of rain days	Wind speed (km/hr)	Cloud cover (tenths)	Humidity (%)
Cox's Bazar	Pre-monsoon	Mean daily temperature (°C)	0.1	0.05	0.06	-0.23	**-0.45**	-0.26	-0.34	**-0.37**
		Maximum daily temperature (°C)	**0.57**	0.32	0.34	0.09	-0.21	-0.34	**-0.62**	-0.39
		Minimum daily temperature (°C)	-0.34	0.37	**0.48**	0.14	-0.25	-0.24	-0.01	-0.01
	Monsoon	Mean daily temperature (°C)	**0.36**	0.18	-0.05	-0.14	-0.2	-0.18	-0.11	**-0.47**
		Maximum daily temperature (°C)	**0.89**	0.13	-0.22	-0.22	0	**-0.5**	**-0.46**	0.02
		Minimum daily temperature (°C)	0.13	-0.02	**-0.44**	**-0.42**	0	0.06	-0.16	-0.33
	Post-monsoon	Mean daily temperature (°C)	0.28	-0.19	-0.08	-0.25	-0.27	-0.09	**-0.49**	-0.23
		Maximum daily temperature (°C)	**0.63**	-0.09	-0.17	-0.4	-0.33	**-0.65**	**-0.72**	-0.22
		Minimum daily temperature (°C)	-0.13	-0.09	-0.06	-0.15	-0.12	0.04	-0.18	0.07
Chittagong	Pre-monsoon	Mean daily temperature (°C)	-0.2	0.24	0.03	-0.41	**-0.62**	0.15	**-0.46**	-0.13
		Maximum daily temperature (°C)	0.44	0.15	0.16	-0.21	-0.42	-0.22	**-0.71**	-0.43
		Minimum daily temperature (°C)	**-0.48**	0.32	0.19	-0.19	**-0.6**	0.23	-0.22	0.28
	Monsoon	Mean daily temperature (°C)	0.11	-0.23	0.17	0.03	-0.44	-0.41	-0.12	-0.03
		Maximum daily temperature (°C)	0.43	-0.22	0.38	0.15	**-0.56**	-0.39	0.05	0.11
		Minimum daily temperature (°C)	-0.46	-0.11	0.35	0.17	-0.47	-0.06	-0.08	0.01
	Post-monsoon	Mean daily temperature (°C)	0.42	-0.1	-0.09	-0.19	-0.09	0.06	-0.08	-0.05
		Maximum daily temperature (°C)	**0.68**	0.15	0.09	-0.15	-0.21	-0.06	-0.15	0.09
		Minimum daily temperature (°C)	-0.26	**-0.47**	-0.18	-0.07	0.12	0.09	0.15	-0.03
Rangamati	Pre-monsoon	Mean daily temperature (°C)	0	-0.16	-0.04	**-0.48**	**-0.68**	0.25	-0.31	**-0.65**
		Maximum daily temperature (°C)	**0.59**	**-0.51**	**-0.43**	**-0.59**	**-0.48**	-0.26	**-0.64**	**-0.56**
		Minimum daily temperature (°C)	**-0.53**	-0.04	-0.18	**-0.4**	**-0.46**	0.24	0.06	-0.21
	Monsoon	Mean daily temperature (°C)	-0.14	0.21	0.07	-0.27	**-0.47**	0.28	-0.09	**-0.57**
		Maximum daily temperature (°C)	**0.71**	-0.1	-0.35	-0.15	**0.43**	**-0.52**	**-0.44**	-0.31
		Minimum daily temperature (°C)	**-0.58**	0.27	0.13	0.04	-0.08	**0.54**	-0.33	-0.09
	Post-monsoon	Mean daily temperature (°C)	-0.33	**-0.41**	-0.16	**-0.36**	**-0.43**	0.35	0.25	**-0.57**
		Maximum daily temperature (°C)	**0.76**	-0.26	-0.29	**-0.37**	-0.22	0	**-0.71**	0.13
		Minimum daily temperature (°C)	**-0.77**	**-0.36**	-0.16	-0.32	**-0.37**	**0.54**	**0.42**	-0.34
Sitakunda	Pre-monsoon	Mean daily temperature (°C)	-0.32	-0.17	-0.33	**-0.54**	**-0.5**	**0.59**	-0.06	**-0.55**
		Maximum daily temperature (°C)	**0.73**	-0.2	-0.2	**-0.4**	-0.33	-0.26	**-0.63**	0.14
		Minimum daily temperature (°C)	**-0.63**	-0.15	-0.33	**-0.49**	**-0.51**	**0.52**	0.13	-0.12
	Monsoon	Mean daily temperature (°C)	-0.04	0.12	-0.08	-0.1	-0.1	**0.46**	-0.06	**-0.69**
		Maximum daily temperature (°C)	**0.95**	-0.09	-0.03	0	0.01	**-0.37**	-0.36	**0.47**
		Minimum daily temperature (°C)	-0.07	-0.11	-0.26	-0.22	0	-0.12	-0.21	0
	Post-monsoon	Mean daily temperature (°C)	-0.27	0.01	0.06	-0.14	**-0.39**	**0.58**	0.16	**-0.78**
		Maximum daily temperature (°C)	**0.89**	-0.25	-0.32	**-0.37**	-0.22	-0.21	**-0.5**	**0.52**
		Minimum daily temperature (°C)	**-0.57**	0.04	-0.01	-0.05	-0.13	0.28	0.14	-0.29

Table 4: Summary of correlations between temperature variables and related climate variables for individual seasons, where the pre-monsoon season is defined as March through May, the monsoon season as June through September, and the post-monsoon season October through February. Numbers in bold are significant at the 95% level.

0.2°C per decade in our study), Chittagong (0.16°C versus 0.3°C per decade in our study) and Rangamati (-0.11°C versus -0.1°C per decade in our study), despite the fact that the analyses cover different periods of time. There is, however, a discrepancy in the direction of the observed trend in mean annual temperature at Sitakunda, which is positive (0.19°C per decade) according to Islam [20] but negative (-0.3°C per decade) according to our analyses. Further diagnosis is required to determine whether this is a reflection of differences in seasonal patterns or a function of the time interval used. Shahid et al. [25] found a year-round increase in diurnal temperature range at three stations from south-eastern Bangladesh (Chittagong, Cox's Bazar, Rangamati) over the period 1961-2008. The rates are comparable to those obtained here (0.17 versus 0.3 in our study for Chittagong, 0.28 versus 0.3 at Cox's Bazar, 0.44 versus 0.8 in Rangamati) and confirm that there is a distinct gradient in the magnitude of the trend between northern and southern stations. More detailed comparisons with earlier studies cannot be made because neither Islam [20] nor Shahid et al. [25] diagnose the relationships between seasonal changes in different components of the temperature regime.

The difficulty in identifying statistically robust trends in precipitation is a common theme of previous analyses [21-24,26] and also in this study. However, we have shown that there is a significant increase in total precipitation and the number of rain days during the monsoon season at Sitakunda, in the number of rain days during the monsoon season at Rangamati, while evidence of an increase in mean daily precipitation on rain days in May at Cox's Bazar suggests an earlier onset of the monsoon season there. Ahasan et al. [21] suggested that, while there was little or no change in rainfall during the monsoon season, there was a trend towards increased precipitation in the pre-monsoon season for Bangladesh as a whole between 1961 and 2010. Shahid [23] also found a significant increase in mean rainfall during the pre-monsoon season at Chittagong and a marginally significant increase at Cox's Bazar, consistent with our findings. In a second paper, Shahid [22] also showed similarly significant trends for the northern stations (Rangamati, Sitakunda) during the pre-monsoon season. Analyses based on the interval 1958-2007 suggest that the number of rain days per year has increased at Cox's Bazar [24], but unfortunately these analyses did not examine the records from Sitakunda and

Rangamati – which show a statistically significant increase in the number of rain days during the monsoon season in our analyses. We are limited in our ability to compare the magnitude of the trends in precipitation characteristics because the earlier studies use different stations (or numbers of stations), examine different intervals of time, and focus on different precipitation variables. Nevertheless, all of the analyses suggest that there are changes in precipitation that could have an important influence on the hydrological regimes of south-eastern Bangladesh.

There are significant decreases in cloud cover and wind speed during the pre- and post-monsoon seasons in south-eastern Bangladesh over the past three decades. These changes are closely correlated with changes in temperature, and particularly maximum daily temperature. This suggests that the observed increases in maximum daily temperature during the pre- and post-monsoon seasons can, at least partly, be explained by reduced cloud cover and reduced wind speed – both of which will led to enhanced surface heating. There is no change in either cloud cover or wind speed during the monsoon season at any of the stations. The absence of a discernible change in cloud cover and wind speed may reflect the fact that precipitation changes during the monsoon season are small (or hard to detect).

This analysis is based on meteorological records covering the interval from 1980 onwards, chosen in order to focus on the time corresponding to the most marked global warming. It is more difficult to identify statistically significant trends from short records. Nevertheless, significant trends exist in multiple seasonal climate variables. We suspect that similar trends at different sites, or similar trends in different seasons at a given site, or coherent trends between different climate variables, are real even when they are not statistically significant. For example, it seems likely that the positive but non-significant trend in maximum daily temperature in the pre-monsoon season at Chittagong is a real feature of the climate, given that similarly positive but significant trends are found in this season at other stations and that similarly positive and significant trends are found in the other seasons at Chittagong itself. Similarly, it seems likely that monsoon precipitation is increasing across the whole of south-eastern Bangladesh, even though the indicators that register a significant change are different at the different stations: change in total precipitation during the monsoon season at Sitakunda, a lengthening of the monsoon season at Cox's Bazar, and an increase in the number of rain days at Rangamati. Nevertheless, changes in all of the precipitation-related variables are coherent across the four stations even when they are not significant. Thus, the relatively short length of the records analysed in this study is not a drawback to detecting climate changes across the region.

The observed trends in temperature and precipitation are small but, if these changes continue in the future, they are likely to have significant impacts on water resources in south-eastern Bangladesh. Diagnosing whether the overall impacts on water availability, river flows and the incidence of flooding will be positive or negative requires forward modelling of the system [29,30]. Our analyses identify both the need for such modelling and provide calibration data sets that make it feasible. There is already considerable concern about water availability and water quality in south-eastern Bangladesh [31,32] and thus an assessment of how these have been affected by recent climate changes and how they will be affected in the near-term future is a matter of some urgency.

Acknowledgements

We acknowledge the Bangladesh Meteorological Department for providing the meteorological data. We thank Sayma Akhter, FONASO Joint Doctorate research Fellow for collecting the data. FR and LG were supported by a Macquarie University International Research Scholarship (iMQRES).

References

1. IPCC (2013) Summary for Policymakers, Climate Change 2013: The Physical Science Basis, Contribution of Working Group I to the Fifth Assessment Report of the Intergovernmental Panel on Climate Change In: Stocker TF, Qin D, Plattner GK, Tignor M, Allen SK et al. (eds.), Cambridge University Press,Cambridge, United Kingdom and new York, NY, USA.

2. Hartmann D, Klein Tank A, Rusicucci M, Alexander L, Broenniman B, et al. (2013) Observations: Atmosphere and surface, Climate Change 2013: The Physical Science Basis. Contribution of Working Group I to the Fifth Assessment Report of the Intergovermental Panel on Climate Change. In: Stocker TF, Qin, D Plattner GK, Tignor M, Allen SK et al. (eds.), Cambridge University Press, Cambridge, United Kingdom and New York, NY, USA.

3. Bindoff N, Stott P, AchutaRao K, Allen M, Gillett N, et al. (2013) Detection and attribution of climate change: from global to regional, Climate change 2013: the physical science basis, Contribution of Working Group I to the Fifth Assessment Report of the Intergovernmental Panel on Climate Change. In: Stocker TF, Qin D, Plattner GK, Tignor M, Allen SK et al. (eds.), Cambridge University Press, Cambridge, United Kingdom and New York, NY, USA.

4. Zhai P, Pan X (2003) Trends in temperature extremes during 1951-1999 in China. Geophysical Research Letters 30: 193.

5. Qian W, Lin X (2004) Regional trends in recent temperature indices in China. Climate Research 27: 119-134.

6. Kothawale D, Rupa Kumar K (2005) On the recent changes in surface temperature trends over India. Geophysical Research Letters 32: L18714.

7. Klein Tank A, Peterson T, Quadir D, Dorji S, Zou X, et al. (2006) Changes in daily temperature and precipitation extremes in central and south Asia. Journal of Geophysical Research: Atmospheres 111: D16105.

8. Su B, Jiang T, Jin W (2006) Recent trends in observed temperature and precipitation extremes in the Yangtze River basin, China. Theoretical and Applied Climatology 83: 139-151.

9. Trenberth KE (2011) Changes in precipitation with climate change. Climate Research 47: 123-138.

10. Sen Roy S, Balling RC (2004) Trends in extreme daily precipitation indices in India. International Journal of Climatology 24: 457-466.

11. Bony S, Bellon G, Klocke D, Sherwood S, Fermepin S, et al. (2013) Robust direct effect of carbon dioxide on tropical circulation and regional precipitation. Nature Geoscience 6: 447-451.

12. Hertel TW, Burke MB, Lobell DB (2010) The poverty implications of climate-induced crop yield changes by 2030. Global Environmental Change 20: 577-585.

13. Alam MS, Hossain MS, Monwar MM, Enamul M (2013) Assessment of fish distribution and biodiversity status in Upper Halda River, Chittagong, Bangladesh. International Journal of Biodiversity and Conservation 5: 349-357.

14. Parthasarathy B, Sontakke N, Monot A, Kothawale D (1987) Droughts/floods in the summer monsoon season over different meteorological subdivisions of India for the period 1871-1984. Journal of Climatology 7: 57-70.

15. Chowdhury M, Debsharma S (1992) Climate change in Bangladesh-A statistical review. In: Proc Report on IOC-UNEP Workshop on Impacts of Sea Level Rise due to Global Warming, NOAMI.

16. Ahmed A, Munim A, Begum Q (1996) El Nino-southern oscillation and rainfall variation over Bangladesh. Mausam New Delhi 47: 157-162.

17. Karmakar S, Shrestha ML (2000) Recent climatic changes in Bangladesh. SAARC Meteorological Research Centre.

18. Ahmed R, Kim I-K (2003) Patterns of daily rainfall in Bangladesh during the summer monsoon season: case studies at three stations. Physical Geography 24: 295-318.

19. Mia NM (2003) Variations of temperature of Bangladesh. In: Proceedings of SAARC Seminars on Climate Variability In the South Asian Region and its Impacts, SMRC, Dhaka.

20. Islam AS (2009) Analyzing changes of temperature over Bangladesh due to

global warming using historic data. Jawaharlal Nehru Centre for Advanced Scientific Research (JNCASR): 15-17.

21. Ahasan M, Chowdhary MA, Quadir D (2010) Variability and trends of summer monsoon rainfall over Bangladesh. Journal of Hydrology and Meteorology 7: 1-17.

22. Shahid S (2010) Rainfall variability and the trends of wet and dry periods in Bangladesh. International Journal of Climatology 30: 2299-2313.

23. Shahid S (2010) Recent trends in the climate of Bangladesh. Climate Research 42: 185-193.

24. Shahid S (2011) Trends in extreme rainfall events of Bangladesh. Theoretical and Applied Climatology 104: 489-499.

25. Shahid S, Harun SB, Katimon A (2012) Changes in diurnal temperature range in Bangladesh during the time period 1961-2008. Atmospheric Research 118: 260-270.

26. Hasan Z, Akhter S, Islam M (2014) Climate change and trend of rainfall in the south-east part of coastal Bangladesh. European Scientific Journal 10: 25-39.

27. Suppiah R, Hennessy KJ (1998) Trends in total rainfall, heavy rain events and number of dry days in Australia, 1910-1990. International Journal of Climatology 18: 1141-1164.

28. Blender R, Luksch U, Fraedrich K, Raible CC (2003) Predictability study of the observed and simulated European climate using linear regression. Quarterly Journal of the Royal Meteorological Society 129: 2299-2313.

29. Arnold J, Fohrer N (2005) SWAT2000: current capabilities and research opportunities in applied watershed modelling. Hydrological Processes 19: 563-572.

30. Gassman PW, Reyes MR, Green CH, Arnold JG (2007) The soil and water assessment tool: historical development, applications, and future research directions, Working Paper 07-WP 443, Center for Agricultural and Rural Development, Iowa State University.

31. Zuthi M, Biswas M, Bahar M (2009) Assessment of supply water quality in the Chittagong city of Bangladesh. ARPN Journal of Engineering and Applied Sciences 4: 73-80.

32. Akter A, Ali MH (2012) Environmental flow requirements assessment in the Halda River, Bangladesh. Hydrological Sciences Journal 57: 326-343.

11

A Comparative Study of the Physico-Chemical Characteristics of Dust-Full Particles in Three Stations of Birjand City in Eastern Iran

Amin Donyaei* and Alireze Pourkhabbaz

Department of Environmental Sciences, University of Birjand, Birjand, Iran

***Corresponding author:** Donyaei A, Department of Environmental Sciences, Faculty of Natural Resources, University of Birjand, Birjand, Iran
E-mail: Amin_donyaei@yahoo.com

Abstract

Dust storms happening frequently in arid and semiarid regions of the world carry huge amount of particulate matters; hence, these are recognized as one of the most important environmental issues at regional and international scales. This study aimed to evaluate heavy metals laden dustfall particles in Birjand, East Iran.

Sampling of dustfall particles was conducted once every 1 month from July 2016 for 6 month at three stations of Agriculture university, Aboozar avenue, and Shokat university simultaneously using Deposit Gauge Method. The concentration of Fe and Mn were measured using Flame Atomic.

Absorption Spectroscopy and Furnace Atomic Absorption Spectroscopy was used for analyzing Cd, Cr, Pb, As, Cu, and Zn. Then, pollution levels for the heavy metals were evaluated using Geo-Accumulation Index (I_{geo}) and Integrated Pollution Index (IPI). Mineralogy and morphology of the dustfall particles were inspected using X-ray diffraction and scanning electron microscopy.

I_{geo} index indicated that levels of pollution for Pb and Zn higher than other among the dustfall particles of the three stations whereas the concentrations of the other heavy metals were not evaluated as contaminant. Moreover, means of IPI values calculated for stations of Agriculture, Aboozar, and Shokat were 1.51, 1.99 and 1.66 respectively. X-ray diffraction revealed abundance of silicate (quartz) and carbonate (calcite) minerals in dustfall particles. Additionally, scanning electron microscopy emphasized on similarity of shape of dustfall particles in the three stations.

Mean of dustfall particles on area unit in Shokat station was less than Agriculture and Aboozar.

Keywords: Dustfall phenomenon; Birjand; Heavy metals

Introduction

The phenomenon of air pollution in urban regions is one of outcomes of the industrial revolution which has been embarked 300 years ago and is daily increased with the development of industry and urbanism. Whatever is interpreted as "world crisis of urban pollution" is in fact derived from disordering the balance between the main fundamental elements of natural environment?

The air pollution is one of the basic problems of modern societies. According to definition: the air pollution is predicated to the existence of one or several contaminants in the free air in an amount, time and features which is dangerous for the life of human, animal and plant and things and appurtenance of life, in a way that causes disorder in the relative welfare of human [1].

The phenomenon of dust is one of the air pollutions with which we were encounter in recent years. The dust is regarded to be the solid material composed of soil, metal anthropogenic mixtures and natural biogenic materials [2].

The phenomenon of dust is one of the serious bioenvironmental problems in the specific regions of the world. Whereas the most dust of atmosphere is occurred with the origin of fine aggregates and these aggregates enjoy high frequency in dried and semi-dried regions. In fact, the dust is a kind of reaction to change in the vegetation of environment that in this regard, the role of human activities should be attended beside the natural conditions of geographical environments. The effects of dust may be kept at a distance of 4000 km from the main source and cause the outbreak of unfavourable environmental effects and plenty damages in agricultural, industrial, transport and transmission systems. In addition to unstable climate, the existence or nonexistence of humidity is more effective in creating dust so that if the unstable climate consists of humidity, it would create rain, storm and thunder and if it has no humidity, it would cause dust storm [3].

The dust particles, with regard to their movement origin and path, enjoy high capacity in transporting the heavy metals. So, the study of amount of heavy metals contamination in dust is of much importance because of the dangers which threat the health of human beings. In general, the heavy metal contamination in the particles of dust is considered to be a serious problem because of toxicity, inseparability and concentration [4,5].

In general, two sources of human activities and detrital patterns of soil have been specified for the pollution of dust particles. This is expected from the human resources to cause the contamination of heavy metals because of further industrial activities in urban regions;

whereas the existence of heavy metals with high concentration in the dust particles in non-industrial regions is generally related to the erosion of geology structures specially soil [6].

The heavy metals which form in the fine and light mixtures stay in the environment in a suspended form and at the time of atmosphere rain, some of these contaminants are dissolved in the rain and return to the earth's surface and the metals which exist in the structure of macro and heavy particles, deposit during the time and reach the earth's surface. Generally, the existence of heavy metals in the atmosphere or dust causes the increase of these metals concentration in the body of citizenship of contaminated regions by swallowing, respiration and cutaneous absorption [7,8]. For example, it has been seen that the concentration of lead has been increased in the blood of children by its increase in the dust of atmosphere. If the children are exposed to more than 10 microgram of lead, the decrease of intelligence quotient would be tangible [9]. The severe effects of cadmium such as lung and Prostate cancer, damaging Kidney, Emphysema and Osteoporosis have been reported in different resources. Also, it has been expressed that the respiration of air polluted with Chromium is resulted in Allergic rhinitis, attenuation of Immune system, severe respiratory problems such as Rhonchus and cough [10].

Of the previous studies about the constitutive mixtures of dust in Asia and Middle East, we can refer to the ones carried out by Krueger et al.; Jaradat et al. in Iran, Rajabi and Suri [6,11,12] carried out some studies about some heavy metals which exist in dust in Sanandaj, Khorramabad and Andimeshk.

The purpose of present research is the study of amount of heavy metals in dust particles of eastern half of Iran by the use of Geo-Accumulation Index and Integrated Pollution Index (IPI) [13]. Also, in this study, the nature and form of sampled dust particles was evaluated by the use of X-ray Diffraction and Scanning Electron Microscope for the purpose of comparing dust particles in two studying seasons.

Materials and Methods

Study area

The Southern Khorasan province, with a land area of about 82864 square kilometer, covers 5/7 percent of total area of country. This province has been located in east of Iran beside eastern north of Lut Desert (Figure 1).

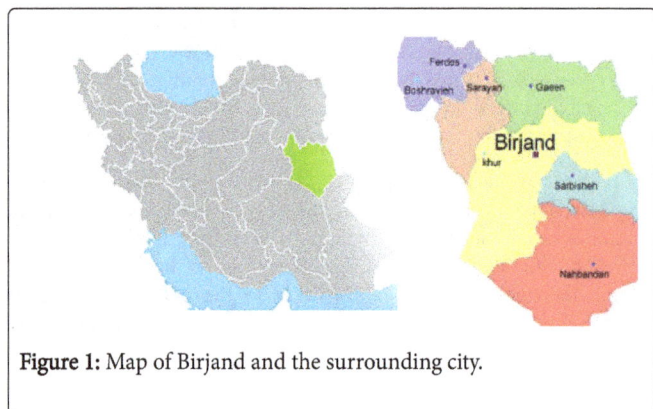

Figure 1: Map of Birjand and the surrounding city.

Birjand, county town of province, has a land area of 14265 square kilometer and with regard to the mathematic location it locates in the coordinates of 3253 north latitude and 5913 east longitude and has been risen 1470 m above sea level. This town has been located at the gradient of Bagheran mountain range and nearly in the center of Birjand Desert and has been surrounded by the mountain range [14].

The climate of Birjand city is of semi-desert kind and has cold winters and dry and hot summers. According to its climate, the rain level of this city is low and the most level of raining is occurred from 21 November to 21 May that this rain is often falls in form of snow in winter. The air of Birjand has severe storm and dust in average in 12 days of year.

Sample preparation and analysis

In present research, the atmosphere dust sampling traps were fitted on rooftops of three builds with 5 m heights and proper locative dispersion for the purpose of sampling atmosphere dust and studying the heavy metals in the atmosphere of Birjand.

The sampling regions are including: Amirabad region, rooftop of Agriculture faculty of Birjand University with the coordinates of 32° 51' 57N latitude and 59° 52' 20E longitude; the second station, the urban region of Aboozar avenue, rooftop of boyish dormitory with the coordinates of 32° 51' 57N latitude and 59° 8' 48E longitude; and the third station, Shokatabad region, rooftop of central build of Birjand University with the coordinates of 50° 37' 32N latitude and 59° 17' 21E longitude (Figure 2).

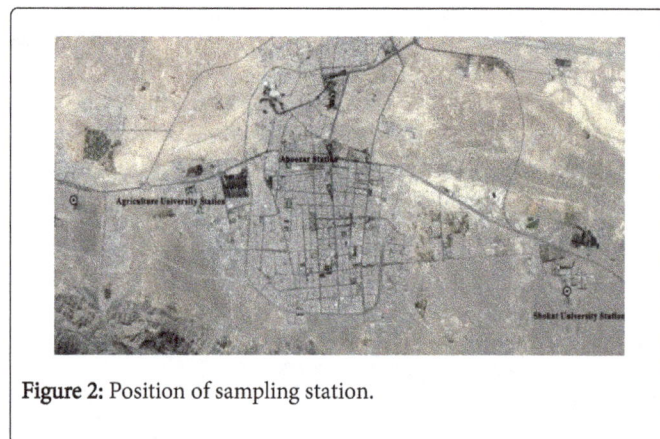

Figure 2: Position of sampling station.

The atmosphere dust sampling traps (Figure 3) consist of one leg, bottle for collecting dust, one funnel put conversely on the bottle and one glass for collecting particles.

The dust samples were collected monthly during 6 months (21 June to 21 December). Six samples at each station and totally 18 samples were collected during these six months. The samples were weighed after transferring to the laboratory.

After transferring the samples to the laboratory, the sampled dust particles were washed with the twice distilled water and were heated for 2 h at the temperature of 105°C for the purpose of decreasing the mass of solution and were refined after being cool by Whatman filter No. 42 and were weighed for determining the amount of dust rained on the area. Then, 100 mg of dissolved containments remained on Whatman filter was digested with 5 ml of 0/01% Nitric Acid and was heated for 2 h at 95°C by the use of Block digest system. In next step, the samples were put in Ultrasonic washroom for 30 min at 50°C and then were again refined by Whatman filter No. 42 and the solution reached the mass of 25 ml [15,16].

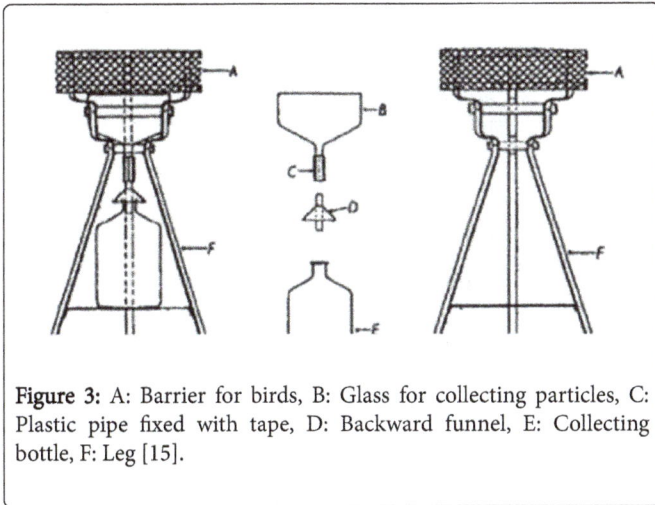

Figure 3: A: Barrier for birds, B: Glass for collecting particles, C: Plastic pipe fixed with tape, D: Backward funnel, E: Collecting bottle, F: Leg [15].

I_{geo}	I_{geo} class	Contamination level
>5	6	Extremely contaminated
4-5	5	Strongly to extremely strongly contaminated
3-4	4	Strongly contaminated
2-3	3	Moderately to strongly contaminated
1-2	2	Moderately contaminated
0-1	1	Uncontaminated to moderately contaminated
<0	0	Uncontaminated

Table 1: Pollution levels classified based on I_{geo} values [18].

IPI	IPI class	Contamination level
>5	3	Extremely contaminated
25	2	Strongly contaminated
12	1	Moderately contaminated
<1	0	Uncontaminated

Table 2: Pollution levels classified based on IPI values [19].

Finally, in the obtained solution, the metals (Cd, Cr, Zn, Cu, Pb and As) were measured by Furnace Atomic Absorption Spectroscopy and the metals (Fe and Mn) were measured by Flame Atomic Absorption Spectroscopy.

Data analysis

The heavy metals measured in the sample were evaluated by the use of following indexes:

1) Geo-Accumulation Index (I_{geo}): The index I_{geo}, according to equation 1, calculates the contamination into heavy metals with regard to the ratio of concentration of each heavy metal in the studied sample into the background concentration of that metal in the earth crust [3].

$$I_{geo} = Log2\left[\frac{C_n}{1.5B_n}\right] \qquad (1)$$

2) Integrated Pollution Index (IPI): The index IPI provides an average of ratio of concentration of several heavy metals into the background concentration of same metals in the sample that is calculated according to equation 2 [13].

$$PI_i = \frac{C_i}{B_i} \qquad (2)$$

Finally, the quantity of IPI is calculated for all metals in the form of average of PI amounts.

The categorization of levels of contamination has been shown according to the amount of I_{geo} and IPI indexes in Tables 1 and 2 [3,17]. Also, the study of nature and form of dust particles was done by x-ray diffraction and electron microscope imaging.

Results

The Figure 4 shows the procedure of change among dust particles rained on the area at three stations in Birjand city during the sampling period. As is observed, the changes of dust amount at three stations follow similar procedure so that the least and most amount of dust at all three stations has been occurred at the end of autumn and middle of summer, respectively. Also, the average of amount of dust particles rained on the area for agriculture faculty, Aboozar Square and Shokat University was measured 5/33, 3/29 and 2/24 g/m² in a period of 30 days.

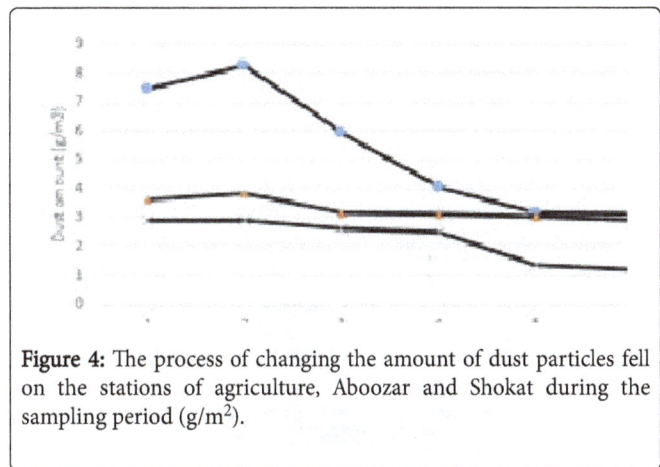

Figure 4: The process of changing the amount of dust particles fell on the stations of agriculture, Aboozar and Shokat during the sampling period (g/m²).

The amounts of maximum, minimum, standard deviation and mean of concentration of measured heavy metals in dust particles of three stations (agriculture faculty, Aboozar Square and Shokat university) have been shown in Table 3 and as is observed, the most and least amounts at all three stations are related to Zn and Cd, respectively.

Station	Parameters	Fe	Pb	Cd	Cr	As	Zn	Cu	Mn
	Minimum	173.4	20.78	0.13	15.35	10.1	212.95	13.6	188.75
	Maximum	246.35	93.28	0.52	65.65	21.95	450.5	60.3	281.25
	Average	199.63	58.78	0.35	34.35	16.52	326.07	33.84	244.15
Agri University	Std. Deviation	26.68	25.32	0.14	17.32	4.33	86.73	17.42	31.99
	Minimum	164.13	27.95	0.21	11.73	2.14	399.5	25.1	150.28
	Maximum	246.3	93.78	0.84	61.28	20.89	700.75	94.08	299.75
	Average	203.1	63.97	0.57	37.2	10.48	498.56	53.69	216.82
Aboozar Avenue	Std. Deviation	31.62	21.78	0.24	17.62	6.56	116.95	24.71	48
	Minimum	171.85	13.28	0.8	8.11	3.79	236.5	21.98	213.2
	Maximum	253	66.28	0.9	89.53	20.26	650.5	75.15	253.5
	Average	202.83	50.35	0.59	36.02	12.86	490.75	37.92	231.71
Shokat University	Std. Deviation	30.62	20.51	0.31	20.63	6.02	158.06	19.31	16.28

Table 3: Values min, max, average and standard deviation of heavy metals in the dust Agriculture University, Abooza Avenue, Shokat University.

The Figures 5-7 and also Tables 4-6 show a little amount of I_{geo} index for the measured heavy metals at three stations during the sampling period.

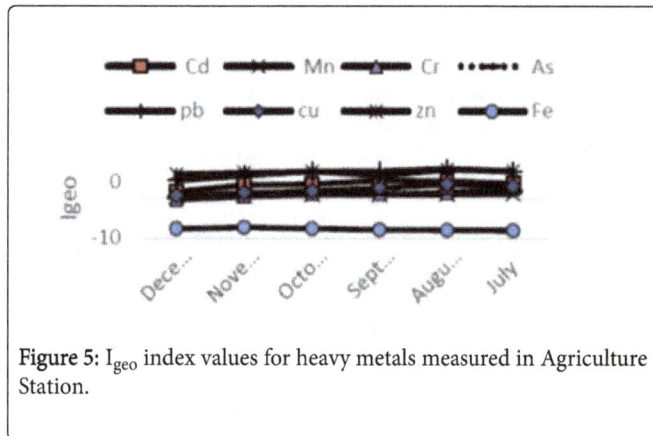

Figure 5: I_{geo} index values for heavy metals measured in Agriculture Station.

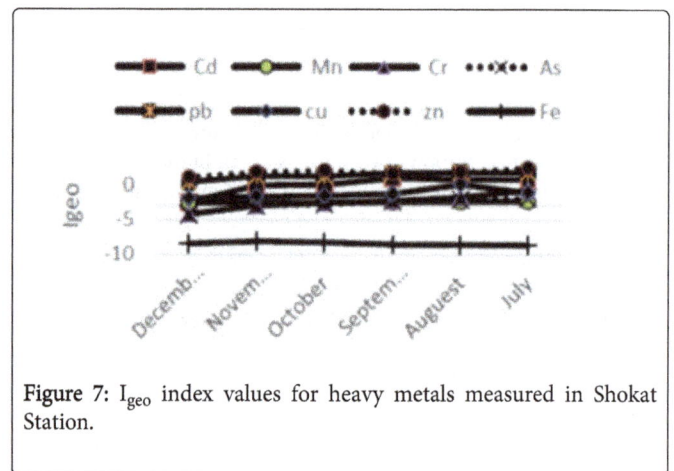

Figure 7: I_{geo} index values for heavy metals measured in Shokat Station.

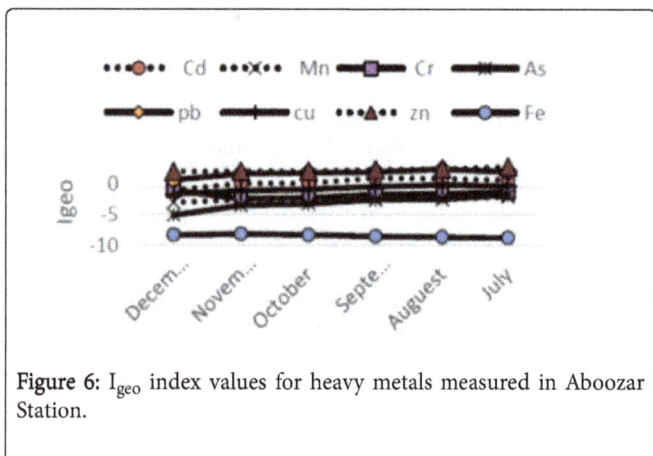

Figure 6: I_{geo} index values for heavy metals measured in Aboozar Station.

As is observed, the amounts of I_{geo} index for Zn and Pb have secured the most level of contamination to themselves at all three stations and after them, Cd and Cu show the lower level of contamination, whereas the amounts of I_{geo} index for the other heavy metals were not evaluated contaminating ($I_{geo}<0$). Also, the comparison of Figures 5-7 shows the more amounts of I_{geo} index, especially for Zn and Pb in Aboozar Square than the other two stations during the summer (20 July to 20 August).

The procedure of changes in IPI index for the studied heavy metals at each station has been shown in Figure 8 that confirms tangibly the higher levels of contamination of dust particles with heavy metals at Aboozar station specially in summer (20 July to 20 August); so that the average of amounts of this index for agriculture faculty, Aboozar Square and Shokat university were calculated 1/51, 1/99 and 1/66, respectively.

A Comparative Study of the Physico-Chemical Characteristics of Dust-Full Particles in Three Stations...

87

Agriculture University								
	Cd	Mn	Cr	As	Pb	Cu	Zn	Fe
I_{geo}	Uncont	Uncont	Uncont	Uncont	Moderately Contaminated	Uncont	Moderately Contaminated	Uncont
IPI	Moderately Contaminated	Uncont	Uncont	Moderately Contaminated	Strongly contaminated	Uncont	Strongly contaminated	Uncont

Table 4: Comparison of heavy metals quality values based on I_{geo} and IPI indices at Agriculture Station.

Aboozar Avenue								
	Cd	Mn	Cr	As	Pb	Cu	Zn	Fe
Igeo	Uncont	Uncont	Uncont	Uncont	Moderately Contaminated	Uncont	Moderately to Strongly Contaminated	Uncont
IPI	Moderately Contaminated	Uncont	Uncont	Uncont	Extremely contaminated	Moderately Contaminated	Moderately to Strongly Contaminated	Uncont

Table 5: Comparison of heavy metals quality values based on I_{geo} and IPI indices at Aboozar Station.

Shokat University								
	Cd	Mn	Cr	As	Pb	Cu	Zn	Fe
Igeo	Uncont	Uncont	Uncont	Uncont	Moderately Contaminated	Uncont	Moderately Contaminated	Uncont
IPI	Moderately Contaminated	Uncont	Uncont	Uncont	Strongly contaminated	Uncont	Extremely Contaminated	Uncont

Table 6: Comparison of heavy metals quality values based on I_{geo} and IPI indices at Shokat Station.

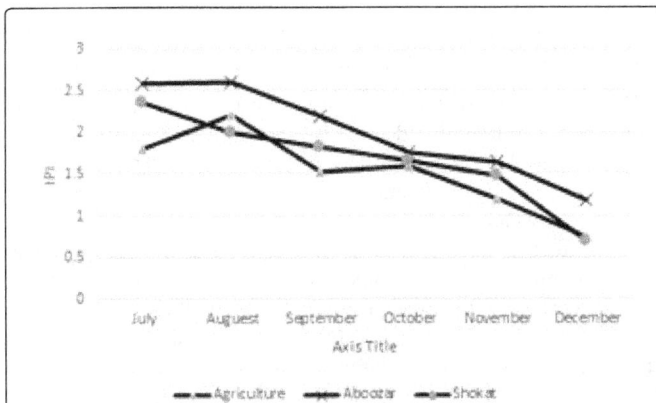

Figure 8: The index changes measured IPI to heavy metals in stations Agriculture, Abooar and Shokat.

Also, the pictures provided by the use of Electron Microscope (SEM) show that the similar particles with irregular forms are more effective in creating dust in both seasons (Figure 9).

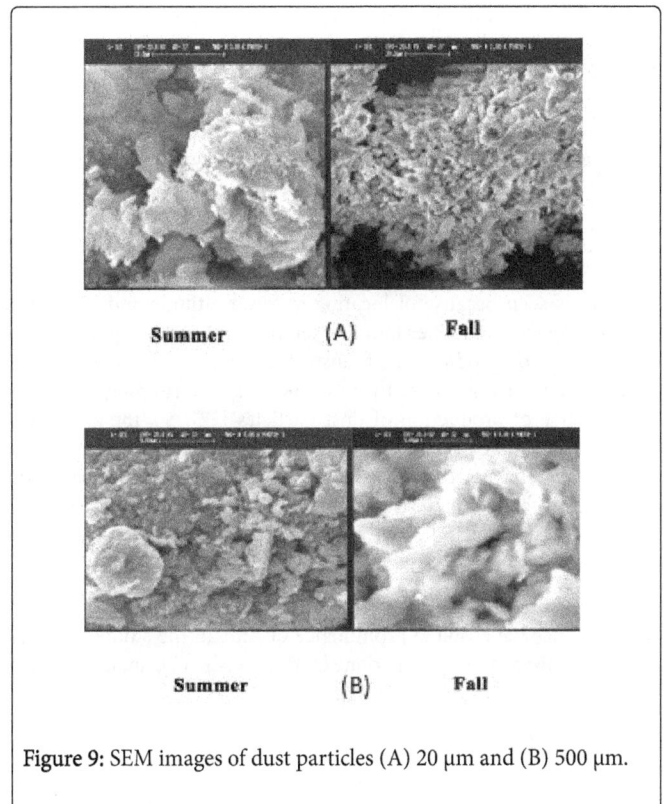

Figure 9: SEM images of dust particles (A) 20 μm and (B) 500 μm.

The study of x-ray diffraction related to the samples of two seasons (Figure 10) confirms the high frequency of Carbonate Minerals (Calcite) and Cilicate (Quartz), although shows the existence of few amounts of the other minerals, too.

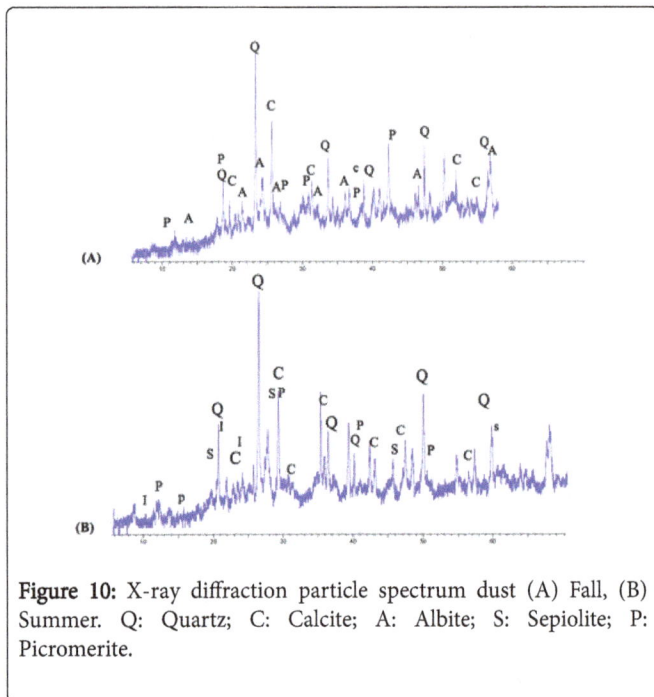

Figure 10: X-ray diffraction particle spectrum dust (A) Fall, (B) Summer. Q: Quartz; C: Calcite; A: Albite; S: Sepiolite; P: Picromerite.

The precedence of entrance of dust and heavy dirt from the Arabian countries adjacent with the west of Iran is not so new matter. With regard to the capacity of these dusts in transporting the metals and the importance of matter of continuous encountering of human with the metals which has harmful effects on the health of all living creatures; the study and more attention to the matter of analysing the harmful metals as a new and important matter, in addition to studying dusts and measuring the amount of particles, is of special importance [20]. Being adjacent with Iraq as main center of development of dust, besides this fact that the dry and semi-dried regions cover more than two-thirds of Iran area; justifies the occurrence and exacerbation of this phenomenon in the country [21]. Of three studied station in present research, Aboozar Square is the one in which more amounts of particles is seen because of locating at lower latitude and also lower raining average and higher monthly temperature and being near to the main center of production of dust. The amount of these particles decreases by moving toward the more latitudes and running away from the recourse of production of dust particles [22]. So, the amount of dust particles rained on agriculture faculty and then Shokat university show a reducing procedure in comparison to Aboozar Square. The similarity of changes pattern of amount of dust rained on three stations (Figure 4) can be the indicative of a similar recourse of dust particles in these three stations.

The comparison of amounts of studied metals (Table 3) demonstrates the absolute prominence of Zinc in the studied samples that corresponds to the study done by Bheravesh et al. about the traffic dusts of Mashhad city in 2013 [23].

As is observable in Figures 5-7 the similarity of changes pattern of I_{geo} index at the studied stations confirms the identic resource of dust particles reached to the agriculture, Aboozar and Shokat stations with regard to the heavy metals. Also, the I_{geo} index shows a high level of contamination of dust particles with Zn and Pb at three stations that corresponds to the study carried out by Salmanzadeh et al. [24]. The levels of contamination evaluated by this index for the studied metals demonstrate a relative reduction in autumn in comparison with summer [25,26]. It seems that the prevalence of Breast cancer in women with more than 30 years old [27] and also increasing Lead in milk tooth of children among the citizens of Birjand city can be related to the toxicity of this heavy metal in dust of eastern hemisphere of country. The I_{geo} index also shows amounts of low to middle contamination of Cd in the sampled dust particles in all stations (agriculture faculty, Aboozar Square and Shokat University) that the results are similar to the ones obtained from the study done by Javidaneh et al. [28]. This index demonstrates a relative reduction for Zn, Pb and Cd in autumn. Furthermore, the application of I_{geo} index didn't evaluate the measured amounts of As, Cr, Mn, Fe and Cu in the dust of three stations contaminating.

The amounts of IPI index show lower oscillation in agriculture faculty and Shokat University during 6 months, whilst follow a similar procedure at these two station (Figure 8); whereas the amounts of this index increase in Aboozar Square in the middle of summer and then descend in autumn. The descending procedure of average of IPI index amounts for the agriculture faculty, Aboozar Square and Shokat university stations is evaluated (1/99, 1/66 and 1/51) respectively and the level of contamination of dust particles with the heavy metals at three stations is evaluated intermediate (Table 2). Kouhzad [29] consider the four main regions including Sudan, some parts of Saudi Arabia, Iraq and Pakistan and some parts of Iran and Afghanistan that enjoy high level of frequency in Sudan, the main resources of dust storms in Middle West that the frequency of its occurrence from 21 May to 21 July is more than the other months and the level of contamination into heavy metals increases by increasing these dust storms that corresponds to the results obtained of this study.

SEM images review show more dust particle in the summer and fall mainly have a spherical and irregular shapes that associated with its natural origin. Also, it can be said in decreasing process of diameter size of the relative distribution of the particle size, initially larger particles come out of suspension during the transportation of dust particles. In other words, the distance from the source of dust particles tend to be thin (smaller) [30]. It seems the smaller dust particles relate with the order of source distance with the mineral sepiolite. Because sepiolite mineral density of dust particles is reduced toward the equator [31]. Also from the minerals observed in the dust, sepiolite mineral having the smallest size [22].

As the main product of erosion in sedimentary environment, the presence of minerals calcite and quartz in dust samples represents the natural and probably sedimentary origin of this phenomenon [32,33].

Relationship of mineral particles contained in dust particles with the particle diameter size can be evaluated. Spherical particles with approximate sizes of 5 µm are more than clay (sepiolite) that create dense cluster [34]. Regular crystal structures with 10-20 µm dimensions are more than Calcite crystals. Coarse crystals in sizes 20-40 μm are with prismatic structure of minerals of the albite. Particles close to spherical or have irregular shapes that are approximately 10-20 μm, are also seen as a single mineral [22,24].

But the presence of clay minerals in the soil sepiolite and particulate matter has always been considered as the most important factors of absorb and transmission. The specific chemical structure of this mineral such that contains van der Waals weak links which the meteorological parameters such as temperature changes and moisture breaks down the links and increase production and stimulate of mineral as dust particles [34]. From types of clay minerals, sepiolite minerals due to its specific morphological structure have special importance. Small size, wide and thin shape and presence of large amounts metal oxides shows its role in attracting, retaining and transport of heavy metals in dust.

Concerning the heavy metals, it can be referred to the role of clay in transferring them. The abundance of Calcite and Quartz in both seasons as main products of Sedimentary Clastic environments is the indicative of totally sedimentary derivation for the dust particles of Birjand. The existence of claying minerals as miner mineral phase plays the absorbent role of some heavy metals. The non- ability of most of metals in transmission and reaction with Quartz and Calcite indicates that these minerals play no effective role in the concentration of heavy metals and as the special heavy metals existed, they were more thrown from the environment around themselves in the structure of dust particles and have unnatural derivation [22].

Also, the study of main oxides which comprise the dust particles in deserts and different regions indicate that Felsic and Tectosilicates mineralogy mixtures are the most essential geochemical structures of these particles [22] that the existence of Albite, which is of Tectosilicates family, corresponds to the foresaid studies. Also, the existence of Pitromerite in two seasons that are of Potassium minerals and the existence of Potassium is one of important factors in the production of dust particles in the world corresponds to the foresaid studies.

The Pearson Correlation of Cr, Cu, Mn, As, Cd, Zn, Pb and Fe was evaluated and has been shown in Table 4. All metals except Mn have positive correlation with each other. The high positive correlation of metals with each other indicates the similar source of these metals in the dust that can be the same nature and Mn has a different source because of non-correlation with these metals. For instance, Pb and Zn exist more in Carbonate phases and Iron (III) Oxide phases that the main minerals of carbonate phase is produced of the dusts derived from the materials of city.

With regard to the present study, it was specified that the level of contamination of heavy metals in the dust is related to the traffic mass and speeds of cars, because the increase of speed of vehicles can be resulted in the increase of fuel consumption and also the erosion of road surface and tires. The Plumbum existed in the urban dust is

	Pb	Fe	Zn	As	Cr	Cu	Cd	Mn
Pb	1	-	-	-	-	-	-	-
Fe	0.824**	1	-	-	-	-	-	-
Zn	0.605*	0.661**	1	-	-	-	-	-
As	0.683**	0.654**	0.449	1	-	-	-	-
Cr	0.656**	0.522*	0.636**	0.774**	1	-	-	-
Cu	0.802**	0.878**	0.691**	0.453	0.551**	1	-	-
Cd	0.721**	0.759**	0.843**	0.498*	0.640**	0.701**	1	-
Mn	0.607**	0.541*	0.398	0.891**	0.678**	0.403	-	-

Table 4: Pearson correlation of heavy metals.

derived from the leaded fuel, vehicle oil and erosion and scrappiness of tires. Also, the erosion of cars tires causes the existence of much Zinc in the environment. The highest level of concentration of Copper is related to Aboozar station because of frequent putting on the brakes. There is a high concentration of Cu and Pb in the places in which there exists traffic light because of exorbitant stopping of waiting cars. The existence of industries and industrial wastewaters and also the treatment plant existing in Amirabad region is of main resources of entering Arsenic into the atmosphere.

Also, according to the study done by the use of I_{geo} and IPI indexes about the amounts of heavy metals in dust particles, there exists rather high level of contamination at the station of Aboozar Square in summer in comparison with the agriculture faculty and Shokat University in summer. These results reveal that the diffusion of heavy metals at first rate is related to the traffic resources and its features such as number of vehicles, kind and speed. With regard to the biological conditions of region, the public approaches of controlling the contaminants can be considered for the purpose of improving the air quality of Birjand city as following: Optimizing the urban traffic or executing methods for controlling traffic and exerting technical and municipal terms and conditions for accelerating the urban traffic, developing the urban greenbelt and regional parks and general education and attracting the attention and cooperation of managers and industrial owners for controlling the air pollution.

References

1. Garrison VH, Shinn EA, Foreman WT, Griffin DW, Holmes CW, et al. (2003) African and Asian dust: From desert soils to coral reefs. Bioscience 53: 469-80.

2. Faiz Y, Tufail M, Tayyeb Javed M, Chaudhry MM, Siddique N (2009) Road dust pollution of Cd, Cu, Ni, Pb and Zn along Islamabad Expressway, Pakistan. Microchem J 92: 186-192.

3. Xuan J, Sokolik IN, Hao J, Guo F, Mao H, et al. (2004) Identification and characterization of sources of atmospheric mineral dust in East Asia. Atmos Environ 38: 52-62.

4. Irabien MJ, Velasco F (1999) Heavy metals in Oka river sediments (Urdaibai National Biosphere Reserve, northern Spain): Lithogenic and anthropogenic effects. Environ Geol 37: 54-63.

5. Naddafi K, Nabizadeh R, Soltanianzadeh Z, Ehrampoosh MH (2006) Evaluation of dust fall in the air of Yazd. Iranian J Environ Health Sci Eng 3: 68-161.

6. Jaradat QM, Momani KA, Jbarah AAQ, Massadeh A (2004) Inorganic analysis of dust fall and office dust in an industrial area of Jordan. Environ Res 96: 44-139.

7. Calabrese EJ, Kostecki PT, Gilbert CE (1987) How much dirt do children eat? An emerging environmental health question. Comment Toxicol 1: 229-241.

8. Hawley JK (1985) Assessment of health risk from exposure to contaminated soil. Risk Anal 5: 289-302.

9. Lanphear BP, Roghmann KJ (1997) Pathways of lead exposure in urban children. Environ Res 74: 67-73.

10. Ahmadizade M (1997) Industrial toxicology (heavy metals). Tehran: Hezaran pp: 1-50.

11. Krueger BJ, Grassian VH, Cowin JP, Laskin A (2004) Heterogeneous chemistry of individual mineral dust particles from different dust source regions: The importance of particle mineralogy. Atmos Environ 38: 61-62.

12. Rajabi M, Souri B (2014) Evaluation of heavy metals among dustfall particles of Sanandaj Khorramabad and Andimeshk cities in western Iran 2012-2013. J Health Environ 8: 11-22.

13. Chen CW, Ka CM, Chen CF, Dong CD (2007) Distribution and accumulation of heavy metals in the sediments of Kaohsiung Harbor, Taiwan. Chemosphere 66: 40-1431.

14. Hossein Zade M (2005) Geomorphological abilities in Birjand urban development. Tehran University, Faculty of Geography, Department of Geography, Iran.

15. Reeve R (2002) Introduction to environmental analysis. John Willey and Sons LTD, UK p: 301.

16. Liu X, Yin ZY, Zhang X, Yong X (2004) Analyses of the spring dust storm frequency of northern China in relation to antecedent and concurrent wind, precipitation, vegetation, and soil moisture conditions. J Geophys Res 109: 1-16.

17. Valdés J, Vargas G, Sifeddine A, Ortlieb L, Guiñez M (2005) Distribution and enrichment evaluation of heavy metals in Mejillones Bay (23°S), Northern Chile: Geochemical and statistical approach. Marine Pollution Bulletin 50: 68-1558.

18. Gonzales-Macias C, Schifter I, Liuch-Cota DB, Endez-Rodriguez L, Hernandez-Vazquez S (2006) Distribution, enrichment and accumulation of heavy metals in coastal sediments of Salina Cruz Bay, Mexico". Environmental Monitoring and Assessment 118: 211-230.

19. Wei BG, Yang LS (2009) A review of heavy metal contaminations in urban soils, urban road dusts and agricultural soils from china. Microchem J 94: 99-107.

20. Ridgwell AJ (2003) Implications of the glacial CO_2 "iron hypothesis" for quaternary climate change. Geochemistry, Geophysics, Geosystems 4: 1-10.

21. Zarasvandi AL (2013) Geochemical composition and source of dust storms particles in Khuzestan Province using REE geochemistry: Concerning on geo-environmental parameter. Proceedings of Conference of Dust Haze, Monitoring, Effects and Solutions, Tehran, Iran (in Persian).

22. Zhang XY, Cao JJ, Li LM, Arimoto R, Cheng Y, et al. (2002) Characterization of atmospheric aerosol over Xian in the south margin of the Loess Plateau, China. Atmos Environ 36: 99-4189.

23. Bheravesh F, Mahmudy Gharaie MH, Ghassemzadeh F, Avaz Moghaddam S (2013) Determination of heavy metals pollution in traffic dust of Mashhad city, and its origin by using Selective Sequential Extraction (SSE) procedure. Geoscience 24: 141-150.

24. Tegen I (2003) Modelling the mineral dust aerosol in the climate system. Quat Sci 22: 1821-1834.

25. Escudero M, Querol X, Pey J, Alastuey A, Perez N, et al. (2007) A methodology for the quantification of the net African dust load in air quality monitoring networks. Atmos Environ 41: 24-5516.

26. DustScan (2004) Dust monitoring and dust consultancy services: Nuisance dust monitoring.

27. Haghighi F, Khodaei S, Sharifzadeh G (2010) Effect of logotherapy group counseling on depression in breast cancer patients. Mod Care J 9: 165-172.

28. Javidaneh Z, Zarasvandi A, Rastmanesh F (2016) Determination of environmental indicators and sources of heavy metals and dust the streets. Masjed Soleiman: Khuzestan Province 9: 155-170.

29. Kouhzad R (2008) Statistical analysis of synoptic phenomenon of dust in the province of Khuzestan, Master's thesis, University of Sistan and Baluchestan, Iran.

30. Udden JA (1998) The mechanical composition of wind deposits. Augustana Library, Rock Island, III.

31. Chester R, Elderfield JJ, Griffin LR, Johnson LR, Padgham RC (1992) Eolian dust along the eastern margins of the Atlantic Ocean. Mar Geol 13: 91-106.

32. Khuzestani RB, Souri B (2011) Evaluation of heavy metal contamination hazards in nuisance dust particles, in Kurdistan Province, western Iran. J Environ Sci 25: 46-54.

33. Wiederkehr P, Yoon SJ (1998) Air qualityindicators. In: Fenger J, Hertel O, Palmgren F (eds.) Urban air pollution European aspects. Kluwer, Dordrecht.

34. Dube A, Zbitniewski R, Kowalkowski T, Cukrowska E, Buszewski B (2001) Adsorption and migration of heavy metals in soil. Polish J Environ Stud 10: 1-10.

12

Environmental Concerns in National Capital Territory of Delhi, India

Shashank Shekhar Singh*, Singh SK and Shuchita Garg

Environmental Engineering Department, Delhi Technological University, Delhi, India

Abstract

After Independence, the city of Delhi became a major center of commerce, industry and education. The rapid urbanization of Delhi along with the level of growth in economic activities in the city and its surrounding areas stressed the natural environment significantly. Among the environmental problems, air pollution, water pollution, loss of biodiversity, municipal waste and noise pollution are major environmental challenges that the city is facing. The city suffers from air pollution caused by transportation, road dust, industries and pollutant emissions. Noise pollution comes mainly from industries, transportation, aircraft etc. Water pollution and lack of adequate solid waste treatment facilities have caused serious damage to the river on whose banks Delhi grew, the Yamuna. Several steps have been taken in the recent past to improve the environment condition which includes massive focus on afforestation, universal use of CNG by commercial vehicles, ban on plastic use, better management of solid waste, treatment of waste water and improvement of sewage system etc. But still many challenges remain to contain the environmental pollution. This paper summarizes the major environmental concerns and the present status of pollution in NCT of Delhi.

Keywords: Environmental concerns; Pollution; Atmosphere; Hazardous wastes

Introduction

Environment of any city is the asset of that city and for a city like Delhi-NCR, the significance of a clean and pleasing environment is as beneficial as it can be. Delhi being hub of political, social, economic and other national/international affairs of India portrays the image of India to the world. Being the national capital, plethora of national and international migration takes place from and to Delhi resulting in increasing pollution stress on natural resources viz. Air Water and Land. The overutilization of these resources makes Delhi prone to all types of pollution making lives of people difficult here. With Delhi expanding its boundaries each day and National Capital Region (NCR) getting to nearby states, the whole effect is quite evident in this area. Proper water use techniques need to be brought in place to make Delhi efficient and making water available to its masses. First step in this direction can be controlling water pollution levels. Similar stress needs to be made to control rising air pollution levels. Noise pollution and Land degradation too makes Delhi unviable and unsustainable. Delhi as a sustainable city needs proper planning and operation to make it as pleasing as cities of the world.

Air environment

The air pollution levels in Delhi are strikingly high and the transport sector is a major contributor. Besides the transport sector, domestic and power sectors are also major sources of air pollution in the capital. Nearly 421.84 tons of CO, 110.45 tons NOx, 184.37 tons HC and 12.77 tons particulate matter is released in Delhi's atmosphere per day (Department of Environment and Forests, 2010).

The CO emission has dipped drastically post-CNG use [1]; SPM and RSPM have increased, SO_2 has declined marginally and NO_2 is still high over 1997-2011 [2]. The composition of pollutants have changed with the introduction of CNG, new pollution standards and phasing out of old vehicles. The concentration of CO, SO_2 and PAHs has declined, while NOx and SPM increased [3-5]. The rise in NOx is attributed to CNG use and SPM to the diesel vehicles' growth.

The Central Pollution Control Board has been monitoring ambient air quality at six locations in Delhi under NAAMP for the past many years. (Figures 1-4) Year-wise annual mean ambient air quality levels in Delhi during 1997 to 2014 is presented in the following table: (Table 1) The values for 1997 to 2010 are of the monitoring stations of CPCB while the values of 2011 to 2014 are of the monitoring station network developed by Delhi Pollution Control Committee. DPCC presently monitors air quality through six online continuous ambient air quality monitoring stations at 6 locations. The stations can be classified in two categories i.e. residential Puram RK, Mandir Marg & Punjabi Bagh and hot spots I.G.I Airport and Anand Vihar. Civil Lines is also influenced by traffic emissions [6] (Table 2).

Increasing levels of air pollution are responsible for higher incidence rate of respiratory diseases, cancer, and heart diseases in the capital. Various studies carried out for Delhi reflect the correlation between air pollution and health impacts. A study by AIIMS reconfirmed the point that respiratory symptoms are more frequent amongst people residing in highly polluted areas. To tackle the problem of air pollution, a number of measures have been taken in the past, such as switching to cleaner fuels, tightening vehicular emission limits, phasing out of old vehicles and maintenance of in-use vehicles, closing or relocating polluting industries, plantation activities etc. However, a lot more still needs to be done if the capital desires to breathe clean air.

Water environment

With the population of Delhi increasing from 0.4 million in 1911 to 18.24 million in 2015, there is an ever increasing pressure on the water resources. Improvement in living standards and access to sanitation

*Corresponding author: Shashank Shekhar Singh, Research Scholar, Environmental Engineering Department, Delhi Technological University, Delhi, India, E-mail: sssinghdtu@gmail.com

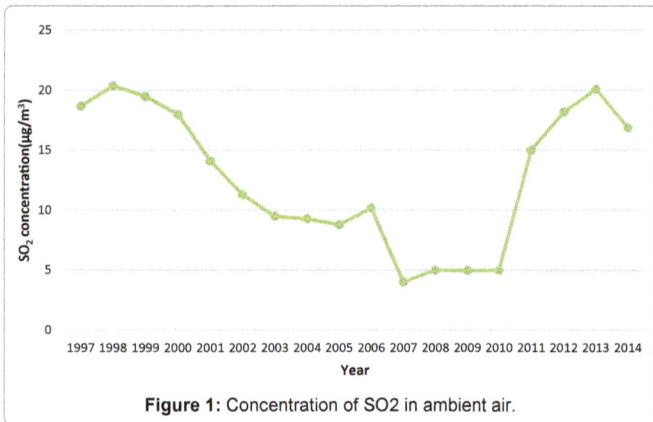

Figure 1: Concentration of SO2 in ambient air.

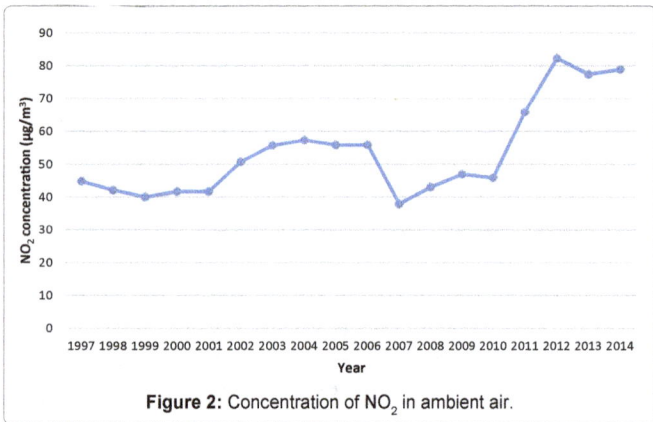

Figure 2: Concentration of NO_2 in ambient air.

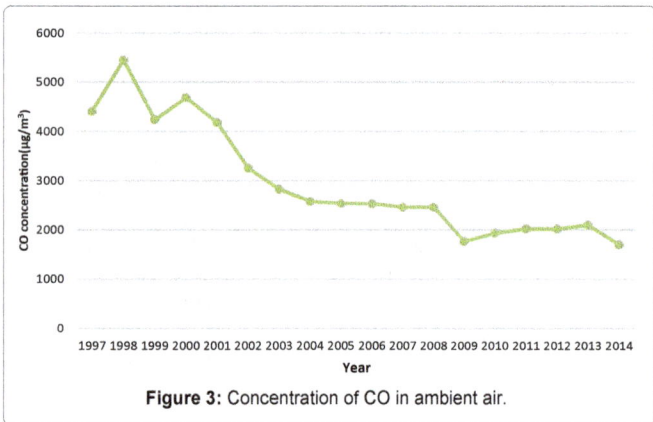

Figure 3: Concentration of CO in ambient air.

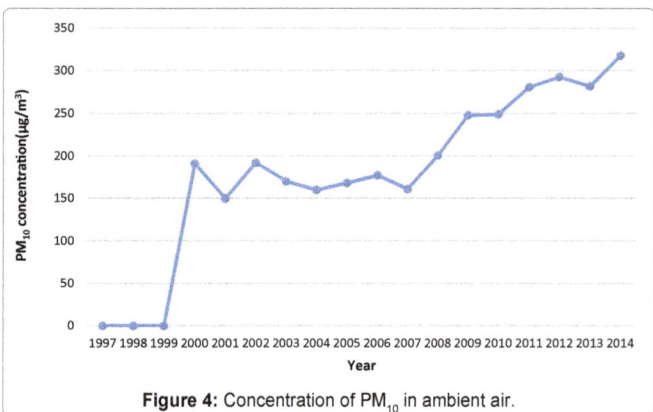

Figure 4: Concentration of PM_{10} in ambient air.

S. No	Years	Ambient Air Quality ($\mu g/M^3$)			
		So_2	No_2	Co	Rspm (Pm_{10})
1	1997	18.7	44.9	4410	--
2	1998	20.4	42.2	5450	--
3	1999	19.5	40.1	4241	--
4	2000	18.0	41.8	4686	191
5	2001	14.1	41.8	4183	150
6	2002	11.3	50.8	3258	192
7	2003	9.5	55.8	2831	170
8	2004	9.3	57.4	2581	160
9	2005	8.8	55.9	2541	168
10	2006	10.2	55.9	2531	177
11	2007	4.0	38.0	2460	161
12	2008	5.0	43.1	2461	201
13	2009	5.0	47.03	1768	248
14	2010	5.0	46.0	1937	249
15	2011	15.0	66.0	2020	281
16	2012	18.2	82.4	2020	293
17	2013	20.1	77.5	2100	282
18	2014	16.9	79.0	1700	318
	Standard	50	40	2000	60

Table 1: Year-Wise Annual Mean Ambient Air Quality Levels in Delhi During 1997 To 2014.

R. K. Puram	13.6	51.6	263	140	41	1.45
Mandirmarg	12.5	87.4	203	125	51	1.28
Punjabi Bagh	17.3	106.4	248	139	39	1.86
Igi Airport	17.7	66.8	289	176	85	1.48
Anandvihar	20.4	84.5	583	191	32	1.73
Civil Lines	19.7	79.4	318	141	96	2.64

Table 2: Annual Average of Critical Pollutants at Six Stations in Delhi (In $\mu g/M^3$) For the Year 2014.

facilities increases the per capita water demand levels. For sustainable development of Delhi, it is essential to ensure adequate supply of water in terms of reliability, quality and quantity. Although Delhi has an average water availability of 225 lpcd, the distribution is not uniform. Some areas get 24 hrs. Water supplies, whereas some get hardly 1-2 hr. water supply in a day.

Delhi depends on river Yamuna and partially on river Ganga for its share of raw water. Surface water contributes to over 86% of Delhi's total water supply. Yamuna, a perennial river, provides the major share of this water supply. Urban agglomeration of NCT Delhi is the major contributor of pollution load in Yamuna followed by Agra and Mathura. The stretch between Wazirabad Barrage and Chambal River confluence is critically polluted and there is significant fluctuation in dissolved oxygen level from nil to critically low levels. This reflects presence of organic pollution load and persistence of eutrophic conditions in the river. Pollution load in the river Yamuna added from various sources like industries and domestic and long dry season, has virtually converted it into a nala. Najafgarh drain along with its 70 sub-drains is the biggest polluter of the river.

DPCC has been conducting monthly water quality monitoring of river Yamuna (at 9 locations) and major drains (24 drains) falling into river Yamuna. Recent water quality monitoring reports of river Yamuna indicate that the water quality parameters, BOD & DO, are in the desirable/prescribed norms, with respect to Water Quality criteria of "C" class, at Palla, which is upstream of Wazirabad Barrage.

However, the water quality of River Yamuna at the downstream of Wazirabad barrage after confluence of Nazafgarh Drain is not meeting the desirable/prescribed norms.

The annual average of DO has ranged from 0.20 mg/l at Shahdara (Downstream) to 8.48 mg/l at Palla. The annual average of BOD has ranged from 1.99 mg/l at Palla to 60.33 mg/l at KhajuriPantoolpul. The water quality standards for DO and BOD as per CPCB norms are 4 mg/l and 3 mg/l respectively for class 'C' of river water. The water quality monitoring results in Delhi stretch clearly indicates river water is grossly polluted (Table 3).

Water quality monitoring results of the drains indicate that most of the drains are not meeting the standards with respect to Bio-chemical Oxygen Demand (BOD), Chemical Oxygen Demand (COD) and Total Suspended Solids (TSS) (Table 4).

S. No.	Locations	Ph (Mg/L)	Cod (Mg/L)	Bod (Mg/L)	Do (Mg/L)
	Water Quality Criteria	6.0- 9.0		3 (Max)	4 (Min)
1	Palla	7.63	14	1.99	8.48
2	Surghat	7.5	22.66	4.51	5.78
3	Khajuri Pantool Pool	7.40	200.33	60.33	Nil
4	Kudesia Ghat	7.45	125.67	37.00	Nil
5	Ito Bridge	7.53	100.67	31.80	Nil
6	Nizamudin Bridge	7.30	88.70	27.10	1.00
7	Agra Canal Okhla	7.50	96.20	29.90	0.90
8	Shahdara (Down Stream)	7.45	138.67	38.80	0.20
9	Agra Canal Jaitpur	7.45	108.83	28.83	0.60

Table 3: Annual Average Water Quality of River Yamuna at Different Locations: April 2014 to March 2015.

S. No	Drains	Ph	Tss (Mg/L)	Cod (Mg/L)	Bod (Mg/L)
1	Najafgarh Drain	7.39	269.67	241.00	70.75
2	Metcalf House Drain	7.53	113.17	85.58	24.00
3	Khyber Pass Drain	7.51	40.17	42	10.30
4	Sweeper Colony Drain	7.33	55.83	100.83	27.42
5	Magazine Road Drain	7.39	212.83	298.33	87.92
6	Isbt Drain	7.40	148.00	283.33	87.92
7	Tonga Stand Drain	7.55	161.33	333.50	114.17
8	Moat Drain	No Flow	No Flow	No Flow	No Flow
9	Civil Mill Drain	7.42	167	302	94.42
10	Power House Drain	7.43	268.33	350.17	117.83
11	Sen Nursing Home Drain	7.48	302	389.33	132.08
12	Drain No. 12a	No Flow	No Flow	No Flow	No Flow
13	Drain No. 14	7.54	58.67	45.67	11.97
14	Barapulla Drain	7.37	163.67	164.50	49.08
15	Maharani Bagh Drain	7.25	454.67	395.50	135
16	Kalkaji Drain	No Flow	No Flow	No Flow	No Flow
17	Saritavihar Drain (Mathura Road)	7.34	272.00	438.00	146.67
18	Tehkhand Drain	7.34	289.67	470.08	150
19	Tuglakabad Drain	7.34	265.67	314.75	98.83
20	Drain Near Lpg Bottling Plant	No Flow	No Flow	No Flow	No Flow
21	Drain Near Saritavihar Bridge	7.43	102.00	130.17	39
22	Shahdara Drain	7.44	376.33	509.67	151.67
23	Sahibabad Drain	7.31	606.33	817.58	271.67
24	Indrapuri Drain	7.42	355.33	476.33	128.42

Table 4: Annual Average Water Quality of Drains at Different Locations in Delhi: April 2014 To March 2015.

As per CPCB, the contribution of pollution load from NCR & non-NCR states are in the proportion of 80:20, i.e. over 3/4th of the pollution load in River Yamuna is contributed by the NCR [7].

Besides surface water sources, groundwater contributes a substantial quantity of water supply in Delhi. Inadequate and intermittent supply of piped water has led to unchecked exploitation of the groundwater resource. A comparison of existing groundwater levels in different administrative blocks with levels in 1960 shows a decline of 2-30 m. Levels in Alipur and Kanjhawala blocks have declined 2-6 m, in the Najafgarh block by 10m, and in the Mehrauli block by 20 m. In addition to quantity, the quality of groundwater is also deteriorating and in several places it has been found to be unfit for human consumption.

Municipal and hazardous wastes

Solid waste includes commercial and residential waste generated in municipal or notified areas. As per the data available with DPCC records, solid waste generation in Delhi was around 8360 MTD. This is expected to increase due to economic and population growth. 700 MGD sewage is also generated, which generates organic sludge. Municipal waste of Delhi is disposed in three landfill sites namely Bhalswa GT Road, Ghazipur and Okhla.

Hazardous waste means any waste which by reason of any of its physical, chemical, reactive, toxic, flammable, explosive or corrosive characteristics causes danger or is likely to cause danger to health or environment. The most critical hazardous waste generated in Delhi is from small-scale enterprises such as pickling units, electroplating units, anodizing units, and sludge from CETPs.

Bio-Medical Waste (BMW) means any waste, which generated during the diagnosis, treatment or immunization of human being or animals or in research activities. With the increase in the number of hospitals and nursing homes in Delhi, hospital waste has become another area of concern. This waste is sent to common biomedical waste facilities in the city. Delhi is having 3 CBWTF operators who collect the waste from HCEs of Delhi and dispose the BMW after its treatment.

Electronic Waste, means any waste, which is generated due to product obsolescence and discarded electronic items, and may include data processing, telecommunications or entertainment in private households and businesses. The quantity of e-waste generated in the city is going to be much higher than hazardous waste and healthcare waste and thus requires proper management.

The most acceptable strategy for solid waste management in Delhi would be to categorize waste streams as biodegradable, recyclables, and inert matter to maximize recovery and minimize the quantity of waste generation. Efforts should also be made towards reclaiming and redeveloping the abandoned and filled landfill sites.

Forest

The vegetation cover is imperative for balanced atmospheric temperature and sustenance of life. As per the reports of Forests Survey of India (2011), total area of forest and tree cover was 40 and 111 km^2 respectively in 2001 that increased to 120 and 176.2 km^2 in 2011 [8]. Total vegetative cover doubled in a decade from 10% to 19.97% on account of substantial increase in tree cover under the Green Action Plan of Delhi Government. Open forests have coverage share of 119.96 km^2 and dense forests are merely 6 km^2 [9]. The National Forest Policy, 1988 provides that a minimum of 1/3rd of the total land area of the country should be under forest or tree cover. Taking this into view, the

Govt. of NCT of Delhi is making all endeavors to meet the national goal as set by the Central Govt. and is constantly adding to the green cover of the State [8] (Table 5).

The forest and tree cover area increased to 297.81 km² in 2013 increasing thereby the share of forests in the total area to 20.08 per cent. Of the total 297.81 km² of forest area in NCT of Delhi, nearly 272 km² has been added during the period 1999 to 2013 [8] (Table 6).

South Delhi district has the highest forest cover area at 79.02 sq. km, South West Delhi has 44.63 sq. km, that of North West Delhi is 16.50 sq. km and New Delhi has 16.31 sq. km. The lowest forest cover is in North West Delhi of 3.75 sq. Km.

Composition of forests in terms of its density is shown in Chart. Out of the total geographical area of NCT of Delhi, very dense forest is spread over 0.45 percent, moderately dense forest is spread over 3.33 percent, open forest is spread over 8.34 percent and scrub is spread over 0.15%, which is almost negligible [8] (Figure 5).

Delhi has 42 city forests. Fifteen city forests are in South-West district, Ten in North-West district, five each are in North-East and South districts, three each in East and North districts and one in West district.

Noise environment

The major contributors to noise pollution are industries, vehicular traffic, festivals, construction activities, diesel generating sets etc. Use of high sound loudspeakers during festivals and many social gatherings in public place directly increases the noise pollution in the affected areas.

Noise levels in Delhi exceed permissible levels in all areas except industrial areas according to a study by Delhi Pollution Control Committee in 1996. Another study carried out by CPCB in Delhi during 2006 revealed that during daytime ambient noise levels exceeded the prescribed residential area standard at all the locations. The ambient

noise levels in commercial and industrial locations were below their respective standard values.

The ambient noise levels permitted by Central Pollution Control Board for different areas:-

(Table 7) [10-14]. Noise levels observed at 40 different residential locations have been tabulated below. The data shows ambient noise levels being exceeded in all the selected residential areas. (Day time standard for residential area 55dB (A), Night time standard for residential area 45dB (A), All values in Leq dB (A)) [14-17] (Table 8).

Discussion and Recommendations

Growing urbanization and migration of population in search for better employment opportunities to Delhi is constantly putting pressure on city's limited environmental resources. Though the green cover has increased in past several years due to massive plantation drives and awareness schemes, other assets like lakes, groundwater, river etc. are under constant threat due to their over exploitation. To make an informed, scientific decision about saving these natural and environmental resources and to retain them to their closest pristine form, urgent measures are required. Measures like easing out transportation services are needed to deal with problems pertaining to air pollution. Initiatives by government of Delhi to only ply those private vehicles on road which have even numbers on dates having even count and same with odd is a noble step which needs to be executed with few exceptions. Waste management is one such area which needs to be dealt with care and urgency. Segregation while collection should be the desired practice for municipal authorities. Different colour bags should be assigned for different kind of wastes which can be directly sent to processing plants. Principle of Recycle and Reuse should be adopted which can help Delhi get rid of tons of pollutants. Zero net waste should be the objective which can only be attained if proper care is taken of the waste. For increasing and maintaining the forest cover, the horticulture department should make proper road map. Large scale plantation drive and maintenance of existing plants should be done. This can be helpful in dealing with the menace of Air pollution. To deal with pollution of river Yamuna in Delhi, one needs to take care of all

S. No.	Year	Forest and Tree Cover	Absolute Increase in Area	% of Total Area
1.	1993	22	--	1.48
2.	1995	26	4	1.75
3.	1997	26	--	1.75
4.	1999	88	62	5.93
5.	2001	151	63	10.2
6.	2003	268	117	18.07
7.	2005	283	15	19.09
8.	2009	299.58	16.58	20.20
9	2011	296.20	-3.38	19.97
10	2013	297.81	1.61	20.08

Table 5: Forest and Tree Cover Area of Delhi 1993-2013.

Sl. No.	Districts	Geographical Area	Forest Cover Area	% of Geographical Area
1.	Central Delhi	25	5.05	20.20
2.	East Delhi	64	3.05	04.77
3.	New Delhi	35	16.31	46.60
4.	North Delhi	59	4.81	8.15
5.	North East Delhi	60	4.02	6.70
6.	North West Delhi	440	16.50	3.75
7.	South Delhi	250	79.02	31.61
8.	South-West Delhi	421	44.63	10.60
9.	West Delhi	129	6.42	04.98
	Total	1483	179.81	12.12

Table 6: District-Wise Forest Cover in Delhi-2013 (Sq. Km).

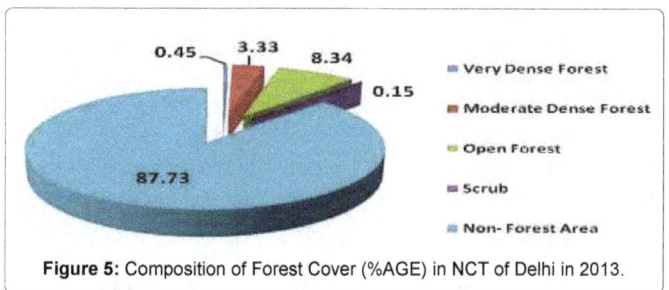

Figure 5: Composition of Forest Cover (%AGE) in NCT of Delhi in 2013.

S. No.	Area	Leq/Db (A)	
		Day Time *	Night Time **
1.	Industrial Area	75	70
2.	Commercial Area	65	55
3.	Residential Area	55	45
4.	Silence Zone***	50	40

Notes: *Day Time-6 Am To 10 Pm
**Night Time-10 Pm to 6 Am
*** Silence Zone is an Area Comprising Not Less Than 100 Meters around Hospitals, Educational Institutions, Courts, Religious Places or Any Other Areas Which Is Declared as Such by Competent Authority.

Table 7: Prescribed Ambient Noise Standards.

Ambient Noise Levels Observed at Different Locations in Delhi

S.No.	Locations	June(2008)		July (2008)		August (2008)		Sep. (2008)		Oct. (2008)		Nov. (2008)		Dec. (2008)		Jan. (2009)	
		Noise (Day Time)	Noise (Night Time)	Noise (Day Time)	Noise (Night Time)	Noise (Day Time)	Noise (Night Time)	Noise (Day Time)	Noise (Night Time)	Noise (Day Time)	Noise (Night Time)	Noise (Day Time)	Noise (Night Time)	Noise (Day Time)	Noise (Night Time)	Noise (Day Time)	Noise (Night Time)
1	Adarsh Nagar	62.3	53	61.6	52.2	62.9	52.4	62.7	55.1	63.4	55.8	59.1	47.1	61.2	52.5	60.5	51
2	Anand Vihar	63.1	57.5	62.5	55.9	63	56	64.3	54.4	62.7	53.9	62	56	61.6	52.7	60.4	53.3
3	Ashok Vihar	61.7	52.6	61.3	62.2	60.2	51.5	61.4	55.1	**	**	61.3	50.9	60.4	51.9	60.1	50.1
4	Badli	58.7	49.1	57.1	50.9	57.4	50.6	63	55.8	**	**	61.4	57.2	59.6	51	60.6	54
5	Braham Puri	59.6	54	57.7	53.6	56.6	51.5	59	54.2	**	**	59.6	52.2	60.4	54.6	59.9	53.7
6	Daryaganj	64	57.4	62.3	52.5	63.7	56.8	62.2	53.4	**	**	62.1	52.5	62.8	56.2	60.6	55.8
7	Defence Colony	60.1	53.8	60.9	55.2	60	54.3	63	57.1	64	57.4	61.6	55.6	60.3	54.3	59.8	53.7
8	Dwaraka	60.9	52.3	59.2	51.3	61.2	52.1	62.9	55.9	**	**	59	51.1	59.3	55.8	58.1	54.5
9	Greater Kailash	63.1	62.2	61.8	56	63.9	62.6	62.7	57	61.5	55.9	61.7	53.1	62.8	55.5	59.2	49.7
10	Inder Puri	59.4	55.9	59.1	53.8	59.6	54.6	62	57.9	**	**	60.6	55.2	61.7	53.3	60.2	53.1
11	Janak Puri	62.7	55.1	61.9	55.6	61.2	53.9	62.9	56.1	**	**	62.5	60.5	62.8	57.6	59.4	53.1
12	Karawal Nagar	62.2	53.2	62.4	52.9	61.4	52.5	63.9	57.7	62.9	56.3	60.5	54.7	63	61.2	61.2	60.8
13	Karol Bagh	62.9	61.3	62.4	60.6	62	61	62.9	54.4			61.8	56.4	62.8	54.9	64	57.4
14	Kondli	60	50.3	60.1	51.1	60.4	54.2	62.4	56.8	62.2	54.7	62.2	54	64.4	54	60.8	48.3
15	Lajpat Nagar	63.6	61.8	64.6	62.3	63.6	61.5	62.5	55.9	61.9	55.5	63.1	53.1	61.3	55.1	61.2	51.9
16	Lawrence Road	59	50.9	59.7	52.2	59.1	52.6	62.4	53.2	**	**	61.4	56.8	62.3	56.5	62.4	49
17	Mandavali	61.7	56.5	63.2	56.6	62.4	55.1	62	57.2	61.6	54.2	63	56.9	59.3	51.3	60.9	52.7
18	Mangol Puri	62.8	56.2	61.6	55.9	62.5	55.9	63.6	56.8	**	**	61.4	56.8	61.3	54.3	62.6	57.6
19	Meera Bagh	60	52	59.4	51.6	60.2	52	61.4	54.8	**	**	61	56.9	62.8	56.2	60.5	52.5
20	Mehrauli	61.3	55.1	62.3	58	62.7	53.4	63.9	54.3	61.1	51.2	60.5	56.8	60.9	57.8	62.7	55.4
21	Moti Bagh	59.3	52.4	58.2	52.2	58.7	51.8	62.1	53	61.6	52.9	62.4	51.3	58.7	52.2	58.6	47.7
22	Moti Nagar	61.2	57.8	60.9	54.2	61.9	57.9	61	52.5			62.9	54.8	62.9	54.8	59.5	53.5
23	Mukherji Nagar	61.7	55.5	61	56.2	62.2	55.9	64.8	56.4	61.6	50.7	61.8	53.8	61.1	55.6	62.8	56.2
24	Nand Nagri	60.6	55.8	59.7	55	60.3	55.4	61.9	53.9	61.4	56.1	62.4	55.7	60.6	55.8	59.4	55.9
25	Naraouji Nagar	60.5	52.6	60.5	55.6	60.9	53.7	61.9	56.1	62.6	51.6	62.9	56.6	60.7	52.8	59.6	51.6
26	New Friends Col	60.1	53.2	62.8	56.5	62.2	53.5	62.9	55.2	63.1	55.6	61.8	56.7	61.3	56.2	59.2	53.5
27	Pahar Ganj	63.6	60.4	63.1	57.6	67	59.1	63.8	59.2	**	**	61.9	55.4	62.7	60.1	62.6	57.6
28	Paschim Vihar	57.4	51.7	57.9	54.1	59.2	51.8	61.3	53.9	**	**	62.2	54.7	57.8	51.4	59.9	48.4
29	Patel Nagar	57.6	48.7	59.6	52.3	58	48.6	61.7	55.4	**	**	62.1	55.5	56.8	48.3	60.2	50.5
30	Prehladpur	63	59.1	59.4	54.5	62.4	57	61.9	56.1	61.6	52.9	62.8	52.8	60.5	57.6	61.7	56.5
31	R.K. Puram	60.2	55.7	60.4	57.1	60.6	55.5	62	55.6	63.1	55	59	51.6	59.6	53.4	58.7	49.1
32	Rajpura Road	62.1	52.5	60.2	52.3	62.4	52.8	63.8	57.2	63.3	53	59.2	52.7	60.2	51.6	57.6	48.7
33	Rana Pratap Bag	61	51.4	60.3	50.8	61.8	52.9	63.6	56.8	**	**	62.4	52.1	60.6	51.2	59.3	52.4
34	Rohini	64	54.3	61.1	53.5	62.3	53.4	63.5	54.7	**	**	61.7	56.8	63.3	53.9	64	57.4
35	Sarita Vihar	69	59.1	60.4	51.7	67.2	58	62.8	54.3	63.3	56.5	60.5	55.5	60.5	52	63.6	54.3
36	Shalimar Bagh	62.6	57.6	63.3	57.6	61.7	54.8	63.8	54.3	**	**	62	54.8	62.4	55.2	61.7	55.7
37	Shanti Vihar	61.3	55.1	65.9	56	59.7	48.3	63.1	59.2	**	**	63	55.5	58.6	47.7	59.1	47.5
38	Tilak Nagar	62.1	54.9	61.5	54.5	62.2	54.9	65	55.6	**	**	60.5	53.9	59.4	53.5	59.2	48.4
39	Tughlakabad	**	**	**	**	**	**	**	**	**	**	62.2	52.6	61	52.5	61.3	48.5
40	Vasant Kunj	62.9	58.6	63.5	58.9	61.7	56.6	61.8	56.8	61.6	55.2	63.5	58.9	61.9	54.9	59.5	55.8
41	Yamuna Vihar	59.6	51.6	59.9	51.3	60.1	52.5	62.7	51.8	63	53.9	63.8	51.3	58.6	51.2	59.9	51.1
	Min	57.4	48.7	57.1	50.8	56.6	48.3	59	51.8	61.1	50.7	59	47.1	56.8	47.7	57.6	47.5
	Max	69	62.2	65.9	62.3	67.2	62.6	65	59.2	64	57.4	63.8	60.5	64.4	61.2	64	60.8
	Average	61.473	54.96	61.02	54.81	61.413	54.47	62.6625	55.53	62.38	54.42	61.69	54.5	60.9	54.1	60.54	60.54

**Data not available

Table 8: Delhi Pollution Control Committee, Government of N.C.T. Delhi.

the drains falling out in Yamuna. Interceptor drains should be made to process waste water reaching Yamuna. Any wastewater reaching Yamuna should be processed beforehand so that sanctity and water quality of Yamuna improves. Recreational models should be developed near Yamuna which will help build connection of Yamuna with residents of Delhi. Conclusively, joint efforts need to be made on part of government machinery and citizenry which will enable overcome the environmental concerns of Delhi.

References

1. Central Pollution Control Board (2006) Water quality status of Yamuna River (1999-2005), Ministry of Environment & Forests, Government of India: New Delhi, India.

2. Department of Planning (2013) Economic Survey of Delhi, 2012-13, Government of India: New Delhi, India.

3. Chelani, AB, Devotta S (2007) Air quality assessment in Delhi: Before and after CNG as fuel. Environ Monit Assess 125: 257-263.

4. Ravindra K, Wauters E, Tyagi SK, Mor S, Grieken RV (2006) Assessment of air quality after the implementation of Compressed Natural Gas (CNG) as fuel in public transport in Delhi, India. Environmental Monitoring and Assessment 115: 405-417.

5. Sindhwani R, Goyal P (2014) Assessment of traffic–generated gaseous and particulate matter emissions and trends over Delhi.

6. Department of Planning, Economic Survey of Delhi, 2014-15, Government of India: New Delhi, India.

7. Aenab AM, Singh SK (2014) Critical Assessment of River water Quality and Wastewater Treatment Plant (WWTP), IOSR Journal of Environmental Science, Toxicology And Food Technology (IOSR-JESTFT) 8: 44-48.

8. Forest Survey of India, State of Forest Report; Ministry of Environment & Forests: Dehradun, India 2013.

9. Department of Planning (2011) Delhi Human Development Report, Government of India: New Delhi, India.

10. Aenab AM, Singh SK (2012) "Evaluation of Drinking Water Pollution and Health Effects in Baghdad", Iraq. Journal of Environment Protection 533-537.

11. Department of Environment and Forests (2010) State of environment report for Delhi, Government of NCT of Delhi, India.

12. Singh RB, Grover A (2015) Sustainable Urban Environment in Delhi Mega City: Emerging Problems and Prospects for Innovative Solutions.

13. Mallick J, Kant Y, Bharath BD (2008) Estimation of land surface temperature over Delhi using Landsat-7 ETM+ J Ind Geophys Union 12: 131-140.

14. McGeehin MA, Mirabelli M (2001). The potential impacts of climate variability and change on temperature-related morbidity and mortality in the United States. Environ Health Perspectives 109: 185-189.

15. Sharma JN, Kanakiya RS, Singh SK (2015) Characterisation Study and Correlation Analysis For Water Quality of Dal Lake, India. International Journal of Lakes and Rivers 8: 25-33.

16. Deepika, Singh SK (2015) Water Quality Index Assessment Of Bhalswa Lake, New Delhi. International Journal of Advanced Research 3: 1052-1059.

17. Singh SK, Katoria D, Mehta D, Sehgal D (2015) Fixed Bed Column Study and Adsorption Modelling on the Adsorption of Malachite Green dye from wastewater using Acid Activated Sawdust. International Journal of Advanced Research 3: 521-529.

13

Modeling of Data from Climate Science Using Introductory Statistics

Mahamad B Pathan*

Department of Statistics, Poona College of Arts, Science and Commerce, Savitribai Phule Pune University, Pune, India

Abstract

The attempt is made for building model(s) to climate data by using introductory statistics. The application of simple statistical methods can expose important insights in climate data. Various statistical tools from introductory statistics help to analyze climate data. The correlation analysis also helps to study interrelations between the variables in given data. The tools such as simple linear regression and multiple regressions are used to fit model for given climate data. Estimation/Prediction of dependent variable is made by using fitted models. The reliability of a model is discussed through residuals. A numerical example consists of three variables such as maximum temperature, minimum temperature and rainfall is considered to illustrate the analysis. The importance of variables in a given data is checked through simple statistical analysis. It is observed that 92% variation in rainfall is explained by range temperature (max-min). The multiple regression models provides better estimate for rainfall rather than two variable analyses.

Keywords: Climate data; Correlation; Simple regression; Multiple regression

Introduction

Climate is a measure of the average pattern of variation in temperature, humidity, atmospheric pressure, wind, precipitation, atmospheric particle count and other meteorogical variables in a given region over long periods of time. Climate is different from weather, in that weather only describes the short-term conditions of these variables in a given region.

Climate is a critical factor in the lives and livelihoods of the people and socio- economic development as a whole. India has to face the challenge of sustaining its rapid economic growth in the era of rapidly changing global climate. The problem has emanated from accumulated greenhouse gas emissions in the atmosphere, anthropogenically generated through long-term and intensive industrial growth and high consumption lifestyles in developed countries. Though, there is need to continuously engage international community to collectively and cooperatively deal with this threat, India needs a strong national strategy to firstly, adapt to climate change and secondly, to further enhance the ecological sustainability of its development path. This path is based on its unique resource endowments, the overriding priority of economic and social development and poverty eradication, and its adherence to its civilization legacy that places a high value on the environment and the maintenance of ecological balance. The national vision is to create a prosperous, but not wasteful society, an economy that is self-sustaining in terms of its ability to unleash the creative energies of our people and is mindful of our responsibilities to both present and future generations. This is in tune with global vision inspired by Mahatma Gandhi's wise dictum - "The earth has enough resources to meet people's needs, but will never have enough to satisfy people's greed". As such, promotion of sustainable production processes along with but equally, sustainable lifestyles across the globe should be the focus point of our efforts.

The climate is a dynamical system influenced not only by immense external factors, such as solar radiation or the topography of the surface of the solid Earth, but also by seemingly insignificant phenomenon. If we know all these factors, and the state of the full climate system (including the atmosphere, the ocean, the land surface etc.), at a given time in full detail, then there would not be room for statistical uncertainty. We do not know all factors that control the trajectory of climate in its enormously large phase space. Thus it is not possible to

map the state of atmosphere, the ocean, and the other components of the climate system in full detail. Also, the models are not deterministic in a practical sense: an insignificant change in a single digit in the model's initial conditions causes the model's trajectory through phase space to diverge quickly from the original trajectory. Therefore, in a strict sense, we have a 'deterministic' system, but we do not have the ability to analyse and describe it with "deterministic" tools. Instead, we use probabilistic ideas and statistics to describe the 'climate' system. The climate is controlled by innumerable factors. Only a small proportion of these factors can be considered, while the rest are necessarily interpreted as background noise. The details of the generation of this 'noise' are not important, but it is important to understand that this noise is an internal source of variation in the climate system.

Many researchers studied various problems related to climate systems. Box and Jenkins [1] suggested time series model for hydrological forecasting. These models include: Auto Regressive Integrated Moving Average (ARIMA), Auto Regressive Moving Average (ARMA), Auto Regressive (AR), and Moving Average (MA). Burlando et al., [2] used ARMA model for forecasting of short-term rainfall [3]. Valipour et al., [4] made comparison of the ARMA, ARIMA, and the autoregressive artificial neural network models in forecasting the monthly inflow of Dez dam reservoir [5]. Number of required observation data for rainfall forecasting according to the climate conditions was studied by Valipour [6,7]. The estimation of parameters of ARIMA and ARMA models studied by Valipour et al., [8]. Mohammadi et al., [9] used goal programming for parameter estimation of an ARMA model for river flow forecasting. Analysis of potential evapotranspiration using limited weather data.Study of different Climatic conditions to assess the role of solar radiation in reference crop evapotranspiration equations

---

Corresponding author: Mahamad B Pathan, Department of Statistics, Poona College of Arts, Science and Commerce, Savitribai Phule Pune University, Pune, 411001, Maharashtra, India, E-mail: must5619@yahoo.co.in

is attempted by Valipour [10,11]. A case study is given by Valipour [12] to see the ability of Box-Jenkins model to estimate of reference potential evapotranspiration.

The paper is arranged as follows. In section 4, the concept of mean and correlation is discussed. In section 5, simple linear regression model is represented. In section 6, the multiple regression model is considered. In section 7, a numerical example depend on secondary data is considered to explain above statistical tools. Conclusions are made in last section [8].

Basic Statistical Tools

The mean climate state

From the point of view of the climatologist, the most fundamental statistical parameter is the mean state. This seemingly trivial animal in the statistical zoo has considerable complexity in the climatological context. The computed mean is not entirely reliable as an estimate of the climate system's true long-term mean state. The computed mean will contain error caused by taking observations over a limited observing period, at discrete times and a finite number of locations. It may also be affected by the presence of instrumental, recording, and transmission errors. In addition, reliability is not likely to be uniform as a function of location [13].

Correlation

In the statistical lexicon, the word correlation is used to describe a linear statistical relationship between two random variables. The phrase 'linear statistical' indicates that the mean of one of the random variables is linearly dependent upon the random component of the other. The stronger relationship indicates the stronger correlation. A correlation coefficient of +1(-1) indicates a pair of variables that vary together precisely, one variable being related to the other by means of a positive (negative) scaling factor.

Simple Regression Model

Let Y be the dependent variable and X be the independent variable. Let $y_1, y_2,....$ y be n- observations recorded on Y variable. Let $x_1, x_2,..... x_n$ be n- observations recorded on X variable. Under the assumptions of linear relationship between Y and X, simple regression model of Y on X is as below:

$$Y = \alpha + \beta X + \varepsilon \qquad (1)$$

The unknown parameters α and β are to be estimated by method of least square (i.e. by minimizing residual sums of squares (S)).The random (noise) factor (ε) is assumed to follow normal distribution with mean zero and unit standard deviation. The estimate of α and β are tain by considering function $S = \Sigma(y_i - \hat{y}_i)^2$, which is to be minimized. The estimate of α and β are obtain by solving partial derivatives $\frac{\partial S}{\partial \alpha} = 0$ and $\frac{\partial S}{\partial \beta} = 0$ respectively, which are given below:

$$\hat{\alpha} = a = \bar{y} - \hat{\beta}\bar{x}, \quad \hat{\beta} = b = \frac{n\sum y_i x_i - \sum y_i x_i}{n\sum x_i^2 - (\sum x_i)^2}$$

The fitted model for equation (1) is Y=a + b X and is used to find estimate of Y ($\hat{Y}$) for given X. The reliability of fitted model to (1) is checked by calculating residual (Y-$\hat{Y}$).

Multiple Regression Model

Multiple regression model is used to study more than two variables. Let $X_1, X_2,....,X_k$ be k-variables under study. The regression model of three variables, by assuming X_1 dependent and other independent variables, can be written as below:

$$X_1 = \beta_1 + \beta_2 X_2 + \beta_3 X_3 + \varepsilon \qquad (2)$$

The unknown parameters β_1, β_2 and β_3 are to be estimated by method of least square (i.e. by minimizing residual sums of squares (S)).The random (noise) factor (ε) is assumed to follow normal distribution with mean zero and unit standard deviation.

Let n-observations are recorded on the variables X_1, X_2 and X_3. The total correlation coefficient denoted as r_{12}, r_{13} and $r_{23},$ are calculated by using following formula,

$$r_{ij} = \frac{n\sum X_{it}X_{jt} - (\sum X_{it}\sum X_{jt})}{\sqrt{n\sum X_{it}^2 - (\sum X_{it})^2}\sqrt{X_{jt}^2 - (\sum X_{jt})^2}}; i \neq j = 1,2,3 \text{ and } t = 1,2,....n.$$

The sample variances s_1^2, s_2^2 and s_3^2 are obtain by using following formula,

$$S_i^2 = \frac{n\sum x_i^2 - (\sum x_i)^2}{n^2}; i = 1,2,3 \text{ and } t = 1,2,.......n.$$

The correlation matrix for three variables is given below:

$$R = \begin{bmatrix} 1 & r_{12} & r_{13} \\ r_{21} & 1 & r_{23} \\ r_{31} & r_{32} & 1 \end{bmatrix}$$

The cofactor of (i, j) th element of determinant of matrix R is defined as below:

$R_{ij} = (-1)^{i+j}$ minor of element (i, j) i= 1,2,3, and j=1,2,3.

The estimates β_1, β_2 and β_3 are obtain by considering function $S = \Sigma(X_1 - \hat{X}_1)^2$, which is to be minimized. The estimates of β_1, β_2 and β_3 are obtain by solving partial derivatives $\frac{\partial S}{\partial \beta_1} = 0$, $\frac{\partial S}{\partial \beta_2} = 0$ and $\frac{\partial S}{\partial \beta_3} = 0$ respectively. The estimates are given as below:

$$\hat{\beta}_1 = a = \bar{X}_1 - \hat{\beta}_2\bar{X}_2 - \hat{\beta}_3\bar{X}_3 \quad \hat{\beta}_2 = b_{123} = \frac{-s_1 R_{12}}{s_2 R_{11}} \quad \text{and} \quad \hat{\beta}_3 = b_{132} = \frac{-s_1 R_{13}}{s_3 R_{11}}$$

The fitted model for equation (2) is $X_1 = a + b_{123} X_2 + b_{132} X_3$ and is used to find estimate of X_1 ($\hat{X}_1$)

for given X_2 and X_3. The reliability of fitted model to (2) is checked by calculating residual (Y-$\hat{Y}$).

A Numerical Example Based on Secondary Data

The secondary data is taken from the India Meteorological Department [14]. The data contains information about mean maximum temperature, mean minimum temperature and mean rain fall for the year 1901 to 2000 (i.e 100 years). The data for Pune city is given below (Table 1).

The correlation coefficient between three variables are calculated and given as below:

$$r_{12} = -0.5189 \; ; r_{13} = 0.7309 \; .$$

It concludes that the variable rainfall and the variable mean minimum temperature has strong correlation than with variable mean maximum temperature. So, the pair (X_1, X_3) will be effective for further analysis. This can be explained by ANOVA analysis (Table 2).

Simple linear regression model of X_1 on X_2 $=353.85107-9.2884X_2+\varepsilon$

As P-value=.0839 >0.05 $\Rightarrow$ Variable mean maximum temperature may not have enough impact on rainfall (Table 3).

Simple linear regression model of X_1 on X_3 $=-131.2993+10.5738X_3+\varepsilon$

As P-value=.006928 <0.05 $\Rightarrow$ Variable mean minimum temperature may have enough impact on rainfall.

Add one more variable as range (difference between max. temp. and min. temp.) in the data, so modified data is given in table below (Table 4).

The correlation coefficient between four variables are calculated and given as below:

$$r_{12} = -0.5189 \; ; r_{13} = 0.7309 \; ; r_{14} = -0.9603 \; .$$

It shows that there is strong correlation between range and rainfall, so it says that the prediction of rainfall with variable range may be more informative than other two variables. Smaller the range says the more chance of rainfall. This can be verified by ANOVA (Table 5).

Simple linear regression model of X_1 on X_4 $=219.3452-11.5981X_4+\varepsilon$

As P-value=7.31326E-07 <<<<0.05 $\Rightarrow$ Variable mean range temperature may have stronger impact on rainfall (Table 6).

Multiple correlation of X1 with X2 and X3=0.960573

Multiple correlation of X1 with X3 and X4=0.960573

Multiple correlation of X1 with X2 and X4=0.960573

Multiple correlation of X1 with X2, X3 and X4=0.960573

From above calculation, it concludes that in multiple regression

Summary Output	
Regression	**Statistics**
Multiple R	0.5188577
R Square	0.2692133
Adjusted R Square	0.1961346
Standard Error	55.443273
Observations	12

ANOVA					
Source of variation	df	SS	MS	F	Significance F
Regression	1	11324.09755	11324.1	3.683883	0.083899789
Residual	10	30739.56495	3073.956		
Total	11	42063.6625			
	Coefficients	Standard error	t Stat	P-value	
Intercept	353.85107	154.8020023	2.28583	0.045322	
X Variable 3	-9.2884044	4.839362702	-1.91934	0.0839	

Table 2: Simple linear regression analysis of X1 and X2.

Summary Output	
Regression	**Statistics**
Multiple R	0.730872
R Square	0.5341739
Adjusted R Square	0.4875913
Standard Error	44.265508
Observations	12

ANOVA					
Source of variation	df	SS	MS	F	Significance F
Regression	1	22469.31083	22469.3108	11.46724	0.006928443
Residual	10	19594.35167	1959.43517		
Total	11	42063.6625			
	Coefficients	Standard Error	t Stat	P-value	
Intercept	-131.2993	57.43645068	-2.2859918	0.045322	
X Variable 3	10.573843	3.1225071	3.38633116	0.006928	

Table 3: Simple linear regression analysis of X1 and X3.

Month	Temperature in centigrade			Rainfall in mm
	Maximum	Minimum	Range	
	X_2	X_3	X_4	X_1
Jan	30.2	11.6	18.6	1.6
Feb	32.3	12.7	19.6	1.1
Mar	35.8	16.3	19.5	2.7
Apr	37.9	20.1	17.8	13.6
May	37.2	22.3	14.9	33.3
Jun	32	22.8	9.2	120.4
Jul	28.1	22	6.1	179
Aug	27.6	21.3	6.3	106.4
Sep	29.2	20.6	8.6	129.1
Oct	31.7	18.9	12.8	78.8
Nov	30.5	14.8	15.7	28.6
Dec	29.3	11.8	17.5	5.3

Table 4: Difference between maximum temperature and minimum temperature and add one more variable as range.

adding functional variable (range) does not change value of multiple correlation coefficient. So, we study multiple regression only by original variables as below:

Multiple regression model between X_1 on X_2 and X_3 $=204.9587-$

Month	Temperature in centigrade		Rainfall in mm
	Maximum	Minimum	
	X_2	X_3	X_1
Jan	30.2	11.6	1.6
Feb	32.3	12.7	1.1
Mar	35.8	16.3	2.7
Apr	37.9	20.1	13.6
May	37.2	22.3	33.3
Jun	32	22.8	120.4
Jul	28.1	22	179
Aug	27.6	21.3	106.4
Sep	29.2	20.6	129.1
Oct	31.7	18.9	78.8
Nov	30.5	14.8	28.6
Dec	29.3	11.8	5.3

Table 1: Monthly mean maximum and minimum temperature & total rainfall based upon 1901 to 2000 data (Place: Pune).

Summary Output	
Regression	**Statistics**
Multiple R	0.96024691
R Square	0.92207413
Adjusted R Square	0.91428155
Standard Error	18.1048262
Observations	12

ANOVA					
Source of variation	df	SS	MS	F	Significance F
Regression	1	38785.81518	38785.82	118.3271	7.31326E-07
Residual	10	3277.847316	327.7847		
Total	11	42063.6625			
	Coefficients	Standard Error	t Stat	P-value	
Intercept	219.345168	15.69816974	13.97266	6.90003E-08	
X Variable 3	-11.598091	1.066214104	-10.8778	7.31326E-07	

Table 5: Simple linear regression analysis of X_1 and X_4.

Summary Output	
Regression	**Statistics**
Multiple R	0.960573218
R Square	0.922700908
Adjusted R Square	0.905523332
Standard Error	19.0072586
Observations	12

ANOVA					
Source of variation	df	SS	MS	F	Significance F
Regression	2	38812.17958	19406.09	53.71543	9.92624E-06
Residual	9	3251.482917	361.2759		
Total	11	42063.6625			
	Coefficients	Standard Error	t Stat	P-value	
Intercept	204.9586912	55.74734782	3.676564	0.005103	
X Variable 2	-11.2616988	1.674400111	-6.72581	8.6E-05	
X Variable 3	11.80349582	1.353187306	8.722736	1.1E-05	

Table 6: Multiple Regression model between X1 on X2 and X3.

Observation	Predicted Y	Residuals
1	73.34125378	-71.7412538
2	53.83560454	-52.7356045
3	21.32618915	-18.6261891
4	1.820539909	11.77946009
5	8.322422988	24.97757701
6	56.62212586	63.77787414
7	92.84690301	86.15309699
8	97.49110521	8.908894786
9	82.62965818	46.47034182
10	59.40864718	19.39135282
11	70.55473246	-41.9547325
12	81.70081774	-76.4008177
	Total	2.55795E-13

Table 7: Comparison of various simple regression models based on Residuals.

$11.2617X_2 + 11.8035X_3 + \varepsilon$

As P-value for mean maximum temperature=8.6E-05< <0.05 $\Rightarrow$ Variable mean maximum temperature may have enough impact on rainfall.

As P-value for mean minimum temperature=1.1E-05<<<0.05

$\Rightarrow$ Variable mean minimum temperature may have stronger impact on rainfall.

Comparison of various simple regression models based on residuals

Comparison of various simple regression models based on Residuals are given in Tables 7-10.

Conclusions and Future Scope

It is observed that the correlation between rainfall and mean minimum temperature is positive and significant than with mean maximum temperature. Also, the correlation between rainfall with

Observation	Predicted Y	Residuals
1	-8.642672914	10.2426729
2	2.988554487	-1.8885545
3	41.05438962	-38.35439
4	81.23499337	-67.634993
5	104.4974482	-71.197448
6	109.7843697	10.6156303
7	101.3252952	77.6747048
8	93.92360508	12.4763949
9	86.52191491	42.5780851
10	68.54638166	10.2536183
11	25.19362498	3.40637502
12	-6.527904296	11.8279043
Total		2.7001E-13

Table 8: Residual output for model X_1 on X_3.

Observation	Predicted Y	Residuals
1	3.620669125	-2.020669125
2	-7.977422226	9.077422226
3	-6.817613091	9.517613091
4	12.89914221	0.700857794
5	46.53360713	-13.23360713
6	112.6427278	7.75727217
7	148.596811	30.40318898
8	146.2771927	-39.87719275
9	119.6015826	9.498417359
10	70.88959896	7.910401036
11	37.25513404	-8.655134045
12	16.37856961	-11.07856961
Total		1.3145E-13

Table 9: Residual output for model X_1 on X_4.

Observation	Predicted Y	Residuals
1	1.775939496	-0.1759395
2	-8.88978255	9.989782546
3	-5.81314333	8.513143329
4	15.39057335	-1.79057335
5	49.2414533	-15.9414533
6	113.7040349	6.695965111
7	148.1818635	30.81813651
8	145.5502658	-39.1502658
9	119.2691007	9.830899326
10	71.04891082	7.751089181
11	36.16861649	-7.56861649
12	14.27216757	-8.97216757
Total		9.5568E-13

Table 10: Residual output for multiple regression model X_1 on X_2 and X_3.

range temperature shows stronger impact than other two variables. By ANOVA, it observed that simple regression model of rainfall on range temperature is more significant than others. The multiple regression model of rainfall on mean maximum temperature and mean minimum temperature gives better estimate. Range temperature factor does not alter the result in multiple regression analysis. Hence, I suggest to estimate rainfall by multiple regression model. It is possible to improve analysis by adding some other factors to improve estimation. Some Greenhouse gases, which are responsible for increment of temperature, may be considered in the analysis.

Acknowledgments

The author would like to express the gratitude to the anonymous reviewers whose constructive and insightful comments have led to many improvements of this paper.

References

1. Box GEP, Jenkins GM (1976) Series Analysis Forecasting and Control. Prentice-Hall Inc London.

2. Burlando C, Rosso R, Cadavid LG, Salas JD (1993) Forecasting of short-term rainfall using ARMA models. Journal of Hydrology 144: 193-211.

3. Witt G (2013) Using Data from Climate Science to Teach Introductory Statistics. Journal of Statistics Education 21: 1-24.

4. Valipour M, Banihabib ME, Behbahani SMR (2012) Comparison of the ARMA, ARIMA, and the autoregressive artificial neural network models in forecasting the monthly inflow of Dez dam reservoir. Journal of Hydrology 476: 433-441.

5. Valipour M, Banihabib ME, Behbahani SMR (2012) Monthly Inflow Forecasting Using Autoregressive Artificial Neural Network. Journal of Applied Sciences 12: 2139- 2147.

6. Valipour M (2012) Critical Areas of Iran for Agriculture Water Management According to the Annual Rainfall. European Journal of Scientific Research 84: 600-608.

7. Valipour M (2012) Number of Required Observation Data for Rainfall Forecasting According to the Climate Conditions. American Journal of Scientific Research 74: 79-86.

8. Valipour M, Banihabib ME, Behbahani SMR (2012) Parameters Estimate of Autoregressive Moving Average and Autoregressive Integrated Moving Average Models and Compare Their Ability for Inflow Forecasting. Journal of Mathematics and Statistics 8: 330-338.

9. Mohammadi K, Eslami HR, Kahawita R (2006) Parameter estimation of an ARMA model for river flow forecasting using goal programming. Journal of Hydrology 331: 293-299.

10. Valipour M (2014) Study of different Climatic conditions to assess the role of solar radiation in reference crop evapotranspiration equations. Archives of Agronomy and Soil Science 61: 679- 694.

11. Valipour M (2014) Analysis of potential evapotranspiration using limited weather data. Applied Water Science.

12. Valipour M (2012) Ability of Box-Jenkins Models to Estimate of Reference Potential Evapotranspiration (A Case study: Mehrabad Synoptic Station, Tehran, Iran). IOSR Journal of Agriculture and Veterinary Science (IOSR-JAVS) 1: 1-11.

13. Yadowsun Boodhoo. Guide to Climatological Practices.

14. India Meteorological Department.

Modeling of Surface and Weather Effects Ozone Concentration Using Neural Networks in West Center of Brazil

Amaury de Souza[1*], Flavio Aristones[1] and Fabio Verissimo Goncalves[2]

[1]Federal University of Mato Grosso do Sul, Institute of Physics, PO Box 549, CEP 79070-900, Campo Grande Mato Grosso do Sul, Brazil
[2]Federal University of Mato Grosso do Sul, Faculty of Engineering, Architecture and Geography, Graduate Program in Environmental Technologies, PO Box 549, CEP 79070-900, Campo Grande Mato Grosso do Sul, Brazil

Abstract

The estimative of the concentration of surface ozone promotes the creation of data for planning forecasting the air quality, useful in the management of public health. The aim of this study was to develop an Artificial Neural Network (ANN) to estimate the concentration of surface ozone due to climate data daily. The ANN, the Feedforward Multilayer Perceptron kind, was trained taking as reference the daily concentration of ozone measured. In the intermediate and output layers we used activation functions type tan-sigmoid and linear, respectively. The performance of the ANN developed was very good, and it can be considered as part of the set of indirect methods to estimate the concentration of surface ozone. The proposed model can be used by the government as a tool to enable the public interventional actions during the period of atmospheric stagnation, when ozone levels in the atmosphere may represent risks to public health.

Keywords: Neural networks; Climate; Surface ozone; Public health; Air quality

Introduction

Surface ozone (O_3) is one of the most important pollutants in troposphere. Its concentration in any given area is the result of the combination of its formation, transport, destruction and deposition. The O_3 sources include: [1] photochemical reactions involving its precursors (volatile organic compounds and nitrogen oxides) with natural or anthropogenic origin; [2] downward transport from stratosphere; [3] long-range transport (intercontinental) of ozone from distant pollutant sources [1,2]. The increase of precursors' emissions due to the economic development of many countries in the world led to the rise of the surface O_3 concentrations [3-6]. Consequently, a public concern about its negative effects on human health, climate, vegetation and materials it has been observed [7-9].

About the human health protection, several studies were implemented to predict the O_3 concentrations [10-12]. The statistical models are the most commonly used ones, due to the complexity of the chemical chain reactions that are associated to O_3 formation and destruction. In this context, linear and nonlinear models have been applied to predict the concentration of this air pollutant. Multiple linear regression, principal component regression, quantile regression, among others, are a few examples of linear models [13-15] and on the other hand, artificial neural networks are the nonlinear models most commonly used [12,16-20]. Evolutionary procedures to determine predictive models were also applied, which include threshold autoregressive models optimized by genetic algorithms (GAs) and genetic programming models [21,22]. Moreover, in several research fields, GAs have been also applied to optimize data division, the weights or the structure of the artificial neural networks [23-26].

Data

Information on daily levels of ozone (O_3) were obtained from the Department of Physics of UFMS. The Ozone Analyzer which was used to perform the measurements has the working principle of the absorption of ultraviolet radiation by ozone molecule. The analyzer is installed near Campo Grande, away from local resources. The measurements are performed continuously 24 hours per day, and

every 15 minutes, values are given of the ozone concentration. Then, when the arithmetic mean was calculated per day, it was assumed that this estimate was representative of air pollution in the city of Campo Grande. Information about rainfall, average temperature and relative humidity were obtained from Embrapa — Gado de Corte — Campo Grande.

In this study, we performed a descriptive analysis of variables which subsequently were associated with ozone concentration data, the rainfall climatic variables, maximum temperature, relative humidity and wind speed, from the period of 2004 to 2010.

Methods

Artificial neural networks

ANN can perform several functions such as classification, regression, association and mapping tasks [27-29]. They have a wide range of applications including adaptive control, optimization, medical diagnosis, decision making, as well as information, signal and speech processing [30]. ANN models are characterized by: (1) a set of processing neurons (also designated by nodes), (2) a pattern of connectivity among neurons, (3) an activation function for each neuron and (4) a learning rule. The processing neurons are distributed in layers: (1) input layer (first layer), (2) output layer (last layer) and (3) hidden layers (layers between the input and the output layers). The neurons in different layers are linked by synapses (each one storing a weight value) and the way which these linkages are done defines the

*Corresponding author: Amaury de Souza, Federal University of Mato Grosso do Sul, Institute of Physics, PO Box 549, CEP 79070-900, Campo Grande Mato Grosso do Sul, Brazil, E-mail: amaury.de@uol.com.br

structure of the network. These models were described in more detail by [29,31].

In this study, a feed forward ANN with three layers was applied to predict surface ozone concentrations with five input variables (O_3, T, RH, speed, precipitation). A linear function was used as activation function of the output neuron. Concerning the hidden neurons, four functions were tested: sigmoid, hyperbolic tangent, inverse and radial basis. The early stopping method (training procedure is stopped when an increase of validation error is observed) was applied to try to avoid the over fitting.

Daily data were stored between January — 2004 and December — 2010, and the total were divided into a training group (2/3) and a test group (1/3). Ozone observed data were necessary for training and validation of the results.

The program for training and testing ANN was developed with Matlab software. Aiming the desired map, a lot of net topologies of the Feed Forward Multilayer Perceptron were tested with variations of the numbers of neurons of the intermediate layers. Since the air temperature, humidity, rainfall, wind velocity and the transport fleet are the main factors that influence the estimative of ozone concentration, its maximum, minimum and average values were used as input data in ANN. In the intermediate layer were used activation functions of tan-sigmoid type and in the output layer were used activation functions of linear type, featuring this neural net as a universal approximator of functions. The data standardization were made depending to the kind of activation function in the output layer of the RNA, this procedure became necessary. The software Matlab offers two forms of data standardization in an interval [-1,1] and with average=0 and variance=1 and finally the total data were divided in 2/3 consecutive for training and 1/3 for validation.

Considering that, in the beginning of training, the free parameters are randomly created and that these initial values could influence in the final result of the training, each net architecture was trained ten times, being selected that one presented the highest value of determination coefficient (R^2). This coefficient was calculated from the data of the observation of the ozone concentration in the test sample and the respective values estimated by ANN.

Aiming the desired map, were trained a lot of net topologies, varying the number of neurons, activation functions in the intermediate layers, as well the numbers of the interactions (Table 1).

The ozone values estimated by the ANN were compared with the numbers calculated by the accumulated percentage error, the Root Mean Square Error (RMSE), the exactitude coefficient of Willmot (d) and the performance index (c).

The RMSE was calculated from the equation1.

$$RMSE = \frac{1}{n}\sum_{i=1}^{n}(\tilde{Y}_i - Y_i)^2 \qquad (1)$$

According to Camargo and Sentelhas [32], the following statistics indicators are considered to correlate the values estimated with the measures: exactitude – index of Willmott "d"; and of trust or performance "c". The exactitude, related to the detachment of the estimated values in relation to the observed ones, is given statistically by the agreement index proposed by Willmott [31]. Its values varies from zero, for no one agreement, to 1, for the perfect agreement. The index is given by the equation 2:

$$d = 1 - \frac{\sum_{i=1}^{n}(P_i - O_i)^2}{\sum_{i=1}^{n}\left(|P_i - O| - |O_i - P|\right)^2} \qquad (2)$$

Being: Pi = estimated value; Oi = observed value; O=average of the observed values.

The performance index "c", presented by Camargo and Sentelhas [32], evaluates the performance of the different methods of estimative. This index gathers the indexes of precision, given by the coefficient of correlation (r) that indicates the degree of dispersion of the obtained data in relation to the average, ie, the random error and of the agreement "d". The index "c" is calculated according the equation 3.

$$C = r.d \qquad (3)$$

Camargo and Sentelhas [32] proposed one criterion to interpret the performance of the estimative methods by the index "c", presented in the Table 2.

After the developing of the training algorithm of the ANN and the realization of analyses of the available climate data and the training algorithms, it was obtained an ANN capable of estimate, in a satisfactory mode, the concentration of surface ozone. This estimate is realized by mapping the relation between the maximum, average and minimum temperature data, maximum, average and minimum related humidity, wind speed, rainfall, the numbers of automotive vehicles that were counted as input and the concentration of reference ozone that is the desired output.

Results and Discussion

The ANN selected presented the best performance with the minimum configuration possible. This configuration is composed of one input layer with three variables, two intermediate layers each one with 4 and 2 artificial neurons, respectively, and one neuron in the output layer. The activation function of Sigmoid Hyperbolic Tangent type was adopted for the neurons in the intermediate layer. Generally, the trained nets presented better performances with smaller numbers of cycles with the ANN selected reaching better efficiency in 200 cycles. Beyond this it was verified that the nets with more than 200 cycles presented "memorization" problems.

The annual average value was c=0.81 with a great performance and an annual monthly average of performance equal to 0.79.

Parameter	Value
Number of neurons in the intermediate layers	1 to 5; 5 to 10
Activation functions in the intermediate layers	Sigmoid Logistic; Sigmoid Hyperbolic Tangent
Number of cycles	50; 100; 200; 500

Table 1: Parameters tested in the training of the RNAs.

C Value	Performance
>0.85	Great
0.76ª 0.85	Very good
0.66ª 0.75	Good
0.61ª 0.65	Average
0.51ª 0.60	Tolerable
0.41ª 0.50	Bad
<0.40	Terrible

Table 2: Criterion of interpretation of the estimative performance of concentration of surface ozone.

Month	[O₃] obs	[O₃] est	RMSE	c
Jan	10.32	13.07	-0.27	0.88
Feb	12.08	13.55	-0.12	0.88
Mar	13.28	14.34	-0.08	0.84
Apr	13,88	14.08	-0.01	0.83
May	14.66	11.40	0.22	0.73
June	14.93	15.54	-0.04	0.69
July	15.54	20.50	-0.32	0.56
Aug	24.29	24.57	-0.01	0.87
Sep	29.69	26.30	0.11	0.86
Oct	26.79	21.19	0.21	0.79
Nov	21.20	19.84	0.06	0.82
Dec	16.10	14.49	0.10	0.68

Table 3: Statistics indicators of the adjust between the values observed of the ozone and the values estimated by then RNA, monthly average relative error, values of "c" from January of 2004 to December if 2010.

Table 3 presents the values of the performance index (c) and of the root mean square error (RMSE) to the ANN's. Lowest values of RMSE associated with highest values of "c" indicate the performance of the methodology in the estimate of ozone concentration from the collected data.

The ANN's developed generally presented a good performance, except in the month of July, when they presented statistics index RMSE of -0.32, presenting values of "c" with a terrible performance. The concentrations of ozone presented four months of great performance, as shown in Table 3.

The ANN's performance was very good, mainly due to lots of data used in its training, making its learning easier. It also contributed to the very good performance the fact that different architectures were tested in the network, i.e., different numbers of layers, algorithms of learning, number of cycles, etc.

Some works like [32,33], evaluated many architectures for the ANN's, obtaining exceptional performances. It was emphasized that the number of cycles used in the training of the ANN's was high, making its learning easier, reducing the possibility of memorization occurrence. The memorization leads ANN's to present a good statistic performance (a high value of "c" and low value of RMSE), because this one is calculated based only in the sample of available data. On the other hand, the memorization would lead to serious distortions in the spatialization of the concentration of ozone extremely high or extremely low.

Analyzing the values with the ANN's (Table 3) we can verify that the memorization didn't occur, because were note evidenced severe deviations of the concentration of ozone estimated.

Analyzing the data of Table 2, we can verify that the average concentrations vary between 10.32 and 29.69 ppb (Table 3), with losing in the months of January, February and March, our rainfall season. The highest values were evidenced in the months of August, September and October, because it's the time to prepare the land for planting the crops.

We observed that high values of R^2 and "d" were obtained. This results were compared with those ones obtained in other studies with previsions of daily concentrations of ozone (Grivas, Chaloulakou [23] (0.60 and 0.86); Nagendra and Khare [34] (0.61 and 0.78)). The average of annual values of R^2 and "d" of this study were (0.8796 and 0.923798).

Figure 1 shows the graphic that compares the values observed and predicted by the model in the phase of validation. Figure 2 presents the histograms of the residues of the model evaluated in the phase

of validation. A good model must have a normal distribution of the residues, i.e., the histogram of the residue must be symmetric, in the shape of a bell. To visualize the performance of the model and of the ANN, the values observed and the simulations were compared as shown in the Figure 3. The graphical shows a good adjust of the model to the observed data, both in the phase of estimation/training and in the phase of validation.

Conclusion

The study of the methods for the estimate ozone concentrations

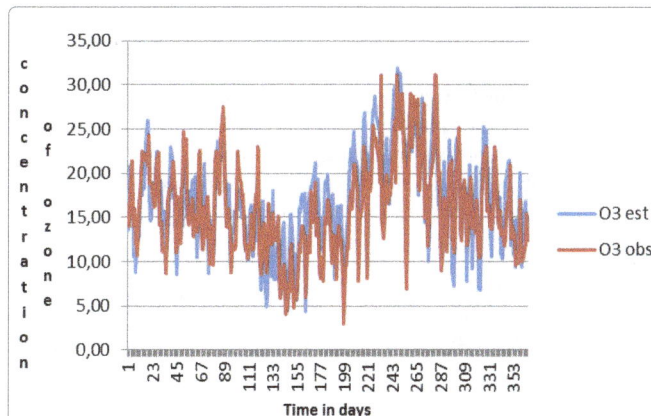

Figure 1: Concentration of ozone observed and predict in the phase of validation: ANN.

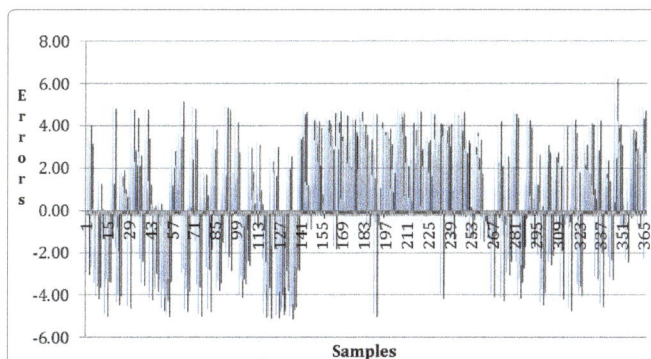

Figure 2: Histogram of the concentration of ozone in the phase of validation: ANN.

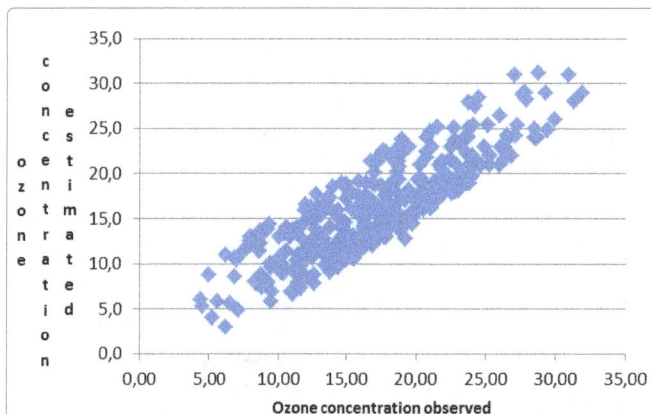

Figure 3: Simulation of the values observed/trained and a validation.

provides the average behavior of the parameters of study, which may be useful in prevention works of air quality, aiming the modeling work. Based on the results obtained in this work, we can conclude that:

1. The ANN's developed for estimating the ozone concentration presented a very good statistic performance.

2. There is a need of more training of the ANN's and variation of its architecture in order to obtain better statistic results.

3. The ANN's developed were capable of spatialize the concentrations of ozone without the presence of greater variances in its estimate.

4. Depending on the number of variables and the complexity of the architecture result the root mean square error may decrease or increase. Correlation values can be adjusted considering the size of the data presented to the network for training, variables that may represent more adequately modeled the environment as well as the development of other network architectures, enabling forecasts for longer periods.

Acknowledgement

We would like to extend our gratitude to the many people who helped to bring this article to fruition. First of all the UFMS. Karita Cristina Francisco Veríssimo Gonçalves for the contribution in English reviewing.

References

1. Derwent RG, Kay PJA (1988) Factors Influencing the Ground Level Distribution of Ozone in Europe. Environ Pollut 55: 191-219.

2. EPA (Environmental Protection Agency) (1993) Air quality criteria for ozone and related photochemical oxidants. Environmental protection agency, pp: 3-06 (EPA-600/ P-93-004aF-cF).

3. Cartalis C, Varotsos C (1994) Surface ozone in Athens, Greece, at the beginning and at the end of the 20th century. Atmospheric Environment 28: 3-8.

4. Lisac I, Grubisic V (1991) An analysis of surface ozone data measured at the end of the 19th century in Zagreb, Yugoslavia. Atmospheric Environment 25: 481-486.

5. Nolle Michael, Ellul Raymond, Ventura Frank, Gusten Hans (2005) A study of historical surface ozone measurements (1884-1900) on the island of Gozo in the central Mediterranean. Atmospheric Environment 39: 5608-5618.

6. Roxanne Vingarzan (2004) A review of surface ozone background levels and trends. Atmospheric Environment 38: 3431-3442.

7. Jack Fishman (1991) The global consequences of increasing tropospheric ozone concentrations. Chemosphere 22: 685-695.

8. Fuhrer J, Skarby L, Ashmore MR (1997) Critical levels for ozone effects on vegetation in Europe. Environmental Pollution 97: 91-106.

9. Lippmann M (1991) Health effects of tropospheric ozone. Environmental Science & Technology 25: 1954–1962.

10. Hanna SR, Chang JC, Fernau ME (1998) Monte Carlo estimates of uncertainties in predictions by a photochemical grid model (UAM-IV) due to uncertainties in input variables. Atmospheric Environment 32: 3619-3628.

11. Robert Vautard, Matthias Beekmann, Jean Roux, Dominique Gombert (2001) Validation of a hybrid forecasting system for the ozone concentrations over the Paris area. Atmospheric Environment 35: 2449-2461.

12. Yi JS, Prybutok VR (1996) A neural network model forecasting for prediction of daily maximum ozone concentration in an industrialized urban area. Environ Pollut 92: 349-357.

13. JCM Pires, Martins FG (2011) Correction methods for statistical models in tropospheric ozone forecasting. Atmospheric Environment 45: 2413-2417.

14. Sousa SIV, Pires JCM, Pereira MC, Alvim-Ferraz MCM, Martins FG (2009) Potentialities of quantile regression to predict ozone concentrations. Environmetrics 20: 147-158.

15. Cannon AJ, Lord ER (2000) Forecasting summertime surface-level ozone concentrations in the Lower Fraser Valley of British Columbia: an ensemble neural network approach. Journal of the Air & Waste Management Association 50: 322-339.

16. Gardner M, Dorling S (2001) Artificial neural network-derived trends in daily maximum surface ozone concentrations. Journal of the Air & Waste Management Association 51: 1202-1210.

17. Inal F (2010) Artificial neural network prediction of tropospheric ozone concentrations in Istanbul, Turkey. Clean-Soil Air Water. 38: 897-908.

18. Latini G, Grifoni RC, Passerini G (2002) The importance of meteorology in determining surface ozone concentrations—a neural network approach. Ecology and the Environment 8: 405-414.

19. Lu HC, Hsieh JC, Chang TS (2006) Prediction of daily maximum ozone concentrations from meteorological conditions using a two stage neural network. Atmospheric Research 81: 124-139.

20. Jose CM Pires, Maria CM Alvim–Ferraz, Maria C Pereira, Fernando G Martins (2010) Atmospheric Pollution Research 1: 215-219.

21. JCM Pires, MCM Alvim-Ferraz, MC Pereira, FG Martins (2011a) Prediction of tropospheric ozone concentrations: application of a methodology based on the Darwin's theory of evolution. Expert Systems with Applications 38: 1903-1908.

22. Bowden GJ, Maier HR, Dandy GC (2002) Optimal division of data for neural network models in water resources applications. Water Resources Research 38: 2-1-2-11.

23. Chaloulakou A, Grivas G (2006) Artificial neural network models for prediction of PM10 hourly concentrations, in the Greater Area of Athens, Greece. Atmospheric Environment 40: 1216-1229.

24. Garcia-Gimeno RM, Hervas-Martinez C, de Siloniz MI (2002) Improving artificial neural networks with a pruning methodology and genetic algorithms for their application in microbial growth prediction in food. International Journal of Food Microbiology 72: 19-30.

25. Hansen JV, Mcdonald JB, Nelson RD (1999) Time series prediction with genetic algorithm designed neural networks: an empirical comparison with modern statistical models. Journal of Computational Intelligence and Electronic Systems 15: 171-184.

26. Corne SA (1996) Artificial neural networks for pattern recognition. Concepts in Magnetic Resonance 8: 303-324.

27. Mukta Paliwal, Usha A Kumar (2009) Neural networks and statistical techniques: a review of applications. Expert Systems with Applications 36: 2-17.

28. Schneider Gisbert, Paul Wrede (1998) Artificial neural networks for computer-based molecular design. Progress in biophysics and molecular biology 70: 175-222.

29. Gupta RR, Achenie Lek (2007) A network model for gene regulation. Computers and Chemical Engineering 31: 950-961.

30. Guoqiang Zhang, B Eddy Patuwo, Michael Y HU (1998) Forecasting with artificial neural networks: the state of the art. International Journal of Forecasting 14: 35-62.

31. Cortj Willmott et al. (1985) Statistics for the evaluation and comparison of models. Journal of Geophysical Research Atmospheres 90: 8995-9005.

32. Camargo AP, Sentelhas PC (1997) Evaluation of the performance of different potential evapotranspiration estimation methods in the state of Sao Paulo, Brazil. Journal of Agrometeorology 5: 89-97.

33. Moreira MC, Cecilio RA, Silva KR (2007) Comparison of methods for estimating the air temperatures in the Brazilian Northeast. In: Brazilian Congress Agrometeorology, 15,Aracaju. Anais. Aracaju : SBAgro, (CD-ROM).

34. Nagendra SMS, Khare M (2005) Modelling urban air quality using artificial neural network. Clean Technologies and Environmental Policy 7: 116-126.

15

Climate Change and its Effect on Social Phenomena

Christopher G A Harrison*

Rosenstiel School of Marine and Atmospheric Science, University of Miami, 4600 Rickenbacker Causeway, Miami FL-33149, USA

Abstract

We now understand that climate has been affected by human activity (the most important being fossil fuel burning). But less is known of the effect of climate on human activity. In this paper, I examine data recently published by Zhang et al. [1] consisting of sixteen data sets lasting between 1500 and 1800 A.D. and in addition another 11 relevant data sets. These data sets consist of information about climate change, bioproductivity, prices and wages, and many socioeconomic factors, such as wars, war fatalities, population, health etc., related to Europe. It is shown that many of the strong correlations between the sixteen data sets claimed by Zhang et al. [1] are in fact not as strong as they claimed. Yet there are still some strong correlations. Multiple linear regression was used to determine which external factors (such as climatic or oceanographic variables) controlled each internal factor (involving human activity). Based on these calculations, some data sets did not seem to have any close correlations, while other data sets were clearly correlated at high accuracy and high confidence. Based on these data sets there is a set of correlations that strongly suggests that climatic variables have affected human activity, with warmer temperatures benefitting society in several ways.

Keywords: Climate change; Environmental science; Social phenomena

Introduction

WH. Beveridge, a distinguished British academic, economist, Member of Parliament, Director of the London School of Economics and Political Science and a principal planner of the British welfare state, gathered together many data bearing on wages and prices in England from 1251 to 1914. Only one volume of the five planned was published [2] because Beveridge became involved in the war effort and after the war in planning for the old age pension system and the National Health Service (the welfare state). It is ironic that most of the copies of this first volume were destroyed by fire during WWII, but the volume has since been reprinted. Gregory Clark (UC Davis) used data collected by Beveridge and stored in the library of the London School of Economics and Political Science, as well as other data, to continue the analysis started by Beveridge. He has produced a yearly record of prices from 1251 until 1914 [3]. Beveridge wanted to investigate the effect of climate on wages and prices during this time interval but had not arrived at a reasonable climatic series with which to compare his data. He wrote "My own interest in the subject arose not from general considerations but from the belief that the study of prices could be used to throw light upon the problem of periodicity of harvests and so of weather". With the publication of a major new paper connecting climate, economics, environment, and social behavior by Zhang et al. [1] it is now possible to make some headway at understanding the connections between these phenomena.

Zhang et al. [1] suggest that there are major correlations between various climatic, physical and environmental measurements and some socioeconomic functions during the period from 1500 to 1800 CE. Sixteen different time series were developed and they showed that the correlations between these time series were generally high and also highly statistically significant. The sixteen different time series are described below and were made available as annual data sets by Dr. HF.Lee, the second author of the paper. All had nominally 301 data points.The data sets from Zhang et al. [1] are as follows.

- Northern Hemisphere temperature anomaly. Annual data averaged from 12 individual data sets.

- European temperature anomaly. Two annual temperature records were combined [4,5] by normalizing and then averaging for each year .

- NH extra-tropical tree-ring widths, which may represent bioproductivity. Annual data obtained from Esper et al. [6].

- Grain yield ratio in relation to seed. Data were obtained from Slicher van Bath [7], covering many different European countries for four different grain types. Individual records were interpolated to make up for missing data. Yield ratios for each country and for each grain type were produced and arithmetically averaged to give one data set.

- Agricultural production index. Obtained from population size divided by inflation corrected (CPI index) grain price. These data sets were detrended.

- Average adult height. Various data on femur size were averaged to give a record which had eleven uniform heights of non-uniform duration in years. One duration only lasted for a year, and one lasted for 5 years. The other nine averaged about 33 years duration. Detrended before analysis.

- Grain price. Derived from European commodity price data using four grain types and data from sixteen major European regions. Note the inverse relationship to data set 5. Detrended before analysis.

- Wage index. Combination of two data sets of farm wages (England, by decade and interpolated) and building craftsmen

*Corresponding author: Christopher G A Harrison Rosenstiel School of Marine and Atmospheric Science, University of Miami, 4600 Rickenbacker Causeway, Miami FL-33149,USA. E-mail: dvdoma@saude.sp.gov.br

and laborers (19 European cities, annual). Detrended before analysis

- Magnitude of social disturbance. Sorokin's [8] data set covering Europe defined magnitude using social area of disturbance, duration, size of the masses involved and other things including intensity. This magnitude was divided by its duration. Annual magnitudes of all social disturbances in Europe were summed and then divided by the number of countries in Europe.

- Number of wars. Data obtained from Brecke [9]. 582 wars causing more than 31 deaths were fought in Europe during this time. Each war had a start year and an end year, and the number is just the number of wars in existence.

- War fatality index. Data also obtained from Brecke [9]. There is one sustained period lasting from 1618 to 1648 (the Thirty Years War), and another from 1791 to the end of the record (French revolution and its aftermath) where the fatality index is much higher than during the rest of the time.

- Plagues. Data obtained from Kohn [10] using only European results. The data come every ten years as plagues per decade, and these were linearly interpolated to give 301 data.

- Population growth rate. Established from the yearly differences in the next data set (#14), resulting in eight growth rates. The slopes in this data set are caused by the logarithmic interpolation in Population Size data set.

- Population size. European data were taken from McEvedy and Jones [11]. Population data are at irregular time intervals. Data were logarithmically interpolated. There were nine data points. Detrended before analysis. See data sets #5 and #13.

- Famine. European famines taken from Walford [12]. The original data set was for ten-year periods (31 data). These were linearly interpolated to give 301 data points.

- Migration. Complex index which comes in 25-year units (13 data), which were linearly interpolated to give 301 data.

Data Analysis

An important factor in Zhang et al. [1] is the correlation between the 16 data sets as represented by the 120 correlation coefficients between pairs of data sets. They produced a table of these 120 correlations, and all but four of them are significant at a level smaller than 0.001 (less than 0.1% chance of coming from randomly generated data set pairs). The other four are significant at a level smaller than 0.002. In addition, the correlation coefficients are large, with 86 absolute values greater than 0.5. Before doing the correlations the data were smoothed with a 40 year Butterworth low pass filter. Five data sets were detrended before filtering and correlation, as noted above.

I suggest that the data have been over-interpreted for three reasons. Firstly, the use of a Butterworth filter removes some of the information, implying that 301 should not be used as the number of independent data pairs. Secondly, the linear interpolations which have been done on some data sets do not produce 301 independent results to be used in the correlation analysis, although this number was in fact used to determine the significance of the correlation. Thirdly, detrending the data can result in addition of an important signal in some cases. Each of these will be discussed in more detail in the following sections.

Butterworth filtering

Butterworth filtering the data involves Fourier transforming

each data set and multiplying the resulting Fourier coefficients by the following factor.

$$\frac{1}{\sqrt{1 + \left(\dfrac{\omega}{\omega_0}\right)^{2n}}}$$

where ω is the frequency of the coefficient & ω_0 is the reference frequency, here 0.025 cpy, and then recreating the signal using the filtered Fourier coefficients. The power (half of the amplitude squared) of a 40 year period (0.025 cpy) is reduced by a factor of 2. I have assumed n to be 1. I subjected the undetrended data to a Butterworth filter and determined the RMS difference between the Butterworth data set and the original data set. I then determined the number of low frequency Fourier spectral estimates that are needed so that the simple low pass filter gives the same RMS difference between the filtered and the unfiltered data as the Butterworth-filtered and unfiltered data. If this number is n then the number of degrees of freedom should be 2n (Table 1).

Degrees of freedom

Zhang et al. [1] used 301 as the number of data pairs when they did their correlation analysis on the 120 pairs of data sets. But six of the sixteen data sets did not have 301 data, and simply interpolating between the existing data does not produce any more data, unless it can be clearly proven that the interpolation accurately presents the real variation of the particular data set. For instance data set #6 (Height) consists of the measurements of femur lengths, which were then translated into heights, which were then averaged over eleven non-uniform time periods; there is no reason to believe that the actual average height stayed constant over these time periods, and hence that there are not 301 independent data points in data set 6. One is forced to believe that it is not correct to use 301 as number of data pairs for any data set that did not start with 301 data. Instead it is more correct to use the smaller number when such a data set is being correlated with another data set. If it were legitimate to interpolate between data points, then this could also be done with the yearly data, increasing the number of data points indefinitely, thus rendering all non-zero correlation coefficients statistically significant! Shown in Table 1 of column 5 is the adjustment of degrees of freedom called for by the real number of data in each data set. This affects six of the data sets.

1	2	3	4	5	6
1	NH Temp.	0.560	169	169	51
2	European Temp.	0.405	122	122	141
3	Tree Ring Widths	0.585	176	176	73
4	Grain Yield	0.804	242	242	25
5	Agricultural Production	0.729	219	219	18
6	**Height**	**0.804**	**242**	**11**	**5**
7	Grain Price	0.590	178	178	31
8	Wage Index	0.524	158	158	28
9	Social Disturbance	0.112	34	34	205
10	Wars	0.624	188	188	40
11	War Fatalities	0.522	157	157	19
12	**Plagues**	**0.750**	**226**	**31**	**2**
13	**Population Growth**	**0.633**	**191**	**8**	**12**
14	**Population Size**	**0.781**	**235**	**9**	**0.007**
15	**Famine**	**0.683**	**206**	**31**	**2**
16	**Migration**	**0.709**	**213**	**13**	**1**

Table 1: Degrees of Freedom.

Another consideration concerning degrees of freedom for serial data such as those considered here is that serial correlation implies that the degrees of freedom should be less than the number of data. For instance Slonosky et al. [13] has suggested that the effective number of samples should be reduced using the following formula.

$$n_{eff} = n \left(\frac{1 - r_1(a) r_1(b)}{1 + r_1(a) r_1(b)} \right)$$

where $r_1(a)$ and $r_1(b)$ are lag-1 autocorrelation coefficients for each time series, n is the number of data and n_{eff} is the corrected number of data to be used in calculating the number of degrees of freedom. If this formula is modified to deal with just one data set then the appropriate formula would be

$$n_{eff} = n \left(\frac{1 - r_1^2}{1 + r_1^2} \right)$$

where r_1 is the lag-1 autocorrelation coefficient of the single data set. If this is done, the effective number can drop greatly, as is shown in column 6 of Table 1. This method of dealing with serial correlation does not work in some cases because if one of the lag-1 autocorrelations is negative, this results in an effective number of samples greater than n. For very high values of the lag-1 autocorrelation coefficients, such as with the population size data set, the effective number of data becomes very small. I have not used this method.

Detrending

The problem with detrending some signals is that this automatically adds new strong signals to the data sets. This is illustrated in (Figure 1), which shows two of the data sets that were detrended [1]. The original signal is shown along with the detrended signal. The process of detrending has produced a significant low in each data set in the middle range of age. It turns out that many of the data sets in this paper have either lows or highs in the middle of the age range, which is basically what causes the good correlations found by Zhang et al. [1]. There are five data sets detrended in [1] before analysis. In the analysis to be described below I have not detrended any of the data sets.

Summary

The large absolute values and the statistical significance of the correlations reported by Zhang et al. [1] are due to three erroneous assumptions. One is that filtering a data set does not remove information. Because filtering does remove information the degrees of freedom to be used when calculating if a correlation coefficient is significant is less than the number of original data. A method of allowing for this has been suggested. This is that the data number should be reduced by the same ratio as the power of the spectrum derived from the filtered profile to the power of the original signal. Secondly, the number of independent observations cannot be increased by linear interpolation unless there is a good argument for doing so. This means that for some data sets the regression coefficients are not as accurate as has been assumed because the number of degrees of freedom should be equal to the number of original data, not to the 301 data produced by interpolation. Thirdly, detrending data may add another variation to the original data set, causing spurious correlations.

New Correlations

To the 16 data sets in [1], I have also considered the following data

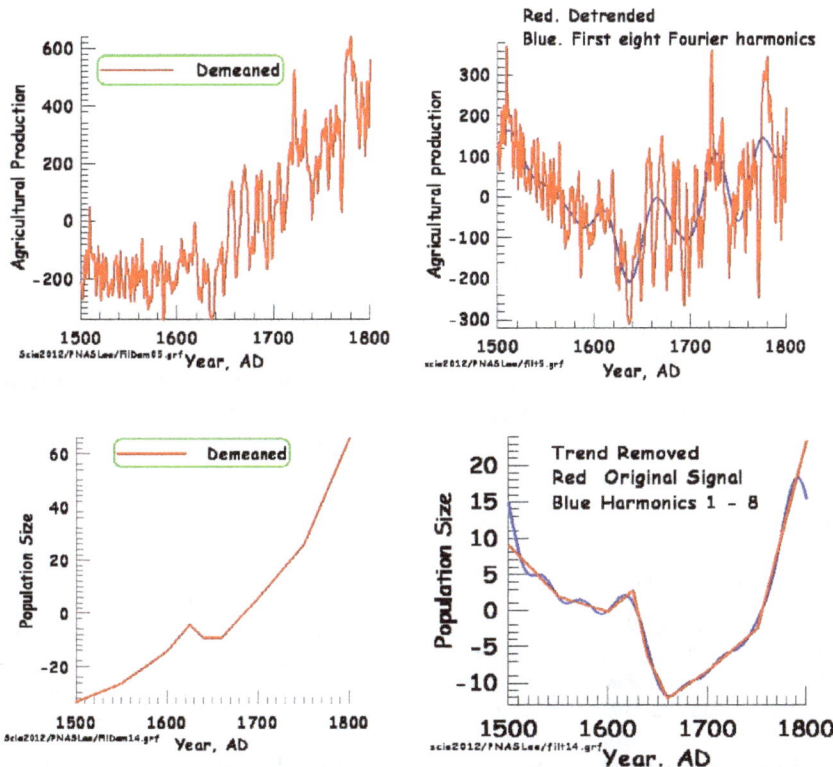

Figure 1: Two examples of how detrending can produce a completely different looking graph. Top left: Agricultural Production Index demeaned. Top Right: Agricultural production index detrended. Bottom Left: Population Size demeaned. Bottom right:. Population Size detrended. Both detrended curves show a large negative signal in the middle of the time period. This is one factor in generating large correlation coefficients with other signals. There were three other data sets that were detrended.

sets, all given on an annual time scale, and all but one from 1500 to 1800.

1. Clark [3] took the raw data gathered by Beveridge [2] and other data to produce an annual price index for England.

2. Phenology consisting of Swiss grape harvest dates, supposed to be controlled to some extent by the warmth of the summer months, warmer summers causing the grapes to ripen earlier [14].

3. Northern European grape harvest dates (see data set #18) coupled with observations of temperature in England later in the period [15].

4. North Atlantic Oscillation. NAO is the variation in atmospheric pressure difference between Azores and Iceland, which may have an effect on European climate as it controls the westerly winds that predominate in northern Europe [16]. Proxy time series are produced for years prior to the availability of barometers. There is much variation in the various different proxy manifestations of the NAO so that it is difficult to be confident that it represents a real climatic phenomenon ("In fact, prior to the twentieth century, the various NAO index reconstructions published in the literature remain inconsistent with no significant correlations among the different reconstructions" [17].

5. Another phenology paper based on harvest dates for Pinot Noir from Burgundy [18].

6. Atlantic Multidecadal Oscillation, from 1567 [19]. This is a measure of sea surface temperatures in the N. Atlantic Ocean and is calculated prior to the time of actual measurement using proxy indicators. It is only available (in proxy form) since 1567, giving 234 data. It measures North Atlantic sea surface temperature variations, and seems to be correlated to air temperatures and rainfall over much of the Northern Hemisphere.

7. English Price Index [20].

8. Idaho temperatures from tree ring analysis. This is somewhat far away from the European results, but it is interesting to see if there is any connection [21].

9. Phenology based on grapes from France, Switzerland, Alsace and the Rhineland [22].

10. ENSO proxy reconstruction [23,24]. There are so many teleconnections from El Niño Southern Oscillation so it is appropriate to use it here. The El Niño portion refers to changes in low latitude currents in the southern Pacific, with warmer equatorial waters during El Niño and cooler equatorial waters during La Niña. The Southern Oscillation refers to the variation of atmospheric pressure difference between Darwin and Tahiti.

11. Wheat index [25]. This data set was subjected to a form of spectral analysis [26] after having been passed through a high pass filter and the predominant period was 15.25 years, which was claimed to correspond to periods seen in some climate signals.

Because the AMO index is lacking 2/3 of a century of data, it has not been considered further, leaving 26 data sets. Correlation coefficients have been calculated for all 325 pairs of data sets without using any filter and without detrending any of the data sets. I have used a method of data selection which requires the correlation coefficients to be both significantly different from 0 and also I require them to explain at least 5% of the variance, meaning that the absolute value of the correlation coefficient has to be greater than 0.2236 (=√0.05). In order to check for correlation coefficients that are significantly different from 0 I used

a transformation given by Fisher [27,28] in which the transformed correlation coefficient is related to the correlation coefficient by the following equation,

$$z = 0.5.\ln\frac{1+r}{1-r}$$

where z is the Fisher transformed coefficient and r is the correlation coefficient. z is approximately normally distributed with a standard error only dependent on the number of pairs of data and given by

$$\sigma_z = \frac{1}{\sqrt{n-3}}$$

where σ_z is the standard error of z and n is the number of pairs of data. This contrasts with r which is not normally distributed, especially when $|r|$ gets close to 1; the standard error for r depends on the true but unknown value of the population coefficient ρ. If z is divided by $2\sigma_z$ then values of this quantity less than 1 will signify that the correlation coefficient is not statistically different from 0 at the 2.5% level. This number will be called f. The number of data for each data set is 301 apart from the six data sets given in Table 1 (in bold), which have smaller numbers as indicated there. When a correlation is done with two data sets with different numbers of data, the number to be used in the calculation of the standard error of z is the smaller of the two numbers.

These calculations show that 264 data pairs passed the barrier of explaining at least 5% of the variance. 226 data pairs passed the barrier of being significantly different from zero. This adds to 490 positive results out of a total of 1300 [2,26,25] or less than 38%. A total of 174 data set pairs passed both tests out of a possible 750 or just over 23% (Table 2). This is in distinct contrast to the results obtained by Zhang et al. [1], because all 240 of their data set pairs had correlation coefficients significantly different from zero by a large margin.

Multiple Linear Correlation Analysis

As a means of investigating further the connection between external and internal data sets, I used multiple correlation analysis. I decided to see how much the various external (climatic) parameters affected the internal (social) parameters. In order to aid in understanding the results of these multiple correlation coefficients, I demeaned all the data sets and then divided by their standard deviations. The result is that in the normal equation for multiple correlation coefficients given here,

$$y_1 = a + b_1 x_{1,i} + b_2 x_2 + ... + b_n x_{n,i}$$

the value of **a** became zero and the values of the slopes **b**$_i$ showed the relative importance of the independent data sets **x**$_1$, **x**$_2$ etc., on the value of the dependent data set **y**. Also, when doing two parameter (i.e., **y** and only one **x**) correlations, the slope **b** will be the same as the correlation coefficient. The results of the multiple correlation analysis are given in (Table 3).

In order to simplify the analysis I decided to combine all four price indices (data sets #7, #17, #23 and #27) by simple averaging and count this as a social, dependent data set. It is illustrated in (Figure 2). I also combined the four phenology data sets (data sets #18, #19, #21, and #25) into a single independent data set by averaging the four data sets. This may indicate temperature change during the summer months, and so is an external data set. These data are illustrated in (Figure 3).

Since I did not use the AMO data set because of its limitation of extension to 1500 CE, I ended up with 20 data sets. These were renumbered with seven independent data sets (NH Temp., European Temp., Tree Ring Widths, NAO, Idaho Temperatures, ENSO and

1	2	3	4	5	6	7	8
21	1	NH Temps.	301	13	14	9	27
2	2	European Temps.	301	15	14	12	29
3	3	NH Tree Ring Widths	301	14	14	12	28
20	**4**	**NAO**	**301**	**0**	**2**	**0**	**2**
24	5	Idaho Temp.	301	3	8	3	11
26	6	ENSO	301	0	3	0	3
18	7	Swiss Wine	301	9	10	8	19
19	7	Engl. Temp. plus Wine	301	4	10	3	14
21	7	Pinot Noir Wine	301	2	5	2	7
25	7	European Wine	301	0	1	0	1
4	8	Grain Yield	301	14	15	12	29
5	9	Agric. Prod.	301	14	12	10	26
6	**10**	**Height**	**11**	**15**	**3**	**3**	**18**
8	11	Wage Index	301	13	14	11	27
9	**12**	**Social Dist.**	**301**	**3**	**0**	**0**	**3**
10	13	Wars	301	14	13	10	27
11	14	War Fatalities	301	15	12	12	27
12	15	Plagues	31	12	8	8	20
14	**16**	**Population Size**	**9**	**13**	**4**	**4**	**17**
15	17	Famines	30	13	9	9	22
16	**18**	**Migrations**	**13**	**13**	**1**	**1**	**14**
7	19	Grain Prices	301	16	16	13	32
17	19	Beveridge/Clark Prices	301	12	14	11	26
23	19	Phelps Brown/Hopkins	301	13	12	11	25
27	19	Beveridge Wheat Prices	301	12	12	10	24
13	**20**	**Population Growth**	**8**	**12**	**0**	**0**	**12**
		Totals		264	226	174	490

Bold entries were eventually rejected. Column 1, Old #; 2, New #; 3, Data Set Name; 4, Number of data; 5, Number with $r^2 > 0.05$; 6, Number with z>2*SE; 7, Number passed both; 8, Total Number = columns 5 + 6.

Table 2: Summary of Correlations between 20 data sets.

Wine Phenology) being given the numbers 1 to 7. The new numbering system is given in (Table 2).

The Social Disturbances data set (#12) had one correlation with Northern Hemisphere Temperature (#1). However the correlation coefficient, although significantly different from zero, was only -0.146, representing less than 2.2 % of the variance. On doing a correlation with each other external data set as well as data set #1, no other data set had a slope statistically different from zero. On doing a multiple correlation with all seven independent data sets the correlation coefficient only went up to 0.175 representing 3% of the variance. So

the Social Disturbances data set will not be considered further.

Table 3 gives the results of these multiple correlations. The Plagues data set (#15) is correlated with the European Temperature data set (#2) with a correlation coefficient of -0.461, which is 2.65 times its standard error (using n=31) and so is significantly different from zero. On adding other data sets none was found that had a slope significantly different from zero. Nevertheless, since the correlation coefficient is of reasonable size, it has been put in Table 3, even though it is not a "multiple" correlation.

The Famines data set (#17) is correlated with the NH Tree Ring Width data set with a correlation coefficient of -0.518. If other external data sets are added serially one by one none of them achieves a slope significantly different from zero. The ratio of the correlation coefficient (or slope) to its standard error is 3.305. So this is a strong and significant correlation and has been put into Table 3, with the same caveat as for data set #15.

The method of multiple linear correlation consisted of taking each dependent data set and doing a multiple correlation with all seven of the external independent data sets. This gives the values of the seven **b**s, plus their standard errors, as well as the correlation coefficient, whose squared value represents the proportion of variance explained by the multiple correlation. If a **b** value for a specific external data set was less than twice its standard error, then that independent data set was considered to have too uncertain a contribution to the explanation of the dependent data set to worry about and another run was made with only those independent data sets left whose slopes were greater than twice their standard errors. It was found that the NAO independent data set (number 4 in the new numbering) never achieved a slope greater than twice its standard error. Thus there were only six independent data sets to be considered.

There were also five dependent data sets that did not have significant correlation with any independent data set to achieve the goal of having the slope be greater than twice its standard error. All of the six data sets that were not considered further are identified in Table 2 as bolded lines. Four of these six data sets had far fewer than 301 independent data and so this increased the standard error which was calculated for any correlation by the factor $\sqrt{\{298/(n-3)\}}$ where n is the number of data in the data set. This factor can be larger than 7. These four data sets had the smallest number of data, all less than 14.

So we are left with six independent data sets and eight dependent data sets. The results for the relationship between the dependent data sets and the independent data sets are given in Table 3. In the last

Indep. Data Sets		1	2	3	5	6	7	Correlation Coefficient
Dep. Data Sets		NH Temp	Eur Temp	Tree Ring	Idaho T.	ENSO	4 Wine	
8	Grain Yield	0.191/2.42	0.212/3.69	**0.219**/2.87	0.154/3.09			.599/.595
9	Agric. Prod.	**0.548**/6.89	0.340/5.61	-0.397/5.28			0.447/5.90	.571/.567
11	Wage Index	-0.401/4.80	0.144/2.26	**0.635**/8.06			0.245/3.08	.507/.502
13	Wars	-0.241/3.86	**-0.302**/4.69				0.285/3.54	.472/.472
14	War Fatal.	0.300/3.53	-0.175/2.80	**-0.525**/6.32	0.115/2.11			.477/.477
15	Plagues		**-0.461**/2.64					.535/.461
17	Famines			**-0.518**/3.305				.580/.518
19	4 Prices	0.342/6.41		**-0.529**/10.07		0.241/7.10	-0.222/4.33	.636/.544

First number in a box is slope of variation of dependent variable with independent variable. Second number is |slope|/[SE of slope] (numbers greater than 2 show slopes significantly different from zero at 97.5% confidence). Blanks (23 out of 48, including 8 from data set 4) are caused by slopes not significantly different from zero. For the correlation coefficients column, the first number shows the correlation coefficient when all seven independent data sets are used (including data set 4) and the second number is the correlation coefficient when only the indicated data sets are used.

Table 3: Multiple Correlation Analysis Between 8 Dependent Data Sets and 6 Independent Data Sets.

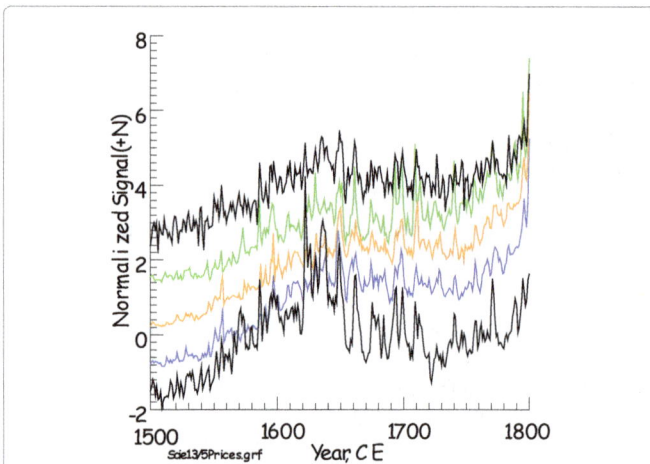

Figure 2: Four different Price data sets from demeaned data divided by their standard deviation. Red-Grain Price, Blue-Beveridge/Moore Price, Orange-Phelps Brown and Hopkins Price, Green-Beveridge Wheat Price, Black-Average. Each data set is displaced one ordinate unit above the data set below it.

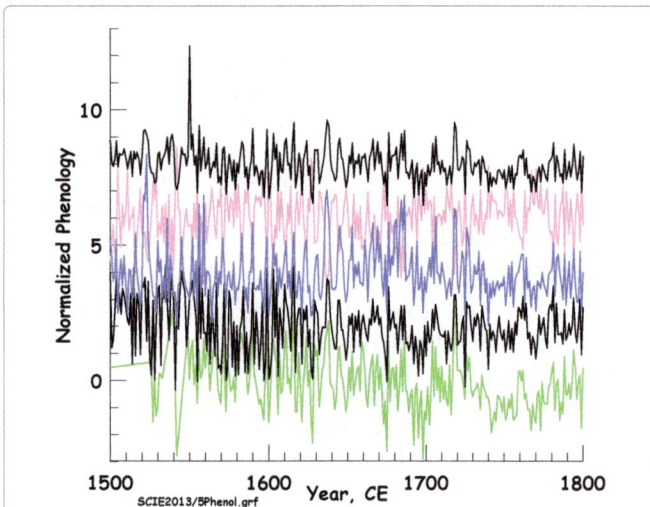

Figure 3: Four different Phenology data sets from demeaned data divided by their standard deviation. Green-Swiss, Red-England and N Europe, Blue-Burgundy, Pink-Europe, Black-Mean. Each data set is displaced two ordinate units above the data set below it.

column, the first number is the correlation coefficient when all seven independent data sets are used (including the NAO data set), and the second number is when only the ones with slopes significantly different from zero are used. These two numbers are very close, two cells having almost the same values. This illustrates that the independent data that were removed had little influence on the ability to reduce the overall variance.

In the boxes showing correlations between independent (horizontal) and dependent (vertical) data sets the left hand number is the value of the slope b_j for the independent variable listed at the top of the table. The right hand number is the ratio of this to the standard error of b_j. This ratio is always greater than 2, by choice. It can be seen that the Northern Hemisphere Temperature record, the European Temperature record and the Tree Ring record have strong correlations to six of the eight dependent variables left after the winnowing process. The phenology data set contributed significantly to four of the data sets,

the Idaho Temperature record was important in two dependent data set correlations and the ENSO data set was strongly correlated with just one dependent data set. The strength of the correlations with the Extra Tropical Tree Ring Widths was considerably greater than for the other five data sets. This is shown by the bolded numbers in Table 3, indicating the independent data set with the largest slope, of which five out of eight are for the extra-tropical tree ring width data set. The average absolute slope for the Tree Ring Width data set was 0.471, that for the NH Temperature was 0.337, that for the European temperature was 0.272. As might be expected, the correlation between the two numbers in each cell of the central part of Table 3 is high. As regards the dependent data sets, five of them had significant correlations with four independent data sets, one with three independent data sets and two with one independent data set (as pointed out above).

In order to illustrate some of these correlations (Figure 4) shows the Grain Yield index plotted against the four independent variables correlated with it. This does not bring out the strength of the correlations as well as (Figure 5), which shows the correlation between Extra Tropical Tree Ring Widths and Grain Yield. Figures 6-9, show plots of four of the eight dependent variables and the independent variable that gives the highest correlation with the dependent variable under consideration, as a function of time. Others are shown in the supplemental information (Figures 10-13). The correlations can easily be seen by eye in these sequential plots.

Discussion

We started off with 27 data sets. Eleven could be classified as external (having to do with climate or external forces of nature, including climate proxies such as the four wine phenology data sets) and sixteen could be classified as social (including four price data sets). The four phenology data sets were averaged and the four price data sets were averaged, and the AMO index was removed from further consideration because it started late in the sixteenth century. This reduced the number of external data sets to seven and the number of social data sets to thirteen.

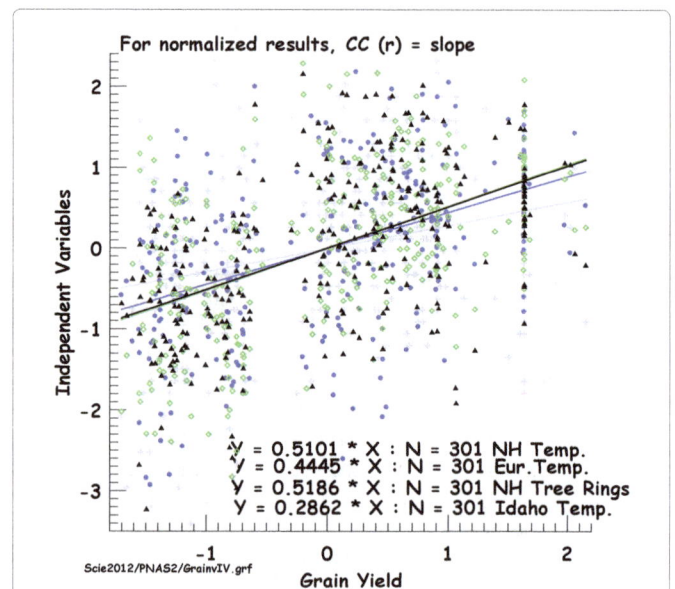

Figure 4: Grain yield (abscissa) plotted against the four independent data sets (ordinate) shown in table 1. All five data sets have been normalized. Red-Northern Hemisphere Temperature. Blue-European Temperature. Green-Northern Hemisphere Tree Ring Widths. Powder Blue-Idaho Temperature.

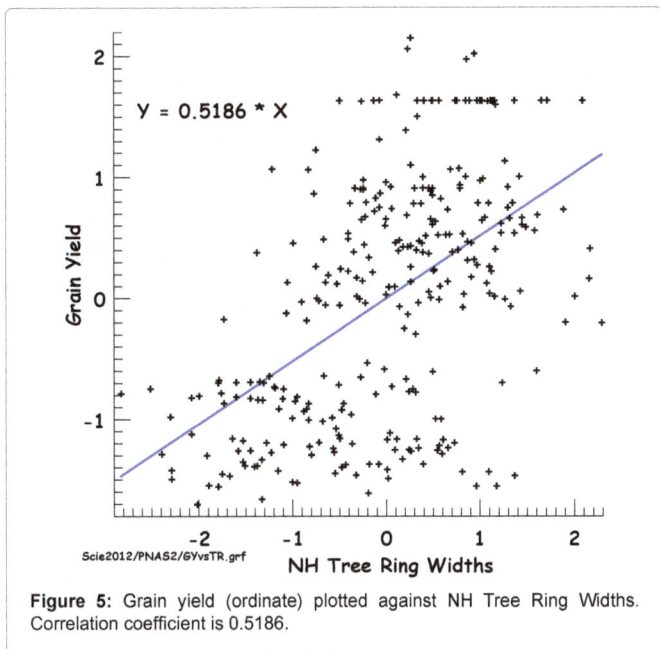

Figure 5: Grain yield (ordinate) plotted against NH Tree Ring Widths. Correlation coefficient is 0.5186.

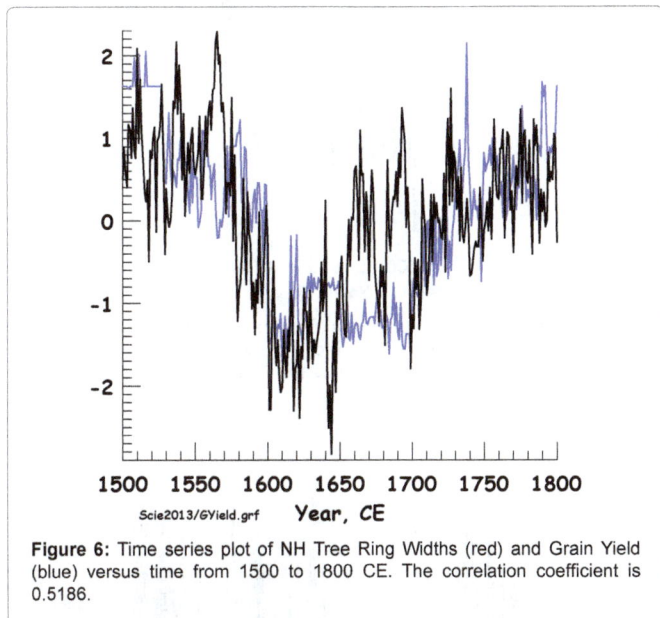

Figure 6: Time series plot of NH Tree Ring Widths (red) and Grain Yield (blue) versus time from 1500 to 1800 CE. The correlation coefficient is 0.5186.

this into later periods although allowance would have to be made for the effect of humans on climate.

As regards the way in which the independent data sets and the dependent data sets are correlated Table 3 shows the following. On the assumption that larger tree ring widths are also showing higher temperatures, then the eight bolded numbers in Table 3 show the following. Grain Yield, Agricultural Production, and Wages are all positively correlated with a Temperature index. Wars, War Fatalities, Plagues, Famines, and Prices are negatively correlated with a Temperature index. A simple conclusion is that positive aspects of human activity in Europe are correlated with high temperatures whereas the negative aspects of human activity are correlated with low

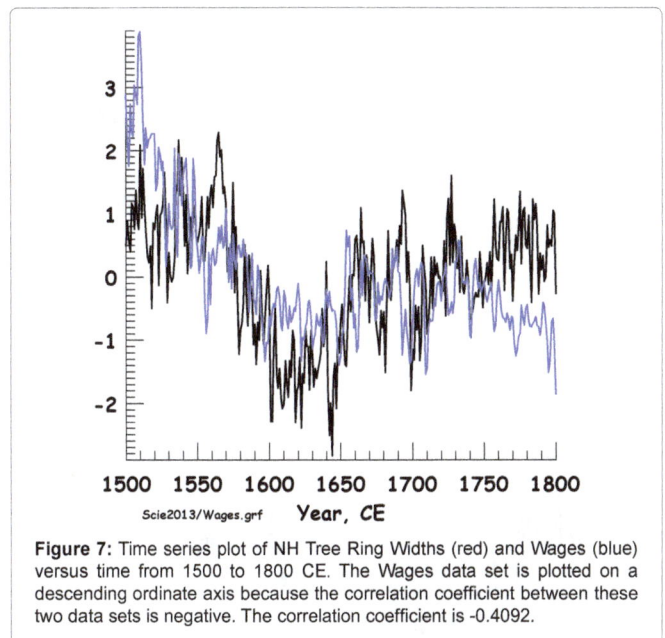

Figure 7: Time series plot of NH Tree Ring Widths (red) and Wages (blue) versus time from 1500 to 1800 CE. The Wages data set is plotted on a descending ordinate axis because the correlation coefficient between these two data sets is negative. The correlation coefficient is -0.4092.

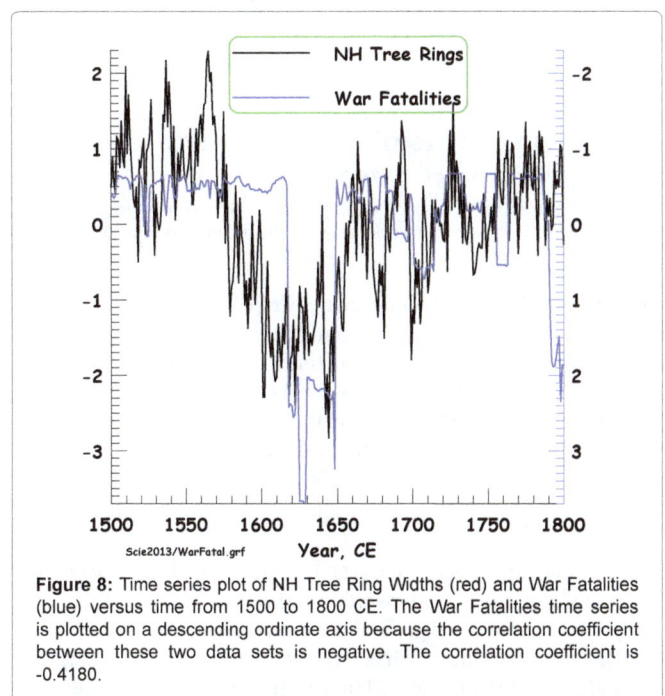

Figure 8: Time series plot of NH Tree Ring Widths (red) and War Fatalities (blue) versus time from 1500 to 1800 CE. The War Fatalities time series is plotted on a descending ordinate axis because the correlation coefficient between these two data sets is negative. The correlation coefficient is -0.4180.

Multiple linear correlations were done between each (dependent) social data set and the seven (independent) external data sets. All four of the 13 dependent (social) data sets in paper (1) with the smallest number of original data were not correlated significantly with any of the seven independent (external) data sets. One additional data set with 301 data (social disturbance) also fell into this category. After removal of these data sets there were eight dependent data sets remaining. All of these data sets had strong correlations with one or more of the external data sets (Table 3), but one of the external data sets did not appear in the list of correlations. So the final set of correlations was between eight dependent (social) data sets and five independent (external) data sets. Hopefully, other social and physical data sets during the time period under consideration (1500-1800) can be constructed to determine if there are any other valid correlations. It would also be useful to expand

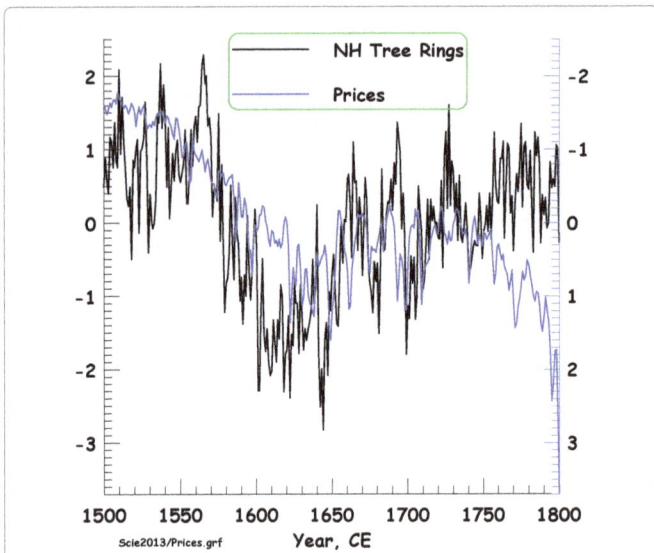

Figure 9: Time series plot of NH Tree Ring Widths (red) and Prices (blue) versus time from 1500 to 1800 CE. The Prices time series is plotted on a descending ordinate axis because the correlation coefficient between these two data sets is negative. The correlation coefficient is-0.3622.

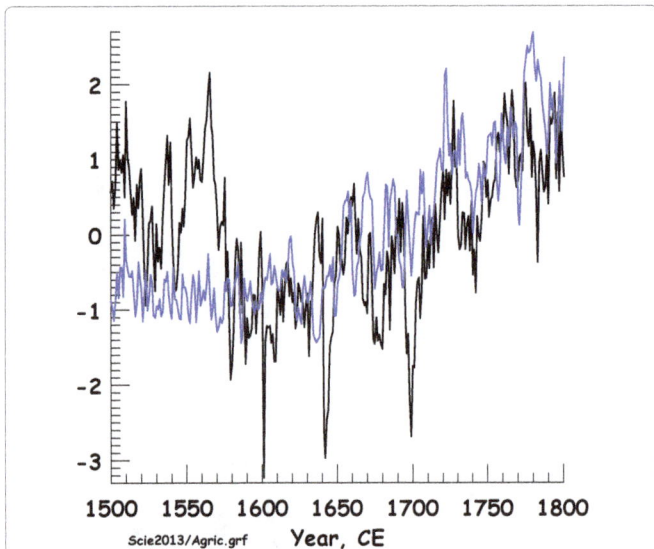

Figure 10: Agricultural Production (blue) and NH Temperature (red) plotted against time. Correlation coefficient between the data sets is 0.3970.

data sets were added. These data sets were divided into two groups, one group representing external variables having to do with weather or climate, and classified as independent variables, and one group representing internal variables of human activity, and classified as dependent variables. After combining four price indices, and four wine phenology indices and other adjustments, six independent variables and eight dependent variables were left and were subjected to multiple linear correlation. The independent data sets had to do with climate and were Northern Hemisphere Temperature, European Temperature, Northern Hemisphere Extra Tropical Tree Ring Widths, Idaho Temperature, the ENSO index, and Wine Phenology as a measure of European summer temperatures. The dependent variables having to do with social phenomena were Grain Yield (significantly correlated with four independent data sets), Agricultural Production

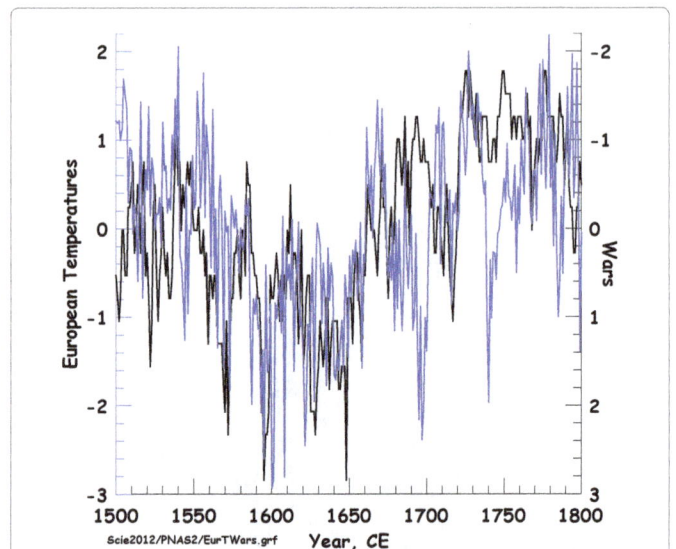

Figure 11: Wars (red) and European Temperatures plotted against time. Wars is plotted along a descending ordinate axis because of its negative correlation with European Temperatures (blue). Correlation coefficient is -0.3864.

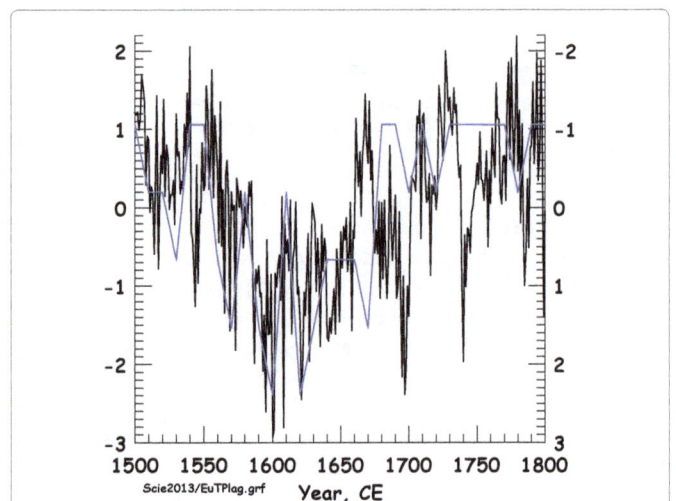

Figure 12: Plagues (blue) and European Temperatures (red) plotted against time. Correlation coefficient between the data sets is - 0.4610. The Plagues data base is plotted inversely because of its negative correlation with European Temperature.

temperature. Higher temperatures are better for humanity in Europe during this time interval apart from those people producing goods, as the price index is reduced during times of higher temperature. The simple correlation coefficients shown in Figures 6-13 correspond in sign to the partial correlation coefficients shown as the left hand number in the columns of cells in Table 3 referring to the six independent variables (1-3 and 5-7).

Conclusions

The original sixteen data sets in Zhang et al. [1] were over interpreted due to (a) Butterworth filtering, (b) detrending five data sets produced spurious signals and correlations, and (c) six data sets had far fewer than the nominal 301 data that were used to estimate correlation significance. In addition to the 16 data sets, another 11

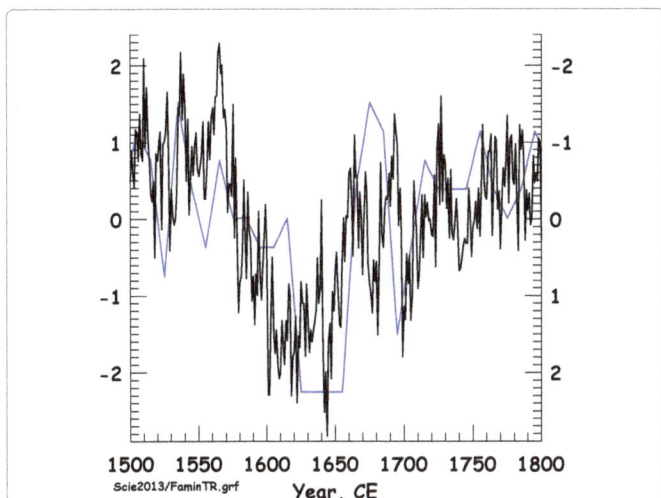

Figure 13: Famines (blue) and NH Tree Ring Widths (red) plotted against time. Correlation coefficient between the two data sets is -0.5180. Famines is plotted inversely because of the negative correlation between the two data sets.

(four), Wage Index (four), Wars (three), War Fatalities (four), Plagues (one), Famines (one), and Price Data (four). Northern Hemisphere Temperature was significantly correlated with six dependent data sets, European Temperature (with six), Tree Ring Widths (six), Idaho temperatures (two), ENSO (one), and Wine Phenology (four). The independent variable Tree Ring Widths was the most important independent variable both from the point of view of the total absolute slopes (2.437, the next highest being 2.17 from the NH Temperature index) and also as an average (0.487, the next highest being 0.362 also from the NH Temperature index). For this limited set of data there were strong correlations between external data sets (those related to climatic measurements) and some of the internal data sets (involving human activity), indicating that some (but not all) of the conclusions arrived at by Zhang et al. (1) were justified. Finally, warmer temperatures seem to benefit Europeans in several aspects such as greater grain yields (positive correlation) and fewer wars and war fatalities (negative correlations).

Acknowledgements

I thank Dr. H. F. Lee for sending the sixteen data sets soon after publication of their interesting paper. I also thank Dr. G. Clark for sending me his prepublication data set on English prices, which helped to get me interested in the connection between weather, climate and social phenomena.

Reference

1. Zhang DD, Lee H, Wang C, Lie B, Pei Q, et al. (2011) The causality analysis of climate change and large-scale human crisis, Proceedings of the National Academy of Science 108: 17296-17301.

2. Beveridge WH (1939) Prices and Wages in England from the Twelfth to the Nineteenth Century (1stedn.), Price Tables, Mercantile Era. Longmans p.756.

3. Clark G (2004) The Price History of English Agriculture. Research in Economic History 22: 41-123.

4. Luterbacher J, Dietrich D, Xoplaki E, Grosjean M, Wanner H (2004) European seasonal and annual temperature variability, trends, and extremes since 1500. Science 303: 1499-1503.

5. Osborn T, Briffa KR (2006) The spatial extent of 20th century warmth in the context of the past 1200 years. Science 311: 841-844.

6. Esper J, Cook ER (2002) Low-frequency signals in long tree-ring chronologies for reconstructing past temperature variability. Science 295: 2250-2253.

7. Slicher von Bath BH (1963) Yield ratios, 810-1820. A. A. G. Bijdragen 10: 1-264.

8. Sorokin PA (1937) Social and Cultural Dynamics, American Book Company, New York.

9. Brecke P (1999) Violent conflicts 1400 A. D. to the present in different regions of the world, 1999 Meeting of the Peace Science Society (International).

10. Kohn GC (2001) Encyclopedia of Plague and Pestilence. Facts on File (3rdedn.) New York.

11. McEvedy C, Jones R (1978) Atlas of World Population History, Allen Lane, London.

12. Walford C (1970) The Famines of the World: Past and Present. Burt Franklin New York.

13. Slonosky V, Jones PD, Davies TD (2000) Variability of the surface atmospheric circulation over Europe, 1774-1995. Int. J. Climatology 20: 1875-1897.

14. Meier N, Rutishauser T, Pfister C, Wanner H, Luterbacher J (2007) Grape harvest dates as a proxy for Swiss April to August temperature reconstructions back to AD 1480. Geophysical Research Letters 34: 20705.

15. Bray JR (1982) Alpine glacial advance in relation to a proxy summer temperature index based mainly on wine harvest dates, A. D. 1453-1973. Boreas 11: 1-10.

16. Cook ER, D'Arrigo RD, Mann ME A (2002) well-verified multiproxy reconstruction of the winter North Atlantic Oscillation Index since A. D. 1400. j. Climate 15: 1754-1764.

17. Schmutz C, Luterbacher J, Gyalistras D, Xoplaki E, Wanner H (2000) Can we trust proxy-based NAO index reconstructions. Geophysical Research Letters 27: 1135-1138.

18. Chuine I, Yiou P, Viovy N, Seguin B, Daux V, et al. (2004) Grape ripening as a past climate indicator. Nature 432: 289-290.

19. Gray ST, Betancourt J, Graumlich LJ, Pederson G (2004) Atlantic Multidecadal Oscillation (AMO) Index Reconstruction. NOAA/NGDC Paleoclimatology Program, Boulder CO, USA.

20. Phelps Brown EH, Hopkins S (1956) Seven centuries of the prices of consumables. Economica, New Series 23: 296-314.

21. Biondi F, Perkins DL, Cayan DR, Hughes MK (1999) July temperature during the second millennium reconstructed from Idaho tree rings. Geophysical Research Letters 26: 1445-1448.

22. Ladurie E Le Roy, Baulant M (1980) Grape harvests from the fifteenth through the nineteenth centuries. The Journal of Interdisciplinary History 10: 839-849.

23. Cook ER (2000) Nino 3 Index Reconstruction. International Tree-Ring Data Bank. IGBP PAGES/World Data Center-A for Paleoclimatology Data Contribution Series. NOAA/NGDC Paleoclimatology Program, Boulder CO, USA.

24. D'Arrigo R, Cook ER, Wilson RJ, Allan R, Mann ME (2005) On the variability of ENSO over the past six centuries. Geophysical Research Letters 32: 3.

25. Beveridge WH (1921) Weather and Harvest Cycles. The Economic Journal 31: 429-452.

26. Fisher RA (1958) Statistical Methods for Research Workers (13thedn.) Oliver and Boyd, Edinburgh p 356.

27. Panofsky HA, Brier GW (1965) Some applications of statistics to meteorology (1stedn.) Mineral Industries Continuing Education, University Park, Pennsylvania p 224.

16

The Challenge of Reducing Food Carbon Footprint in a Developing Country

Oyenike Mary Eludoyin*

Department of Geography and Planning Sciences, Adekunle Ajasin University, Akungba-Akoko, Nigeria

Abstract

Reducing the contribution to the greenhouse gases through modification of lifestyles has been the recent focus of geoengineering and climate change discussion. However, many developing countries still have to face the challenge of ensuring food security, and increasing energy availability. This study is an expository review of the impact of food production and peoples' attitudes to food in the face of climate change. The main objective of the study was to examine the sources and dimensions of food carbon footprints, with a focus on Nigeria, and with the view of making recommendations for reduction of greenhouse gasses from food sources. The study argued the need for infrastructural and welfare improvement, and rural development to channel the focus of the people towards environmental preservation and sustenance.

Keywords: Food carbon footprint; Climate change; Food production

Introduction

Sources of CO_2 and other greenhouse gases are diverse, and are both natural and anthropogenic. Anthropogenic sources of greenhouse gases are generally becoming a threat to humankind as the earth's temperature increases [1]. For a long time, Increasing industrialization and urbanization were often the targeted anthropogenic sources of greenhouse gases [2]. Recently however, approach to meeting the Climate Change Act target of 80% cut in greenhouse gases (GHGs) by 2050, in many developed countries, especially in Europe and Americas has shifted to commitment on farming and food [3], alternative energy use, especially as global population rises [4,5]. Convery and Redmond [6] noted that increase in food consumption in the European Union alone may account for about a third increase in carbon footprint in the region by consumers. In most developing countries, records of consumptions are scarce because of poor carbon accounting systems in these countries (if any awareness). Most developing countries are known to rapidly increase in population and contribute significantly to the greenhouse emission through high carbon emitting activities such as use of domestic fuel wood, gas flaring, transportation among others. The study focuses on Nigeria, where most people (at least 60% of both rural and urban settlers) are meat lovers. Meat lovers emit the largest (3.3 t CO_2 e/person) per diet (Figure 1).

Aim and Objective

The aim of this study is to contribute to discussion on the importance of food production and consumption on climate change. Specific objective is to examine the sources and dimensions of food carbon footprints, with a focus on Nigeria, and with the view of making recommendations for reduction of greenhouse gasses from food sources.

Information about Nigeria

Nigeria is a sub-Saharan African country with population of more than 150 million and population growth of 2.5% (1.8% in the rural areas, and 3.8% in urban areas) [7]. It is located on 4–14°N and 3–15°E in the southeastern edge of the West African region, with a land area of about 923 800 km² (14% of West Africa). Nigerian climate is diversified (Figure 2), and supports growth of many agricultural products, especially tree crops, including fruits, most of which are characterised by high carbon contents.

The tropical rainforest climate, designated by the Koppen climate classification as 'Af', characterises the southern Region, and it is sub-grouped into the tropical wet and tropical wet and dry climates as with distance away from the Atlantic Ocean. The tropical rainforest is characterized by small temperature range (26-27°C) throughout the year, and usually convectional storms, as a result of its proximity to the equatorial climate. This region is known for agricultural practices, especially crop production. The tropical savanna climate exhibits a well-marked single peak rainy season and a dry season. Mean temperature in the savanna is between 26 and 28.2°C throughout the year. Dairy farming and vegetable cultivation are dominant agricultural practices in the savanna, especially in the Guinea and Sudan savanna [8,9]. Food cultivation in Nigeria is essentially limited by poor technological infrastructure, and this accounts for large proportion of subsistence agriculture practices in the country.

Food Production Chain and Associated Carbon Sources

Figure 3 is a conceptualized form of food production processes. The first stage of food production involves cultivation of food crops and rearing of livestock.

Farming activities and release of carbon

Energy is consumed during the activities involved in agriculture, hence they contribute substantially to carbon footprint. Livestock farming releases methane (CH_4), and nitrous oxides are often emitted from fertilised fields [3]. Other sources of carbon is the emission from tractor based machines used for tillage, and the soil carbon turnover during tillage and crop planting. The soil carbon stock is large (mean ± standard deviation in Europe; 20 ± 12, 57 ± 34 and 10 ± 9 g Cm^{-2} yr^{-1} from forest, grassland and cropland, respectively) [10], and the distribution is often affected by changes in vegetation and plant growth,

---

*Corresponding author: Oyenike Mary Eludoyin, Department of Geography and Planning Sciences, Adekunle Ajasin University, Akungba-Akoko, Nigeria
E-mail: baynick2003@yahoo.com

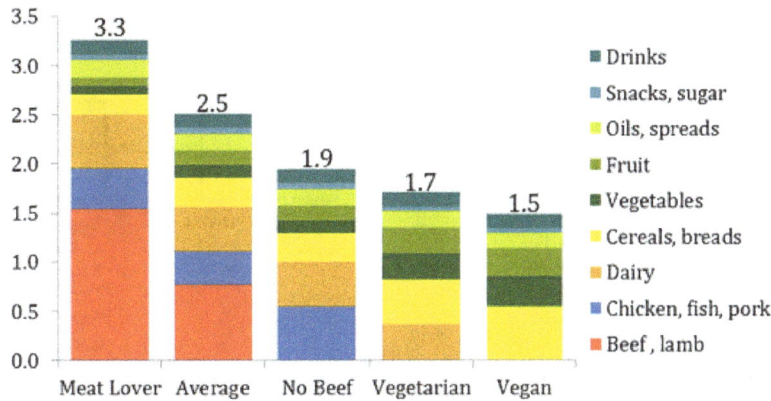

Figure 1: A graphical representation of the amount of CO_2 (t CO_2 e) emission per person by diet type (http://www.csag.uct.ac.za/2014/02/17/sweating-for-carbon-redemption/).

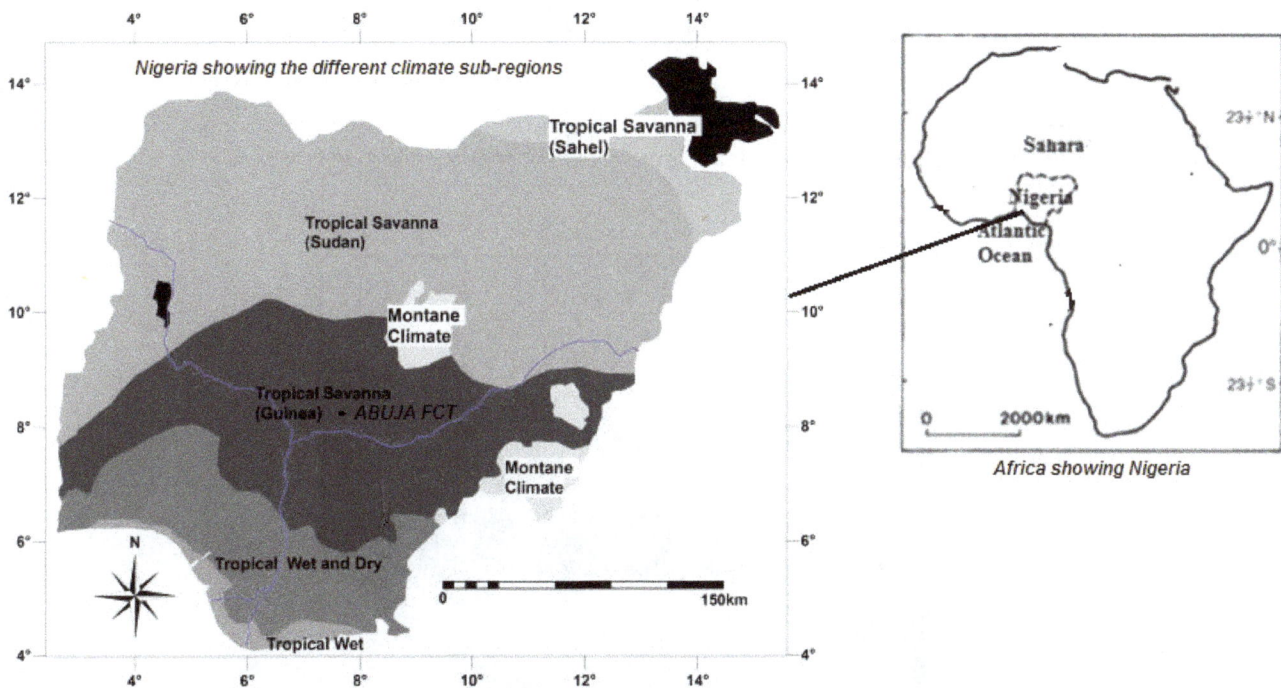

Figure 2: Map showing the different climate sub-regions in Nigeria. Inset is a map of Africa showing the location of Nigeria.

removal of biomass by harvest, and mechanical soil disturbances such as plowing [11].

Food processing

Activities involved in food processing or refining into supermarket and/or consumption-ready products are also tangible sources of carbon into the environment. Often times, meat are frozen or oven-dried. In West African countries, and importantly in Nigeria, smoked fish and 'Suya', a spicy kebab-like skewered meat (beef, ram and chicken) are popular street foods [12] Preparation of Suya and other skewered meat involves significant use of charcoal, coal or fuel wood, which are important sources of carbon.

A number of studies have shown that fuel wood is an important energy source for food processing in developing countries, including Nigeria, especially in the rural areas [13-17]. Fuel wood consumption in Nigeria is estimated at about 87% of the total energy, and it is mostly used irrespective of the economic status of the people [16,18] (Figure 4). The Nigerian Bureau of Statistics, NBS (2007) [19] estimated that the fuel wood use as energy varied from 55% in the southwest to about 96% in the north east (95% in Northwest, 86% in north central, 78% in southeast and 73% in the south-south) an in direct correlation with the poverty rate in the regions.

Food distribution

Food distribution involves transportation of food crops from the farms to where they are processed. In general, transportation and wastages from food distribution are main sources of carbon release. In Nigeria, and developing countries in general, food transport can occur in different forms:

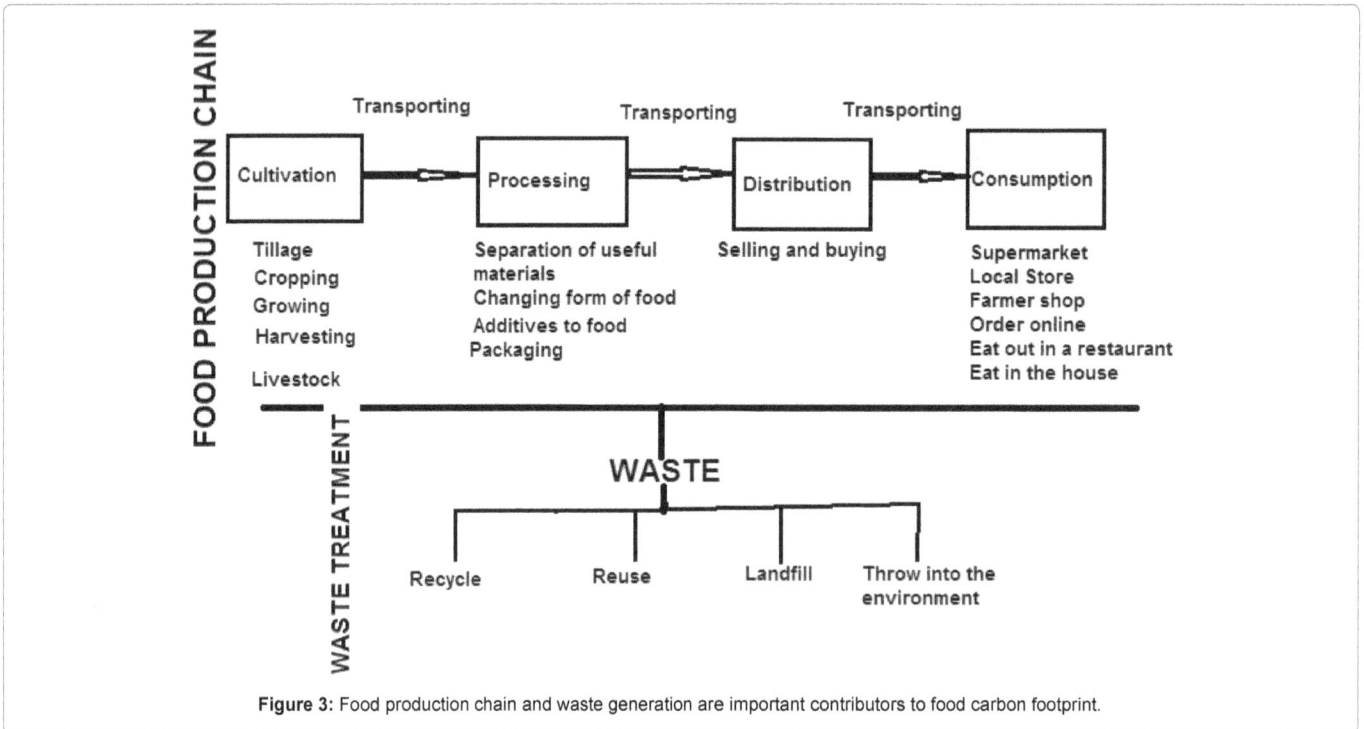

Figure 3: Food production chain and waste generation are important contributors to food carbon footprint.

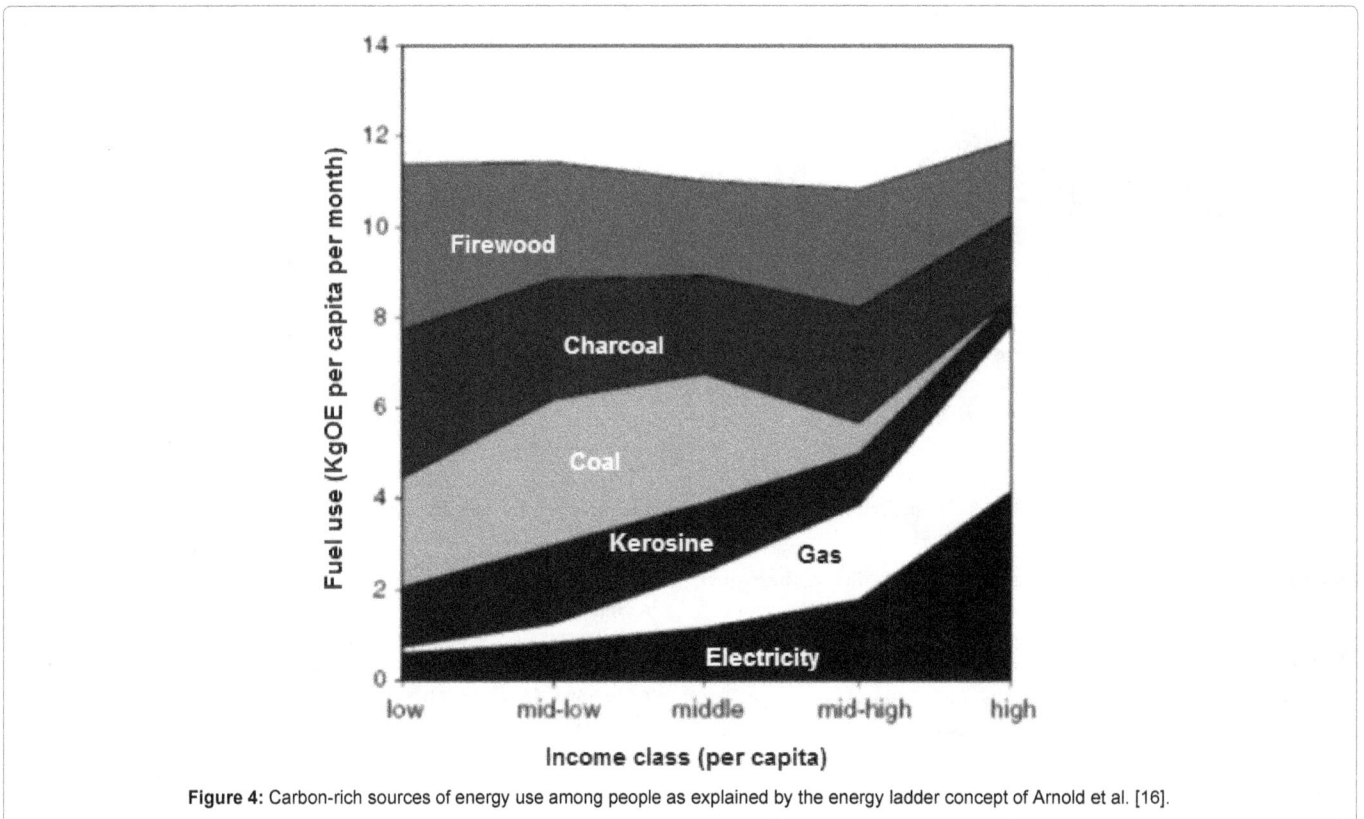

Figure 4: Carbon-rich sources of energy use among people as explained by the energy ladder concept of Arnold et al. [16].

(1) Food products can be sold in open rural food markets. This often involves sales of raw, unpackaged food, and are usually in the rural areas. It also involves sale of vegetable, meat (usually bush-meat such as squirrels, rabbits, grasscutters) and fruits. Food items offered for sale in these markets are usually degradable items because of poor energy and technology for preservation. The markets are often organized daily or four to five days intervals. Unsold degradable items are often left in the market or thrown away to rot in the environment, where they release their carbon contents.

(2) Larger and more organized markets are transacted at longer day interval, weekly or fortnightly, and are often located within or close to urban settlements. Processed, packaged (local and imported) food stuffs are often sold in these urban markets. Unlike the rural markets, these markets are usually provided with basic amenities, including toilets and water supply, although studies [20] have shown that the infrastructure are often dysfunctional. This kind of market generates a large amount of waste, emerging from sale remnant or packaging.

(3) 'Sabo' is characteristically northern Nigerians food market structure in other parts of Nigeria (south, west and east), where food crops (such as vegetable, yam, potatoes) and livestock are offered for sale. Food and animal products into Sabo markets in the southern Nigeria from the northern parts often involve a long journey in diesel or petrol-powered-trucks.

Food consumption

Food type and rate of consumption in Africa vary with location, socio-economic status, believe and interaction of individual or community with the western culture. Due to poor food technology, most food and fruits are seasonally consumed. Based on location, rural dwellers generally tend to consume less processed (and more of natural) food than the urban dwellers, and some fruits are more consumed in a season than another season (large quantities of fruits are wasted immediately after their harvest time in the wet season, than in the dry season) because of poor technology for preservation. Electricity supply in most rural areas in Nigeria is low, and almost nonexistence [20].

Most people in the high socio-economic status in Africa often prefer to live in the urban areas, and a large proportion of these people consume significant amount of processed foods, such as noodles and processed drinks. Fast food joints such as 'Mr Biggs', 'Chicken Republic', 'Captain Cook', etc. are major characteristics of the Nigerian urban area, and the frequency of visit to these 'joints' is often directly related to the socio-economic status of an average urban Nigerian. Visitors to these places also often do so in their cars because typical Nigerians' socio-economic status is often considered as equal to the number of cars they flaunt.

Waste

Food wastes in Nigeria and most developing countries is a major source of carbon to the environment. From cultivation or rearing of livestock, to the period of consumption, wastes are generated. The lifestyle of most Nigerians is also a contributing factor in the amount of wastes generated. Processed foods in many capital cities in Nigeria, especially in the Federal Capital City, Abuja and among the very affluent people often contradicts the law of demand (the higher the price, the lower the demand), as consumers tend to consume more at higher price because of the general perception that such attitude can show their economic 'status'. Unlike many developed countries where unused household materials can be donated to Gift and charity shops or packaged for recycle, most unused materials in Nigeria are thrown away without proper management. Most existing landfills are also rarely protected [21], and they are therefore sources of environmental pollution, releasing CO_2 and methane. Poor electricity power supply, very low re-use, preservation and recycle innovations, poor or non-existent charity shops are major encouragers of food waste in Nigeria, and these often apply to most developing countries.

Conclusion and Recommendations

The main objective of this study was to examine the sources and dimensions of food carbon footprints, and made specific reference to Nigeria as a typical example of most developing countries. The study indicated that large sources of carbon occurred at the various stages of agriculture production and consumption, and that climate education, peoples' lifestyle, technology and approach to waste management are main determinants of the extent of carbon footprints. The study concluded that although food security is a major problem in developing countries, variations in the rural-urban and poor-rich dichotomies in infrastructure and lifestyle or tastes are important challenges facing the carbon footprint reduction in Nigeria. The study recommends change in the feeding habit of people from carbon-rich foods to less of carbon rich foods. It also recommends processes or innovations for reducing food wastages.

References

1. Le Treut H, Somerville R, Cubasch U, Ding Y, Mauritzen C et al. (2007) Historical overview of climate change. Earth 1: 93-127.

2. Boko M, Niang I, Nyong A, Vogel C, Githeko A et al. (2007) Climate change 2007: impacts, adaptation and vulnerability. In: Parry ML et al. (eds.), Contribution of working group II to the fourth assessment report of the intergovernmental panel on climate change. University Press, Cambridge, UK, pp. 433-467.

3. Soil Association (2009) Soil carbon and organic farming, Scotland, pp. 212.

4. Thies PR, Flinn J, Smith GH (2009) Is it a showstopper? Reliability assessment and criticality analysis for Wave Energy Converters, Proceedings of the 8th European Wave and Tidal Energy Conference, Uppsala, Sweden.

5. Thies PR, Johanning L, Smith GH (2011) Towards component reliability testing for marine energy converters. Ocean Engineering 38: 360-370.

6. Convery FJ, Redmond L (2007) Market and price developments in the European Union emissions trading scheme. Review of Environmental Economics and Policy 1: 88-111.

7. United Nations Statistics Division (2013) World Statistics Pocketbook, Country Profile, Nigeria.

8. Ileoje NP (2001) A New Geography of Nigeria. New Revised Edition, Longman Publishers: Ibadan, Nigeria.

9. Eludoyin OM, Adelekan IO, Webster R, Eludoyin AO (2014) Air temperature, relative humidity, climate regionalization and thermal comfort of Nigeria. International Journal of Climatology 34: 2000-2018.

10. Schulze ED, Luyssaert S, Ciais P, Freibauer A, Janssens IA (2009) Importance of methane and nitrous oxide for Europe's terrestrial greenhouse-gas balance. Nature Geoscience 2: 842-850.

11. Schrumpf M, Schulze ED, Kaiser K, Schumacher J (2011) How accurately can soil organic carbon stocks and stock changes be quantified by soil inventories? Biogeosciences 8: 1193-1212.

12. Ekanem EO (1998) The street food trade in Africa: safety and socio-environmental issues. Food Control 9: 211-215.

13. Gustafon D (2001) The role of wood fuels in Africa. In: N Wamukonye (ed.), Proceeding of a high level regional meeting on energy and sustainable development. Food and Agriculture Organization of the United Nations, Rome, pp. 99-101.

14. Girard P (2002) Charcoal production and use in Africa: What Future? Unasylva 211: 30-34.

15. Ogunsawa OY, Ajala OO (2002) Firewood crises in Lagos- implication on the suburban and rural ecosystem management. In: Abu JE, Oni PO and Popoola L (eds.), Proceeding of the 28th annual conference of Forestry Association of Nigeria at Akure, Ondo State, pp. 257-264.

16. Arnold MJE, Kohlin G, Persson R (2006) Woodfuels, Livelihoods, and Policy Interventions: Changing Perspectives. World Dev 34: 596-611.

17. Zaku SG, Kabir A, Tukur AA, Jimento IG (2013) Wood fuel consumption in Nigeria and the energy ladder: A review of fuel wood use in Kaduna State. Journal of Petroleum Technology and Alternative Fuels 4: 85-89.

18. The Solar Cooking Archive (2011) Fuel wood as percentage of energy consumption in developing countries.

19. National Bureau of Statistics (2007) Annual abstracts of statistics, Abuja, Nigeria.

20. Allen AA (2013) Population dynamics and infrastructure: meeting the millennium development goals in Ondo State, Nigeria. African Population Studies 27: 224-229.

21. Oyeku OT, Eludoyin AO (2010) Vulnerability of groundwater resources to heavy metal contamination in a Nigerian urban settlement. African Journal of Environmental Science and Technology 4: 201-204.

17

The Role of Tibetan Plateau Snow Cover in the 1978 and 2001 Western North Pacific Typhoon Seasons

Yan T*, Pietrafesa LJ, Gayes PT and Bao S

School of Coastal & Marine Systems Science, Coastal Carolina University, Conway, SC 29528, USA

Abstract

An inverse correlation has been identified between the annual number of landfall typhoons along East China Seaboard (ECS) and the Tibetan Plateau snow cover (TP-SC) during the preceding winter and current spring. This correlation suggests that the Tibetan Plateau snow plays a key role in the East Asia–West Pacific regional climate system, and the system further influences typhoon track pattern off ECS. In this paper, major climatic factors accounted for the startling contrast in the number of landfall typhoons along ECS during the 1978 (2) and 2001 (7) western North Pacific (WNP) typhoon seasons were investigated. Among other climate/ocean conditions associated with typhoon activity in WNP, the preceding winter and spring TP-SC plays a crucial role to modulate the ECS landfall frequency in 1978 and 2001.

Keywords: Typhoon; Islands; Snow; Subtropical

Introduction

The Tibetan Plateau acts as a heat sink in winter and a heat source in summer. The heating effect of Tibetan Plateau plays an important role in the intensity and progress of East Asian monsoon. For example, anomalously extensive snow cover over the Tibetan Plateau during the preceding winter and current spring causes a slow progress of East Asian monsoon and a weak summer monsoon by reducing the heating over the Plateau. Correspondingly, the subtropical high over the western Pacific is intensive, but is located to the south of its normal position. There was a tendency for the WNP typhoon season to be less (more) active when the snow cover (SC) over the Tibetan Plateau (TP) was above (below) normal in the preceding winter and current spring [1]. Similarly, fewer (more) typhoons made landfall in China (including Hainan and Taiwan Islands) when the preceding winter had an above (below) normal snow cover on the TP. It also appears that heavy TP snow cover not only suppressed the overall WNP typhoon activity, but also delayed the onset of typhoon landfall in China. Historical records show that, there was, in general, a negative correlation between the number of landfall typhoons in China and the TP-SC in the preceding winter and spring [2]. The regions with the most significant correlation appear to be near 103°E over the Eastern TP throughout the winter and near 92°–95°E and 80°E during the spring.

Inter annual variability of WNP typhoon activity is also known to be influenced by anomalous sea surface temperature (SST) and other climatic conditions [1,3-8]. Active WNP typhoon seasons are often associated with above-normal local SST and vice versa [5]. El Niño and Southern Oscillation (ENSO) events are known to strongly modulate the annual hurricane frequency in North Atlantic [9], but the influence of ENSO on the WNP typhoon activity shows a more complex picture. The tropical cyclogenesis in the WNP basin as a whole does not show a significant dependence on ENSO [10-12]. However, there is a strong ENSO signal at sub-basin scales [13]. In El Niño years, fewer tropical cyclones form to the west of 160°E, while more form in the region between 160°E and the dateline. The opposite was found to occur during periods of La Niña. Previous study showed that the WNP TC activities from July to December are noticeably predictable using preceding winter–spring Niño3.4 SST anomalies, while the TC formation from March to July is exceedingly predictable using preceding October–December Niño3.4 SST anomalies [13].

Data

The WNP typhoon data is from National Climatic Data Center (NCDC) of the United States National Oceanic and Atmospheric Administration (NOAA), which contains track and intensity of WNP tropical cyclones since 1945. For each storm, the data contains 6-hourly (0000, 0600, 1200, 1800 UTC) center locations (latitude and longitude in tenths of degrees), intensities (maximum 1-minute surface wind speeds in knots and minimum central pressures in hPa). Typhoon landfall along ECS is defined as a TC with a maximum sustained wind speed of at least 32 m/s as crossing of the storm center over the coastline.

Snow cover data is derived from the satellite-estimated percentage monthly snow cover within each 1° × 1° grid cell on the Tibetan Plateau from 1976 to 2012.In this study, 1° × 1° resolution snow cover data for the region (20°–40°N, 75°–115°E) covering the entire TP for the period of 1976–2012 is utilized. Geo potential height fields from 1950-2014 are derived from the National Center for Environmental Prediction (NCEP)/National Center for Atmospheric Research (NCAR) reanalysis data, archived at NCDC.

Other data sources used in this study include:

1) Japan Meteorological Agency (JMA) ENSO

2) SST from NCEP/NCEP Reanalysis Project

3) Trade wind index: It is the averaged daily 850 hPa wind anomalies in the tropical western Pacific (135°E-180°W, 5°N-5°S), Note that positive values of the 850 hPa zonal wind indices imply easterly anomalies.

4) QBO indices: computed from the zonal average of the 30mb

---

*Corresponding author: Yan T, School of Coastal & Marine Systems Science, Coastal Carolina University, Conway, USA, E-mail: tyan@coastal.edu

zonal wind at the equator as computed from the NCEP/NCAR Reanalysis.

The 1978 and 2001 North Western Pacific (WNP) Typhoon Seasons

Follows [2], the Pearson correlation coefficient between the ECS landfall count and the spring TP-SC is recalculated using the 37-year data from 1976 to 2012, which yields a value of -0.47 (p<0.005), snow cover index is delivered from domain: (92° – 95°E, 34°-38°N). The negative correlation keeps significant. The 1978 and 2001 WNP typhoon seasons showed a startling contrast in the number of typhoons that made landfall along East China Coastline (ECS) (including landfalls on Taiwan and Hainan Islands) (Figure 1). Only two typhoons made landfalls in China in 1978, 47% below the 55-year (1950-2004) climatological average of 3.8. However, seven typhoons made landfalls in 2001 almost doubled the climatological mean. It is evident that the difference between 1978 and 2001 was mainly in the intensity of the tropical cyclone. The number of tropical cyclones including tropical storms and hurricane-intensity typhoons was 32 for 1978 and 33 for 2001, showing no significant difference. However in 1978 Figure 1 only 47% (15) of the total 32 TCs reached typhoon intensity, below the 55-year (1950-2004) average of 64% (17.5 typhoons). Only three typhoons reached category 3 or higher comparing with a climatological mean of 9.5. Only one storm was able to become a super typhoon (category 5) comparing with a climatological mean of 2.7. In contrast, in 2001, 21 of the 33 (61% of the total) TCs reached typhoon intensity, and over half (11) reached category 3 or higher, with 2 became super typhoons.

It is important to note the difference in the spatial and temporal distribution of the TCs between the 1978 and 2001 typhoon seasons. All but only one typhoon in 1978 developed from July to October took a northwestward track and curved northeastward when approaching China coast. During the whole season only two typhoons (#14 in August and #21 in September) made landfall on ECS. In contrast, twenty typhoons developed in a seven month period in 2001, with a total of seven made landfalls. It is also notable that six typhoons developed in early summer (June and July), and five of these six typhoons made landfalls on ECS. A question is raised: what caused such a difference in the early season? The potential causes for the differences between 1978 and 2001 are discussed below.

ENSO and regional SST anomalies

ENSO in 1978 and 2001 seasons both displayed a neutral phase. There is no evidence suggesting that neutral ENSO events and the western North Pacific regional SST pattern are significantly correlated [10-11]. Thus, the differences between the typhoon activity during the 1978 and the 2001 typhoon seasons could not be attributed to the influence of ENSO. Local SST anomalies in the WNP are known to affect the tropical cyclogenesis in WNP [13]. A significant feature of SST distribution in the first half of 1978 was the presence of a persistent below normal SST field centered around 20°N and extended west of (Figure 2) 130°E (Figure 2a). This below normal SST in the western part of the tropical North Pacific south of 22°N continued into the peak typhoon season from July to October 1978 Figure 2b.

In accordance with this SST anomaly pattern, the majority of the typhoons in 1978 formed north of 20°N, avoiding the cooler than normal water located to the south. In the first half and during the period of August-October, 2001, the WNP was featured by an above normal SST while the eastern North Pacific (ENP) was featured by a below normal SST Figure 3a,3b. Tropical cyclone mainly formed and

developed in above normal SST region (Figure 3). The differences in the anomalous SST distribution could be one possible cause to result in differences in typhoon activity between 1978 and 2001.

Snow cover on tibetan plateau (TP): (Figure 4) show an above normal SC in 1978 4 and a below average SC in 2001, respectively. The snow cover (January-March) over the Tibetan Plateau demonstrated a strong negative correlation with the number of typhoons landed on the ECS [3].

The TP-SC influences WNP TC track patterns primarily through the large-scale WNP-EASM circulation, which is associated with

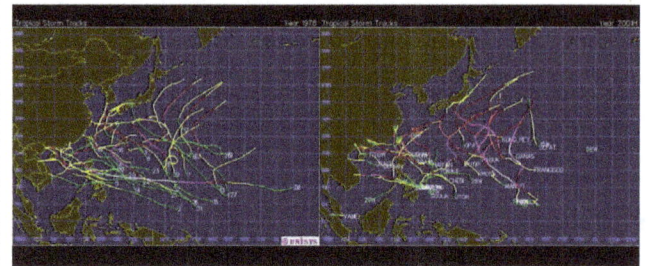

Figure 1: Typhoon tracks in 1978 and 2001 in WNP.

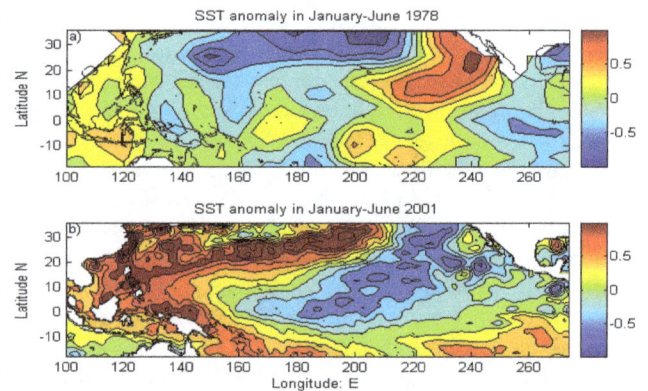

Figure 2: SST anomalies from January to June in: (a) 1978. (b) 2001 (Climatological base: 1971-2000).

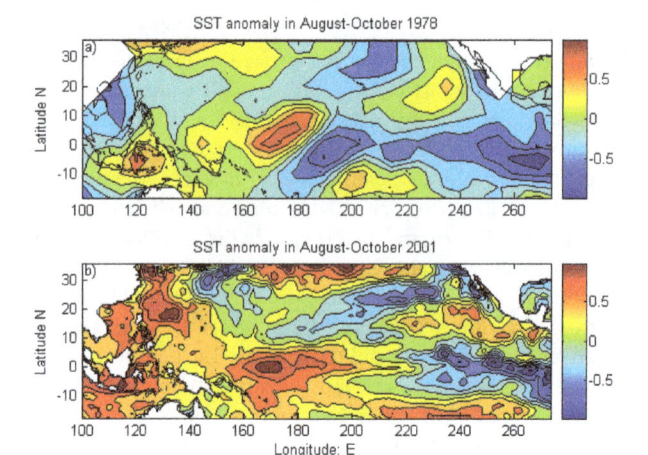

Figure 3: SST anomalies during August-October in: (a) 1978, and (b) 2001.

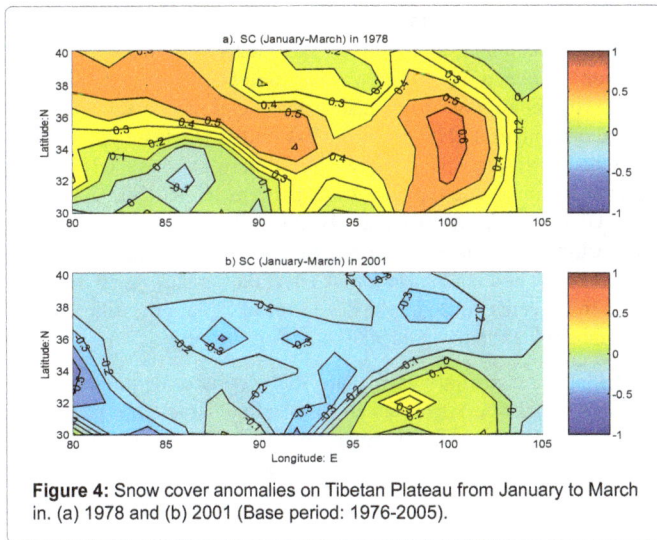

Figure 4: Snow cover anomalies on Tibetan Plateau from January to March in. (a) 1978 and (b) 2001 (Base period: 1976-2005).

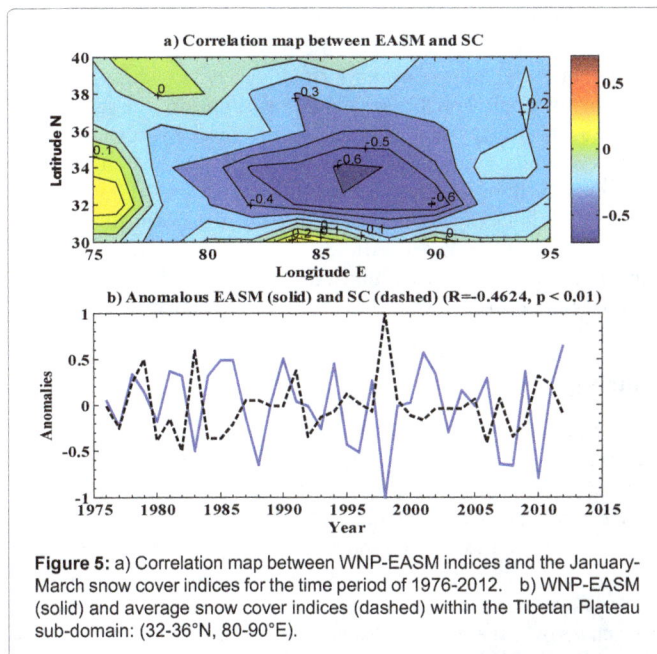

Figure 5: a) Correlation map between WNP-EASM indices and the January-March snow cover indices for the time period of 1976-2012. b) WNP-EASM (solid) and average snow cover indices (dashed) within the Tibetan Plateau sub-domain: (32-36°N, 80-90°E).

shift of the East Asian and WNP subtropical high (EASH) position [1,14,15]. High (low) TP-SC is associated with East Asian summer drought (flood) Figure 5a, 5b. This is especially true for 1978/2001 cases. High snow over the TP changed atmospheric circulation of land-atmosphere-ocean interaction and restricted the EASM from development, typhoon activity, as a consequence, was suppressed. Chan and Gray indicted that the mid-tropospheric 500 hPa flow is associated with TC steering flow [16]. Since the TC track associated with the large-scale atmospheric conditions vary with seasons, it is necessary to examine the variations in the 500 hPa flow in a monthly base (Figure 5).

The June-July 5880 m GHT contours at 500 hPa between 1978 and 2001 demonstrated a substantial distribution pattern Figure 6a, 6b The East Asian/western North Pacific subtropical High (EA/WNPSH) situated at a more southwesterly position in June-July 2001. This pattern is usually associated with low TP snow cover. The southwesterly displacement of EA/WNPSH (Figure 6) prevents typhoons from

making a northward curve; therefore typhoons took a more westerly track approaching the ECS. WNPSH position during the same period in 1978 displayed a significant northeastward shift, however, typhoons likely recurve northward instead of propagate further westerly. This condition prevents typhoons from making landfalls on the ECS. In July the 500 hPa GHT anomaly shows that tropical WNP east of the ECS dominated by a strong negative trough Figure 7a, 7c in 1978; however a southward displacement of subtropical high dominated that area Figure 7b, 7d in 2001. Five typhoons made landfalls in July 2001, however no typhoon made landfall in July 1978 though three typhoon developed during this time period. The above comparison demonstrated that monthly variability of typhoon tracks are strongly tied to GHT distribution pattern (Figure 7).

850hPa trade wind indices (TWI): The 850hPa TWI was identified significantly associated with the WNP typhoon activity. Westerly (easterly) phase of trade wind usually corresponds to strong (weak) TC activity. Trade winds in the peak typhoon season (July-October) in 1978 and 2001 were both in an easterly phase (0.2631, 0.4750), which is generally suppressed typhoon development. TWI was, therefore, not a key factor to influence the overall TC activity in 2001. However it might be one of the factors to suppress typhoon activity in 1978.

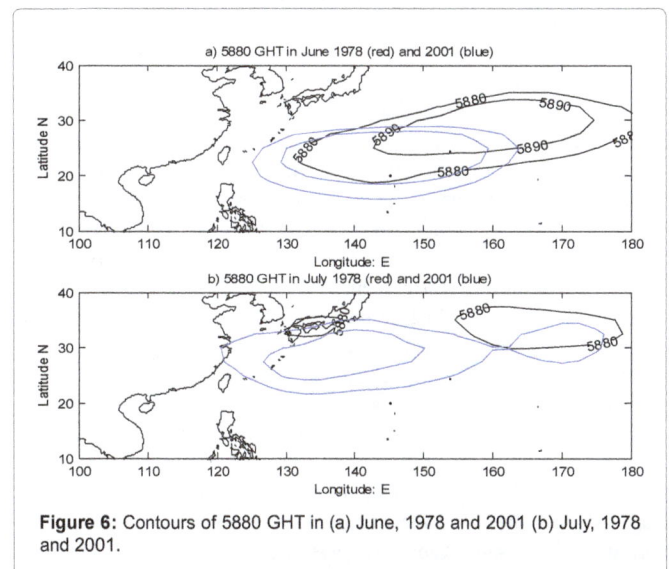

Figure 6: Contours of 5880 GHT in (a) June, 1978 and 2001 (b) July, 1978 and 2001.

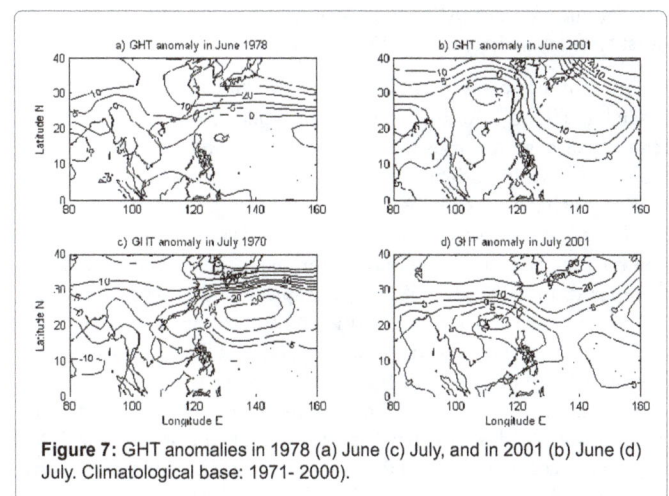

Figure 7: GHT anomalies in 1978 (a) June (c) July, and in 2001 (b) June (d) July. Climatological base: 1971- 2000).

The stratospheric quasi-biennial oscillation (QBO): Easterly QBO from January to February demonstrates a strong association with increased TC activity in the WNP [1]. This correlation is especially robust during the neutral ENSO years. The QBO during the period from January to February in 1978 demonstrated a strong westerly phase, which implied an inactive typhoon season. However QBO during January-February in 2001 displayed a strong easterly phase, which signaled a favorable condition for TC formation. Therefore, QBO may attribute to the overall typhoon activity in those years.

Arctic oscillation (AO): AO indices demonstrated a negative phase (-0.52) in 1978 and a positive phase (0.48) in 2001. As discussed in Yan [1], the positive phase of AO in March is associated with more TC tracks accumulated in eastern portion of the WNP. The majority of typhoon tracks in 1978 stayed off the ECS. The positive AO phase in 2001 is favorable for typhoons to propagate with a more westerly track. However, typhoon tracks east of 130°E may be dominated by other climate factors.

The 2001 WNP typhoon season featured seven landfall typhoons on ECS (climatological average is 3.8). This is one of the largest numbers of landfall typhoons observed in any typhoon seasons (June-December) since 1961, and the twenty typhoons observed were 2.5 times the long-term average. Of these seven ECS landfall typhoons, five occurred in June and July - the most observed during this period in the WNP typhoon history. Typhoon tracks in 2001 split into two clusters: one occupied west of 130°E and the other accumulated east of 130°E. The majority of the east group typhoons are typhoons of category-3 or higher, all those intense typhoons developed from beginning of August while all except one of the west group typhoons were typhoons of category-1 or category-2, seven out of the total nine west group typhoons landed on ECS, five of those total six June-July west group typhoons and all the two September west group typhoons made landfalls on ECS. The east group typhoons demonstrated a strong curving track pattern, which resulted from a systematic southeastward displacement of the mid-tropospheric subtropical low positioned east of Japan during August and September. The west group typhoon track patterns in June and July were associated with a southeasterly extension of the subtropical ridge in the WNP, which strongly linked to the low TP snow cover in early of the year and the active East Asian Summer Monsoon. The development of southerly flow to the west of the mean ridge axis over the WNP and a tropical low centered on (115°E, 20°N) dominated the South China Sea west of Taiwan created a perfect steering condition for typhoons to approach ECS. This process is more evident during the early typhoon season, consistent with the observed burst typhoon activity in July 2001.

In contrast to seven landfall events in 2001, only two typhoons made landfall on ECS during the entire typhoon season of 1978. The majority of named storms that formed in the tropical WNP exhibited a strong curving tendency when passing by northeast Taiwan. The majority of typhoons in south of Taiwan that propagated along the northwestward tracks weakened dramatically as they approached the South China Sea. Super typhoon #9 formed in July was an exception. Compared to 2001, the WNP subtropical high during June-July in 1978 retreated to a more northeasterly position compared to its counterpart in 2001. This distribution pattern is likely a cause for the northward curving typhoon tracks, which prevented typhoons from making landfall along ECS. The displacement of subtropical high position surrounding the ECS has been identified associated with the high snow cover over the TP through large-scale atmospheric flow fluctuation that greatly influenced by the East Asian Monsoon system

[1,3] [14-15]. The variability of local SST distribution and the upper-level winds that tend to steer tropical cyclone movement and inhibit its development over a region may vary without the influence of El Niño. This is why the number of landfall tropical cyclones formed over the WNP during neutral ENSO years varies dramatically from year to year. Combined with local SST patterns, the TP snow cover is likely the key factor to dominate the typhoon track patterns in 1978 and 2001. The sparse landfall events in 1978 may also be attributed to the negative AO phase, which is associated with more easterly tracks. The neutral ENSO events in 1978 and 2001 have minor effect on the East Asian Summer monsoon; it seems not to be the key factor to explain the difference in the number of landfall events between those two years.

Perhaps the primary factor for the increased typhoon activity during 2001 can be attributed to a favorable, large-scale pattern of extremely warm sea surface temperature (SST) distributed throughout the WNP. This above normal SST reflects an abundant thermal energy condition that is favorable for typhoon formation and development and also associated with an abundant rainfall in East Asian and the WNP. Observations of typhoon intensity and tracks coincided well with SST distribution pattern. Compared to 1978, the warm SST in 2001 was well developed prior to the typhoon season and persisted throughout the typhoon season, which ultimately provided a favorable condition for tropical cyclogenesis and intensification. The abnormally warm SST also helps to lower surface pressures hydrostatically by directly warming the lower troposphere. The reduce meridional pressure gradient of modified SLP acts to reduce low level trade winds, thereby contributes to a further warming of the ocean. In addition to the large-scale abnormal warm SST, the enhanced WNP typhoon activity in 2001 has also been linked to the negative phase of the QBO in January and February which intensified the TC activity and favorable for cyclogensis.

Conclusions

Analysis on the 1978/2001 WNP typhoon seasons supports the discovery of a strong negative correlation between the extent of snow cover in the preceding winter (DJF) and spring (MAM) on the Tibetan Plateau and the annual landfall frequency on ECS. This inverse correlation can be explained by the response of the WPSH to the snow-modulated land surface thermodynamic processes over the Tibetan Plateau and the anomalous SST in the WNP. Increased winter snow cover over the Tibetan Plateau leads to reduced heating during the melting period in the spring and summer, which is followed by a weak summer monsoon and a weak WPSH, which, in turn, leads to a reduced number of landfall typhoons along ECS, particularly during the early summer (May-July). Comparison of major climatic factors in 1978 and 2001 showed that the difference in the ECS landfall numbers between 1978 and 2001 typhoon seasons may result from TP-SC and its associated climatic conditions.

The observed seasonal lag correlation between TP-SC and landfall typhoon frequency on ECS is -0.47 based on 37-year period of snow and typhoon data (1976-2012), which suggests that ECS landfall typhoon can be predicted using the amount of winter and/or spring TP-SC as one of the primary predictors.

Acknowledgements

This study is supported by School of Coastal & Marine Systems Science, Coast Carolina University.

References

1. Yan, Tingzhuang (2006) Interannual variability of climatology and tropical

cyclone tracks in North Atlantic and Western North Pacific. Ph.D. Dissertation, North Carolina State University.

2. Xie L, Yan T, Pietrafesa LJ (2005) Relationship between Western North Pacific typhoon activity and Tibetan Plateau winter and spring snow cover. Geophysical Research Letters 32: 16703.

3. Xie L, Yan T (2007) North Pacific typhoon track patterns and their potential connection to Tibetan Plateau snow cover. Natural hazards.

4. Liu KS, Chan JCL (2003) Climatological characteristics and seasonal forecasting of tropical cyclones making landfall along the South China coast. Mon. Wea. Rev. 131:1650-1662.

5. Chen TC, Weng SP, Yamazaki N, Kiehne S (1998) Interannual variation in the tropical cyclone formation over the western North Pacific. Mon. Wea. Rev. 126: 1080-1090.

6. Clark MP, Sereze MC (2000) Effects of variations in East Asia snow cover on modulating atmospheric circulation over the North Pacific Ocean. J. of Climate 13: 3700-3710.

7. Chen L, Xu X, Luo Z, Wang J (2002) Introduction to tropical cyclone dynamics. China Meteorological Press, Beijing 319.

8. Xu X, Zhou M, Chen J (2002) A comprehensive physical pattern of land-air dynamic and thermal structure on the Qinghai-Xizang plateau. Science in China 45: 577-594.

9. Gray WM (1984) Atlantic seasonal hurricane frequency: Part I. El Nino and 30 mb quasi-biennial oscillation influences. Mon. Wea. Rev 112: 1649-1668.s

10. Chan JCL (1985) Tropical cyclone activity in the northwest Pacific in relation to the El Niño/Southern Oscillation phenomenon. Mon. Wea. Rev 113: 599-606.

11. Chan JCL (2000) Tropical cyclone activity over the west North Pacific associated with El Niño and La Niña Events. J. of Climate 13: 2960-2972.

12. Lander MA (1994) An exploratory analysis of the relationship between tropical storm formation in the western north Pacific and ENSO. Mon. Wea. Rev 122: 636-651.

13. Wang B, Chan JCL (2002) How Strong ENSO Events Affect Tropical Storm Activity over the Western North Pacific. J. of Climate 15: 1643-1658.

14. Kripalani RH, Kulkarni A, Sabade SS (2003) Western Himalayan snow cover and Indian monsoon rainfall: A re-examination with INSAT and NCEP/NCAR data. Theor. Appl. Climatol. 74: 1-18.

15. Zhang S, Tao S (2001) A diagnostic and modelling study of the effect of Tibetan Plateau snow cover on the Asian summer monsoon. Chinese Journal of Atmospheric Sciences 25: 372-390.

18

Evaluation of Future East Asia Drought Using Multi-Model Ensemble

Jae-Won Choi*, Yumi Cha and Jeoung-Yun Kim

National Institute of Meteorological Sciences, 33, Seohobuk-ro, Jeju 63568, Korea,

*Corresponding author: Jae-Won Choi, National Institute of Meteorological Sciences, 33, Seohobuk-ro, Jeju 63568, Korea
E-mail: choikiseon@daum.net

Abstract

We analyzed the changes in precipitation and drought climatology over East Asia by global warming using the daily precipitation data from 14 coupled atmosphere-ocean general circulation model simulations under the Special Report on Emission Scenarios (SRES) A1B scenario at the end of the twenty-first century. The models were consistent in predicting an increase in the mean precipitation over East Asia. However, the increase was less significant in Southeast Asia, and was accompanied by even larger increase in precipitation variability. This predicted precipitation climatology was translated into a change in drought climatology using the effective drought index (EDI). According to the increased precipitation, East Asia tends to be wetter with a decreased frequency and duration of drought. However, because of the enhanced precipitation variability, extreme droughts are predicted to be more frequent, especially over Southeast Asia.

Keywords: Precipitation; Drought; East Asia; Global warming; Effective drought index

Introduction

A report from the Intergovernmental Panel on Climate Change [1] indicated that East Asian water resources are threatened by an enhanced variability in the precipitation under global warming. However, few studies have tried to estimate quantitatively the hydrological disasters that we should expect. This study has an interest on drought which caused by precipitation deficits over a prolonged period.

Several modeling studies have shown that over Asian monsoon regions, increases in greenhouse gas concentrations lead not only to an increase in mean precipitation but also to a significant enhancement in precipitation variability on sub-seasonal to inter-annual timescales [2-4]. The significance of these findings was verified by recent studies using the Multi-Model Ensemble (MME) method [3,5,6]. The results showed that the frequency of non-precipitation increases in a way that is similar to the frequency of heavy rainfall [3]. Monsoon excesses and deficiencies are also projected to intensify [5]. However, it remains to be understood how drought patterns are affected by the enhanced variability of precipitation.

Some of the studies on future drought were interested in the global scale dryness revealed by changes in soil moisture conditions [7-9] or Palmer Drought Severity Index [10,11]. However, the predictions for the magnitude and extent of dryness in East Asia are considerably different in each study. For example, Manabe et al. [8] predicted dryer conditions while Burk et al. [10] predicted wetter conditions. One reason for these different results could be the use of different Global Circulation Models (GCMs). Due to the strong model dependence of the hydrological response to a greenhouse gas increase, different models may predict changes with different signs, even for the same region and the same variable. Moreover, climate sensitivity also differs substantially among models. The MME averaging approach can be very useful in reducing the uncertainties related to model dependence.

Although this method is widely used to investigate the future climate, only a single study [12] has attempted to evaluate the likelihood of future drought. That study used 15 state-of-the-art GCMs. Furthermore, there is another limitation. Most of them derived their results by comparing the climatologically averaged values of present-day and future. This simple comparison can only measure the climatological dryness and/or wetness but fail to catch the actual drought change, which is an extreme natural phenomenon with very irregular time scales.

This study examined the impact of greenhouse gas warming on East Asian drought by comparing the projected climate (2081-2100) in the Special Report on Emission Scenarios (SRES) A1B experiment with the present-day control climate (1981-2000). The projected daily precipitation data were translated into drought climatology by using the effective drought index (EDI) [13], which quantifies the drought intensity in daily time steps. We used the MME average from 14 GCMs and assessed its roughness.

Models and Methods

Models

The 14 GCMs used this study are a part of the IPCC's data archives at the Lawrence Livermore National Laboratory. All of the models begin their integration from the "20th Century Climate in Coupled Model" run, in which the level of anthropogenic forcing is based on historical data from the late 19th century through the 20th century. From the end of the 20C3M run, SRES A1B conditions were imposed and integrated through the year 2100. The SRES A1B assumes rapid economic and population growths that peak mid-century and decline thereafter. Two time periods of twenty years each were chosen for analysis: the late 20th century (1981–2000; hereafter 20C3M) and the late 21st century (2081-2100; hereafter A1B). The analysis based on the 14 GCMs and their MME average (average of 14 GCMs). The following model data were used in this study: CCSM3, CGCM3.1 (T47), CGCM3.1 (T63), CNRM-CM3, CSIRO-Mk3.0, ECHAM5/MPI-

OM, FGOALS-g1.0, GFDL-CM2.0, GFDL-CM2.1, GISS-AOM, INM-CM3.0, MIROC3.2 (hires), MIROC3.2 (medres), and MRI-CGCM2.3.2. Model characteristics for all components and other details are available at http://www.pcmdi.llnl.gov/ipcc/model_documentation/. In addition, characteristics on advantages or disadvantages of these models can be found some previous studies [7,14,15].

In addition to the differences in the parameterization of the physical and dynamical processes, the models also differ from each other in their spatial resolution. The resolutions of the 14 models range from coarse (e.g., $4° \times 5°$ for the GISS-AOM) to fine (e.g., $1.4° \times 1.4°$ in CCSM3). To obtain the MME pattern, the original model outputs were converted to the same resolution ($2.5°$ longitude/latitude) by employing the bi-linear interpolation technique. Several studies have demonstrated that these models are capable of reproducing the temporal and spatial features of the East Asian precipitation climate [7,14,15].

Effective Drought Index (EDI)

The EDI was applied to measure the drought. Unlike many other drought indices, the EDI is calculated with a daily time step.

$$EP_i = \sum_{n=1}^{i} \left[\left(\sum_{m=1}^{n} P_m \right) \middle/ n \right] \qquad (1)$$

$$DEP = EP - MEP$$

$$EDI = DEP/ST(DEP)$$

Where P_m is the precipitation m days before and the index i represents the duration of summations in days. Here i=365 is used; that is, the summation is equal to a year, which is the most dominant precipitation cycle worldwide. EP is the summed value of daily precipitation with a time dependant reduction function. DEP represents the deviation of EP from MEP (30-year average EP for the calendar date). ST (DEP) denotes the standard deviation of each day's DEP. EDI expresses the standardized deficit or surplus of stored water on a daily basis. It enables one location's drought severity to be compared to that of another location, regardless of climatic differences. The "drought range" of EDI indicates extreme drought at EDI<-2.5, severe drought at -1.5>EDI>-2.49, and moderate drought at -0.7>EDI>-1.49. Near normal conditions are indicated by 0.69>EDI>-0.69. The use of EDI has been tested in several drought studies [16-20]. When using EDI to explore the changes in drought as a result of future climate scenarios, the calibration factors were set at present-day values.

Results

Future precipitation climatology

Figure 1a shows the area-averaged ($10°-50°N$, $100°-140°E$) percentage changes (Eqn. 2) of the mean and the standard deviation of precipitation data set for 14 GCMs.

$$(A1B-20C3M)/20C3M \times 100 \qquad (2)$$

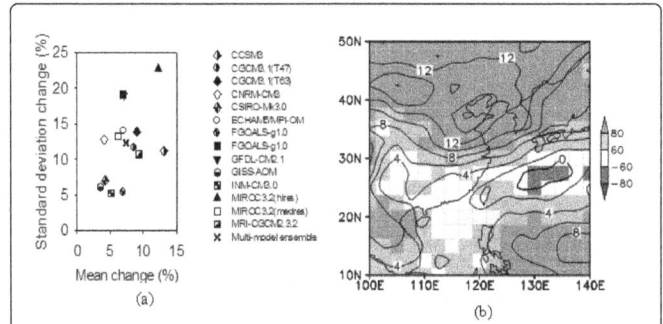

Figure 1: (a) Scatter plot of the area-averaged (East Asia; $10°-50°N$, $100°-140°E$) percentage change in the mean and the standard deviation of the precipitation data for 14 GCMs. The percentage change is defined as $100 \times (A1B-20C3M)/20C3M$. A multi-model ensemble value is denoted by the X symbol. (b) Multi-model ensemble percentage change in mean precipitation from the 20C3M to the A1B experiments. Shading denotes the consistency level (%) of the 14 models in predicting the direction of change in mean precipitation.

It is confirmed that the increase of precipitations in East Asia, mentioned in the introduction, is shown in all the 14 models used in this study. While the minimum increase is projected by GISS-AOM (3.56%), the maximum increase is projected by CCSM3 (13.06%). The increase of the MME average is 7.41%. The increases in the standard deviation vary from 5.26% (INM-CM3.0) to 22.73% (MIROC3.2 (hires)), and the MME average is 12.28%. In 12 out of 14 GCMs, the increase rate in the standard deviation is higher than that in the mean value. In the 4 GCMs (CNRM-CM3, MIROC3.2 (medres), GFDL-CM2.1, GFDL-CM2.0), the increase rate in the standard deviation is over two times that in the mean. These mean that the increase in changeability of precipitation is more distinct than the increase in the average precipitation.

Figure 1b shows the spatial distribution of the percentage changes of the MME average precipitation from the 20C3M to the A1B experiments. To investigate its robustness, consistency level among models is calculated (shading). Here the consistency is defined as a fraction of the number of models with either positive or negative change: That is, the value is +100% if all model have projected an increase, and is -100% if all models have projected a decrease in the future compared to the present. The absolute value of consistency level is always larger than or equal to 50%. In Northeast Asia above 30°N, the increase rate of precipitation is 8-12%, and the model consistency level is high. On the other hand, in the Southeast Asia below 30°N, the increase rate is relatively low (0-8%), and their model consistency level is low too.

Figure 2a shows the time-latitude cross-section of monthly mean precipitation averaged for $100°-140°E$ for the MME of 20C3M experiments. Figure 2b shows the MME percentage changes from the 20C3M to the A1B experiments and the corresponding consistency index (shading). The comparison of these two figures shows that the increase rate of precipitation is bigger as the latitude is higher and the season is colder. It is confirmed that the precipitation in the entire East Asia increases during summer, implying the strengthening of the East Asian summer monsoon. From the region around 30°N during winter to Southeast Asia during spring, the precipitation decreases a little.

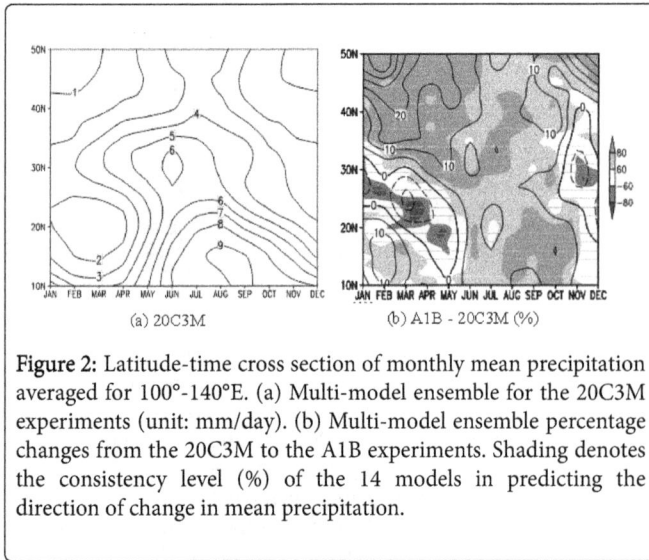

Figure 2: Latitude-time cross section of monthly mean precipitation averaged for 100°-140°E. (a) Multi-model ensemble for the 20C3M experiments (unit: mm/day). (b) Multi-model ensemble percentage changes from the 20C3M to the A1B experiments. Shading denotes the consistency level (%) of the 14 models in predicting the direction of change in mean precipitation.

Future drought climatology

Figure 3 shows the difference in the frequency distribution of the EDI values over East Asia for 14 GCMs between the A1B and 20C3M experiments. EDI, originally, is a standardized index, follows normal distribution with zero mean. The frequency decrease of the negative values which means dryness and the frequency increase of positive values which means wetness are definitely shown in Figure 3. Hence, the center of EDI value frequency distribution moved toward wetness in the A1B experiments. In other words, frequency of wetness increases according to the overall precipitation increase in East Asia, and the frequency of dryness decreases. However, there is seen a special feature here. That is the increase of frequency of extreme values: below -2 and over +2. This means that the hydrologic variability increases greatly in East Asia. That is, the frequencies of extreme flood increase at the same time, and the frequencies of extreme drought increase as well.

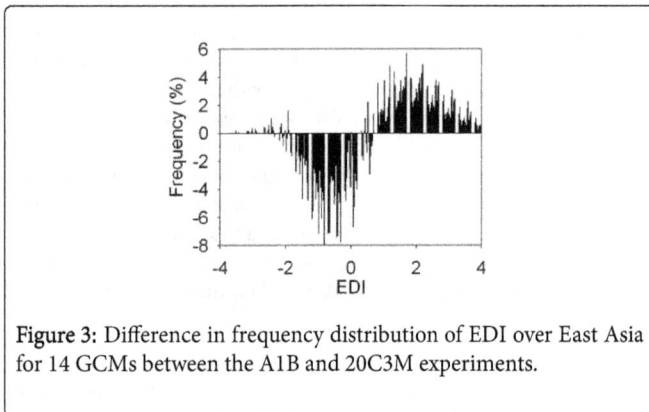

Figure 3: Difference in frequency distribution of EDI over East Asia for 14 GCMs between the A1B and 20C3M experiments.

Figure 4 shows the relationships between the intensity and duration of drought simulated by 14 GCMs in the 20C3M and the A1B experiments. The drought duration herein is the consecutive days of negative EDI, and the drought intensity, the minimum EDI during the duration. The regression coefficients (c) which show the relationship of the two variables are indicated in the bottom of each panel. The percentage change of the number of total drought events from the 20C3M to the A1B experiment is indicated in the top of the A1B panel.

As analyzed above, the frequency of drought decreases from 14.0% (GFDL-CM2.1) to 42.6% (CCSM3) according to the increased precipitation in East Asia. In 13 out of 14 models, on the other hand, the slope of the linear regression line is steeper in the A1B experiment than that was in the 20C3M experiment. This means that the droughts in the future have tendency of intensive precipitation lack although the frequency and duration of drought decrease.

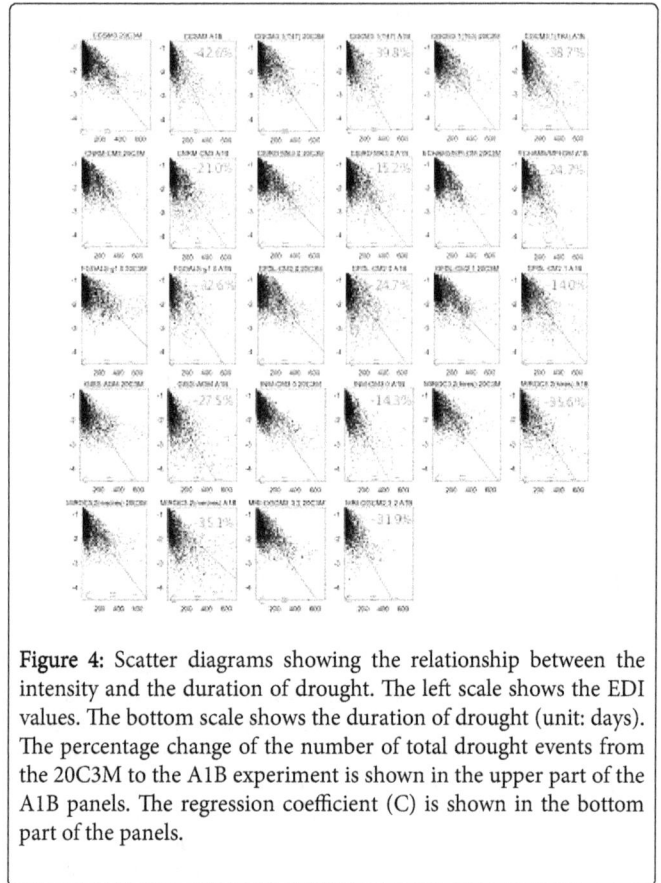

Figure 4: Scatter diagrams showing the relationship between the intensity and the duration of drought. The left scale shows the EDI values. The bottom scale shows the duration of drought (unit: days). The percentage change of the number of total drought events from the 20C3M to the A1B experiment is shown in the upper part of the A1B panels. The regression coefficient (C) is shown in the bottom part of the panels.

Drought intensity

Figure 5 displays the difference in MME total drought days (extreme<-2.5<severe<-1.5<moderate<-0.7) between the A1B and the 20C3M experiments. The panels in the left are the spatial distribution (days/decade), and the panels in the right show the latitude-time cross sections of monthly total drought days (days/decade) averaged for 100°-140°E. In the horizontal spatial distribution, moderate drought days decrease in the entire East Asia, corresponding to the increased mean precipitation. The magnitude of the decrease is large in the Northeast Asia where the model consistency exceeds 80%. Both the magnitude of the decrease and the model consistency level are relatively small in Southeast Asia. The drought days in Northeast Asia show a noticeable decrease during spring while in the Southeast Asia, the decrease is weak. At the latitude of 25°N, there is a weak increase of drought days in April and May.

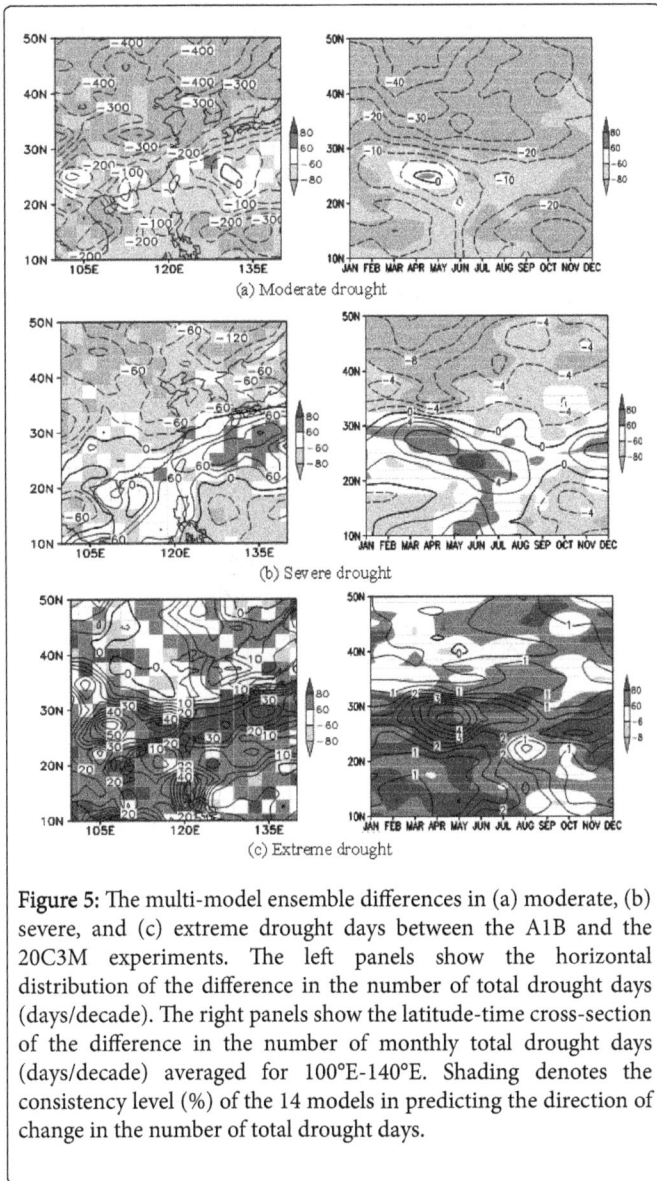

Figure 5: The multi-model ensemble differences in (a) moderate, (b) severe, and (c) extreme drought days between the A1B and the 20C3M experiments. The left panels show the horizontal distribution of the difference in the number of total drought days (days/decade). The right panels show the latitude-time cross-section of the difference in the number of monthly total drought days (days/decade) averaged for 100°E-140°E. Shading denotes the consistency level (%) of the 14 models in predicting the direction of change in the number of total drought days.

In the difference map of severe drought days, although, decreases in the drought days are still dominant over Northeast Asia, the model consistency levels become lower compared with that in the moderate case. In the some regions of the Southeast Asia, there is shown an increase of drought days, which are predicted by approximately half of the models. These increases are noteworthy in the areas of 25°-30°N during winter to spring, and in all Southeast Asia during summer.

Extreme drought days are predicted to increase in all Southeast Asia by the majority of the models. In Northeast Asia, the same reactions appear in some regions. Such an increase of drought days is shown in all seasons in Southeast Asia, especially nearby the regions of 25°-30°N in spring.

Summary and Conclusion

We analyzed the changes in precipitation and drought climatology over East Asia by global warming using daily precipitation data from 14 coupled atmosphere-ocean General Circulation Model (GCM) simulations under the SRES A1B scenario at the end of the twenty-first

century. The Effective Drought Index (EDI) was applied to measure the drought.

The increase in the mean precipitation was predicted in East Asia by the majority of the models. This is outstanding in Northeast Asia, and in Southeast Asia, the magnitudes of the increase and model consistency levels are relatively small. These increases in the mean precipitation are accompanied by a bigger increase in the precipitation variability. The seasonal precipitation increase is the biggest in Northeast Asia during winter, and shown commonly in East Asia during summer. From the regions around 30°N during winter to Southeast Asia during spring, there are weak precipitation decreases.

All GCMs predicted that the frequency of droughts decreases in East Asia according to the increased mean precipitation, and the frequency of wetness increases greatly. However, the frequency of extreme drought also projected to increase due to the greatly increased precipitation variability. That is, the frequency and duration of droughts showed a tendency of decrease, but the precipitation shortage is greater during the drought period. This is because water vapours at troposphere increase according to global warming.

We analyzed the spatial and seasonal changes in the three categories of drought according to the intensity (moderate, severe, and extreme drought). Moderate drought days are predicted to decrease in all East Asia except for regions of 25°-30°N during short period in spring. This weakening of drought is especially noteworthy in Northeast Asia during winter and spring when precipitation increases greatly. However, severe droughts show almost no change in Southeast Asia, but an increasing tendency during spring and summer in those regions. Extreme droughts projected to increase in all Southeast Asia during all seasons. This strengthening of extreme drought intensity is also shown in some regions of Northeast Asia. This regional difference of drought is because global warming continues to more accelerate over mid-latitude of East Asia.

Acknowledgments

This work was supported by the R&D Project of the Korea Meteorological Administration "Development and application of technology for weather forecast".

References

1. McCarthy JJ, Canziani OF, Leary NA, Dokken DJ, White KS (2001) Climate Change 2001: Impacts, Adaptation and Vulnerability, Cambridge University Press, Cambridge, UK.

2. Hu ZZ, Latif M, Roeckner E, Bengtsson L (2000) Intensified Asian summer monsoon and its variability in a coupled model forced by increasing greenhouse gas concentrations. Geophys Res Lett 27: 2681-2684.

3. Kimoto M (2005) Simulated change of the East Asian circulation under global warming scenario. Geophys Res Lett, p: 32.

4. Loo YY, Billa L, Singh A (2015) Effect of climate change on seasonal monsoon in Asia and its impact on the variability of monsoon rainfall in Southeast Asia. Geosci Front 6: 817-823.

5. Kripalani RH, Oh JH, Chaudhari HS (2007) Response of the East Asian summer monsoon to doubled atmospheric CO_2: Coupled climate models simulations and projections under IPCC AR4. Theor Appl Climatol 87: 1-28.

6. Zhi X, Qi H, Bai Y, Lin C (2012) A comparison of three kinds of multimodel ensemble forecast techniques based on the TIGGE data. Acta Meteorol Sin 26: 41-51.

7. Wetherald RT, Manabe S (2002) Simulation of hydrologic changes associated with global warming. J Geophys Res 107: 4379-4394.

8. Manabe S, Wetherald RT, Milly PCD, Delworth TL, Stouffer RJ (2004) Century-scale change in water availability: CO_2-quadrupling experiment. Clim Change 64: 59-76.

9. Kiem AS, Austin EK (2013) Drought and the future of rural communities: Opportunities and challenges for climate change adaptation in regional Victoria, Australia. Global Environ Change 23: 1307-1316.

10. Burke EJ, Brown SJ, Christidis N (2006) Modeling the recent evolution of global drought and projections for the twenty-first century with the Hadley Centre climate model. J Hydrometeol 7: 1113-1125.

11. Zhou TJ, Tao H (2015) Projected changes of Palmer Drought Severity index under an RCP8.5 scenario. Atmos Oceanic Sci Lett 5: 273-278.

12. Wang G (2005) Agricultural drought in a future climate: Results from 15 global climate models participating in the IPCC 4th assessment. Clim Dyn 25: 739-753.

13. Byun HR, Wilhite DA (1999) Objective quantification of drought severity and duration. J Clim 12: 2747-2756.

14. Kitoh A, Uchiyama T (2006) Changes in onset and withdrawal of the East Asian summer rainy season by multi-model global warming experiments. J Meteorol Soci Jpn 84: 247-258.

15. Sperber KR, Annamalai H, Kang IS, Kitoh A, Moise A, et al. (2013) The Asian summer monsoon: an inter comparison of CMIP5 vs. CMIP3 simulation of the late 20th century. Clim Dyn 41: 2711-2744.

16. Yamaguchi Y, Shinoda M (2002) Soil moisture modeling based on multiyear observations in the Sahel. J Appl Meteor 41: 1140-1146.

17. Morid S, Smakhtin V, Moghaddasi M (2006) Comparison of seven meteorological indices for drought monitoring in Iran. Int J Climatol 26: 971-985.

18. Smakhtin VU, Hughes DA (2007) Automated estimation and analyses of meteorological drought characteristics from monthly rainfall data, Environ. Model Software 22: 880-890.

19. Akhtari R, Morid S, Mahdian MH, Smakhin V (2009) Assessment of areal interpolation methods for spatial analysis of SPI and EDI drought indices. Int J Climatol 29: 135-145.

20. Kimoto M, Yasutomi N, Yokyama C, Emori S (2005) Projected changes in precipitation characteristics around Japan under the global warming. SOLA 1: 85-88.

19

Greenhouse Gas Emissions from Landfills: A Case of NCT of Delhi, India

Singh SK, Anunay G*, Rohit G, Shivangi G and Vipul V

Department of Environmental Engineering, Delhi Technological University, Delhi, India

***Corresponding author:** Anunay G, Assistant Professor, Department of Environmental Engineering, Delhi Technological University, Delhi, India
E-mail: anunaygour@live.in

Abstract

The quantity of municipal solid waste (MSW) generated in Delhi is increasing at an alarming rate. Presently Delhi generates 8360 tons per day of MSW, which is projected to rise up to 18,000 tons/day by 2021.This would place immense pressure on the existing infrastructure and soon become a challenge for the local and municipal bodies responsible for waste management. The paper surveys the present state of the solid waste management in the NCT of Delhi and the propagation of greenhouse gas from the landfill sites. The bulk of the waste generated in Delhi is disposed at the three landfill sites viz. Bhalswa, Ghazipur, and Okhla. Waste at landfills is acted upon chemically and biologically to yield stabilized solids, liquid leachate and gases. The degradable organic carbon in the waste is broken down by microorganisms into methane gas which is released as a major contributor to global anthropogenic CH_4 emissions. The paper further calculates the GHG emission potential for the three landfill sites in Delhi using IPCC Methodology - Default Method and First Order Decay Model. The results obtained from both the methods are compared and it is found out that Default Method gave higher GHG emission values than the First order decay model. MSW in Delhi has been found to have enormous waste-to-energy potential, which if employed may cater to Delhi's energy needs and simultaneously reduce the GHG emissions.

Keywords: Municipal solid waste; Landfill; Greenhouse gas; GHG emissions; IPCC - default method; IPCC - FOD method

Introduction

Solid waste management have become a worldwide problem and is getting increasingly complicated day by day mostly due to the rise in population, industrialization and the consequent changes introduced in the lifestyle of people. Hence the per capita generation of solid waste bears strong correlation with the economic development at global as well as local scale. There are six functional elements complementing solid waste management (SWM) practices. It takes off with waste generation, followed by storage and handling of waste at source, waste collection, transfer and transport, followed by treatment and transformation, and at last the disposal [1]. Presently majority of the waste generated is disposed in open dumps or ordinary landfills in the developing rural areas or in sanitary landfills in the developed ones. Of the various SWM techniques, landfilling is found to be the cheapest and easiest to dispose municipal solid waste (MSW) all over the globe [2]. With the increasing urbanization and evolving life styles, Indian cities are generating eight times more MSW compared to that generated during 1947. Annually about 90 million tons of solid waste is produced as by-products of industrial, mining, municipal, agricultural and other activities. While the average collection efficiency in metropolitan cities (population > 1 million) is below 70% and that for smaller cities, it is less than 50% [3-5]. Waste at the landfills have been known to be one of the major sources of anthropogenic greenhouse gas (GHG) emission and a key contributor to global warming [6]. Methane (CH_4) emission from landfill is estimated to account for 3% - 19% of the anthropogenic sources in the world [6]. India, one of the world's largest emitter of CH_4 from landfills, currently produces about 16 tons of CO_2 equivalent per year which is predicted to increase to almost 20 tons of CO_2 equivalent per year by 2020 [7]. CH_4 alone constitutes

about 29% of the total GHG emissions in India which is nearly twice the worldwide average of 15%. Moreover the emission from wastes is also twice (6%) than the global average of (3%) [8]. Major part of MSW constitutes of biodegradable organic materials, which undergo anaerobic decomposition in landfills generating a variety of gases collectively called landfill gas (LFG). It is composed of approximately 60% methane (CH_4) and 40% carbon dioxide (CO_2) in concert with low amounts of non-methane organic compounds and other trace gases [9-11]. Both CH_4 and CO_2 are GHGs. Over a time period of 100 years, the global warming potential (GWP) of CH_4 is 25 times of the GWP of CO_2 and has an atmospheric residence time of 12 ± 3 years [12]. Moreover percolating rainwater through the landfills produces leachate which contaminates the ground water [13,14]. Presently, approximately 8360 tons of MSW is generated daily in Delhi, out of which 87% is collected which amounts to 7273 tons per day (TPD). Moreover of the collected 7273 TPD of waste only 28%, i.e., 2000 TPD is treated [15]. As only one waste to energy plant at Timarpur - Okhla with a capacity of 1950 MTD is operational.

State of Municipal Solid Waste Management in Delhi

Population of Delhi is increasing at an alarming rate and it is estimated that the waste generation is expected to touch 18,000 TPD by 2021 [16]. State of the Delhi's landfills is not very encouraging with all the landfills being used way beyond their design life. Bhalswa landfill (BL), Ghazipur landfill (GL) and Okhla landfill (OL) are the three operational landfill sites in Delhi while two landfill sites in Jaitpur and Bawana are proposed. MSW is treated at three composting plants at Bhalswa, Narela / Bawana and Okhla; two incineration plants at Okhla which also generates 16 MW of power from the waste; one RDF plant at Narela / Bawana; and one construction and demolition waste dump at Burari [17]. Also all the three landfill sites have crossed the 30 metre height ceiling. Since landfills are major contributor of methane emissions [16], emissions from waste-water treatment plants [18,19] and other waste management practices have not been considered in scope of the present study.

Study area

The NCT of Delhi is spread over an area of 1483 sq. km., located in the northern part of India. It is divided into three parts Delhi Ridge, Yamuna flood plains and plains. The air quality is poor and highly variable [20,21]. Hence periodic critical assessment of air quality by analytic and statistical tests is necessary [22]. It is the hub for employment for millions from the periphery, i.e., the NCR. As a result of which a huge mass of population survive in Delhi, which produces huge quantity of waste. Literature related to MSW management in Delhi shows that wastes are handled by three municipal agencies namely, Municipal Corporation of Delhi [23], New Delhi Municipal Corporation (NDMC) and Delhi Cantonment Board. All the MSW generated in the city is transported to landfill sites at Ghazipur in East Delhi, Bhalswa in North Delhi and Okhla in South East Delhi [16]. Brief descriptions about these landfills are given in (Table 1).

Characteristics	Bhalswa (BL)	Ghazipur (GL)	Okhla (OL)
Starting year	1992	1984	1996
Location	28 44'27.16" N, 77 9'27.92" E	28 37' 22.4" N, 77 19' 25.7" E	28 30'42" N, 77 16' 59" E
Area (Hectare)	26.22	29.62	16.89
Slope ()	60-70	60-70	70 80
Average height (m)	18	25.5 - 30.5	27 - 40
Dumping quantity (TPD)	1500	2200	1200
Type of waste	Household, vegetable market, C and D waste	Household, animal waste from poultry, fish market and slaughter house	Mainly household with C and D waste
Zones supplying waste	Civil Lines, Karol Bagh, Rohini, Narela, Najafgarh and West	Shahdara (North andWest), City zone, Sadar Pahargunj, NDMC	Central, Najafgarh, South Delhi and Cantonment Board
Depression, below ground level (m)	4	3	4

Table 1: Salient features of Delhi's landfills [16].

All the landfills have been receiving waste from more than 15 years; with GL being the oldest site operating since 1984; followed by BL which started in 1992 and OL became operational in 1996. Based on the municipal records, it has been estimated that the cumulative waste quantities reached at GL, BL and OL landfills till 2008 - 2009 are 11 million tons, 9.2 million tons and 6.1 million tons respectively [16]. Figure 1 shows a five-point summary and comparison of the quantity of solid waste received at BL, GL and OL landfill sites on a logarithmic y-axis of the box plot.

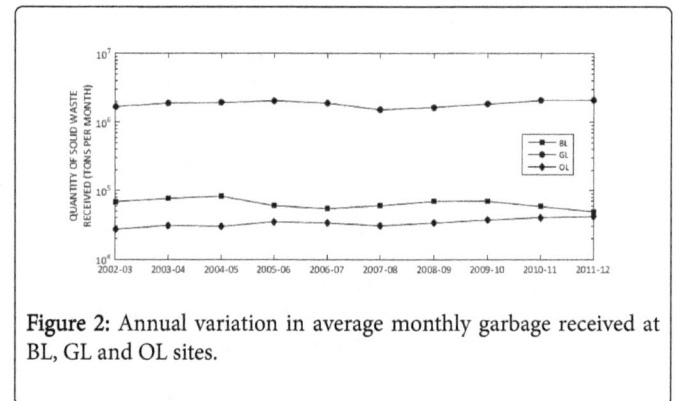

Figure 2: Annual variation in average monthly garbage received at BL, GL and OL sites.

Composition

The composition of MSW is the key to estimate the GHG emissions from landfills. Table 2 shows the relative content of various parameters of Indian MSW. It reveals that the amount of recyclables is very less in the waste owing to the rag-pickers who pick up recyclable matter (paper, plastic, glass, and metal) at dhalaos (garbage bins) before the waste reached the landfill [24].

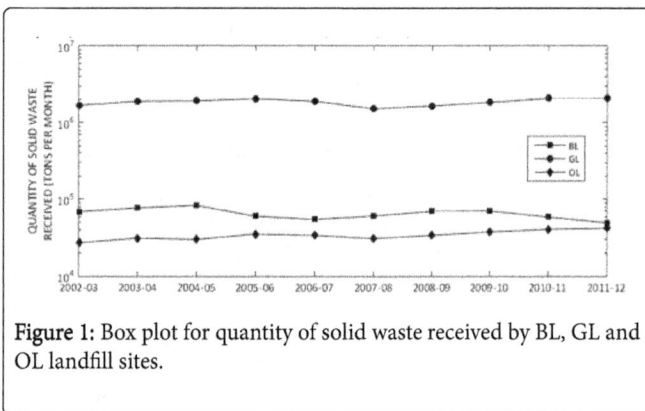

Figure 1: Box plot for quantity of solid waste received by BL, GL and OL landfill sites.

Overall trend of quantity of solid waste received at the landfill sites have been found to be increasing (except OL). Figure 2 shows the annual variation in average monthly garbage received at BL, GL and OL sites from year 2002 to 2012.

Methane Emission from Landfills: Methodology for Inventory Estimation

Due to disposal of solid waste to landfills, emissions of methane take place due to anaerobic decomposition of organic matter. IPCC suggests two methods for calculating methane emissions from landfills.

The Default IPCC methodology based on the theoretical gas yield (a mass balance equation).

Theoretical first order kinetic methodologies, through which the IPCC Guidelines introduces the 'First order decay (FOD) model'.

Sr. No	Parameter	Content (%)
1	Biodegradables (includes green matter, kitchen waste, paper, textiles, dry leaves)	73.7
2	Recyclables (includes glass, rubber / leather, metal, plastic)	9.2
3	Inert (include concrete, sand, brick, stone)	17.1
4	Others (includes dead animals, etc.)	6.3
5	Moisture	47

Table 2: Composition of waste reaching the landfill site in Delhi [23].

The main difference between the two methods is that method IPCC-Default method does not reflect the time variation in solid waste disposal and the degradation process as it assumes that all potential methane is released the year the solid waste is disposed whereas the effect of timing of the actual emissions is reflected in FOD method [25].

IPCC Default method

The method assumes that all the potential methane emissions are released during the same year the waste is disposed. The method is simple and emission calculations require only input of a limited set of parameters, for which the IPCC Guidelines provide default values, where country-specific quantities and data are not available.

The Default Method is based on the following equation:-

Methane emissions (Gg/yr) = (MSWT * MSWF * MCF * DOC * DOCF * F * 16/12-R) * (1-OX)

Where,

Symbol	Parameter	Value
MSWT	Total municipal solid waste (MSW) generated (Gg yr/yr.)	MSWT * MSWF = Waste disposed at landfill site (see Tables 4-6)
MSWF	Fraction of MSW disposed of at the disposal sites.	
MCF	Methane correction factor	0.6 (IPCC default value)
DOCF	Fraction DOC dissimilated	0.77 (IPCC default value)
DOC	Fraction of degradable organic carbon in the waste	0.5 (IPCC default value)
F	Fraction of methane in the landfill gas	0.5 (IPCC default value)

R	Recovered methane (Gg/yr)	Recovery of LFG is not adopted in India, hence the value is zero
OX	Oxidation factor	0 (IPCC default value)

Parameters and values of the equation are tabulated above.

Methane generation by first order decay method

The amount of methane generated from the landfill site is calculated based on a First Order Decay Model. In this method, methane generation in the landfill is described as a function of time. Nationally adjusted FOD model has been used for calculating yearly emissions from the waste dumped in landfills.

$$QT, x = k * MSWT(X) * MSWF(X) * MCF(X) * Lo(X) * e - k (T-X) * F$$

Where,

Symbol	Parameter		
QT, x	The amount of methane generated in the current year from waste disposed in the year X		
X	The historical year of the disposal of the relevant national MSW quantities		
K	Ln (2)/$t^{\frac{1}{2}}$. (1/yr) and $t^{\frac{1}{2}}$ is half-life period for degradation process		
MSWT(X)	Total MSW generated (Gg/yr) in year X		
MSWF(X)	Fraction of MSW disposed to solid waste disposal sites in year X		
MCF(X)	Methane correction factor (fraction) for year X		
Lo(X)	DOC x DOCF for the year X (Gg CH_4 / Gg waste) Where:		
		DOC	Fraction of degradable organic carbon in the waste
		DOCF	Fraction DOC dissimilated
T	The current year (year of the emission estimate) (Gg/yr)		

Parameters of the equation are tabulated above.

The value of is taken as 7 years (default value) for tropical countries and rest all parameters have same values as that in IPCC default method.

Results and Discussion

IPCC- Default methodology and IPCC-FOD methodology has been utilized to yield the values of net methane emission from the three landfills at Bhalswa, Ghazipur and Okhla. Table 3 shows the descriptive statistics, geometric increase / decrease occurred in the average monthly solid waste received at the landfills over the years 2002 to 2012. It also presents the correlation of quantities of solid waste received, which indicates stronger correlation between GL and OL than others. This may be due to the geographic proximity of the landfills, similarity in generation, logistics and transport of wastes. Bhalswa Landfill is the fastest growing landfill as the solid waste received here is growing at a rate of 9.61% per year. Whereas the solid waste received at Okhla Landfill is declining i.e. at a negative growth rate. The quantities of solid waste received at the respective landfills are presented in Tables 4-6 along with the calculated methane emission by Default Method and FOD Method.

	Correlation			Mean	Std. Dev.	Geometric Increase (%)
	BL	GL	OL			
BL	1			65215.8	9789	9.61
GL	-0.26	1		1867657	180209	5.76
OL	-0.62	0.64	1	34185.7	4450	-7.75*
* % decrease is observed						

Table 3: Correlation between solid waste reaching landfill sites and their descriptive statistics [17].

Various values and parameters required for calculation of methane emission potential are described in the methodology. The discussion of the results is discussed here.

Bhalswa landfill (BL)

Bhalswa Landfill was started in 1992 and receives household, vegetable market and Cand D waste. It receives around 1500 tonne per day of wastes from Civil Lines, Karol Bagh, Rohini, Narela, Najafgarh and West Delhi Area. It is a major source of ground water pollution. The quantity of solid waste received at BL and the respective methane emission by IPCC-Default method and IPCC-FOD Methods are given in Table 4. The emissions have been found to gradually decrease from 127.66 Gg/yr to 91.23 Gg/yr by Default method and 86.71 to 61.97 Gg/yr by FOD method as the waste reaching this landfill has decreased over the years.

Year	MSW received at BL (tons per month)	Methane emission by Default Method (Gg/yr)	Methane emissions by FOD Method (Gg/yr)
2002-2003	69072	127.66	86.71
2003-2004	77009	142.30	96.67
2004-2005	82773	153.00	103.91
2005-2006	60236	111.32	76.62
2006-2007	54566	100.84	68.49
2007-2008	60674	112.13	76.17
2008-2009	69617	128.65	87.39
2009-2010	70134	129.61	88.04
2010-2011	58711	108.50	73.70
2011-2012	49366	91.23	61.97

Table 4: Methane emissions from Bhalswa landfill (BL).

Ghazipur landfill (GL)

Ghazipur Landfill was started in 1984 and receives household, animal waste from poultry, fish market and slaughter house. It receives around 2200 tonne per day of wastes from Shahdara (North and West), City zone, Sadar, Pahargunj, NDMC municipal areas. It receives the highest quantity of solid waste to be disposed. The quantity of solid waste received at GL and the respective methane emission by IPCC-Default method and IPCC-FOD Methods are given in Table 5. The

quantity of waste reaching GL has increased; hence the CH_4 emissions have also been observed to increase gradually from 3125.121 Gg/yr to 3845.20 Gg/yr by default method and increased from 2122.85 Gg/yr to 2611.99 Gg/yr by FOD method. GL shows the maximum emission as it received the largest quantity of solid waste. Also the methane emissions calculated by First order decay method are less than of Default method.

Year	MSW received at GL (tons per month)	Methane emission by Default Method (Gg/yr)	Methane emissions by FOD Method (Gg/yr)
2002-2003	1691083	3125.12	2122.85
2003-2004	1907599	3525.24	2394.65
2004-2005	1929465	3565.65	2422.10
2005-2006	2061538	3809.72	2587.89
2006-2007	1903583	3517.82	2389.61
2007-2008	1524059	2816.46	1913.18
2008-2009	1659741	3067.20	2083.51

2009-2010	1845896	3411.21	2317.20
2010-2011	2072873	3830.67	2602.12
2011-2012	2080736	3845.20	2611.99

Table 5: Methane emissions from Ghazipur landfill (GL).

Okhla landfill (OL)

Okhla Landfill was started in 1996 and receives mainly household waste with some C and D waste. It receives around 1200 tonne per day of wastes from Central, Najafgarh, South Delhi and Cantonment Board. The quantity of solid waste received at OL and the respective methane emission by IPCC-Default method and IPCC-FOD Methods are given in Table 6. The emissions have been found to gradually decrease from 350.29 Gg/yr to 77.42 Gg/yr by default method and increased from 34.16 Gg/yr to 52.60 Gg/yr as the waste reaching this landfill has increased but the rate at which waste is received at OL is reducing over time. OL shows a geometric growth rate of -7.75% of solid waste received.

Year	MSW received at OL (tons per month)	Methane emission by Default method (Gg/yr)	Methane emissions by FOD Method (Gg/yr)
2002-2003	27216	50.29	34.16
2003-2004	31180	57.62	39.14
2004-2005	30004	55.48	37.66
2005-2006	35249	65.14	44.25
2006-2007	33736	62.34	42.35
2007-2008	30836	56.98	38.71
2008-2009	33807	62.48	42.44
2009-2010	37379	69.08	46.92
2010-2011	40554	74.94	50.91
2011-2012	41896	77.42	52.60

Table 6: Methane emissions from Okhla landfill (OL).

Figure 3 shows the methane emission from years 2002 to 2012 calculated by IPCC- Default method (DM) and FOD Method at BL, GL and OL landfill sites. It is observed that Default Method yields higher value of emission than FOD method because any change in the amount of MSW disposed is reflected immediately in the results of IPCC-Default method whereas the FOD method responds slowly to the changes. The IPCC-Default Method is easy to implement and the result are also lucid. The FOD method requires adequate knowledge of the decay process at solid waste disposal sites and comparatively more data is required (Table 7).

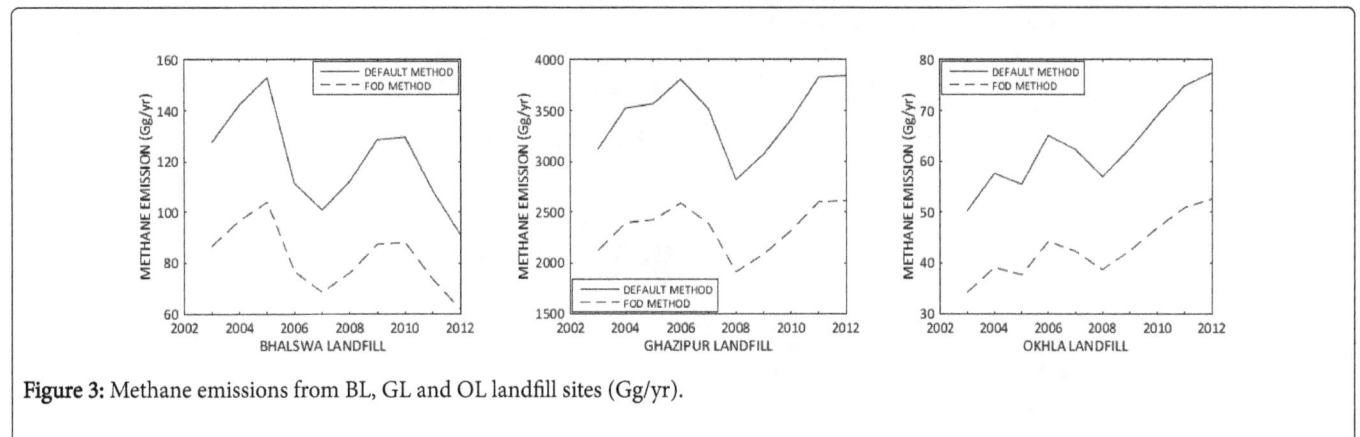

Figure 3: Methane emissions from BL, GL and OL landfill sites (Gg/yr).

In consonance with the total quantity of solid waste received at the landfills, Ghazipur emits the largest share of methane emission from both methods. The variation in methane emissions is found to be fluctuating the most at GL with a standard deviation of 333 in Default Method CH_4 emission results and 226.2 in FOD method CH_4 emission results.

Landfill	BL		GL		OL	
IPCC Method	DM	FOD	DM	FOD	DM	FOD
Mean	120.5	82.0	3451.4	2344.5	63.2	42.9
St. Dev.	18.1	12.2	333.0	226.2	8.2	5.6

Table 7: Descriptive statistics of methane emission from BL, GL and OL landfills.

Uncertainty in quantity and composition of waste received at landfill sites, oxidation mechanism and rate of methane generation in upper crust of landfill, fraction of DOC and DOCF contribute to the uncertainties in GHG emission estimation from landfills. IPCC model that predict the amount of methane generated (mass balance, first order decay) need to be precise for accurate results. Therefore, even a small variation in the DOC or methane generation rate constant may lead to large variations in CH_4 emission estimates.

Conclusion

Municipal solid waste disposal is a significant source of methane emissions globally. The increasing pressure of waste from the population needs an alternative for waste management and disposal other than landfilling in topographic depressions. Present study revealed that the methane emissions from Bhalswa, Ghazipur and Okhla Landfills for the year 2011-2012 was 91.23 Gg/yr, 3845.20 Gg/yr and 77.42 Gg/yr by Default Method; and 61.97 Gg/yr, 2611.99 Gg/yr and 52.60 Gg/yr by First Order Decay method. Hence there emerges an urgent need to initiate mitigation steps for controlling GHG emission from landfill areas. Moreover the quantity of methane emission has been increasing as a result of increased consumption. The first step of waste management practice to be employed is the segregation of organic component from solid waste at the source, which will be effective for composting. Secondly, construction of scientifically planned sanitary landfill site with provision of tapping landfill gas for its fuel value [11]. Hence the huge quantity of MSW in Delhi has enormous waste to energy potential [3,19] which if utilized can cater to Delhi's increasing energy needs.

References

1. Tchobanoglous G, Theisen H, Vigil SA (1993) Integrated Solid Waste Management: Engineering Principles and Management Issues, McGraw Hill International editions, pp: 978.

2. Jhamnani B, Singh SK (2007) Solid Waste Management by land filling in new Urban Growth Centres. National Seminar on Integrated Development of Towns as new growth Centers. Indian Buildings Congress 17-19.

3. Kumar A, Sharma MP (2014) Estimation of GHG emission and energy recovery potential from MSW landfill sites. Sustainable Energy Technologies and Assessments 5: 50-61.

4. Singh SK, Bisht A (2015) Environmental Management in Mass Gatherings, Int J of Engg Sci and Mgmt 5: 130-138.

5. Sahai S, Sharma C, Mitra AP, Singh SK, Gupta PK (2011) Assessment of trace gases, carbon and nitrogen emission from field burning of agricultural residues in India. Nutrient Cycling in Agroecosystems 89: 143-157.

6. Intergovernmental Panel on Climate Change (IPCC) (1996) Report of the Twelfth Season of the Intergovernmental Panel on Climate Change, Mexico City.

7. Singh SK, Singh N (1998) Solid Waste Management of Bathinda City. Proceedings of Thirteenth International Conference on Solid Waste technology and Management. Philadelphia, USA.

8. Siddiqui FZ, Khan ME (2011) Landfill gas recovery and its utilization in India: current status, potential prospects and policy implications. J Chem Pharm Res 3: 174-83.

9. Hegde U, Chang TC, Yang SS (2003) Methane and carbon dioxide emissions from Shan chu-ku landfill site in northern Taiwan. Chemosphere 52: 1275-1285.

10. Singh J, Ramanathan AL (2015) Solid Waste Management: Present and Future Challenges. In: Singh SK, Mahour M (eds18.) Clean Development Machanism: An Opportnity in Solid Waste Perspective, pp: 178-185.

11. Singh J, Ramanathan AL (2015) Solid Waste Management: Present and Future Challenges In: Singh SK, Mahour M (eds2.) Methane Estimation from Landfill Site, pp: 19-28.

12. Solomon S, Qin D, Manning M, Chen Z, Marquis M, et al. (2007) Climate Change 2007: The Physical Science Basis, IPCC Fourth Assessment Report.

13. Singh SK (2007) Ground water Modeling-Trends and Practices. Journal of American Society of Civil Engineers-IS 4(1).

14. Chawla A, Singh SK (2014) Modelling of Contaminant Transport from Landfills. International Journal of Engineering Science and Innovative Technology 3: 222-227.

15. Govt. of Delhi, Department of Environment (2015) Waste Management.

16. Chakraborty M, Sharma C, Pandey J, Singh N, Gupta PK (2011) Methane Emission Estimation from Landfills in Delhi: A comparative Assessment of Different Methodologies. Atmospheric Environment 45: 7135-7142.

17. Kumar A (2013) Existing Situation of Municipal Solid Waste Management in NCT of Delhi, India, Research Forum: International Journal of Social Sciences 1: 6-17.

18. Gupta D, Singh SK (2012) Green House Gas Emission from Waste Water Treatment Plants: A Case Study of Noida, India. Journal of Water Sustainability 2: 131-139.

19. Gupta D, Singh SK (2015) Energy use and Greenhouse Gas emissions from Waste Water Treatment Plants, Journal of Environmental Engineering 7: 1-10.

20. Gour AA, Singh SK, Tyagi SK, Mandal A (2013) Weekday/Weekend Differences in Air Quality Parameters in Delhi, India. International Journal of Research in Engineering and Technology 1: 69-76.

21. Gour AA, Singh SK, Tyagi SK, Mandal A (2015) Variation in Parameters of Ambient Air Quality in National Capital Territory (NCT) of Delhi (India). Atmospheric and Climate Sciences 5: 13-22.

22. Aenab AM, Singh SK, Lafta AJ (2013) Critical Assessment of Air Pollution by ANOVA Test and Human Health Effects. Atmospheric Environment 71: 84-91.

23. Municipal Corporation of Delhi (2004) Feasibility study and master plan for Optimal Waste Treatment and Disposal for the Entire State of Delhi on Public Private Partnership Solution. Volume 6: Municipal Solid Waste Characterization Report.

24. Sastry DBSSR. Composition of Municipal Solid Waste- Need for Thermal Treatment in the present Indian context.

25. Frøiland-Jensen JE, Pipatti R. CH_4 emissions from Solid Waste Disposal. Good Practice Guidance and Uncertainty Management in National Greenhouse Gas Inventories 419-439.

20

Variability of Precipitation regime in Ladakh region of India from 1901-2000

Shafiq MU, Bhat MS, Rasool R, Ahmed P*, Singh H and Hassan H

Department of Geography and Regional Development, University of Kashmir, J & K India

***Corresponding author:** Ahmed P, Department of Geography and Regional Development, University of Kashmir, J & K India, E-mail: pervezku@gmail.com

Abstract

Ladakh region being a high elevation cold desert of India is marked by extreme aridity with acute moisture deficit throughout the year. Annual precipitation is extremely low due to rain shadow effect caused by Karakoram ranges on one side, mighty Greater Himalayas and Zanskar ranges on the other side. The study of precipitation variability is extremely important for a region like Ladakh, in which the lifestyle of inhabitants, agriculture, live stock rearing and water resources are dependent on the nature and magnitude of precipitation. Despite experiencing low precipitation, extremely low temperature enables Ladakh to contain some of the largest alpine glaciers in the world. These glaciers feed a number of river systems including Indus being the largest and the most important. The precipitation regime however has shown a changing trend over the period of time. In this backdrop, the present study attempts to analyze, core summer and core winter trends to ascertain the temporal variability in the precipitation regime from 1901-2000. Mann-Kendall test (non-parametric test) has been used to analyze the significance levels. Results for summer season show non-significant results with a test statistic of -1.102 with decrease of the order of 0.127 mm per year while for winter season Mann-Kendall test shows a rise of 0.04 mm per year showing significant trend with test statistic of 1.92 at 0.10 significance levels. The results are indicative of decreasing precipitation during summers and increasing precipitation off late during winters which will have a profound impact on the glacial environment of Ladakh region.

Keywords: Mann-Kendall test; temporal variability; precipitation; Ladakh; Himalayas

Introduction

Mountain systems account for roughly 20 per cent of the terrestrial surface area of the globe and are found on all continents. They are usually characterized by sensitive ecosystems and enhanced occurrences of extreme weather events and natural catastrophes; they are also regions of conflicting interests between economic development and environmental conservation [1]. Once, regarded as hostile and economically nonviable regions, mountains have attracted major economic investments for industry, agriculture, tourism, hydropower, and communication routes [1]. The mountains provide the direct life-support base for about a tenth of humankind and indirectly affect the lives of more than half of the global population [2].

Mountains especially Himalayas, are widely recognized as areas containing highly diverse and rich ecosystems, and thus, they are key elements of the global geosphere-biosphere system. At the same time, mountains contain ecosystems that are quite sensitive and highly vulnerable to natural risks, disasters, and ecosystem changes, be it through the occurrence of rapid mass movements, such as landslides, or via slow land degradation due to human activities, with all the attendant socioeconomic consequences [3].

The Himalayan Mountains of snow have also been called the third pole, since they are the third largest body of snow on our planet after the Antarctic and Arctic [4]. Almost 9.04% of the Himalaya is covered with glaciers, with 30-40% additional area being covered with snow [5]. They feed the giant rivers of Asia, and support half of humanity [4]. They abode highly fragile and sensitive ecosystems which provide vital clues about the impact of global warming and other man induced ecological imbalances.

Prevalence of varied climatic conditions that are similar to those of widely separated latitudinal belt, within a limited area, make the high mountain areas such as Himalaya, the Alps, the Andes, the Rockies etc. the ideal sites for the study of climate change [5]. The high mountains of South Asia covering the Hindukush, Karakoram Himalayan belt have reported warming trend in the past few decades [6,7]. A study of precipitation data for a fairly long period is essential for estimating its distribution and other characteristics like irrigation, availability of ground water, vegetative cover.

Leh (Ladakh), which lies in the north-western Himalayas also exhibit the impact of global environmental changes as has been reported by a number of studies conducted all over the world. Ganjoo in his study on 114 glaciers of the Nubra sub-basin came to the conclusion that 39 glaciers (34%) have shown gain in the area, 43 glaciers (38%) have vacated the area and 32 glaciers (28%) do not show any change in their area [8]. The monitoring of snouts of 2018 glaciers in Himalaya and Karakorum for the period 2001-2010/11 with the help of satellite imageries has further revealed that 1752 (86.8%) glaciers are stable in their position, 248 (12.3%) glaciers have shown retreat in their snout positions whereas 18 (0.9%) glaciers have advanced confirming the erratic behaviour of glaciers with respect to latitude, elevation and climatic/weather variation from region to region [8].

In Ladakh, the northern most region of India, all life depends on snow. Ladakh is a high elevated cold desert [9] with only 50-70 mm of average annual rainfall. Ladakh's water comes from the snow melt both the snow that falls on the land and provides the moisture for farming and pastures, as well as the snow of the glaciers that gently melts and feeds the streams that are the lifeline of the tiny settlements [4].

Thus the present study aims to analyze the trends of changing precipitation regimes in the cold desert of Leh Ladakh.

Study Area

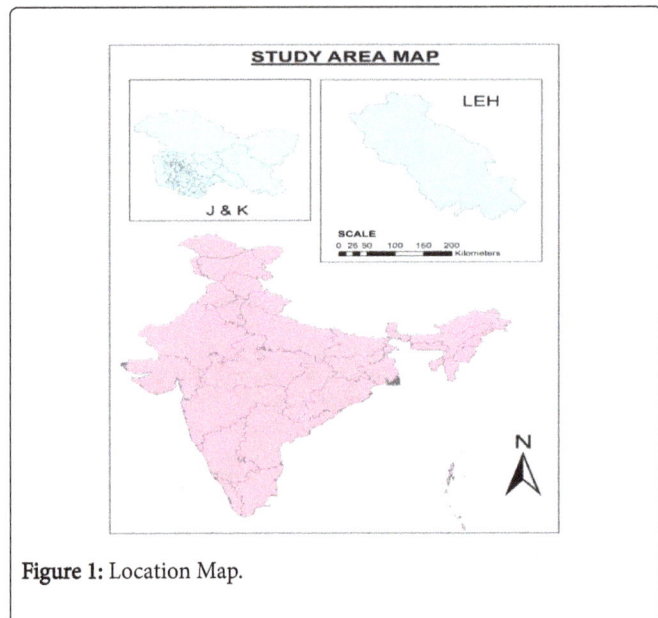

Figure 1: Location Map.

Ladakh region of Jammu and Kashmir state situated in the western Himalayas is a low rainfall zone situated on the leeward side of the Zanskar range in the western Himalayas with Leh (34.09 N / 77.34 E, 3514 m asl) as the main town in the region. Leh with an area of 45110 Sq Km makes it largest district in the country in terms of area. It is at a distance of 434 Kms from Srinagar and 474 Kms from Manali.

Physiographically, Leh is bounded by the Ladakh range to its north and the east and by the Zanskar range to its south and west. Greater Himalayan range lie further southwest of the Zanskar range. The town is situated on the right bank of river Indus which traverse the ladakh region from southeast to northwest.

Ladakh lies on the rain shadow side of the Himalayas, where dry monsoon winds reach Leh after being robbed of its moisture in plains and the Himalayan Mountain. The district combines the condition of both arctic and desert climate. Therefore Ladakh is often called "COLD DESERT" [9]. Wide diurnal and seasonal fluctuations in temperature with -40°C in winter and +35°C in summer are observed. Precipitation is very low mainly in the form of snow. Air is very dry and relative humidity ranges from 6-24%. Due to high elevation and low humidity the radiation level is very high. The global solar radiation is as high as 6000-7000 W/m2 (which is among the highest in the World). Dust storms are very common in the afternoon. Soil is thin, sandy and porous. The entire area is devoid of any natural vegetation. Irrigation is mainly through channels from the glacier-melted snow.

The greater Himalayan ranges are the highest among the three and act as a barrier for the eastward moving systems. Precipitation decreases considerably towards east of this range. Leh being east of Zanskar range receives very little annual precipitation as most of the weather systems move from west to east across the region. The station receives precipitation in form of both rainfall and snowfall. The total

annual precipitation of Leh is about 50-70 mm with July and august being the rainiest months with mean rainfall of 15.2 and 15.4 mm, respectively. The average number of rainy days for both the months is two each.

Materials and Methods

Records of Monthly precipitation (mm) of Meteorological station Leh available from 1901 to 2000 have been obtained from the National data center of the Indian Meteorological Department (IMD), Pune and Meteorological centre (IMD) Srinagar. The same have been used to compute the seasonal variability of precipitation for Leh. Due to the profound impact of only two seasons in the Ladakh region, only summer (June-august) and winter months (November-February) were taken into consideration depending upon climatic conditions prevailing over the region. The summer and winter precipitation data was plotted against time, and trends were examined by fitting the Time series analysis to the data. Linear trends were also drawn to analyze the pattern of precipitation over the last 100 years of the study period.

For detecting trend in a variable, the data series is smoothened. In the present study, 5-year moving average was applied. Basically, it performs low pass filtering in time domain. By doing this smoothening, the year-to-year variations get suppressed and dominant behaviour, if any, emerges. If there is increasing trend in the climatic series, then the slope coefficient should be positive and statistically significant. Conversely, for decreasing trend in behaviour of a variable the slope value should be negative.

For this study, the widely used Mann-Kendall test was run on time series data for the time period, 1901 to 2000. The Mann-Kendall test is a non-parametric test for identifying trends in time series data. Non-parametric Mann-Kendall method is a distribution-free method, more resistant to outliers, can usually be used with gross data errors, and can deal with the missing data values unlike the parametric method [10]. Mann-Kendall rank test is widely used in environmental monitoring for its simplicity and the focus on pair-wise slopes [11]. However, non-parametric methods are fraught with more uncertainty in the statistical estimates than the parametric method [12]. The test was suggested by Mann and has been extensively used with environmental time series. This test detects the presence of a monotonic trend within a time series [13] with different significance levels p: 0.1, 0.05, 0.1 and 0.001. The Mann-Kendall S Statistic is computed as follows:

$$MKT = [(4\Sigma ni / N (N-1)] -1$$

With

$$var (MKT) = [(4N+10) / (9N (N-1)]$$

Where ni is the no. of values larger than the ith value in the series subsequent to its position in the series of N values.

Results and Discussions

Long term trends have been utilized to detect changes in the pattern of winter as well as summer precipitation in the study area by utilizing time series analysis and applying Mann-Kendall test for showing the monotonic trend in the series. Winter precipitation shows rising trend although of a little order and is statistically significant at (0.1 significance level) which is a healthy sign with respect to the glacial environment of study area. Summer precipitation shows a decreasing trend over the last century in the area.

The complex topographic setting, altitudinal variation from glacier basin to glacier basin, impact of monsoon in high altitudes, variation in various parameters of micro-climate [11], increase and/or decrease in the cloudiness, encroachment of human population to the interiors of high altitude of Karakorum-Himalaya [12] can be some of the factors in the erratic movement of the glaciers. A slight variability or fluctuation in climate could prove detrimental to the agriculture, horticulture and fast emerging hydel power sector in the Leh, Ladakh [13].

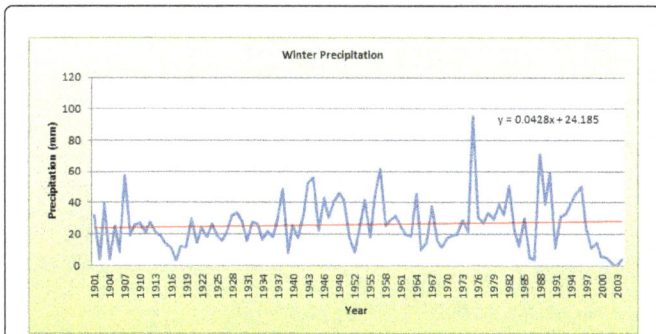

Figure 2: Time series analysis of winter precipitation in Leh from 1901-2000.

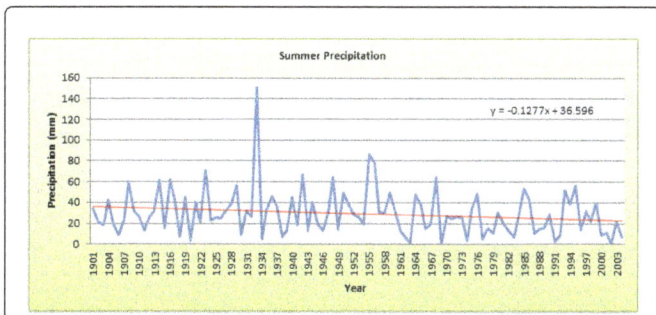

Figure 3: Time series analysis of summer precipitation in Leh from 1901-2000.

Trends in winter precipitation

Although average winter precipitation in Leh is only 27.38 mm, it is highly erratic and fluctuating. Occasionally Leh has received precipitation much higher than normal (1975, 1988) and on the other hand it has experienced a deficit, much below than the normal winter precipitation like in 1902, 1904, 1952, 1987.

If we analyze the monthly pattern of precipitation during four winter months, January comes out as the most precipitous month during the study period with average 11.3 mm precipitation followed by February (7.8 mm), December (5.8 mm) and November with 2.3 mm precipitation respectively.

Overall the winter precipitation has been on a slight rise during last century. The rise is of the order of nearly 0.04 mm per year which accounts to 0.40 mm per decade (Figure 2). The major characteristic of it being the precipitation in the form of snowfall. The linear slope analysis of the winter season in Leh reveals a positive trend with a slope of 0.042 when drawn on y=mx+c.

Trends in summer precipitation

The summer months of have shown an average precipitation of 30.67 mm during 1901-2000 with August (14.4 mm) being the rainiest month followed by July and then June with 12.3 mm and 3.8 mm respectively. The precipitation in the summer is usually in the form of rainfall with linear trend showing an overall decline during past century with recent decades showing some positive trend in summer season rainfall. A reduction of 0.127 mm per year in the summer precipitation is observed over the time period 1901-2000 in Leh (Figure 3).

Decadal trend analysis

The decadal analysis reveals that Leh received 68.09 mm rainfall in 4 months during 1901-2000 which showed negative trend up to 1940. The winter precipitation showed a positive trend during 1941-1950 which was 95.8 mm and thereafter for two decades there was decrease in winter precipitation upto 1970. The 1971-1980 decade showed a positive trend with 86.4 mm precipitation and again showed a decrease upto 66.5 mm during 1991-2000. The summer precipitation during 1901-2000 was 101.9 mm which continuously increased for next 60 years which touched 140 mm during 1951-1960. The period from 1961-1990 showed a slight decrease in summer precipitation but it again started to rise reaching 92.1 mm during 1991-2000.

Applying Mann-Kendall test statistic (Z) indicates how strong the trend in precipitation is and whether it is increasing or decreasing.

Technique	Season/Month	Z value	Trend Magnitude	Trend	Significance (0.10)
Mann-Kendall test	Summer Precipitation	-1.102	-0.127	Decreasing	Insignificant
	Winter Precipitation	1.92	0.042	Increasing	Signficant

Table 1: Mann-Kendall test.

Applying Mann-Kendall test statistic (Z) indicates how strong the trend in precipitation is and whether it is increasing or decreasing.

For average summer precipitation, Mann-Kendall test shows very low value of Z-statistic (-1.102) indicating an insignificant trend. The non-significant trend of Mann-Kendall test is indicative of decreasing precipitation during summer months, as indicated by time series.

For average winter the value of S obtained on positive side indicating increasing trend and is statistically significant at 99% significance level in Mann-Kendall test. The Z statistic for the average winter is 1.92 indicating an increasing trend in the winter precipitation during the study period.

The Mann-Kendall test was also applied for the winter and summer months taking into consideration their average precipitation during the study period. The test reveals a statistically insignificant trend for the all the winter months. Out of the summer months only June shows a statistically significant trend at 90% significance level, whereas the months of July and August follow an insignificant trend.

Conclusion

This study has been done for the study period of 100 years i.e. 1901-2000. The study area comprises of a mountainous topography with its water coming mainly from the snow melt. It reveals that there are discernable variations in the core summer and core winter precipitation in different months for different years with the variation extending across decades. From the trend analysis of precipitation data, it can be concluded that there is a decrease in summer precipitation in the study area, which is confirmed by Mann-Kendall Statistics. However, the study also reveals increasing trend of precipitation during the winter months. This erratic behavior of precipitation regime can have far reaching consequences and can influence the socio-economic conditions of the people living in the region.

References

1. Beniston M, Douglas G, Adhikary S, Andressen R, Guisan A, et al. (1996) Impacts of climate change on Mountain regions. Scientific-Technical Analyses 191-213.

2. Grab SW (2000) Mountains of the world: Climate Change in the Himalayas.

3. Singh SP, Singh V, Kutsch MS (2010) Rapid warming ib the Himalayas: Ecosystem responses and development options. Climate and Development 2: 1-13.

4. Viviroli D, Durr HH, Messerli B, Meybeck M, Weingartner R (2007) Mountains of the world, water towers for humanity: typology, mapping and global significance. water resources research 43: 1-9.

5. Immerzeel WW, Van Beek LPH, Bierkens MFP (2010) Climate change will affect the Asian water towers. Science 328: 1382-1385.

6. Ganjoo RK, Koul MN, Bahugna IM, Ajai (2014) The complex phenomena of glaciers of Nubra valley, Karakorum (Ladakh), India. Natural Science 6: 733-740.

7. Negi SS (1995) Cold deserts of India, Indus Publishing company, New Delhi.

8. Wilcox RR (1998) A note on the Theil-Sen regression estimator when the regressor is random and the error term is heteroscedastic. Biometrical Journal 40: 261-268.

9. Gibbons RD, Coleman DE (2001) Statistical methods for detection and quantification of environmental contamination. Wiley-IEEE.

10. Alexander LV, Zhang X, Peterson TC, Caesar J, Gleason B, et al. (2006) Global observed changes in daily climate extremes of temperature and precipitation. Journal of Geophysical Research 111: D05109.

11. Sicard (2013) Decrease in surface ozone concentrations at Mediterranean remote sites and increase in the cities. Atmospheric Environment 79: 705-715.

12. Koul MN, Ganjoo RK (2009) Impact of inter- and intra-annual variation in weather parameters on mass balance and equilibrium line altitude of Naradu Glacier (Himachal Pradesh), NW Himalaya, India. Climatic Change 99: 119-139.

13. Owen AL, Dortch MJ (2014) Nature and timing of Quaternary Glaciation in the Himalayan-Tibetian Orogen. Quaternary Science Reviews 88: 14-54.

21

Assessment of Runoff Changes under Climate Change Scenarios in the Dam Basin of Ekbatan, Hamedan Iran

Nazari P*, Kardavany H, Farajirad P and Abdolreza A

Department of Geography, Science and Research Branch, Islamic Azad University, Tehran, Iran

*Corresponding author: Nazari P, Department of Geography, Science and Research Branch, Islamic Azad University, Tehran, Iran
E-mail: hnpoya@yahoo.com

Abstract

In this study, the uncertainty of the effects of climate change on temperature, rainfall and runoff in the watershed done using the output of models MPEH5, HADCM3 and IPCM4 under two scenarios for 2045-2065 period. After performance of LARS-WG model for downscaling of rainfall and temperature variations, the monthly change rainfall and temperature evaluated for the 2065-2045 period relative to 2010-1983 base period. Results showed that all three models based on two scenarios reduce the amount of rainfall and increases in average temperature in the region. The average annual temperature rise according to the scenario A2, 2/12°C and scenario B1, 1/12°C. The amount of rainfall decreases for the period 2045-2065 under the scenario A2, -6/1 and the scenario B1, -1/4 per cent. To study the effects of climate change on monthly runoff regime, production variables was used to the model of rainfall- runoff IHACRES after that runoff was predicted for the period 2065-2045. The results showed that annual river flow reduced under the A2 scenario -17/2 percent and the B1 scenario -19/4 per cent. The overall results showed that despite the uncertainty in climate models MPEH5, HADCM3 and IPCM4 under the A2 and B1 in the amount of temperature increase and rainfall decreases and negative effects of these changes on the runoff area.

Keywords: Climate change; Runoff; Uncertainty; Hamedan; LARS-WG; IHACRES

Introduction

In recent decades, increasing greenhouse gases, consumption of fossil fuels, especially gas CO_2 as a result, an increase in the gas concentration of 280 ppm in 1750 to 380 ppm in 2005. IPCC1 reports indicate that, if current trends continue using these fuels, the concentration of gas before the end of the twenty-first century may reach more than ppm 600. If emissions not reduced, average earth surface temperature 1.1 to 6.4°C by 2100 will be reached. Therefore, according to reports IPCC, climate change causes changes in the hydrological regime in recent decades globally [1]. Increased surface temperatures and changes in rainfall patterns are dominant phenomena of climate change which affects almost all other sectors of the water cycle. All models AOGCMs predict increase in global temperature and increased precipitation intensity and its value as a result of increased concentrations of greenhouse gases in this century. Several studies by Hamlet and others in 2007 have been done about the potential impact of climate change on water resources, including the impact on water quantity, hydrology and water demand. Chang and Jung [2] examined annual runoff, seasonal and minimum and maximum values of runoff and uncertainty in the 218 Sub basin Willamette river in Oregon. The results showed that increased in seasonal changes runoff during the winter and decreased during the summer and change in temporal and spatial of runoff in the future. Gaussian studied impact of climate change scenarios on discharge in Indian River basin for the period 2041-2060. Their results indicate that the flow and intensity of floods and droughts is increasing. Perovskite using output of monthly rainfall HadCM3 model, investigated the effect of rainfall changes on runoff in eight regions in the United States. They reported that annual rainfall changes from 9.6 % to 1.6 and

runoff from 42.5 % to 14%. By the Senator and others studied, effects of regional climate models based on A2 and A1B scenarios for river basin Kraty in southern Italy. They predicted that in the period 2077-2099 mean temperature increases, between 3.5 to 3.9°C while precipitation will decrease 9% to 12%. This causes reduction annual cumulative snow 28% to 29%. The amount of subsurface water decrease between - 6.5 to 41.4 % and surface runoff decrease between -52.4 percent to -14.2 percent. Yi et al. [3] used IHACRES model in Australia for the some of basin with an area of 1582 and 517 square kilometers. Its performance was compared with 22 parameters LASCAM model and parameter 8 GSFB model. Absolute error of assessment in daily flow was calculated using IHACRES, 10 mm per day. However, according to Nash factor, performance has been good, thus model has done excellent. Post to simulate daily flow in watersheds without data relationship established between model parameters and characteristics of watersheds in south eastern Australia and have used to predict daily stream by temperature data, daily rainfall and basin characteristics. Some of these relationships were determined with acceptable accuracy. While others were relatively weak, so the predicted daily flow of current observations showed little difference in quality. Crooke and Little wood 2005 showed that IHACRES model in seven water shed in Wales so CMD edition model has better performance than the previous version for basins larger than 1,000 square kilometers. Anderson for reduced forest cover in order to show the possible impact on river discharge at the lake in the north of the United States, created regression among the factors affecting river flow. But could not gain a regional model for flow routing parameters. Groce used IHACRES for 42,000 square kilometers area in the southern basin in Wales, Australia. They used a version of IHACRES which includes a soil moisture was used and integrated way to the daily data for each subbasin. IHACRES parameters values were estimated using average values for the subbasins with no data from the subbasins of statistics or by local relations. They established relationship between the parameter values and percentage of forest cover and watershed area. Results showed that in two subbasin, model performance was not satisfactory due to the lack of meteorological stations but overall performance is suitable for use in runoff. Bovany evaluated the effects of climate change in the two periods 2070-2099 and 2010-2039 under two climate change scenarios A2 and B2 in the river flow Zayandehrood. Their findings indicate that reduced in both periods between 10 and 16 percent in rainfall and increase in average temperatures between 4.6 and 3.2°C, respectively A2 and B 2 scenarios. They were using a neural network to simulate rainfall runoff basin, findings show decrease to 5.8% and increased runoff coefficient of variation up to 3 times in the next period. Zarghami et al. [4] to predict the effects of climate change using outputs HadCM3 and LARS-WG model in Azerbaijan. The results using artificial neural network showed that reducing runoff of rivers. In 2009 Abbaspoor concluded that using CGCM climate model scenarios A1B, B1, A2 for 37 meteorological stations in Iran in the period 2070-2100, rainfall in arid regions decreases and increases in wet areas. Modaresi investigated the impact of climate change on the annual discharge of the river in Gorgan and downscaled outcome of various scenarios of general circulation models ECHAM, GFDL-R30, CGCM2, CSIRO, HadCM and the CCSR. They concluded that Gorgan river discharge decrease in the 50 and 100-year return periods, 1.83 and 1.33 percent, respectively. The impact of climate change on the basin runoff Gharehsou in the northwestern basin floodplains with respect to uncertainty in hydrological models were studied by Kamal and Bovany 1389. The results determined that reducing runoff for the fall season and it increased in other seasons during the period 2040-2069. Ghorbanizadeh studied the impact of climate change on the distribution of runoff from snowmelt in the basin of Karoon using the scenario of climate model ECHAM4 outputs in the 25-year period from 2000 to 2050. They found that the maximum flow will be transferred from spring to winter also increase in the basin discharge 10% in winter but decreases spring and summer. According to studies, we can conclude that the status of surface runoff of rivers affected by climate change in future periods will be some changes compared to the baseline. Over all Temporal and spatial variations in each region are different. So it is essential to study climate changes in different basins separately. Therefore, in this research effort to study the impact of climate change on variables such as, temperature, precipitation and surface runoff Dam Basin, located in Hamedan province in the period 2045-2065 under A2 and B1 emissions scenarios of future greenhouse gas.

Materials and Methods

Study Area

A schematic of the study area, along with catchment boundaries and gauging sites, is given in (Figure 1). Dam Basin of Ekbatan with an area of 160 square kilometers located in the center of the Hamedan province 48°41' and 48°30'east longitude and 34°36' and 34°46' north latitude on the northern slopes of Alvand. Surface soil texture of basin is 13.5 percent light and 55 percent average. About 22.3 percent of the area under cultivation is rainfed and 17.1 percent irrigated and 60.6 percent used for grassland. The maximum and minimum height of basin range between 1970 and 3467 meters and average altitude of basin is 2524 meters. About 26.5 percent of the watershed area have 0-10% slope, 3.6 percent watershed area has slopes greater than 30% and 67.3% of the watershed area also has a slope of between 10% and 30%. During the period 1983-2009 the total volume of the Dam Basin

of Ekbatan is 44.15 million cubic meters per year and average annual discharge 1.4 cubic meters per second. The lowest monthly discharge 0.008 cubic meters per second is in September and highest average monthly discharge 9.4 cubic meters per second in April.

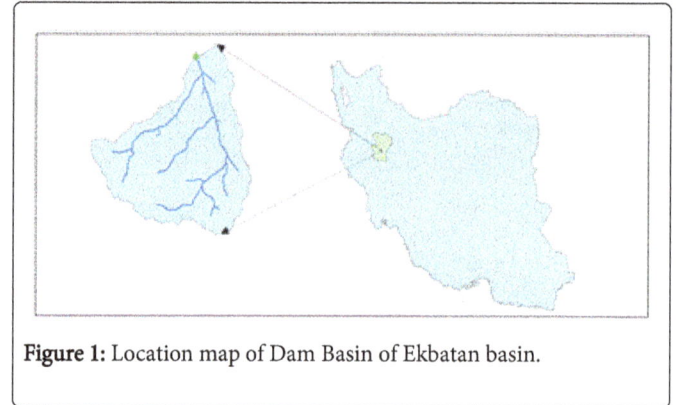

Figure 1: Location map of Dam Basin of Ekbatan basin.

Generation Weather Data Using AOGCM Model

One of the main uses of stochastic weather generators is in the generation of daily weather data representing scenarios of climate change. Most climate change scenarios are derived from the output of global climate models (GCMs). For assesment of climate change effect on runoff first were simulated, temperature and precipitation parameters for the next decade 2046-2065, using results of model outputs coupled ocean atmosphere (AOGCM) under two emission scenarios. Hydrological studies are often associated with small scale process and basins which scale is much smaller than the scale atmospheric circulation models give us. The GCM models should be downscaled for use in hydrological studies. There are two ways to downscaling data on a local or regional scale using global climate scenarios, include dynamical and statistical methods. In this study to predict climatic parameters, weather generatore, LARS-WG is used to downscaling of general circulation models. Therefore Climatic variables such as daily maximum temperature, minimum and precipitation were produced for study area during period 2065-2045 by three general circulation models MPEH5 'HADCM3 and IPCM4 for scenario A2, B1. Then using simulated daily precipitation and temperature data, the upcoming runoff was assessed using hydrological IHACRES model.

LARS-WG model

LARS-WG model used for generating synthetic weather data and to simulate meteorological data in one place, present, and future climatic conditions. Data generated on a daily time series suitable for a range of variables, for example, precipitation, minimum and maximum temperatures and radiation (MJ per square meter). The remarkable thing is that the artificial generation of climate data are not tools to predict that can be used to predict weather but they are ability to produce time series data for climate monitoring. The data produced by the Simulation and artificial climate can be done on a local scale in a regional climate change research. In this model, the process of production of artificial weather data is done in three sections: 1-Site Analysis 2-Model verification 3-Production and simulation of climatic variables for the future period. So the implementation of the model include: 1-Observation data 2-Analysis 3-Calibration 4-verification 5-Select appropriate General circulation model 6- Select emission scenarios 7-Simulation daily data 8-Analysis Results.

Rainfall–runoff modelling

IHACRES is a lumped metric conceptual water balance model developed by Jakeman and Hornberger [5]. IHACRES has been successfully applied to a large number of catchments varying in scale, time step and climate. Originally, IHACRES was developed to model stream flow in humid regions. The first application was realized in two small humid upland catchments in Wales [5]. The first application in semiarid to arid catchments was accomplished by Ye et al. [3] who applied a modified version of IHACRES to three low-yielding ephemeral catchments in western Australia.

The IHACRES model includes a non-linear loss module to estimate the effective rainfall and a linear routing module to model the conversion of the effective rainfall into stream flow using the total unit hydrograph incorporating both quick and slow flow components [5]. The model estimates stream flow from rainfall and temperature inputs calibrated against observed stream flow on a daily basis.

The IHACRES model is a hybrid conceptual-metric model, using the simplicity of the metric model to reduce the parameter uncertainty inherent in hydrological models while at the same time attempting to represent more detail of the internal processes than is typical for a metric model. Figure 2 shows the generic structure of the IHACRES model. It contains a non-linear loss module which converts rainfall into effective rainfall (that portion which eventually reaches the stream prediction point) and a linear module which transfers effective rainfall to stream discharge. Further modules can be added including one that allows recharge to be output. The inclusion of a range of non-linear loss modules within IHACRES increases its flexibility in being used to access the effects of climate and land use change. The linear module routes effective rainfall to stream through any configuration of stores in parallel and/or in series. The configuration of stores is identified from the time series of rainfall and discharge but is typically either one store only, representing ephemeral streams, or two in parallel, allowing base flow or slow flow to be represented as well as quick flow. Only rarely does a more complex configuration than this improve the fit to discharge measurements [5].

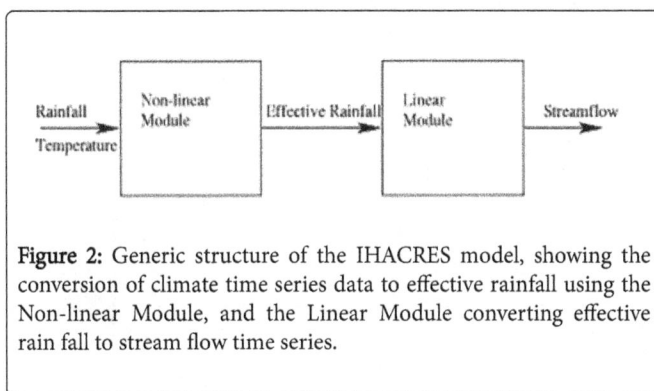

Figure 2: Generic structure of the IHACRES model, showing the conversion of climate time series data to effective rainfall using the Non-linear Module, and the Linear Module converting effective rain fall to stream flow time series.

Results

Uncertainty of climate change scenarious

Uncertainty is the absence or lack of information about the situation or the results of a process. Uncertainty cannot be completely removed but its scope will be reduced with further investigation. The greatest uncertainty in climate modelling, which features in all climate downscaling techniques, stems from the unpredictability of future anthropogenic greenhouse gas emissions and their resultant atmospheric concentrations. The IPCC Special Report on Emissions Scenarios [6] discusses several factors that impact on the atmospheric greenhouse gas concentrations projected over the present century: population growth, economic and social development, the development and utilization of carbon-free energy sources and technology and changes to agricultural practices and land use. The four storylines on which the SRES scenarios are based capture just some of the ways in which these driving forces might change [7]. Uncertainty in climate science is a case of 'imperfect knowledge and what Gershon [8] identifies as 'causes of imperfect knowledge are all present. However, due to the complexity of the climate system and the modelling process, the relationships between uncertainty types must also be considered. Uncertainty in the climate system has two main sources. First, there is uncertainty over human action, including uncertainty due to unknown future emission concentrations of greenhouse gases and aerosols. This uncertainty is largely due to unknowable knowledge, and is inherently irreducible [9]. Second, there is uncertainty over how the climate system is likely to respond to our actions. Further research may reduce this uncertainty, but may also uncover previously unknown processes, thereby increasing uncertainty. Additionally, in a complex, non-linear system the existence of unknown states or the occurrence of 'surprise' events is also possible [10].

So in this study changes in rainfall and temperature have been investigated over the period 2045-2065 by three general circulation models MPEH5 'HADCM3 and IPCM4 and scenario A2, B1 with the implementation of the LARS-WG model and finally evaluated uncertainty climate change models and A2, B1scenarious.

Monthly mean precipitations

The results of the uncertainty of estimates by taking three general circulation models MPEH5 ' HADCM3 and IPCM4 models, A2 and B1 scenario for the period 2045-2065 indicate that there is significant uncertainty in estimating the monthly, quarterly and annually in the basin. Box plot charts used in order to assess the uncertainty. Accordingly, minimum, 25% or first quartile, median or 50%, 75% or third quartile and maximum calculated then Box charts produced to determine the uncertainty climate models [11-13].

Over the period 2045-2065 the highest and lowest uncertainty of monthly precipitation in the basin under two scenarios occurred in April and October. The results showed that the monthly precipitation changes in future conditions in A2 scenario decreases percent of monthly precipitation -45.3% in October (-10.8 mm) and according to the B1 scenario, decreases -34.5 percent in October (-8.2 mm).

Due to the high and low monthly precipitation changes, based on A2 scenario percent of monthly precipitation will be changed, -35 percent in October to 25.4 percent in November (-11.8 to 12.6 mm) (Figures 3 and 4).

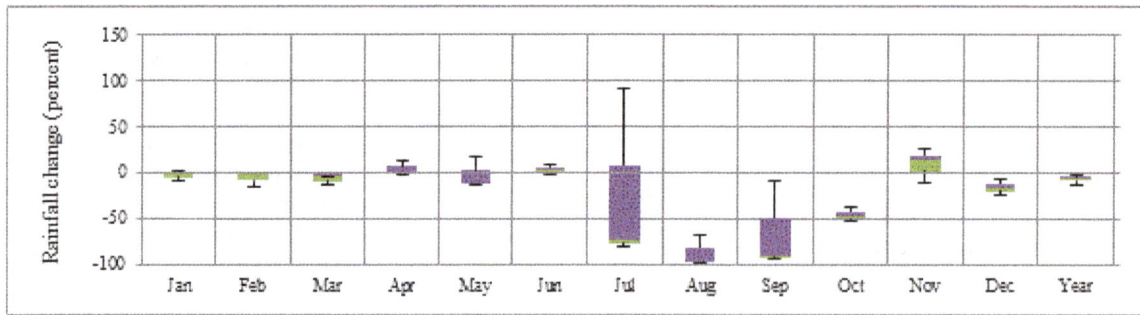

Figure 3: Boxplots showing uncertainty climate models (MPEH5 'HADCM3 and IPCM4) on monthly rain fall Change under the A2 scenarios for the future periods 2045-2065 relative to reference periods 1983-1999.

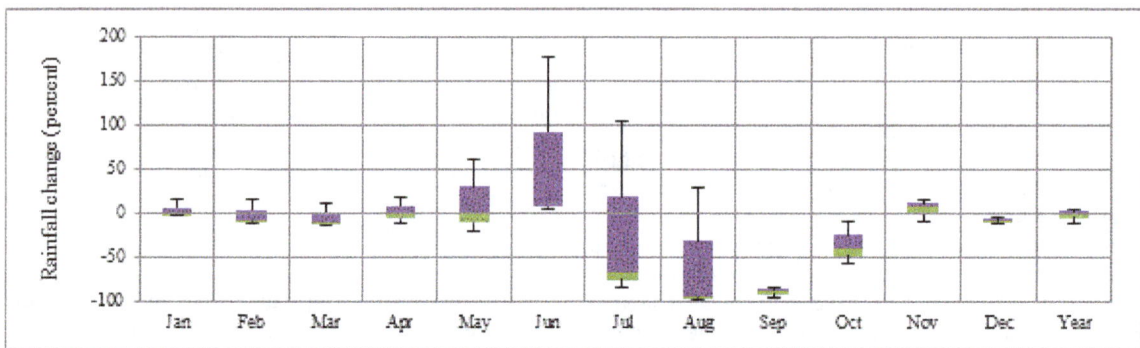

Figure 4: Boxplots showing uncertainty climate models (MPEH5 'HADCM3 and IPCM4) on monthly rain fall Change under the B1 scenarios for the future periods 2045-2065 relative to reference periods 1983-1999.

The maximum and minimum uncertainty seasonal rainfall in the basin during the period 2065-2045, according to two scenarios occurred in the spring and fall. Review on seasonal rainfall results shows that in the worst case future, in A2 scenario percent of seasonal rainfall decreases -11.4 percent in the fall (-13.1 mm) and the scenario B1, -8.1 percent in the fall (-10.4 mm). In this period the amount of monthly rainfall decrease all of the months in the year exception of the months of April, June and November. Also except spring, seasonal rainfall will decrease in other seasons. Annual rainfall will be reduced according to A2 and B1, respectively, -6.1% and -1.4% (Figures 5-7) (Table 1).

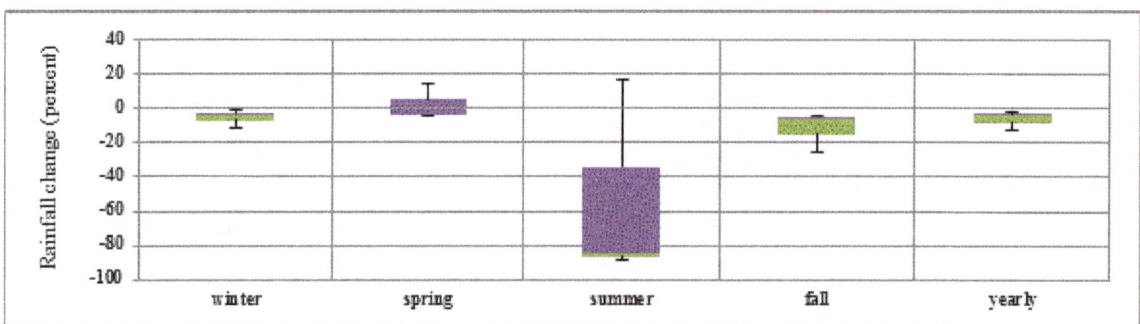

Figure 5: Boxplots showing uncertainty climate models (MPEH5 'HADCM3 and IPCM4) on seasonal rain fall Change under the A2 scenarios for the future periods 2045-2065 relative to reference periods 1983-1999.

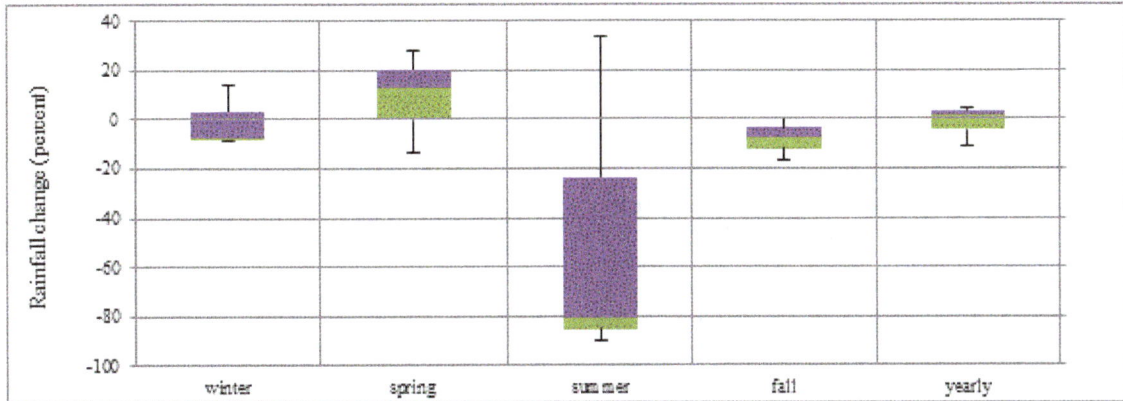

Figure 6: Boxplots showing uncertainty climate models (MPEH5 'HADCM3 and IPCM4) on seasonal rain fall Change under the B1 scenarios for the future periods 2045-2065 relative to reference periods 1983-1999.

Season	B1						A2					
	Min	Q1	Median	Mean	Q3	Max	Min	Q1	Median	Mean	Q3	Max
Winter	-9	-8.3	-7.6	-0.8	3.4	14.3	-10.9	-7.2	-3.5	-5	-2.1	-0.7
Spring	-13.3	-0.3	12.6	9.2	20.4	28.2	-4.6	-4	-3.4	2.3	5.7	14.7
Summer	-90.1	-85.2	-80.3	-45.7	-23.5	33.3	-88.8	-86.6	-84.4	-52.1	-33.7	16.9
Fall	-17	-12.1	-7.2	-8.1	-3.6	0	-24.8	-15.2	-5.7	-11.4	-4.8	-3.9

Table 1: Quartiles change in future (2045-2065) rainfall (%) relative to 1983-2009 historical period for Sub basin for three GCM models and the A2, B1 emissions scenario.

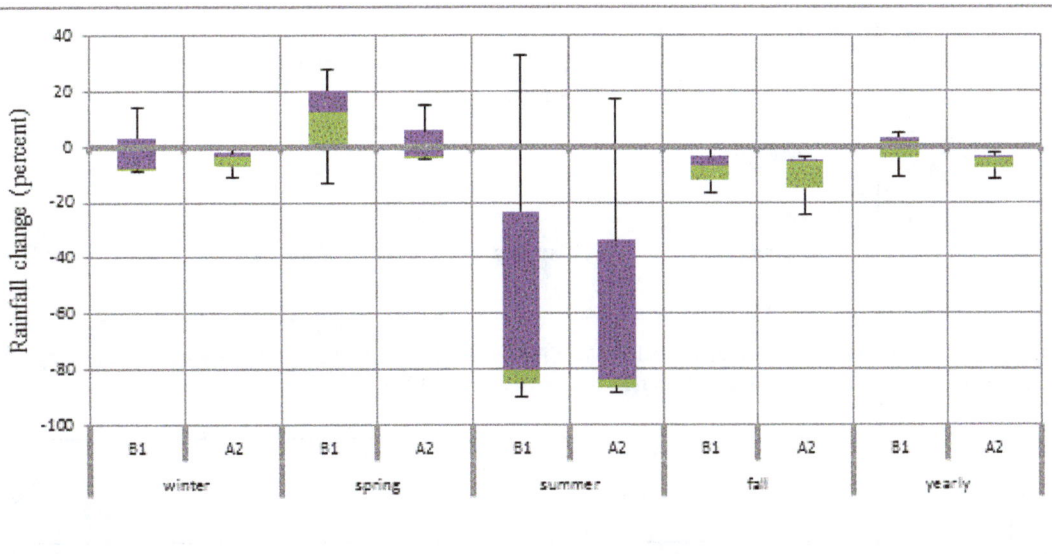

Figure 7: Boxplots comparing uncertainty climate models (MPEH5 'HADCM3 and IPCM4) on seasonal rainfall Change under the A2 and B1 scenarios for the future periods 2045-2065 relative to reference periods 1983-1999.

Average temperatures

Uncertainty of climate models under A2 and B1 for temperature estimate Show an increase in temperature of the region in the future. Survey results show that the seasonal average temperature under the scenario A2, increase 2.6°C and Under the Scenario B1, 2.2°C in summer. The average annual temperature will increase under two scenarios A2 and B1, respectively, 2.12°C and 1.79°C.

During the period 2045-2065 the highest and lowest uncertainty average monthly temperature under two scenarios in the region are in May, June and July. The results showed that temperature changes coming in the future under the worst scenario A2, average monthly temperature will increase 2.92°C and the scenario B1, 2.51°C in July.

Due to the lower and upper limits of mean monthly temperature, according to the A2 scenario will increase 1.32°C in February to 3.44°C in July.

The highest and lowest average seasonal uncertainty in the region during the period 2045-2065 under the two scenarios is higher in spring than other seasons. Upper and lower limit average seasonal temperature changes in the region based on the scenario A2 varies from 1.54°C in winter to 2.96°C in summer and based on the B1 scenario varies from 2.21°C in autumn to 2.62°C in spring. Summer temperature increases are greater than for other seasons under the two emmision scenarious (Figures 8 and 9) (Table 2).

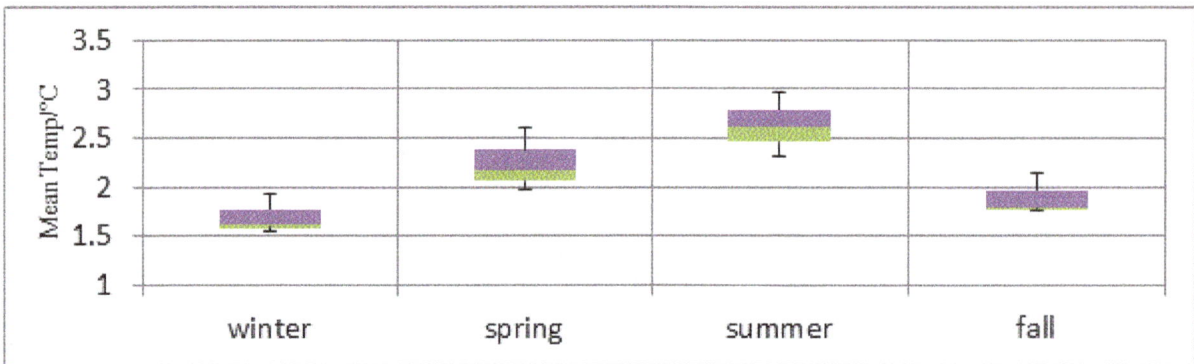

Figure 8: Boxplots showing uncertainty climate models (MPEH5 'HADCM3 and IPCM4) on seasonal Mean Temp Change under the A2 scenarios for the future periods 2045-2065 relative to reference periods 1983-1999.

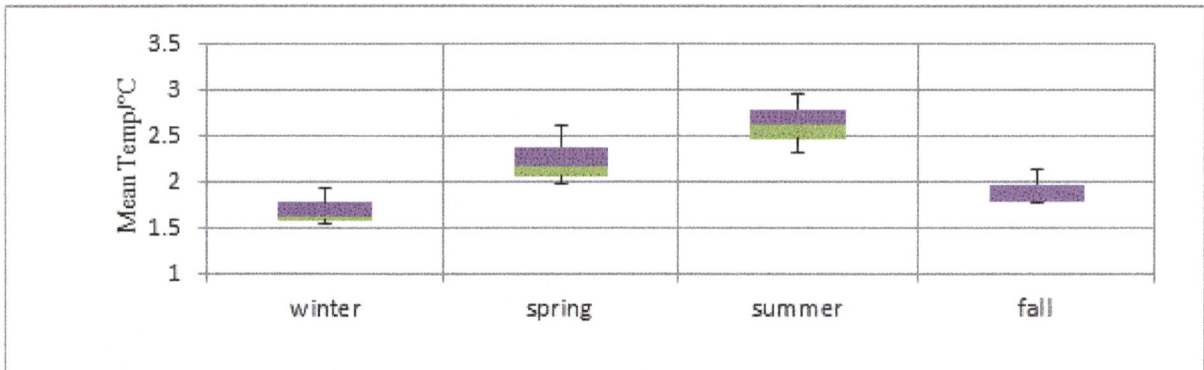

Figure 9: Boxplots showing uncertainty climate models (MPEH5 'HADCM3 and IPCM4) on seasonal Mean Temp Change under the B1 scenarios for the future periods 2045-2065 relative to reference periods 1983-1999.

Season	B1						A2					
	Min	Q1	Median	Mean	Q3	Max	Min	Q1	Median	Mean	Q3	Max
Winter	1.41	1.42	1.43	1.50	1.55	1.66	1.54	1.59	1.63	1.70	1.77	1.92
Spring	1.79	1.84	1.88	2.10	2.25	2.62	1.97	2.06	2.16	2.24	2.38	2.61

Summer	1.97	2.06	2.15	2.16	2.26	2.37	2.32	2.47	2.62	2.63	2.79	2.96
Fall	1.21	1.36	1.51	1.42	1.52	1.52	1.76	1.78	1.79	1.90	1.97	2.14

Table 2: Quartiles change in future (2045-2065) Mean temp (%) relative to 1983-2009 historical period for Sub basin for three GCM models and the A2, B1 emissions scenario.

Model Calibration and Validation

The objective of calibration and validation was to maximizing the model efficiencies and finally using the parameter values obtained through those calibration techniques. This study uses IHACRES model. The calibration was performed using observed data from the year 1983-1999 period [14]. The calibrated model was validated using independent data 2000-2009, on monthly time step at Dam Basin of Ekbatan river gauge.

The time series plot of measured monthly runoff simulated during calibration period for best simulations are shown on Figures 3 and 4, respectively and the summarized objective function results are also shown on (Table 3). And to check the validity of the model, computed runoff have been compared with field observed flow data 1983-1999. The time series plot of measured monthly flow simulated during validation period are shown on Figures 5 and 6, respectively [15].

For determining the ability of the rainfall - runoff model to simulate runoff used Nash-Satcliffe coefficient. Results showed that IHACRES model suitable for the simulation runoff in the watershed. In Table 1 monthly Nash-Satcliffe factors are 0.68 for calibration period 1983-1999 and 0.72 validation period 2000-2009.

Sub Basin	Operate	Period	R^2	Nash-Satcliffe	Observed	Simulate	MAE
Yalfan	Calibration	1983-1999	0.68	0.68	1.4	1.43	0.69
	Validation	2000-2009	0.72	0.72	1.17	1.14	0.54

Table 3: Model performance evaluation coefficients for calibration and validation of Runoff.

The time series plot of measured monthly flow simulated during validation period are shown on Figure 9 and the summarized model predictive performance values are on Table 4 a quit good match is found during monthly time period. Thus, according to Nash coefficient can be concluded that the model is capable for runoff simulation and gives acceptable results in the study basin. In general, the modelled runoff agrees quite well with the observed data (Figure 10 and 11).

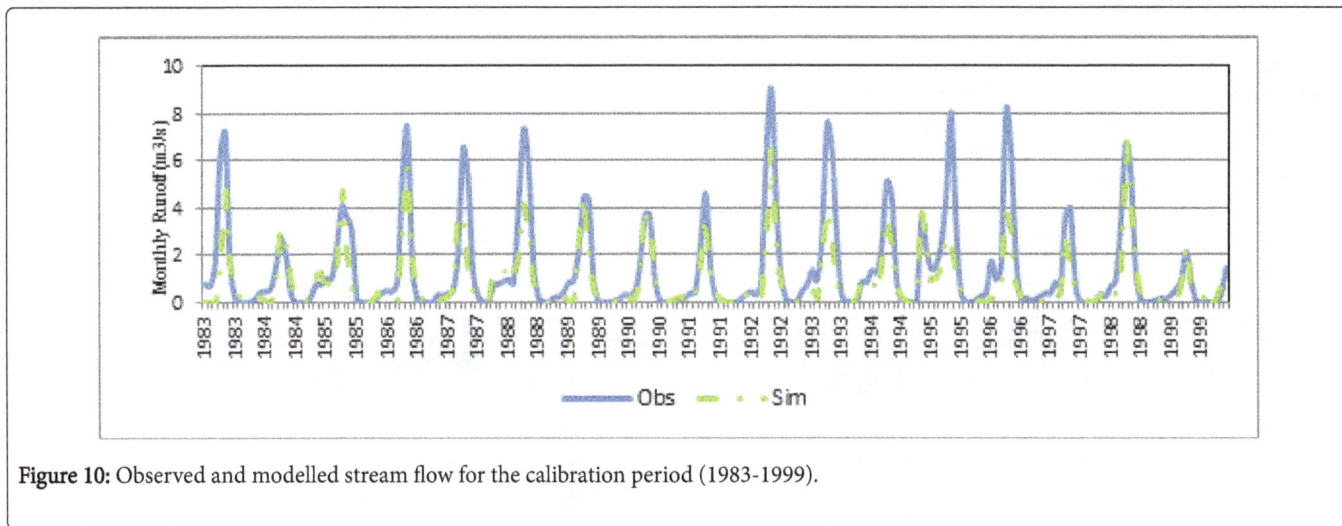

Figure 10: Observed and modelled stream flow for the calibration period (1983-1999).

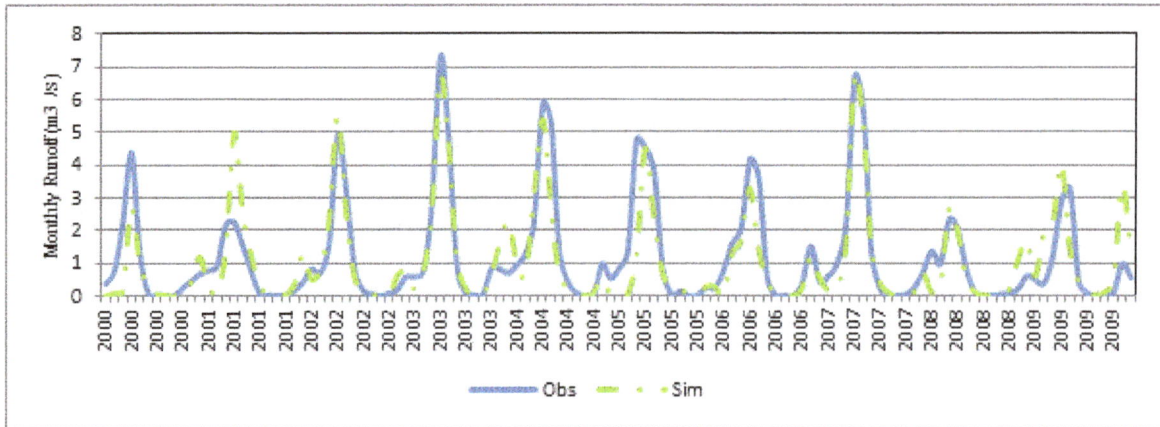

Figure 11: Observed and modelled stream flow for the validation period (2000-2009).

Hydrologic impacts of climate change

The impacts of climate change on hydrology at the catchment were quantified based on runoff simulated with the hydrological model IHACRES.

The results and the estimation of surface runoff basin provided using model IHACRES according to climate models HADCM3, IPCM and MPEH5 under climate change scenarios A2 and B1 in Figure 3 and Table 2 for 2045-2065 period. The results show that there are great uncertainty in the estimation of runoff, according to climate models HADCM3, IPCM and MPEH5 under climate change scenarios A2 and B1. According to Figures 12 and 13 graphs show the greatest uncertainties in summer and autumn and the least uncertainty in winter. The results showed that Based on the Figure14 and Table 4 runoff decreases in most months of the year according to most climate models in the basin. The climate models predict decreases in fall,

summer spring seasons flows exception winter. Consequently with larger increase in winter temperatures the liquid winter precipitation rapidly contributes to runoff instead of being accumulated in the snow cover. Thus, there is not very much snowmelt in spring to contribute to peak discharge.

According to climate models HADCM3, IPCM, MPE5 under climate change A2 and B1 scenarios increases the amount of runoff in the basin during the winter season. The average annual runoff will be reduced according to climate models. Percentage runoff changes based on climate models and emission scenarios A2 and B1 are shown in Table 4. Winter runoff increases under the tow emission scenarios. The mean annual runoff is expected to reduce by 17.2% and 19.4% in 2045-2065 period for the A2 and B1 scenarios, respectively relative to base period.

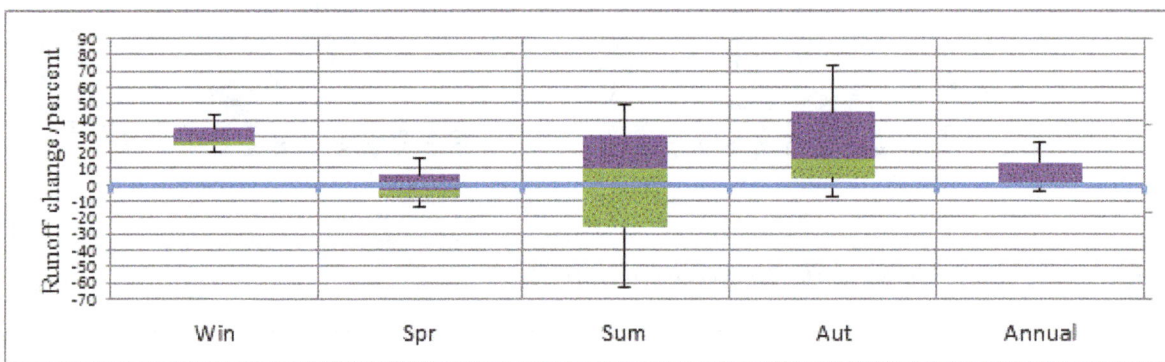

Figure 12: Uncertainty in climate models under the emissions scenario A2 for Runoff change in 2045-2065 period relative to base 1983-2009 period.

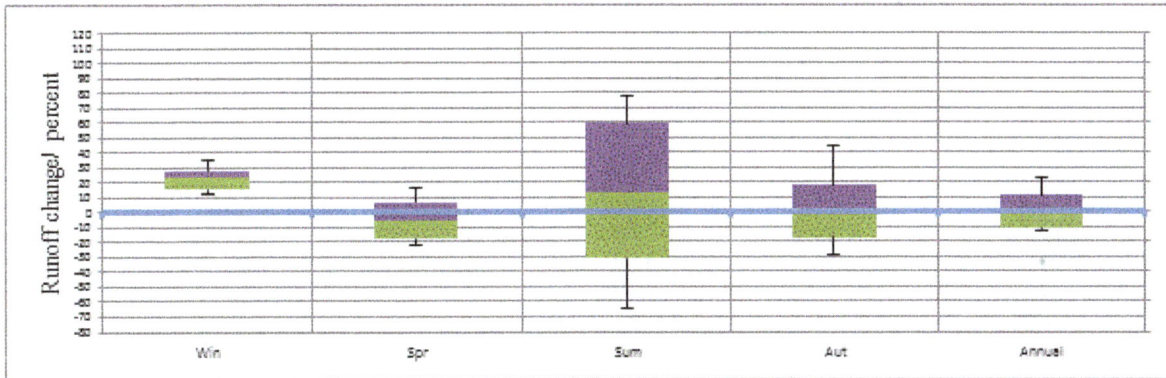

Figure 13: Relative uncertainty in climate models under the emissions scenario B1 for Runoff change in 2045-2065 period relative to base 1983-2009 period.

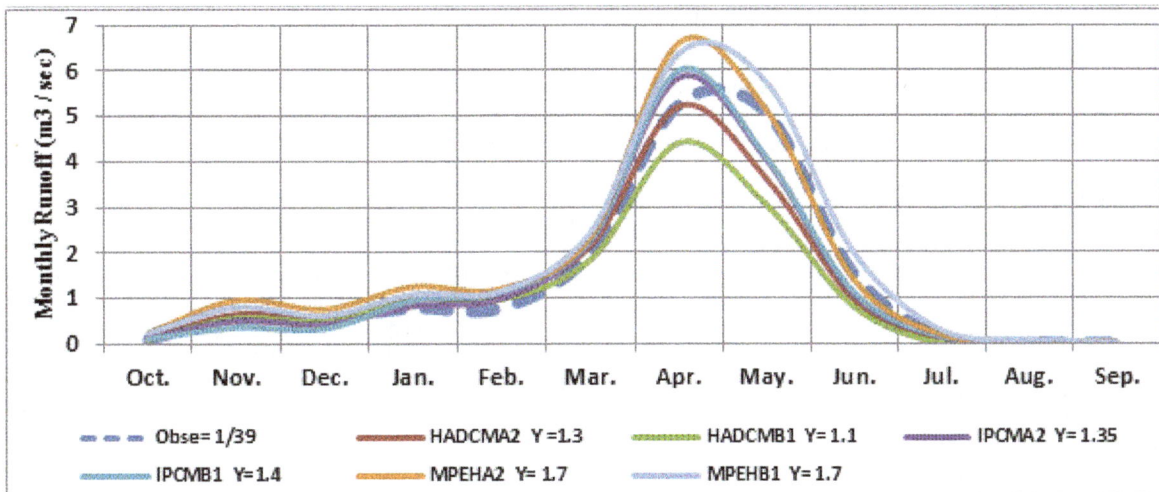

Figure 14: Monthly runoff hydrographs, 2050s time horizon for climate models under the A2 and B1 emissions scenarious.

Season	HADCA2	HADCB1	IPCMA2	IPCMB1	MPEHA2	MPEHB1
Winter	-4.9	22.1	22.7	23.2	69.4	46
Spring	-23.6	-49.4	-35.1	-29.7	-31.8	-17.1
Summer	-38.8	-70.5	-53.7	-30.7	-44.3	40.3
Fall	-43	-34.9	-53.7	-75.3	31.3	-15
Yearly	-21.2	-33.5	-24.4	-21.8	-6	-3

Table 4: Runoff change for climate models under the A2 and B1 emissions scenarious on 1945-2065 period relative to 1983-2009.

Discussions and Conclusion

The contributions of GCMs and hydrologic model parameters to prediction uncertainty are investigated from an ensemble of monthly runoff time series. The relative uncertainty to variability in hydrologic time series is demonstrated with 27 years of monthly runoff predictions for the Dam Basin of Ekbatan.

The effects of climate change, changes in temperature, precipitation and runoff area. For assessing the impact of climate change on climate parameters and runoff, there are various uncertainties which include downscaling techniques, climate models and climate scenarios. According to assessing the impact of climate change on water resources for various uncertainties affect the final results and to ignore any of them do not have the final results ensure. So for this reason, the uncertainty of climate models IPCM4, MPEH5, HADCM3 and A2, B1 scenario for the period 2045-2065 were calculated for the studied area [16]. Review monthly rainfall during the period 2045-2065 in the area showed that under the A2 scenario the greatest reduction in rainfall will happen -45.3% in October (-10.8 mm) and under the B1 scenario -3.5% in October (-8.2 mm). Evaluation of seasonal rainfall in the same period in the designated area showed that the most significant reduction in the A2 scenario -11.4% in the fall (-13.1 mm) and according to the B1 scenario, -8.1 percent in the fall (-104 mm). Annual precipitation decrease in the A2 scenario -6.1% (-12.2 mm) and according to the B1 scenario -1.4% (-0.5 mm).

The results of the seasonal average temperature over the period 2045-2065 showed that, under the scenario A2, average monthly temperature increase 2.6°C in summer and under the the scenario B1, 16/2 in summer and the average annual temperature increased under the tow emission scenarious A2 , B1 2.12°C and 1.79°C, respectively.

The results of the hydrological model simulations and estimates suggest that in the coming period, the monthly runoff hydrograph in the study basin will be changed compared to the baseline and shows move towards the winter and will be shift peak runoff in the winter (left).

On the other hand predicted hydrologic responses increased winter runoff special March and decreased fall, spring and summer. The shift in seasonal runoff has been attributed to warmer air temperatures forcing greater rain precipitation compared to snow precipitation. Subsequently less winter snow precipitation results in decreased snow melt runoff in the spring and subsequently lower base flow in summer. This shift caused by increased river flow in March and reduce in the spring. March runoff increases caused by rising temperatures and changing precipitation from snow to rain in the winter and early melting snow in the upper basin reservoirs in mountainous areas. The results of this study show that ignore climate model uncertainty huge impact on runoff and precipitation basin will be affected by climate.

Proportional to the increase in average temperature between 1.4 to 2.9°C in most months will increase evapotranspiration and Crop water requirement in the region. Therefore, it is essential to plan for the proper management of water resources by altering the pattern of plants and should be considered modification of cropping patterns and irrigation methods to adapt and mitigate the effects of climate change.

References

1. IPCC (2007) Summary for Policymarkers in: Climate Change 2007, Climate Compatibility Comprehensive Program. The Grand Karun Watershed 1.

2. Chang H, Jung IW (2010) Spatial and temporal changes in runoff caused by climate change in a complex large river basin in Oregon. J Hydr 388: 186-207.

3. Ye W, Bates BC, Viney NR, Sivapalan M, Jakeman AJ (1997) Performance of conceptual rainfall-runoff models in low-yielding ephemeral catchments. Water Res 33: 153-166.

4. Zarghami M, Abdi A, Babaeian I, Hassanzadeh Y, Kanani R (2011) Impacts of climate change on run off sin East Azerbaijan, Iran. Global and Planetary Change 78: 137-146.

5. Jakeman AJ, Hornberger GM (1993) How much complexity is warranted in a rainfall - runoff model? Water Resources Research 29: 2637-2649.

6. Nakicenovic N, Alcamo J, Davis G, de Vries B, Fenhann J, et al. (2000) Special Report on Emissions Scenarios: A Special Report of Working Group III of the Inter governmental Panel on Climate Change. New York: Cambridge University Press.

7. Foley AM (2010) Uncertainty in regional climate modelling. Progress in Physical Geography 34: 647-670.

8. Gershon ND (1998) Visualization of an imperfect world. Computer Graphics and Applications 18: 43-45.

9. Hulme M, Carter TR (1999) Representing uncertainty in climate change scenarios and impact studies. In: Carter TR, Hulme M, and Viner D (eds) Representing Uncertainty in Climate Change Scenarios and Impact Studies. ECLAT-2 Report No 1, Helsinki Workshop, 14-16.

10. Massah Bovani A (2006) Evaluation of Climate Change Risk and its Effect on Water Resources, Case Study: Zayandeh Roud, Isfahan, Ph.D. Final Thesis Report, Water EngineeringResearch Center, Tarbiat Modares University, Tehran.

11. Pruski FF, Nearing MA (2002) Runoff and soil-loss responses to changes in precipitation: A computer simulation study. J Soil and Water Cons 57: 7-6.

12. Gosain A, Rao S, Basuray D (2006) Climate change impact assessment on hydrology of Indian river basins. Current Science 90: 346-353.

13. Hamlet AF, Lettenmaier DP (2007) Effects of 20th century warming and climate variability on flood risk in the western U.S. Water Resour 43: W06427.

14. Senatore A, Mendicino G, Smiatek G, Kunstmann H (2011) J Hydr 339: 70-92.

15. Littlewood IG (2002) Improved unit hydrograph characterisation of the daily flow regime (including low flows) for the River Teifi, Wales: towards better rainfall-streamflow models for regionalisation. Hydrology and Earth System Sciences 6: 899-911.

16. Croke BFW, Jakeman AJ (2004) A catchment moisture deficit module for the IHACRES rainfall-runoff model. Environmental Modelling & Software 19: 1-5.

22

Analyzing Periodicity in Remote Sensing Images for Lake Malawi

Alinune Musopole*

University of Malawi, The Polytechnic Blantyre, Malawi

Abstract

Climate change is one of the biggest challenges that we are fighting in the 21st century. One of the indicators of climate change is lake surface water temperature (LSWT)-LSWT is expected to be periodic and a move away from periodicity verifies change in climate. With surface temperature of water on a lake obtained at high frequency both spatially and temporally, the volume of data is high. One of the ways used in reducing dimensionality of data is by approaching the data as functional data- functional principal components (fPCs) reduce dimensionality by giving modes of variation that are dominant in the data. In this paper we apply a method called principal periodic components (PPCs) that is capable of separating variability in the data into that which is nearly-periodic and that which is non-periodic, on LSWT data for Lake Malawi. We also carry out a test to check whether there is any exact annual variation in the data or not. The data are remote sensing images. The analysis has shown that there is no any exact annual variation in LSWT data for Lake Malawi- LSWT for Lake Malawi, though with strong periodicity, is not strictly periodic.

Keywords: Climate change; Remote sensing; Precipitation; Evaporation

Introduction

Malawi, just like the rest of the world, is facing several challenges relating to climate change. Changes in precipitation are one of the problems that Malawi has been facing. Reduction in precipitation that the country has faced in some growing seasons led to scarcity of water (Population Action International and the African Institute for Development Pol- icy, 2012), which in turn led to food crisis. In 2001/2002 and 2004/2005 growing seasons about 2, 829, 435 and 5, 100, 000 Malawians, respectively, were affected by drought and food assistance was required [1]. In 2006 Malawi was forced to import food to distribute in the lean months of January to March due to 2004/2005 drought which affected food production [1]. The areas that are mostly hit by droughts are Karonga, Salima, Zomba, Nsanje, and Chikwawa [1]. Apart from droughts, Malawi is also frequently hit by floods, with the lower Shire being the most affected area. Droughts and floods leave no time for farmers to recover [2]. The frequency of floods and droughts has been increasing, with the number of individuals affected increasing too [2,3].

Changes in growing seasons that we are currently facing are due to climate change [2]. Growing seasons no longer start at the time they used to start [2]. We are also currently experiencing short growing seasons, which are characterized by prolonged dry spells [4]. Due to unpredictable changes in growing seasons, farmers are unsure of when to plant [2]. In a quest to increase production, farmers have resorted to expensive hybrid crops [2] making crop production expensive.

The drying up of Lake Chilwa in 1995/1996 and 2011/2012 [5], simply confirms that we are experiencing a different climate to what it used to be. Temperatures in the lower Shire were noticed to be increasing [6]. Change in temperature is one of the indicators of climate change. Climate change, specifically change in temperature, affects many aspects of aquatic life. For example, water temperature controls growth, development, and survival of cold-blooded animals [7]. Change in temperature, above or below the preferred range, is likely to lead to decrease in number of many aquatic organisms [8]. The rise in temperature has an effect on metabolism, life cycle, and behavior of some marine species [9]. The migration of fish and other species from shallow to deep waters (to areas where they can survive) is associated with change in temperature [8]. Change in water temperature is associated with many changes in the aquatic life, hence deep waters are considered to be good indicators of climate change. Deep waters are sensitive to changes of the environment [10].

Lake surface water temperature (LSWT) data, in form of remote sensing images, pro- vided by the Arc lake Project are obtained through the use of satellites. Surface water temperature observations, within a lake, are obtained at various locations and for each location at different times (the data are obtained both spatially and temporally). Advanced along Track Scanning Radiometers (AATSRs) are used to derive these observations. With LSWT observations obtained at high frequency both spatially and temporally, the volume of data to handle is high. One of the ways through which computational burden can be reduced in this case is by approaching the data as functional data and transforming the data into principal components [11] principal components reduce the dimensionality of data by giving modes of variation that are dominant in the data.

Most climate data, for example LSWT, are characterized by periodicity due to annual effects. With the climate changing, there is a possibility that LSWT recordings of a particular location on a lake, for each specific time of the year, may not be similar for different years. Thus one of the ways through which climate change can be observed is by examining the periodicity in LSWT observations. Liu et al. [12] provides a method called principal periodic components (PPCs) that are capable of separating the components that explain the variability in the response into those that are nearly-seasonal and those that are not.

The purpose of this study is to find out whether there is any exact annual variation (component of LSWT that occur repeatedly each

*****Corresponding author:** Alinune Musopole, University of Malawi, The Polytechnic Blantyre, Malawi, E-mail: alinunemusopole@gmail.com

year/that is seasonal) in LSWT observations for Lake Malawi or not. Elements of climate such as precipitation, temperature and wind conditions have an influence on LSWT. We expect these elements to be periodic; hence their effect on LSWT should also be periodic. If there is no any exact annual variation in LSWT, then the elements of climate that affect LSWT are changing- which is an indication of climate change. LSWT observations for the study are remote sensing images which are processed into continuous data.

This study demonstrates one of the possibilities on the utilization of deep waters in carrying out climate change-related studies. Apart from looking at climate change using LSWT, this study looks at LSWT of Lake Malawi in relation to the lake being a home to biodiversity and a global heritage [13]. In the quest to protect biodiversity in Lake Malawi, there is a need to monitor the temperature of water which is an important component when it comes to survival of various living organisms in the lake- this paper addresses just part of such studies.

Data description

LSWT data for Lake Malawi is considered in this study. The data are provided by the Arc lake Project. The data are in form of remote sensing images which are then processed in to continuous data. The images were taken at a spatial resolution of one kilometer. Advanced along Track Scanning Radiometers (AATSRs) were used to derive these observations of lake surface water temperature. The observations were made at different locations within the lake, and at different times. We will consider monthly observations from January 1992 to December 2011 (228 months) in order to have only complete cycles for easy application of some of the techniques that will be used in the analysis. We have 1061 spatial locations whose time series are under consideration (Figure 1) shows these spatial locations.

Bootsma and Jorgensen (n.d.) provide some details on Lake Malawi. Lake Malawi, which is the ninth largest lake in the world, the third deepest freshwater lake in the world, and a home to greater diversity of fish species than any other lake, is a global heritage. The lake contains about 7% of the available surface water and it covers an area of about 29, 500Km2, with a volume of about 7, 775Km3. The maximum depth of the lake is about 700m, with a mean depth of about 264m. The largest catchment area of the lake is in Malawi, followed by Tanzania, then Mozambique. The level of water on the lake is greatly influenced by precipitation and evaporation; hence it is likely to be affected by climate change. The benefits from the lake include fisheries, water for irrigation, transportation, and generation of hydroelectricity.

Methodology

The surface temperature of water on Lake Malawi was obtained at high frequency both spatially and temporally, hence the volume of data is high. We will approach the data as functional data in order to reduce dimensionality- functional principal components (fPCs) and principle periodic components (PPCs) reduce dimensionality by giving modes of variation that are dominant in the data [14]. Since the data will be approached as functional data and we intend to study the periodicity in the data, the method of principal periodic components (PPCs) which is capable of separating variability in the data into that which is nearly-periodic and that which is non-periodic will be used. The method also involves checking whether there is periodic variation in the data or not. The details for the method are provided by [12].

Data representation

Continuous observations collected at multiple times and assumed

to be from a smooth process can be approached as functional data [15]. With functional data, an entire time series function is viewed as a single datum [14]. Given N time series each with T observations, the jth observation for the ith time series, yij , is represented in functional form as

$$y_{ij} = x_i(t_{ij}) + \varepsilon_{ij} \tag{1}$$

ij s are errors which are assumed to have a 0 mean and a constant variance. xi (tij) is the jth functional observation for the ith time series. Thus the ith time series is given as yi = xi (ti) + i. The functional curve xi (ti) is a linear combination of basic functions. With Φ (t) = (1, Φ1(t), Φ2(t), · · ·, ΦK (t)) a set of basic functions and Φj (t) the jth basis function,

$$x_i(t_i) = \sum_{j=1}^{K} c_j \Phi_j(t_i) \tag{2}$$

With correlated data, both ordinary cross-validation and generalized cross-validation methods of choosing the smoothing parameter lead to under-smoothing [16] considering the relationship between λ and degrees of freedom, which are used to smooth the data, can be used in selecting the smoothing parameter in such a case. λ controls the degrees of freedom in the sense that increasing the smoothing parameter reduces the degrees of freedom, and vice versa [14]. λ and degrees of freedom are related in the following way:

$$df(\lambda) = trace[S(\lambda)] \tag{3}$$

Where S(λ) is the smoothing matrix and is given as

$$S(\lambda) = \Phi[\Phi^T\Phi + \lambda R_L]^{-1}\Phi^T \tag{4}$$

RL is the penalizing matrix.

Functional principal components (fPCs)

Functional principal components are used in summarizing variation in the data [15] they give the structure of variability in the data. fPCs, γi(t), define strong modes of variation by maximizing the variance

$$var\left[\int \gamma k(t)x_i(t)dt\right] \tag{5}$$

Subject to the constraints $\int \gamma k(t)^2 dt = 1$ and $\int \gamma k(t)\gamma_d(t)dt = 0$, with k≠d.

VARIMAX rotation described by Ramsay and Silverman [15] provides a separation between the most common modes of variation and those that are not common [14]. VARIMAX rotation is not capable

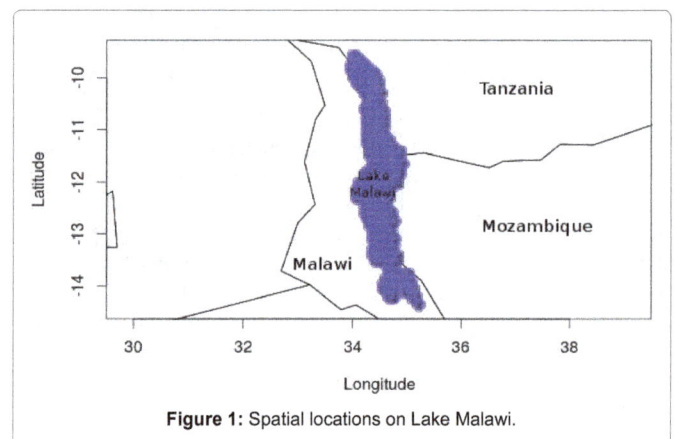

Figure 1: Spatial locations on Lake Malawi.

of separating fPCs into those with strong annual periodicity and those without strong annual periodicity, hence we will adopt the method of principal periodic components (PPCs) described by [12]. Using a small number of fPCs to explain variability implies having less flexibility in explaining the variability [12] after rotating the fPCs, each fPC will have a higher variability to explain than if many fPCs were used [14]. On the other hand, using many fPCs, despite that it offers flexibility, means much concentration is put on individual time points [12], and each fPC has less variability to explain (after rotation).

Principal periodic components (PPCs)

The method of principal periodic components starts by defining benchmarks [12]. Given a vector of basic functions, fk with $1<k<P$ (P is limited to the set of periodic coefficients used to smooth the data), a space FP with functions that are strictly periodic is spanned from the basic functions. This space contains benchmarks. Benchmarks are linear combinations of basic functions. Given a vector of fPCs, γ of M dimension, a space ΓM is spanned. The method involves rotating the two spaces (space spanned by benchmarks and that spanned by fPCs) and then trying to align them in the same direction. Rotated fPCs are compared to their associated benchmarks. The standardized inner-product of the fPC and the associated benchmark provides a measure of periodicity of the fPC that is to say the method orders rotated fPCs depending on their periodicity. Full details of the method of principal periodic components are provided by Liu et al. [12].

Consider U and V as rotation matrices on γ, vector of M fPCs, and f , vector of annual functions which are linear combinations of basic functions, respectively. The rotation matrices result from singular decomposition of the inner product of the spaces ΓM and FP .With utj the jth row of U and vt the jth row of V , the jth principal periodic component (PPC), ξj which is a rotated fPC, is given as

$$\xi_j = u_j' \gamma \qquad (6)$$

And the jth benchmark, θj which is the rotated annual function, is given as

$$\theta_j = v_j' f \qquad (7)$$

Since we are trying to align the two functions in the same direction, the closeness measure can be looked at in terms of the angle between ξj and θj which is given as

$$\rho_j = \frac{<\xi_j, \theta_j>}{\| \xi_j \| \| \theta_j \|} \qquad (8)$$

Liu et al. [12] uses annual information (cumulative variation explained by benchmarks expressed as a proportion of the cumulative variation explained by PPCs) to separate the PPCs into those that are nearly-annual and those that are not. The annual information is given as

$$AI_j = \frac{\sum_{k=1}^{J} \lambda_k^\theta}{\sum_{k=1}^{J} \lambda_k^\xi} \qquad (9)$$

K and λk are the proportions of variation explained by the kth benchmark and PPC respectively. An elbow on the plot of annual information is used as a cut off. PPCs before the cut off are described as nearly-annual while those after the cut off are non-annual. Assuming nearly-annual variation and non-annual variation in the data, the curves can be decomposed as

$$z_i(t) = \hat{\mu}(t) + \sum_{j=1}^{J} s_{i1}^\xi \xi_j(t) + \sum_{j=J+1}^{m} s_{ij}^\xi \xi_j(t) + \sum_{j=m+1}^{K} s_{ij}^\gamma \gamma_j(t) \qquad (10)$$

Where $1 \le i \le N$, N is the number of spatial locations; $1 \le t \le T$, T is the number of times observations have been taken; m is the total number of nearly-annual and non-annual components; J is the number of nearly-annual components; K is the total number of fPCs (K = M in this case); $\mu\hat{}(t)$ is the sample mean function; $\sum_{j=1}^{J} s_{i1}^\xi \xi_j(t)$ is the nearly annual part; $\sum_{j=J+1}^{m} s_{ij}^\xi \xi_j(t)$ Is the non-annual part; $\sum_{j=m+1}^{K} s_{ij}^\gamma \gamma_j(t)$ Part with components that explain a very small part of variation; ξ_j are PPCs and scores of zi(t) (standardized time series curves) on ξj are given by $s_{ij}^\xi = \int T \overline{Z}_i(t) \xi_j dt$, where $\overline{z}_i(t) = z_i(t) - \hat{\mu}(t) \gamma_j$. γ_j are the fPCs; s_{ij}^γ given by $\int T \overline{Z}_i(t) \gamma_j dt$ is the fPC score of $z_i(t)$ on γ_j

Test for periodic variation

Liu et al. [12] provides a procedure for testing whether there is any exact annual variation in ΓM (space spanned by the M fPCs) or not- with the first PPC having a strong correlation ($\rho 1$) with its associated benchmark, the procedure involves testing the null hypothesis that this correlation is 1, against the alternative hypothesis that the correlation is less than 1. That is:

$H_0: \rho_1 = 1$, against

$H_1: \rho_1 < 1$. $\qquad (11)$

To formulate the null hypothesis, either M, the number of fPCs that span the space ΓM, or the percentage of variation to be explained can be fixed. The test involves checking if there is an intersection between leading fPCs and predefines Subspaces in the space that is nearly-annual [12].

In generating the null distribution for $\rho 1$, data are slightly altered and least-favorable covariance is found- then bootstrap is applied. The hypothesized curves that are used to approximate the functional covariance under null hypothesis are given by

$$z_i^*(t) = s_{i1}^\xi \xi_1(t) + \sum_{j=2}^{m} s_{ij}^\xi \xi_j(t) + \sum_{j=m+1}^{K} s_{ij}^\gamma \gamma_j(t) \qquad (12)$$

We set K to be the total number of fPCs and m the number of PPCs returned by PPCs method- $1 \le i \le N$ and $1 \le t \le T$ where N is the number of spatial locations and T is the number of times observations have been made at each spatial location. $\xi 1$ is the first PPC. Given that the correlation between ξj and its associated benchmark is strong, the first PPC can be replaced by its benchmark $\theta 1$ and zi* is expected not to change much. Replacing the first PPC by its benchmark, we get the replacement curves

$$\tilde{z}_i(t) = s_{i1}^\xi \theta_1(t) + \sum_{j=2}^{m} s_{ij}^\xi \xi_j(t) + \sum_{j=m+1}^{K} s_{ij}^\gamma \gamma_j(t) \qquad (13)$$

Bootstrap procedure is applied by sampling z~i with replacement (and denote z~b(tij) as the results), and residuals obtained from pre-smoothing are also sampled with replacement (denoted ebij) to accommodate the effect of pre smoothing. The bootstrap curves and residuals are added (new observations are Wb ij = ~zb i (tij) + ^ebij with $1 \le b \le B$ where B is the number of times sampling has been done) and then re-smoothing is done. PPCs and first correlations are computed. The process is done several times. With the correlation between the first PPC and the associated benchmark not sufficiently large enough, Liu et al. [12] provides an approach for handling such a case. If the correlation between the first pair of the PPC and the benchmark is less than 0.05 critical value of the null distribution for $\rho 1$, the null hypothesis is rejected in favour of the alternative hypothesis.

Analysis

We are considering 1061 different locations on Lake Malawi on which LSWT observations were made. For each location, 228 observations were made. (Thus we have 1061 time series, each with 228 time points.) We are considering monthly observations from January 1992 to December 2011. The observations were taken in form of remote sensing images, and then processed into continuous data. The maximum temperature observed is 29.599°C and the minimum recording is 21.486°C (Figure 2) suggests that LSWT observations of a particular location on Lake Malawi, for each specific time of the year, are similar for different years- suggesting the presence of periodicity in the data. The figure shows that maximum temperatures on the lake are experienced around the months of September to April while minimum temperatures are experienced in the months of May to August in all years Malawi experiences a cool season from May to mid-August; a hot season is experienced from mid-August to November; the rainy season is experienced in the months of November to April. Thus the changes in LSWT from month to month can be attributed to change in seasons. With time plots in Figure 2 showing the presence of peaks and troughs, LSWT recordings may have periodicity. Despite LSWT observations of a specific time of the year being similar for different years (what seems to be strong periodicity in the data), the temperatures (for a particular time of the year on a particular location) seem not to be very similar for different years as observed from minimum and maximum temperatures through the years- the periodicity seems not to be strict.

We are looking at repeated measures of the same process (temperature). The observations are continuous and Figure 2 shows that LSWT observations for each location change smoothly. With these characteristics in mind, we represent the data in functional form. The time series for the various locations seem to share a sinusoidal shape. Thus we use Fourier basis to smooth the data. 229 functions are used (fitting a saturated Fourier basis), then we penalize the total curvature. A smoothing parameter, λ, of 10^{-1} is used The focus of analysis is not on the overall mean and the total variability but on the direction of variation, hence we consider standardized time series [10] mean centering is done to remove vertical variation while dividing by the square-root of the variance makes the mean-centered time series have a variance of one. Figure 3 gives the standardized and smoothed time series.

We intend to apply the PPCs method on the LSWT data (standardized and smoothed LSWT time series) for Lake Malawi. We fix the number of fPCs, M, that span the space TM. We set M=166, we use a large number of fPCs in order to put much focus on individual time points since we are interested in the direction of variation. Using a less number of fPCs will offer less flexibility. Approximately 99.998% of variation in LSWT for Lake Malawi is explained by leading 166 fPCs (explained by the space T166). 229 Fourier basis functions are used to span the space of benchmarks. We have monthly observations from January 1992 to December 2011, thus the frequency is 20 and the period, P, is 12. Figure 4 gives the total variation explained by a specific number of fPCs and PPCs. The plot for fPCs (blue plot) simply shows the decrease in variation explained by fPCs as we move from one leading fPC to the next. The plot for PPCs does not have a clear sinusoidal shape (sinusoidal shape depicts the presence of periodicity) due to the large number of fPCs we have decided to keep. For a specific number of principal components, the variation explained by PPCs is much lower compared to that explained by fPCS which shows low annual variation.

P=12 and M>P, hence the PPCs method used only the first 12 fPCs

in ΓM   and it (PPCs method) returned only 12 PPCs. Only the first 12 fPCs have associated bench- marks to compare to. The measure of periodicity for each PPC is provided by the co-relation between the PPC and the associated benchmark. The PPCs are arranged in descending order based on their periodicity. The correlations between the PPCs and their associated benchmarks are provided in Figure 5. For the first six PPCs, the co-relations between PPCs and their associated benchmarks are strong. The drop in correlations is noticeable after the

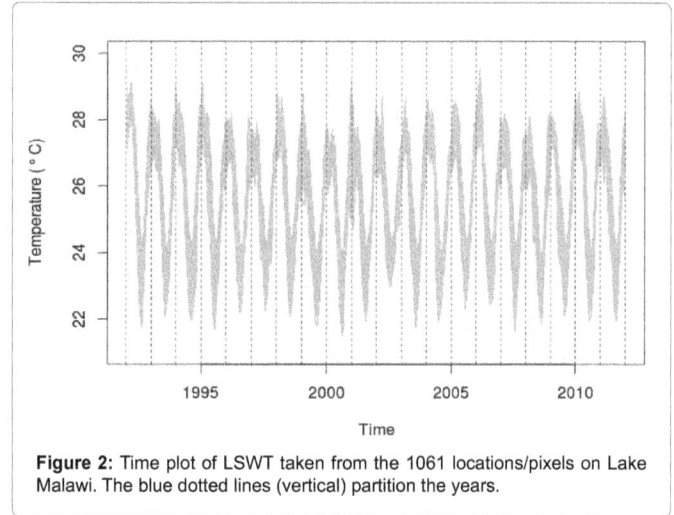

Figure 2: Time plot of LSWT taken from the 1061 locations/pixels on Lake Malawi. The blue dotted lines (vertical) partition the years.

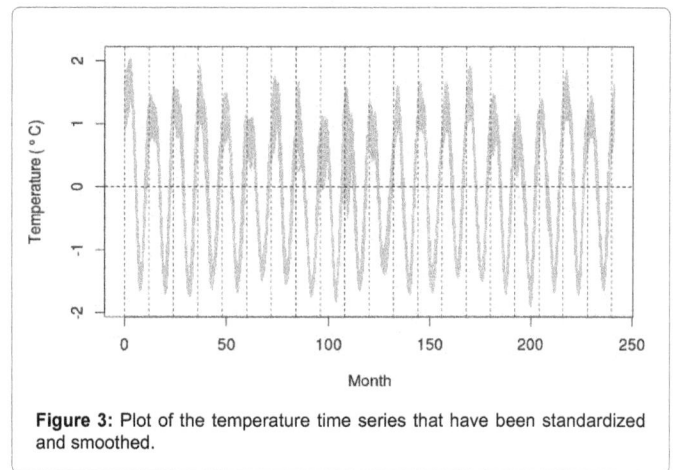

Figure 3: Plot of the temperature time series that have been standardized and smoothed.

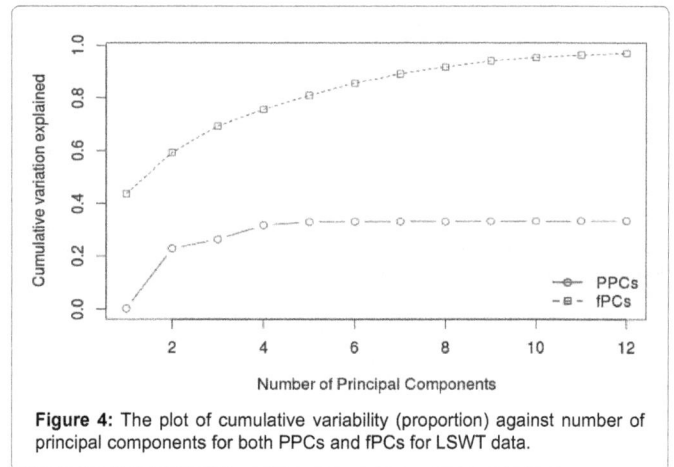

Figure 4: The plot of cumulative variability (proportion) against number of principal comportents for both PPCs and fPCs for LSWT data.

sixth PPC. The strong correlations between the first six PPCs and their associated benchmarks suggest that there is periodicity in LSWT data for Lake Malawi. The periodicity is as a result of annual factors such as seasons. There are also PPCs with less strong and weak correlations with their associated benchmarks these PPCs suggest that LSWT for Lake Malawi is also affected by non-annual factors.

Figure 6 shows 4 of the 12 PPCs together with their associated benchmarks. Plot of the first pair (Pair 1) which has the strongest correlation (approximately 0.999) among all the pairs shows that there is much resemblance between the PPC and its associated benchmark which is strictly periodic. There is also strong resemblance between the second and third PPCs and their associated benchmarks. For the first three PPCs, though there is much resemblance, there seem to be some slight departure from being strictly periodic. The twelfth PPC clearly departs from being strictly periodic.

Figure 7 (Scree plot), which is a plot of annual information (AI), separates the PPCs into those that are nearly annual and those that are not. There is a very sharp drop on how good the PPCs are at explaining annual variation in LSWT data for Lake Malawi after the sixth PPC- thus the cut off is 6. In Figure 5 we also notice a slight big drop in correlation between PPCs and benchmarks after the sixth PPC. The first six PPCs are nearly-annual, while the rest (the other six) are non-annual. Thus there is strong annual variation (periodicity) in LSWT data for Lake Malawi (evidenced from the six PPCs which are nearly-annual). The six non-annual PPCs show that there are also non-annual factors that are making the data less periodic.

There are both nearly-annual variation and non-annual variation in LSWT data for Lake Malawi. Hence the LSWT curves for Lake Malawi can be decomposed into three parts- seasonal, unseasonal,

and noise. Figure 8 gives plots of the noise (top plot) remaining after the effect of the 166 fPCs specified is removed; the middle plot gives the nearly- annual (seasonal) part of the data; while bottom plot gives the non-annual (unseasonal) part of the data. It Figure 8 shows that the patterns of LSWT time series for Lake Malawi are influenced by both annual and non-annual factors. By looking at the plots, a big proportion of variation is nearly-annual (annual factors have a great influence on LSWT).

Given that the correlations between the first six PPCs and their respective benchmarks are strong (there is strong periodicity in LSWT data for Lake Malawi), we are interested in finding out whether there is/are any component/s in LSWT data that occur repeatedly each year (that is/are cyclic) or not we want to find out if there is any exact annual variation in T166. We follow the procedure described in Subsection

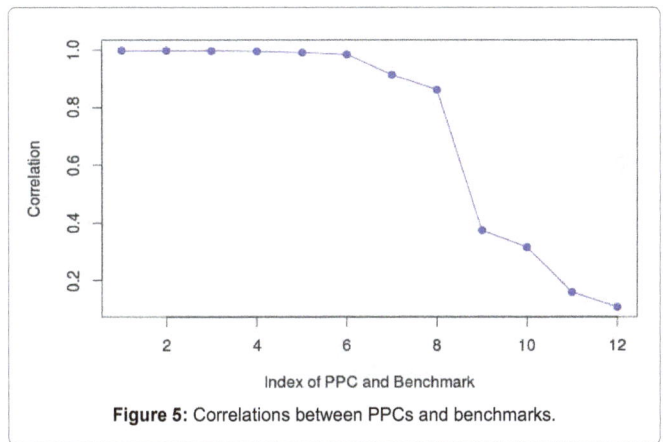

Figure 5: Correlations between PPCs and benchmarks.

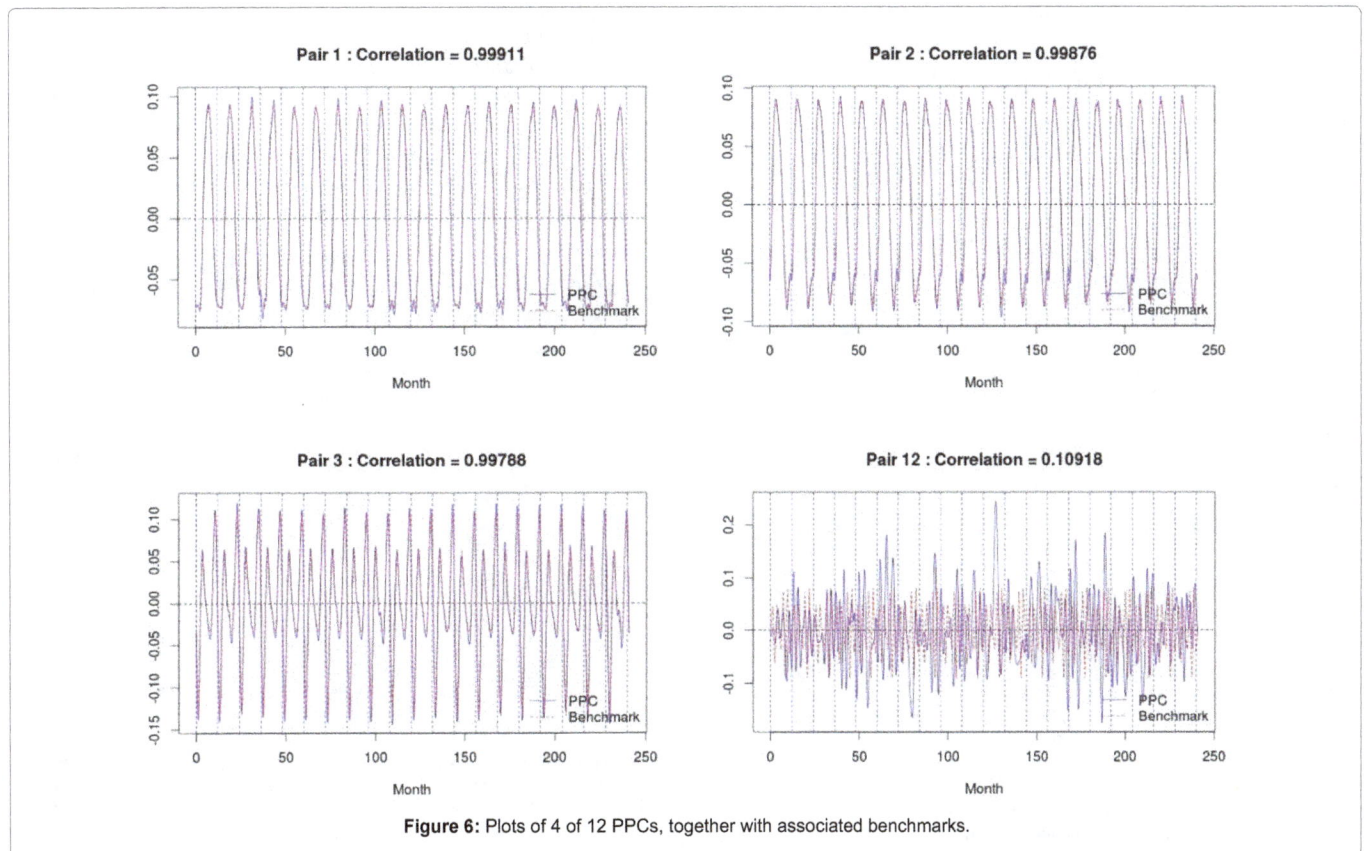

Figure 6: Plots of 4 of 12 PPCs, together with associated benchmarks.

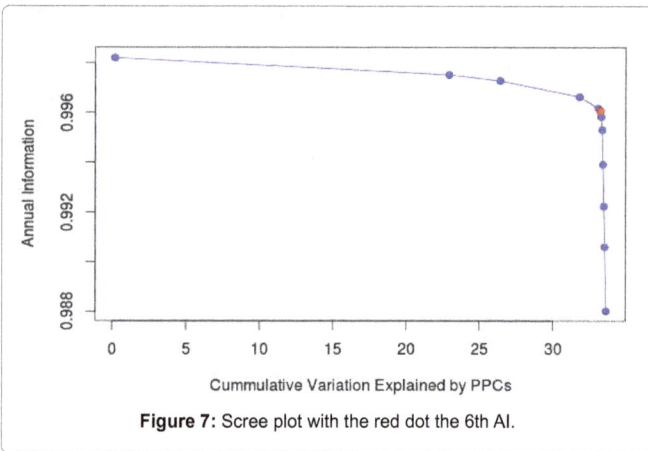

Figure 7: Scree plot with the red dot the 6th AI.

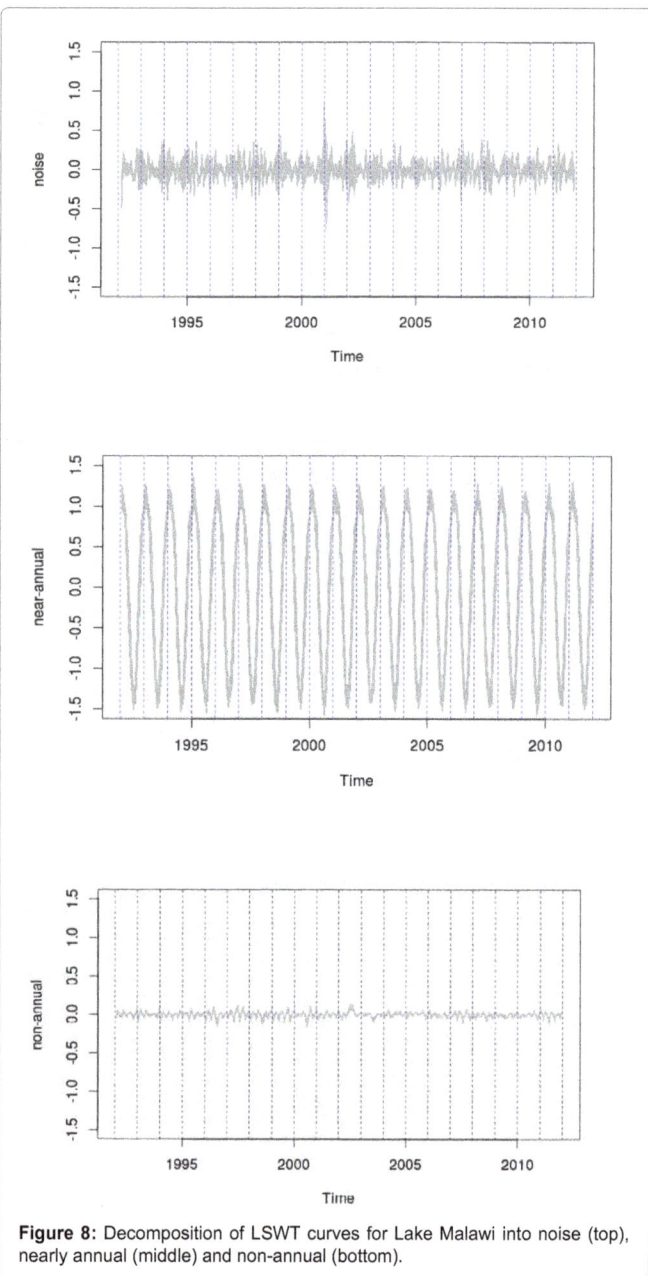

Figure 8: Decomposition of LSWT curves for Lake Malawi into noise (top), nearly annual (middle) and non-annual (bottom).

4.4. Despite that the periodicity in LSWT data for Lake Malawi is strong; Figure 9 shows that it is not strict. There are no components in the data that occur repeatedly each year. This could be as a result of climate change. With no any exact annual variation in LSWT data for Lake Malawi, we can conclude that LSWT for Lake Malawi is not strictly periodic.

Conclusion

Both annual and non-annual factors affect LSWT on Lake Malawi. Annual factors have a stronger influence than non-annual factors- this may be the result of most elements of climate being strongly periodic. Climate change may also be a cause of lack of strict periodicity in most elements of climate, resulting in LSWT lacking strict periodicity. The study has not identified specifically which factors are annual and which ones are non-annual. Identifying and studying changes in such factors is important in combating climate change and protecting biodiversity. We propose future research work to consider looking into specific factors that affect LSWT. The study has also not looked into the direction in which the changes are taking if there are extreme temperatures recorded on the surface of Lake Malawi.

Despite that LSWT for Lake Malawi has strong periodicity, it (periodicity) is not strict. Thus LSWT recordings for a particular time of the year are expected not to be very similar for different years. This is evidence that the temperature that the aquatic life in Lake Malawi is exposed to is changing. There are many aquatic organisms in Lake Malawi whose survival, growth and development depend on the temperature of the water. An example of such organisms is the various species of fish in the lake. There are millions of Malawians whose everyday survival depends on fisheries as source of employment, food and raw material. Each of the species of fish (cold-blooded animals in general) in Lake Malawi has a preferred range of temperature. Many species are likely to die when the temperature of water gets outside their preferred range. If the temperature of water in the lake (LSWT) will keep changing, one day we will wake up only to find that the surface of Lake Malawi is covered by a carpet of dead fish- which is a blow to biodiversity and a nightmare to those whose livelihoods heavily depend on fisheries. Some fish species (cold- blooded animals in general) are already scarce in places where they used to be abundant.

LSWT or water temperature of the lake in general, is greatly influenced by the elements of climate. Thus most changes that happen to the climate also affect the aquatic life, making deep waters a tool that

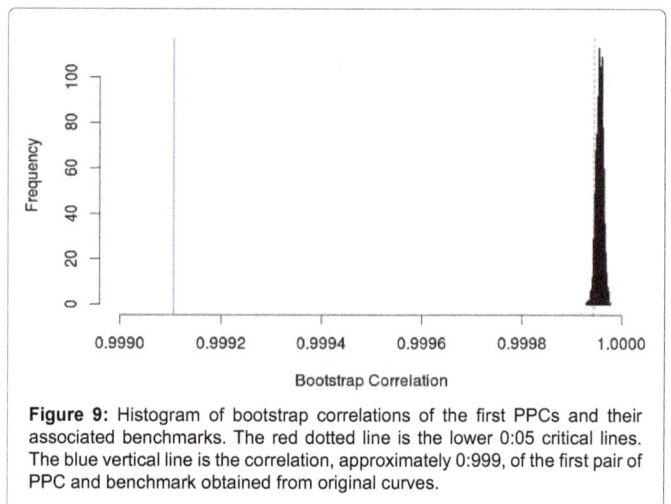

Figure 9: Histogram of bootstrap correlations of the first PPCs and their associated benchmarks. The red dotted line is the lower 0:05 critical lines. The blue vertical line is the correlation, approximately 0:999, of the first pair of PPC and benchmark obtained from original curves.

can be used in detecting and measuring climate change. There is no any exact annual variation in LSWT for Lake Malawi; hence the elements of climate that affect LSWT are changing which is an indication of climate change.

Human activities have been associated with much of climate change and deterioration of the environment. Lake Malawi has only been spared from the horrors faced by Mudi and Lilongwe rivers in Blantyre and Lilongwe respectively for the fact that no major towns and cities are close to the lake. If we are to save our biodiversity in Lake Malawi, we need to stop climate change. The temperature of water on the surface of Lake Malawi is changing, an indication that water temperature in the lake is changing- this is likely to affect the aquatic life, which in turn is likely to affect millions of people. Climate change is taking place in Lake Malawi.

Acknowledgement

The Commonwealth Scholarship Commission (CSC) the University of Glasgow, and African Institute for Mathematical Sciences (AIMS).

References

1. World Bank (2010) Malawi: Economic Vulnerability and Disaster Risk Assessment in Malawi and Mozambique, Washington D.C.

2. Action Aid (2006) Climate Change and Smallholder Farmers in Malawi: Understanding poor people's experiences in climate change adaptation.

3. Chabvungma SD, Mawenda J, Kambauwa G (2006) Drought Conditions and Management Strategies in Malawi.

4. Population Action International and the African Institute for Development Policy (2012) Population, Climate Change, and Sustainable Development in Malawi.

5. Jamu D, Chapotera M, Likongwe P, Chiotha S (2012) Is Lake Chilwa Drying? Evidence from Past Trends and Recent Data.

6. Phiri MG, Ibrahim Saka, Alex R (2005) The Impact of Changing Environmental Conditions on Vulnerable Communities of the Shire Valley, Southern Malawi, Lilongwe 545-559.

7. Nye J (2010) Climate Change and its Effects on Ecosystems, Habitats and Biota. State of the gulf of maine report 1-17.

8. Learmonth JA, Macleod CD, Santos MB, Pierce GJ, Crick HQP, et al. (2006) Potential Effects of Climate Change on Marine Mammals, Oceanography and Marine Biology: An Annual Review 44: 431-464.

9. Doney SC, Rucklshaus M, Duffy JE, Barry JP, Chan F, et al. (2011) Climate Change Impacts on Marine Ecosystems. Annual Review of Marine Science 4: 11-37.

10. Finazzi F, Haggarty R, Miller C, Scott M, Fasso A (2014) A Comparison of Clustering Approaches for the Study of the Temporal Coherence of Multiple Time Series 29: 463-475.

11. Ullah S, Finch CF (2013) Applications of functional data analysis: A systematic review 13: 1471-2288.

12. Liu C, Ray S, Hooker G, Friedl M (2012) Functional Factor analysis for Periodic Remote Sensing Data, The Annals of Applied Statistics 6: 601-624.

13. Bootsman HA, Jorgensen SE (n.d.) Lake Malawi/Nyasa: Experience and Lessons Learned Brief 260-276.

14. Musopole A (2015) Investigating Temporal Patterns and Spatial Correlation in Remote Sensing Images (unpublished masters thesis), University of Glasgow, Glasgow.

15. Ramsay JO, Silverman BW (1997) Functional Data Analysis, Springer Verlag, New York.

16. Fernandez MF, Opsomer JD (2003) Smoothing Parameter Selection Methods for Nonparametric Regression with Spatially Correlated Errors, the Canadian Journal of Statistics 6: 279-295.

23

Temporal Variations in the Distribution of Interstitial Meiofauna along the Southwest Coast of India

Anila Kumary KS[*]

Kuriakose Gregorios College, Pampady, Kottayam, Kerala, India

[*]**Corresponding author:** Anila Kumary KS, Kuriakose Gregorios College, Pampady, Kottayam, Kerala, India, E-mail: ksanilakumary@yahoo.co.in

Abstract

The present paper depicts the dynamics of meiobenthic assemblages in relation to environmental variables along the coast of Thiruvananthapuram, Kerala on the southwest coast of India. Meiobenthic community consisted of 11 taxa in which nematodes, foraminifers and copepods dominated. Considerable temporal variations are identifiable for all meiofaunal taxa. The ambient physico-chemical conditions of water and physical changes in sediment are responsible for the temporal distribution of meiobenthos. Vertically a downward migration of meiofauna has been observed in the sand column due to better drainage, high atmospheric temperature and exposure.

Keywords: Meiofauna; Seasonal variation; South west coast of India; Macrofauna

Introduction

Studies on benthic populations have been widely accepted as a tool for assessing the health of the environment because of certain unique qualities displayed by benthic invertebrates. Being an important link in the food chain between bacteria and macrofauna of sediments, meiofauna populations are probably suitable indicators of the benthic ecosystem balance. High sensitivity, rapid turnover rate, quick response, life cycles entirely spent in sediments and relative pollution stability makes meiofauna a valid tool to assess the impact of environmental stress. Monitoring of coastal environment is one of the key tools in scientific management of coastal resources.

Studies on interstitial meiobenthic assemblages in relation to environmental variables from the Indian beaches are scanty. Initial meiobenthic studies reported from the Indian coast were from the mud bank region of Kerala coast [1]. Following those a few more studies on the distribution and abundance of meiobenthos have been made off the Indian coast [2-8]. Thiruvananthapuram, the capital district of Kerala on the southwest coast of India, is the southernmost district of the state. Considering the role of meiobenthos as key indicators of environmental stress, the present paper explores the distribution of meiobenthos and its temporal variation in Thiruvananthapuram coast of Kerala in relation to the prevailing environmental parameters as structuring factors of interstitial meiofauna in the sandy beaches.

Materials and Methods

The study was carried out along the Thiruvanandapuram coast of Kerala at two selected beaches, station I, located at Poonthura coast and station II, at Adimalathura coast, lying between latitudes 8020'-8030; North and longitudes 76055'-77003' east. Samples were taken monthly from the 2 stations up to a depth of 25 cm using a graduated steel cover having a length of 25 cm and diameter 5.5 cm. The sediment core was then divided into 5 cm and each segment was immediately removed intact into separate polythene bags. The samples were anaesthetized with 7% MgCl and preserved in 4% buffered formalin 0.1% Rose Bengal was added to the sample for efficient extraction of the fauna and was separated by suspension decantation method [9]. Separated benthic sample were then processed through a set of two sieves with 500 mm and 42 mm mesh sizes for the separation of meiofauna. Meiobenthos was then counted on a higher taxonomic level using a binocular microscope.

Observations of physic-chemical characteristic of sea water were made according to standard methods [10,11]. Bottom sediment was subjected to the analysis of geochemical variables temperature, pH, organic carbon and texture [12]. Monthly values of all parameters analysed were pooled to obtain the seasonal values as pre monsoon (Feb-May), monsoon (June-Sep) and post monsoon (Oct-Jan).

Results and Discussion

The interstices of sandy beaches are profusely inhabited by meiobenthic invertebrates which are of great ecological significance. The taxonomic composition, density and distribution of meiobenthic fauna vary considerably from space to space depending on a wide variety of factors. Exposure, predation, competition, grain size, organic matter and oxygen largely determine the distribution of meiofauna in Indian beaches [13]. Numerically meiobenthic abundance varied slightly in the two beaches. The overall density variation was from 1288 to 8386/100 cm^2 at the Poonthura coast and from 1151 to 10795/100 cm^2 at the Adimalathura coast (Table 1).

Faunal composition of meiobenthos obtained from the two sandy beaches of Thiruvananthapuram coast consisted of 11 taxa coprising foraminifera, Turbellaria, Kinorhyncha, Nematoda, oligochaeta, Polychaeta, Archiannelida, Ostracoda, Copepoda, Amphipoda and Arachnida. Over all abundance of meiobenthos has been in the order Nematoda, Copepoda, Foraminifera, Oligochaeta and Ostracoda at the Poonthura coast and in the order Nematoda, Foraminifera, Copepoda, Archiannelida and Oligochaeta at the Adimalathura coast (Figure 1). In general nematodes dominated the meiobenthic community of Thiruvananthapuram coast. Nematodes are the most abundant meiofaunal community of Indian beaches which often represents more than 80% of benthic meiofauna [5,14-18].

Season	Poonthura coast	Adimalathura coast
Pre monsoon	Range 3465-6073	Range 2579-4631
	Mean 4945	Mean 3592
Monsoon	Range 1288-4590	Range 1511-5677
	Mean 2511	Mean 2508
Post monsoon	Range 1559-8386	Range 4540-10795
	Mean 6617	Mean 7884

Table 1: Temporal variations (No/100 cm^2) in meiofaunal density along Thiruvananthapuram coast, Kerala.

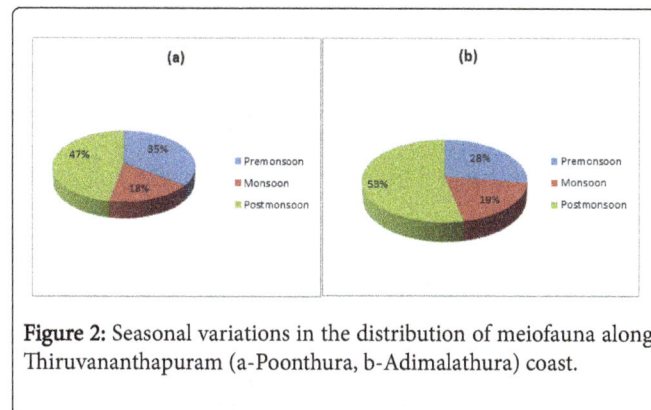

Figure 1: Composition of meiofauna along Thiruvananthapuram (a-Poonthura, b-Adimalathura) coast.

The study revealed distinct temporal variation in the interstitial meiofaunal components along the coast of Thiruvananthapuram (Figure 2). Faunal abundance was higher during the post monsoon period followed by pre monsoon. A distinct feature of the Indian beaches is the influence of monsoon rains that adversely affect the density of the fauna. During the monsoon period (June-Sept) the beach configuration changes drastically at short term intervals due to severe erosion or heavy deposition. Strong wave action during the monsoon has the capacity to completely remove or deposit the substratum. During the high turbulence period sediment particles get rearranged affecting the interstitial spaces and the living space available for the organisms that get shifted continuously. This

phenomenon might uproot the benthic fauna and expose them to the risk of predation [19,20].

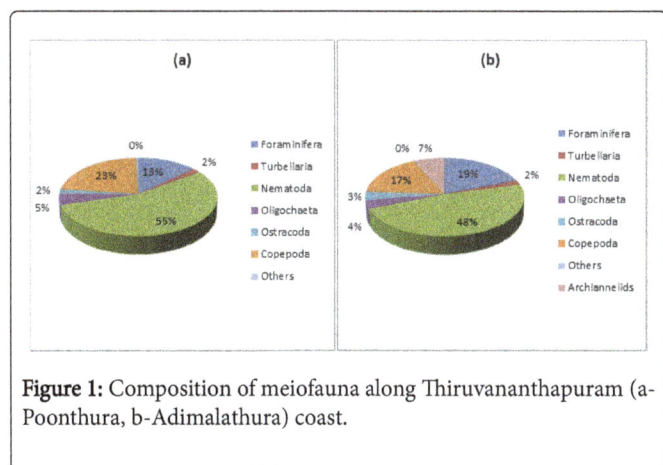

Figure 2: Seasonal variations in the distribution of meiofauna along Thiruvananthapuram (a-Poonthura, b-Adimalathura) coast.

The ambient physico-chemical conditions and the physical change in the sediments are responsible for the temporal distribution of meiofauna. There were considerable fluctuations in the density of all taxa from month to month. Increased temperature, high salinity, stable beach conditions and the probable greater food availability favored the rich post and pre monsoon populations. Seasonal variations in hydrobiological and geological variables are presented in Tables 2 and 3 respectively. The periods of high density of interstitial meiobenthic community in the present study is coincided with increased water and sediment temperature, pH, dissolved oxygen and increased organic carbon in the sediment together with higher proportion of silt and clay. Seasonal breeding is characteristic of meiofauna [21] and the increased meiobenthic density is coinciding with intense breeding activities of meiofauna during the high temperature period [22,23]. Temperature may also influence population increase indirectly by controlling supply of bacterial and diatom food. Size of sand grains was reported to be a major factor influencing meiofaunal abundance [5,13,24-27]. Interstitial fauna develop best in sands with medium diameter [9] and moderate organic enrichment [21]. Sandy beaches of Kerala in general have extremely low organic matter in sediment. Faunal abundance was higher during the post and pre monsoon months (Figure 2) with nematodes recording maximum abundance followed by copepods in the Poontura beach and foraminifers at the Adimalathura beach. On average nematode contributed 57.96% (pre-monsoon), 47.4% (monsoon) and 49.25% (post monsoon) of the total meiobenthic fauna. Prevalence of nematode fauna in meiofaunal community of Indian beaches was reported earlier [15,16,28]. Abundance of foraminifera in sandy substrata was also reported from Indian beaches [1,15].

Variable	Poonthura coast			Adimalathura coast		
	Pre monsoon	Monsoon	Post monsoon	Pre monsoon	Monsoon	Post monsoon
Temperature (°C)	30.30 ± 0.337	24.10 ± 0.447	28.20 ± 0.561	29.80 ± 0.719	23.80 ± 0.137	28.40 ± 0.666
pH	8.21 ± 0.088	7.642 ± 0.026	7.91 ± 0.018	8.21 ± 0.031	7.54 ± 0.073	8.14 ± 0.003
Salinity (S.10-3)	33.88 ± 0.113	32.14 ± 0.831	32.21 ± 0.192	34.24 ± 055	32.18 ± 0.512	33.90 ± 0.522
Dissolved oxygen (ml/l)	5.64 ± 0.696	4.32 ± 0.768	5.98 ± 0.632	5.72 ± 0.631	4.92 ± 0.731	6.01 ± 0.610
Nitrite-nitrogen (µmol/l)	0.211 ± 0.018	0.182 ± 0.016	0.202 ± 0.022	0.198 ± 0.022	0.21 ± 0.003	0.23 ± 0.061
Nitrate-nitrogen (µmol/l)	0.31 ± 0.086	0.23 ± 0.014	0.28 ± 0.136	0.33 ± 0.046	0.28 ± 0.029	0.31 ± 0.14

Phosphate-phosphorus (µmol/l)	0.21 ± 0.07	0.161 ± 0.072	0.27 ± 0.056	0.28 ± 0.008	0.21 ± 0.013	0.29 ± 0.017
Silicate-silicon (µmol/l)	0.23 ± 0.03	0.106 ± 0.008	0.181 ± 0.06	0.21 ± 0.061	0.091 ± 0.007	0.198 ± 0.012

Table 2: Temporal variations (mean) in hydrological variables along Thiruvananthapuram coast, Kerala.

Parameter	Poonthura coast			Adimalathura coast		
	Pre monsoon	Monsoon	Post monsoon	Pre monsoon	Monsoon	Post monsoon
Temperature (°C)	30.00 ± 0.61	25.10 ± 0.88	29.10 ± 0.58	29.70 ± 0.701	24.2 ± 0.515	29.50 ± 0.412
Organic carbon (%)	0.31 ± 0.28	0.23 ± 015	0.56 ± 0.13	0.70 ± 0.14	0.45 ± 0.09	0.64 ± 0.09
Sand (%)	77.76 ± 5.03	96.45 ± 3.65	96.62 ± 4.32	73.61 ± 3.12	98.75 ± 1.01	84.86 ± 6.15
Silt (%)	14.16 ± 2.92	2.76 ± 1.22	1.33 ± 0.92	16.1 ± 4.46	0.62 ± 03	3.51 ± 1.25
Clay (%)	8.08 ± 3.30	0.79 ± 0.63	2.05 ± 1.16	10.29 ± 3.50	0.62 ± 0.03	11.63 ± 1.25

Table 3: Temporal variations (mean) in geological parameters along Thiruvananthapuram coast, Kerala.

Vertically, the majority of meiobenthic organisms are confined to the upper 10 cm depth (Figure 3). Mostly nematodes are found to penetrate the deep layers and found in the entire 25 cm depth. Foraminifera are the other group in the deeper layers of the sediment. 21% of foraminifera and 25% nematoda penetrate the deeper (more than 10 cm) layer. Of all the other groups only oligochaeta and archiannelida were recorded from the deepest (20-25 cm) layer. Decrease in faunal density in the deeper layers has been attributed to the reduction in interstitial space, oxygen content and food material [29]. One of the reasons for the successful penetration of nematodes into deeper layers could be attributed to their capacity of anaerobic existence [1]. Seasonal variations were evident in the vertical distribution meiobenthos with highest density in the surface section (0-5 cm) during the monsoon period at both the beaches and maximum density in the 5-10 cm layer during other seasons (Figures 4 and 5). Meiofauna in sandy sediments generally appear to be concentrated at those levels where desiccation is not too severe and oxygen availability is not too low [30,31]. Because of better drainage of sand and higher temperature at the surface layer the fauna found penetrated to the deeper layer during pre and post monsoon seasons.

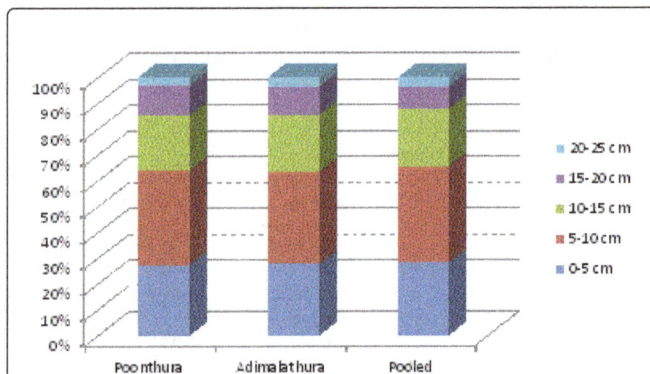

Figure 4: Seasonal variation in the vertical distribution of meiofauna (Poonthura coast).

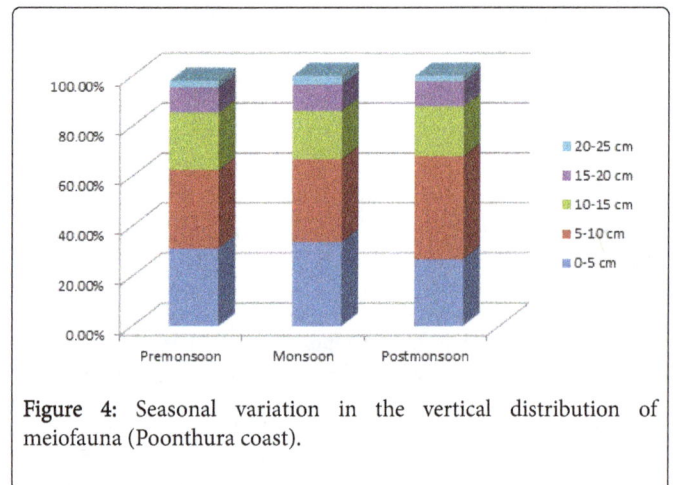

Figure 5: Seasonal variations in the vertical distribution of meiofauna (Adimalathura coast).

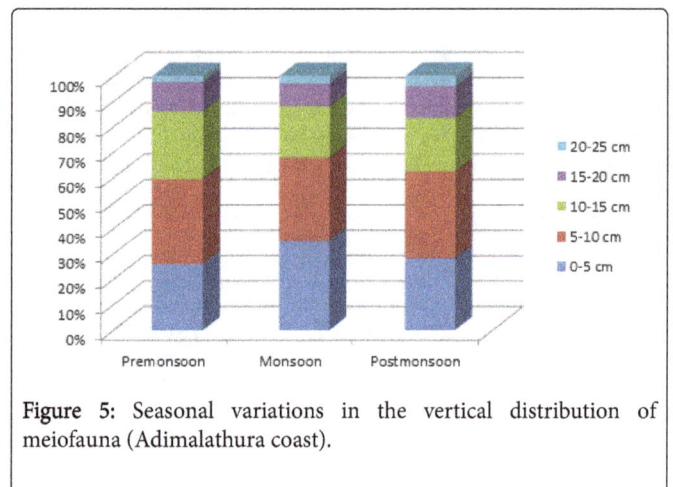

Figure 3: Vertical distribution of meiofauna (annual%) along Thiruvananthapuram coast.

The study revealed distinct seasonal variation in interstitial meiobenthos along the coast of Thiruvananthapuram, Kerala on the southwest coast of India. The meiobenthic abundance in general was found in accordance with sediment granulometry and physico-chemical characteristics of water prevailing along the coast. Temperature, pH, salinity, dissolved oxygen, sediment composition and % of organic carbon in the sediment are proved to be important descriptive parameters related to the abundance and distribution of meiobenthos. Different meiofaunal components showed difference in vertical movement. A downward migration of meiofauna has been observed in the study area due to better drainage, high temperature and exposure.

References

1. Damodaran R (1973) Studies on the benthos of the mud banks of the Kerala coast. Bull Dept Mar Sci univ Cochin pp: 1-126.

2. Ansari ZA, Parulekar AH, Jaytap TG (1980) Distribution of sublittoral meiobenthos of Goa coast, India. Hydrobiologia 74: 209-214.

3. Ansari ZA, Chatterji A, Parulekar AH (1984) Effect of domestic sewage on sand beach meiofauna at Goa, India. Hydrobiologia 111: 229-233.

4. Ingole BS, Ansari ZA, Parulekar AH (1987) Meiobenthos of Saphala salt marsh, west coast of India. Indian J Mar Sci 16: 110-113.

5. Sajan S, Damodaran R (2007) Faunal composition of meiobenthos from the shelf regions off the west coast of India. J Mar Biol Ass India 49: 19-26.

6. Sajan S, Joydas TV, Damodaran R (2010) Meiofauna of the western continental shelf of India, Arabian Sea. Estuar Coast Shelf Sci 86: 665-674.

7. Priyalakshmi G, Menon NR (2014) Ecology of interstitial faunal assemblages from the beaches along the coast of Kerala, India. Int J Oceanogr pp: 1-9.

8. Sinu J Varghese, Miranda MTP (2015) Meiobenthic diversity and abundance along Arthunkal coast in Kerala, southwest coast of India. J Mar Biol Ass India 57: 78-83.

9. Wieser W (1960) Benthic studies in Buzzards Bay II. The meiofauna. Limnol Oceanogr 5: 121-137.

10. Strickland JDH, Parsons TR (1972) A manual for sea water analysis (2nd edn.) Bull Fish Res Brd Canada 167: 310.

11. Grasshoff K, Ehrhartt M, Kremling K (1983) Methods of sea water analysis, Verlag, Chemic Germany p: 419.

12. Krumbein WC, Pettijohn FJ (1938) Manual of sedimentary petrography, Appleton-Century Crafts, NewYork p: 549.

13. Ansari ZA, Prita Ramani C, Rivonker V, Parulekarv AH (1990) Macro and meiofaunal abundance in six sandy beaches of Lakshadweep islands. Indian J Mar Sci 19: 159-164.

14. Rao DG (1987) Ecology of meiobenthos of Rambha bay in Chilka lagoon. J Mar Biol Ass India 29: 74-85.

15. Ansari ZA, Parulekar AH (1994) Meiobenthos in the sediments of sea grass meadows of Lakshadweep atoll, Arabian sea. Vie Milieu 44: 185-190.

16. Sarma ALN, Wilsand V (1994) Littoral meiofauna of Bhitarkanika mangrove of Mahanadi system, East coast of India. Indian J Mar Sci 23: 221-224.

17. Ansari KGM, Lyla TPSI, Ajmalkhan S (2012) Faunal composition of metazoan meiofauna from the south east continental shelf of India. Indian J Geo-Marine Sci 41: 457-467.

18. Anupama C, Srinivasa Rao M, Vijaya Bhanu CH (2015) Distribution of meiobenthos off Kakinada Bay, Gaderu and Coringa estuarine complex. J Mar Biol Ass India 57: 17-26.

19. Anita Patnaik, Rao MVL (1990) Composition and distribution of interstitial meiofauna in the sandy beaches of Gopalpur, south Orissa coast. Indian J Mar Sci 19: 165-170.

20. Suresh K, Shafiq Ahamed M, Durairaj G (1992) Ecology of interstitial meiofauna at Kalpakkam coast, east coast of India. Indian J Mar Sci 21: 217-219.

21. Harris RP (1972) Seasonal changes in population density and vertical distribution of harpacticoid copepod on an intertidal sandy beach. J Mar Biol Ass UK 52: 493-505.

22. Sarma ALN, Ganapati PN (1975) Meiofauna of the Visakhapatnam harbor in relation to pollution. Bull Dept Mar Sci Cochin 7: 243-255.

23. Sarma NSR, Mohan PC (1981) On the ecology of interstitial fauna inhabiting the Bhimilipatnam coast (Bay of Bengal). Mahasagar-Bull Natn Inst Oceanogr 14: 257-263.

24. Harkantra SN, Parulekar AH (1989) Population distribution of meiofauna in relation to some environmental features in a sandy intertidal region of Goa, west coast of India. Indian J Mar Sci 18: 202-206.

25. Palacin C, Martin D, Gile JM (1991) Features of spatial distribution of benthic infauna in a Medeterranian shallow water bay. Mar Biol 100: 315-321.

26. Somerfield JDH, Gee JM, Warwick RM (1994) Soft sediment meiofaunal community structure in relation to a long heavy metal gradient in the Fal estuarine system. Mar Ecol Prog Ser 105: 79-88.

27. Ansari ZA, Mehta P, Furtado R, Aung C, Pandiarajan RS (2014) Quantitative distribution of meiobenthos in the gulf of Martaban, Myanmar Coast, North east Andaman sea. Indian J Geomarine Sci 43: 189-197.

28. Rao KK, Sivadas P, Narayanan B, Jayalakshmi KV, Krishnankutty M (1987) Distribution of foraminifera in the lagoons of certain islands of the Lakshadweep Archipelago, Arabian sea. Indian J Mar Sci 16: 161-178.

29. McIntyre AD (1969) Ecology of marine meiobenthos. Biol Rev 44: 245-290.

30. McLachlan AP, Winter ED, Bhota L (1977) Vertical and horizontal distribution of sublittoral meiofauna in Algoa bay, South Africa. Mar Biol 40: 355-364.

31. Fernando OJ (1987) Studies on the intertidal fauna of the Vellar estuary. J Mar Biol Ass India 29: 86-103.

24

The Impact of University Student's Green Awareness Purchasing on Green Marketing in Egypt

El Sakka S[*]

Department of management, School of Business, Future University in Egypt, 5th Settlement, New Cairo, Egypt

[*]**Corresponding author:** El Sakka S, Department of management, School of Business, Future University in Egypt, 5Th Settlement, New Cairo, Egypt
E-mail: sherinesakka@fue.edu.eg

Abstract

Egypt as one of the developing countries which confront a lot environmental challenges, environmental issues and green products culture is not a priority, spreading the ideas of going green and expanding the culture of green product purchase (GPP) is very important step which it need awareness campaign for green marketing to clarify for new generation the benefit of purchasing green products.

Our paper will investigate how much young Egyptians university students are aware about green product (GP), if the awareness impacts their green product purchase (GPP) and how it might influence their consumer behaviour; we will discover what kind of factors could affect Egyptian university students buying behaviour of green products and green services.

Keywords: Green product purchasing (GPP); Awareness purchase (AP); Green product (GP)

Introduction

Climate changes, global warming, air and water pollution, Eco friendly products, and green marketing are some environment issues popped up internationally and nationally nowadays.

These environmental issues have impact on consumer health, which made producers to think green and to promote for green products and services.

Egypt like other developing country, confront a lot of environmental problems, from water pollution, air pollution, agriculture land problem, and unsuccessful waste management problem, due to economic problems there is community lack awareness about the environment its problem and its products actually there is no priority of the environmental issues in front of the economy.

Our study will try to discover how much young Egyptians are aware about green product and its consumption benefits as a way to highlights the importance of green products consumption.

This paper will be structured as follows, a brief literature review on green marketing and student green product awareness will be discussed, research methodology, statistical analysis result, finally conclusions and recommendations.

Literature Review

Green marketing

According to American marketing association it defines green marketing "as marketing of products that are believed to be environment friendly which organize into various activities such as product adjustment, modification of production processes, packaging, labeling, advertising strategies as well as increases awareness on compliance marketing among industries" [1].

Ruth Dettie [2] define green marketing as "the application of marketing tools to facilitate exchanges that satisfy organizational and individual goals in such a way that the preservation, protection and conservation of the natural environment are upheld.

Arita khare [3] state that green marketing relates to a holistic management approach for identifying, anticipating and satisfying the needs of consumers and society in a profitable and sustainable way [3]

According to Leslilelu [4] green marketing is defined as "a strategic effort made by firms to provide customers with the environment friendly merchandise."

Green marketing tools

Eco label, eco brand and environmental advertisement are green marketing tools which can increase green products features and help consumers to buy green products.

Eco Label: it's a tool which allows consumers to easily distinguish environmentally green product between the ordinary products.

Eco brand: applying it aspects could help consumers to recognize green products between other products which could harm the environment.

Environmental advertisement: it's a way to redirect the attention of customer to The benefit of purchasing green product [4].

Benefit of green marketing

Green marketing can help building a brand value

Green marketing can eliminate the bad effect of production and better the image of the product.

In green marketing the cost of raw materials are low and the price of green product is low on the long term [5].

Green consumer

The consumer who support business that operate in the environmental friendly ways according to Jen Mei [6].

Green consumption behavior controlled by several factors such as changing consumption value and awareness of green product benefits, demographic factors like age, education and income the more the consumer is exposed the more his consumer behavior change; also factors like price and quality of green products has an impact.

According to Schuhwerkan individuals who engage in environmental activities are devoted to purchase green goods.

Green consumers not only consuming green products but also deals with Company's has engagements green practices and activities such as recycling and energy efficiency.

According to Vermillion Consumers are not likely to purchase a product only for it's environmentally attributes.

Green consumer is define as "products that are likely to endanger the health of the consumer or others, cause significant damage to the environment during manufacture, use of disposal, consume a disproportionate amount of energy, cause of harming waste, use materials derived from threatened species or environments" [7].

According to previous studies children and teens influenced by their family's green product and service consumption, consumers are influenced as well by the product price they post pone their choices to buy green products to a time when they can afford it [8].

Green consumer categories

Roper Organization, categories consumers to five segments

True blue green: when consumer believe that their green consumption have an impact on the environment, they have a will to invest on green products and engage in ecological activities.

Green back green: when consumer invests in more expensive green products but is not willing to engage in environmental activities.

Sprouts: green consumer who support environmental regulations, but they are less to spend money on green products.

Grousers: believe that it's not his responsibility to solve environmental issues; they use regular products and avoid consumption of green products.

Basic browns: believe that environmental problems are very complicated no efforts on personal, commercial or political levels can solve any ecological problem [9].

Hunger stated customers could have an attitude towards green purchases but this doesn't guarantee that they will purchase green products

According to Gatersleben 2012 he clarify that people who consume green products are the young generation [10].

Green product

A product which has ecological attribute the use of green products its aim to prevent pollution and reserve resources, the problem of green product use is its cost and that no many consumers could afford it specially in developing countries when the income is low [11].

Consumer purchasing behavior

Is the decision process and acts of people involved in buying and using product [12].

The history of going green idea and its implementation

The force of going green idea started in western marketing at the millennium and then expanded in Asia and Africa, from 1990 consumers worldwide started to become socially and environmentally aware of the importance of the environment issues.

The concept of ecofriendly or going green approach was pushed by the developed countries to initiate international green marketing in order to expand their market and to take advantage of the positive image of their green brand established in their domestic markets.

Lately there is an increase towards green behaviour due to the population increase and resources drain more can Egypt handle it as a developing country, and due to its economic situation it rely on some industries which could harm its environment like ceramic and cement, so we have to discover what is Egyptian consumers awareness situation considering green products.

Consumers environmental knowledge can be improved through education, some of previous studies stated that many consumers fail to understand the connection between their buying decision and environmental consequences.

After the United Nations conference in 1972 the 21st century become a green consumption Era all over the world, the word green gained attention according to previous studies they showed that knowledgeable consumers of ecological problems purchase environmentally friendly products, consumers with high knowledge are more willing to pay higher prices to support green products [13].

The four criteria for achieving the principles of green consumptions are known as the 4 Rs, (reduction, reuse, recycling and recognition) [14].

Socioeconomic environment of Egypt

According to the World Bank classifications, Egypt is a lower middle income country, the 2010 millennium development goal report highlights that poverty is one of the most critical areas of deficit in Egypt; Egypt has a lot of problematic issues concerning the environment, climate change, water sacristy, coastal problem and air pollution, Egypt has a mismanagement of natural resources which make the awareness process for the population about consuming green product a kind of luxury [14,15].

Environmental Awareness

The more people are knowledgeable about green products and its practices the more they will act positively (Roberts 1996), environmental awareness considered as knowledge and general concepts relating to the ecosystems.

Government initiative in Egypt

The government's role is a predictor to green purchasing behavior claiming Punitha & Rahman (2011) according to them that government should play a role in building green purchasing, Egypt pursue long term developments goals taking into consideration at its policies social and environmental dimension.

Research Methodology

Research objectives

To investigate the awareness level of Egyptian university students about green marketing

To measure the green value for Egyptian university students

To understand what factors could persuade Egyptian university students to buy green products or use green services

To discuss if students environmental responsibility could impact their consumption behaviour

Research questions

Is there awareness of green marketing between university students?

Is there a green value between Egyptian university students?

What are the factors which could influence university students to buy green products?

Are students environmental responsibility could impact their consumption behaviour?

Population and sampling

The populations researched were a random sample of future, British and the German universities students in Egypt aged between 18 to 24 years, a quantitative study, and self-administrated questionnaires had been conducted .

The main reasons for choosing this sample was as follows.

To explore the awareness of green product for young generation.

The sample chosen expected to represent consumer's behaviour for green product in the future.

Hypothesis

H0: There is no significant relation between Egyptian university student's awareness and green consumption purchase

H1: There is a significant relation between Egyptian university student's awareness and green consumption purchase

Findings and discussion

Respondent's surveys were not aware of green products, consumption benefits and practices

Data collection and analysis

Green marketing awareness	Mean	Std Deviation
Time spent discussing with your friends green products importance	2.06	1.441
Time spent with your friends discussing environmental problems issue	2.38	1.575
Time spent sharing information regarding green products with your friends	2.36	1.474
Sharing information about the benefit of green product consumption	4.09	889
The importance of raising green products awareness among university students	4.48	703

Table 1: student's perception about green products awareness.

As revealed from the previous table (Table 1) the awareness of the students about green products it's not high, the data standard deviation measured are not concentrated around the mean. Awareness about green product should be raised between universities students and more protection works its need it (Table 2).

Cronbach's alpha	Cronbash alpha based on standardized items	No of items
310	277	5

Table 2: Green products students' awareness scale reliability analysis.

The alpha coefficient for the 5 items have relatively low internal consistency as it is, 277 (Table 3).

Green value between students	Mean	Std deviation
It's essential to promote green living concept in Egypt	4.09	889
Green products consumption is a waste of money and resources	2.67	1.4
Green products consumption issues are not of my interest	2.45	1.00
Consumption of green products is meaningless	2.47	1.02
Worry about worsening of the quality products in Egypt	4.00	865
Using green products as a step to change consumer behavior one of my major concern	3.6	1.00
Involved in environmental protection issues in Egypt	3.21	1.14
the green products improvement in Egypt should be a priority	3.43	1.28
Green products market expansion in Egypt need to be dealt	4.02	710
Think that the green products problem in Egypt is worsening	4.21	656

Usage Non green products threat our health	4.30	577
Non usage of green products problem threat Egypt reputation	4.11	853

Table 3: Green value between Egyptian university students.

Green value is not that high between Egyptian university students, standard deviation show values that data set is not around the mean, the more the standard deviation is smaller the more the data are concentrated around the mean and this is shows in the students opinion that the green living should be promoted in Egypt, it shoes std, 886 while the mean is 4.09, and how they are worry about the worsening the quality of Egypt environment std of 0.865 while the mean is 4.00, they think that problem in Egypt should be dealt std of 710 while the mean is 4.02, they think that environmental problem threat their health std of 577 with a mean of 4.30, they are concerned also about Egypt reputation std of 853 with a mean of 4.11(Table 4).

Cronbach's alpha	Cronbash alpha based on standardized items	No of items
499	542	12

Table 4: Green value scale reliability analysis between Egyptian university students.

The alpha coefficient for the 12 items has relatively low internal consistency as it is, 542 (Table 5 and 6).

Factors could influence students to buy green products	Mean	Std deviation
Green product is Expensive in Egypt	3.46	1.40
Non availability of green product everywhere create a consumption barrier	3.66	1.29
Its unwise to spend a big amount for green product consumption	2.89	1.17
Green products and green services need high income	3.38	1.03
Brand image could persuade to buy green products	3.94	824
Awareness of green product price could persuade to change consumption behavior	3.82	687

Table 5: Factors could influence university students to buy green products.

Cornbrash's alpha	Cronbash alpha based on standardized items	No of items
120	202	6

Table 6: Factors scale reliability analysis could influence university students to buy green products.

The alpha coefficient for the 6 items have relatively low internal consistency as it is, 202 (Table 7).

Environmental students responsibilities	Mean	Std deviation
You should be responsible for protecting our environment	4.01	759
Environmental protection starts with buying green products	3.21	1.14
How much responsibility do you think you have in protecting the environment in Egypt	3.91	1,102
Do you have the will to buy green products as a part of your responsibility to protect the environment in Egypt	3.67	766
Environmental protection is the responsibility of Egypt government not you	3.44	1.402
Environmental protection is the responsibility of environmental organizations	3.53	1.123
If you carry out some pro environmental behaviors by buying green products in your everyday life you would contribute a lot in changing consumption behavior	4.01	659
Your participation in buying green products would encourage your family and friends to participate in changing their consumption behavior	3.86	817
Environmental quality in Egypt will stay the same even if you engage in some pro-environmental behaviors	2.76	1.357
Even if you buy green product recycle and reuse stuff, the consumption behavior will remains as it is	2.97	1.314

Table 7: Student's environmental responsibility could impact their consumption behaviour.

Responsibility for protecting community environment shows a std of 759 and mean of 4.01, willing to take responsibility to protect the environment in Egypt shows Std of 766 and mean of 3.67, carrying pro-environment behavior with std, 569 and mean of 4.01, encouraging family and friends to protect the environment std of ,817 with a mean of 3.86 (Table 8).

Cronbach's alpha	Cronbash alpha based on standardized items	No of items
543	570	10

Table 8: Student's environmental responsibility scale reliability analysis could impact their consumption behaviour.

The alpha coefficient for the 10 items has relatively low internal consistency as it is 570

Conclusion

There is no high awareness for green marketing between private Egyptian universities students; also there is no high green value between them as well, they don't differentiate between green products and non-green products, they don't know the importance of buying green products and how they will participate in saving and protecting the Egyptian environment by going green , the factors like green product price, availability of green product in the markets and brand image could influence university students to buy green products, student's environmental responsibility could impact their consumption behaviour if they aware of the benefit of the consumption of green products and if the green product has a reasonable price and available everywhere.

The relationship between awareness and green products consumption is positive, which mean the more an individual's awareness the more his consumption behaviour towards green products

Recommendation

Awareness campaigns about green products importance and its benefits, is recommended between Egyptian universities students

Environmental issues should be included in education program to help students to know the relation between going green and how consumer behaviour changes could solve environment problems.

Students involvement in finding solution for community environmental problems

Government should highlight the importance of green product awareness campaign in public universities as well as in private as a priority in its agenda

Reinforcement of environmental laws in Egypt will help to highlight the importance of the environment as well as using green products.

Contribution

This study contributes to the theory by adding current literature on green buying behaviour awareness among young consumer in some of private universities in Egypt; it contributes to the practice by shading the light on the importance of the awareness of green products for young generation in Egypt as one of the developing countries.

Limitations of the research

This study was limited to students in some private universities in Egypt and did not cover all the universities in Egypt, in future studies it would be ideal to major the awareness of the students in all Egypt universities on both sectors public and private.

Secondly, this study concentrated on green product awareness in general, further studies could focus on specific process of green product awareness which can generate a more reliable response.

References:

1. Global journal 2015 p. 2.

2. Ruth Dettie, kevein Burchell, Debra Riely (2012) Normalizing green behaviors a new approach to sustainability marketing. J of marketing management 28: 423.

3. Arita khare, Sourjo mukerjee, Tanuj goyal (2013) Social influence and green marketing an exploratory study on indian consumers. J customer behavior 12: 362.

4. Leslilelu, Dora Bock, Mathew joseph (2013) Green marketing: what the millennials buy, Journal of business strategy 34: 3-10.

5. Wong Fuiyeng, Rashad Yaz dnifard (2015) Green marketing a study of consumers buying behavior in relation to green products. Global journal 4-5.

6. Ooi jen Mei, Kwek choon Ling, Tan Hoi (2012) Piew The antecedents of green purchase intention among Malysian consumers 8: 249.

7. Elham rahbar, Nabsiah abul wahid (2007) Investigation of green marketing tools effect on consumers purchase behavior 12: 73-83.

8. Jesicca Ascheman-witzel, Emilie marie niebahr (2014) Elaborating products: young Danish consumer and in store food choice. International journal of consumer 38: 552.

9. Kaman Lee (2008) Opportunities for green marketing: young consumers. Journal of consumer marketing 26: 573-586.

10. Muntaha Anwar, Marike Venter (2014) Attitudesand purchase Behavior of green products among generation Y consumers in south Africa. Mediteranian Journal of social sciences 5.

11. Violeta, Sima (2014) Green behavior of the Romanian consumers, Economic insights trends and challenges 111: 66-77.

12. Hisham el din bin, Ismail, Mohamed fateh ali khan (2008) Consumer perception on the consumerism issue and its influence on their purchasing behavior, A view from Malaysian food industry pp 11: 43.

13. Lynsey Scott, Debbie Vigar-Ellis (2014) Consumer understanding perceptions and behavior with regard to environmentally friendly packaging in a developing nation, International journal of consumer studies 38: 2-4.

14. Cheng-juri Tseng, Shuo chang Tsaj (2011) Effect of consumer environmental attitude of green consumption decision making. pack J statist 27: 1-4.

15. Iman F Abou-El naga (2015) Demographic, socioeconomic and environmental changes affecting circulation of neglected tropical diseases in Egypt. Asian pac j Trop med 8: 881-888.

25

GRACE, Climate Change and Future Needs: A Brief Review

Sandeep N Kundu*

Department of Geography, National University of Singapore, Singapore

***Corresponding author:** Sandeep N Kundu, Department of Geography, National University of Singapore, Singapore, E-mail: geosnk@nus.edu.sg

Abstract

Gravity Recovery and Climate Experiment (GRACE) is a satellite based observation of the Earth's gravitational field through time. Earth's gravitational field has long been studied to infer the densities of underlying rocks for geological characterization at various scales. Convention gravity measurements were done using spring gravimeters on ground. However continuous observations of temporal variations in the Earth's gravity field have become available at an unprecedented resolution of a few hundreds of kilometres through satellite based sensors. With the launch of GRACE in 2002, the study of the exchange of mass both within the Earth and at its surface in the short temporal interval has become possible. This has huge implications in studying the impacts of earth's surface processes involving the interaction of the atmosphere, hydrosphere, cryosphere and biosphere. GRACE gravity, therefore, has gained relevance for Earth scientists as an important tool to study the complex dynamics of the Earth system and climate change. The current article researches on the principles behind grave gravity variations and its applications to infer climate change and proposes the advances required to overcome the limitations of GRACE for climate change forecasting.

Keywords: GRACE; Gravity Field; Climate Change; Ocean mass; Land mass

Introduction

Newton's law of universal gravitation states that a particle attracts every other particle in the universe using a force that is directly proportional to the product of their masses and inversely proportional to the square of the distance between them. Denoted by g, the gravitational field strength is the acceleration that the Earth imparts to objects on or near its surface. This acceleration is measured in metres per second squared (in symbols, m/s^2) is a constant approximately 9.81 m/s^2. This constant is not uniform around the globe because it is a product of its mass distribution, variations in the density of the Earth's interior and undulating topography, thereby causing regional deviations to a scale of few tens of a millionth of g. There are challenges in modelling these variations owing to the poor understanding of our Earth's structure and dynamic nature. Gravity Recovery and Climate Experiment (GRACE), provides a 9 day repeat data over the same region allowing the study of the gravity field's variations in the short temporal scale. As mass transport in the Earth's interior is a slow process, the variations in the short temporal scale can attributed to mass transport in the near surface with high relevance to hydrological processes, surface water changes and ground water [1].

The short temporal variations of the Earth gravity field are dominated by the mass exchange processes between the atmosphere, cryosphere, hydrosphere and biosphere. Climate patterns drive these interactions and therefore variations in mass distributions are a direct impact of climatic variations. Availability of GRACE gravity since 2002 facilitates the study of this variability at different temporal scales providing information on sub-annual to decadal variations interpretations of which provides insight on climatic changes.

Before the availability of GRACE, hydrological applications aimed at studying Total Water Storage (TWS) changes over land could only be done at local watershed scale. Moreover these could only focus on one component i.e., groundwater, soil moisture or surface water, instead of TWS. Gravity field variations and its correlation with precipitation and run-off at regional scale has realised the potential of GRACE in many environmental and climate change applications. The findings of such studies have influenced the water management policies of large countries for secure a sustainable land and water management practice. We discuss the principles behind GRACE data and its characteristics and how it has been used in various applications which infer climate change and finally propose the future requirements which could overcome the limitations of GRACE gravity data.

About Grace

GRACE satellites

GRACE consists of two satellites in a low, near circular, near-polar orbit with an inclination of 89°, at an altitude of about 500 km, separated from each other by a distance of roughly 220 km along-track. The satellites have a repeat imaging over the same area once in 9 days and roughly a month of coverage data is needed to produces a seamless model for the globe (Figure 1)

Figure 1: The twin satellites of GRACE with the earth's gravity field.

Gravity from GRACE

The mission does not measure gravity directly with an active sensor, but is based on the satellite-to-satellite tracking concept, which tracks variations in the inter-satellite distance and its derivatives to recover gravitational information. As the two GRACE satellites are separated by a certain distance, and the gravitational pull of mass on the earth's surface is inversely proportional to the squared distance to each satellite, the orbit of each of the satellites will be perturbed slightly differently when it encounters a sizable dense mass on earth's surface. The leading satellite will be pulled slightly more toward the mass than the trailing one increasing the separation between the satellites. These minute changes-in the order of a few micrometres-can be measured by means of a dual-one way microwave ranging system (KBR). Other non-gravitational forces, such as atmospheric drag will also alter the relative distance, and are accounted for using on-board accelerometer measurements. Two star-cameras measure the orientation of these satellites in space. The satellites are equipped with Global Positioning System (GPS) receivers so that their location is known. Thus from the instantaneous orientation, ranging and GPS data (called the level-1 data), variations in the Earth's gravity field can be estimated. An iterative procedure involving a priori model of the Earth's mean gravity field is used in combination with other background force models (planetary interactions and tidal components) are used to determine the orbit of both satellites. The gravitational effects of ocean and atmosphere mass variations are removed from the measurements at this step using numerical models else their high-frequency contributions would influence the results [2].

Noise Reduction and Smoothing

The gravity field is sampled using the variations in the long-track distance between the two satellites, which circle the Earth in a near-polar orbit. As a result, the observations bear a high sensitivity in the North-South direction, but little in the East-West direction. Errors in the instrument data, shortcomings in the background noise models and other processing errors results in unusual gravity gradients in the East-West direction. This results in stripes along the North-South direction [3]. Several methods have been developed to reduce the effect of this noise in the GRACE data which include spatial and frequency based smoothing methods (Figure 2). These methods reconcile the noise but unfortunately damp the geophysical signals thereby hindering a direct quantitative interpretation from GRACE observations.

The processed GRACE data products are then used to produce computations for land mass and ocean mass grids for integrated studies with other ground based information on climate. Most available as Monthly global 1° grids. Different grids for land mass and ocean mass are available for research. A Global Land Data Assimilation System (GLDAS) has been developed jointly by scientists at the National Aeronautics and Space Administration (NASA) Goddard Space Flight Center (GSFC) and the National Oceanic and Atmospheric Administration (NOAA) National Centers for Environmental Prediction (NCEP), which makes use of new generation of ground and space based observation systems for land surface models aimed at supporting improved forecasting and hydro-meteorological investigations [4].

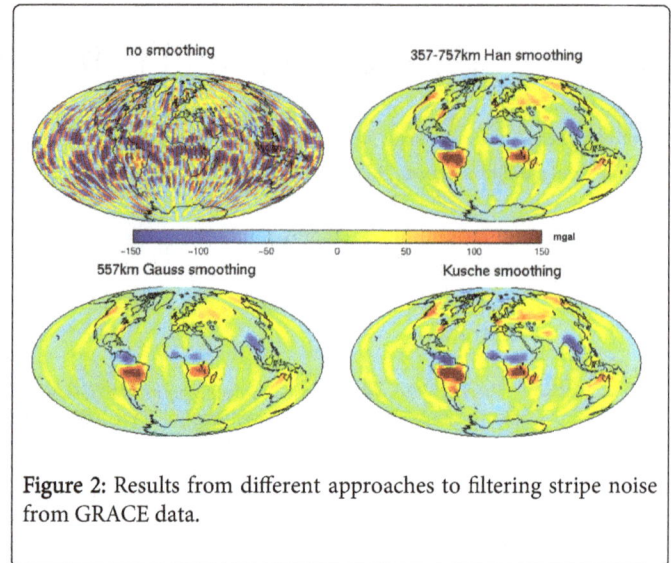

Figure 2: Results from different approaches to filtering stripe noise from GRACE data.

Mass Concentration blocks (Mascons) are another form of gravity field basis functions to which GRACE's inter-satellite ranging observations are fit, which supports the implementation of geophysical constraints [5]. Monthly grids for GLDAS, Mascon, Land mass and Ocean mass can be downloaded from GRACE Tellus server (ftp://podaac-ftp.jpl.nasa.gov/allData/tellus/L3/).

Grace and Climate Change

Climate influences the distribution of mass on earth's surface in regular consistent patterns. Climate change disrupts these patterns which feedback in different forms like shifts in Earth's axis of rotation. Though gradual, a consistent shift over decades could drive further shifts in climate patterns that would add another level of complex mass distribution patterns. Climate change interpretations from GRACE are not direct and are purely based on mass transfer studies related to water and ice on land and oceans at a monthly temporal scale. The below are brief summaries of the different applications trending in literature which involves GRACE data for understanding our Earth's response from the changing climate in the last decade.

Total Water Storage (TWS)

Continental water storage and snow pack at high latitudes and their changes over time are reflected in the gravity field variations. Continental water storage is a key component of the terrestrial and global hydrological cycles, having an important control over water, energy and biogeochemical fluxes, thus playing a major role in the Earth's climate system [6]. However totals continental water storage remains completely unknown at regional and global scales because of the lack of direct hydrological measurements. GRACE satellite gravimetry offers a very interesting alternative remote sensing technique to measure changes in TWS (ice, snow, surface waters, soil moisture, and groundwater) over continental areas, representing a new source of information for hydrologists and global hydrological modellers [7].

Over a drainage basin, the water mass balance equation to solve is

$$\frac{dW}{dt} = P - E - R$$

Where dW/dt is the TWS variations that are directly provided by GRACE over the considered region [8]. If P (precipitation) is known and R (runoff) is modelled then variations in E (Evapotranspiration) can be estimated which essentially represents vertical water fluxes. GRACE TWS data with observed precipitation and stream flow Rodell et al. [4] can be used to identify patterns of changes which would be critical in forecasting drought episodes in a region. Several applications on large river basins have reflected the changes in seasonal changes in TWS [9-11] through integrated studies using field based data. Cazenave and Chen [12] calculated the rates of mass changes in water column equivalent for 2002-2009 (Figure 3). The observations included the loss of Ice sheet thickness in the Himalayas and increasing loss of water (ground water) around the New Delhi region attributed to excessive pumping of water by an population that have increases over the time period.

However there are several limitations to using GRACE gravity especially when the study requires separation of TWS into its hydrologic constituents without the help of field collected data. Integration of similar information through disparate temporal scales and precision results in complicated combination with other hydrologic products, all of which have their own limitations and errors. The differing spatial and temporal scales between GRACE (a global, monthly product) and in situ data such as river or well gauges (point-source measurements which are non-uniform in space and time) makes exact comparisons and combinations difficult.

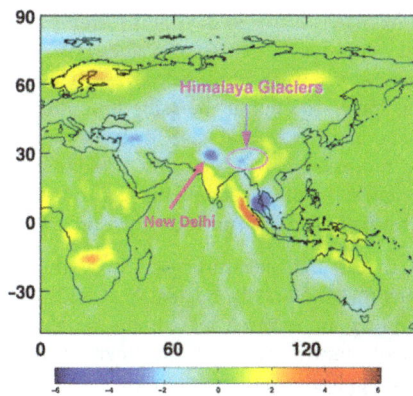

Figure 3: Global mass rate changes (cm/year of equivalent water height rate) estimated from GRACE time-variable gravity data from period Sep 2002–Aug 2009. Modified from Cazenave and Chen [12].

Ice Mass Balance

A giant leap forward in our understanding of the cryosphere was made by the advent of satellite remote sensing. Despite the lack of missions specifically dedicated to observing the mass balance of the cryosphere, estimates of volume and mass changes were already made in the 1990s using satellite radar altimetry. GRACE has taken this forward by providing new insights on the mass balance of smaller ice caps and glacier systems. Direct observations of glaciers are sparse, both in space and in time, because of the labour intensive nature and tend to be biased toward glacier systems in accessible, mostly maritime, climate conditions. GRACE-derived mass balance estimates

of Antarctica and Greenland (Figure 4) have been determined from the gravity fields of the GRACE project [13]. Mass changes of ice sheets have been resolved by using Mascons form various drainage basins.

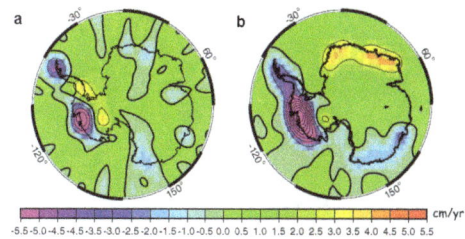

Figure 4: Antarctic ice mass changes a) Jan 2003 to Dec 2006) and b) Dec 2006 to Nov 2012. Sourced from Velicogna and Wahr [13].

Velicogna and Wahr [13] estimated an extreme decrease of ice mass of -152 ± 80 km^3/year for Antarctica. Although GRACE-based estimates are not corrected for long-term Post-Glacial Rebound (PGR), the mass balance of Antarctica appears to be close to zero. Uncertainties in such studies are primarily because of the inherent North-South striping noise and inaccurate modelling of PGR from GRACE data although it there has been reports of possibilities to reduce these from combination of different satellite based techniques and data. Studies have found that approximately 60% of the in situ glacier mass balance records are from the smaller European Alps, Scandinavia and north-western America.

Continuous and uninterrupted observation time series for very large glaciers were earlier unavailable, an issue which are overcome by GRACE, yet at the smaller spatial scale noise is larger required stringent validation of the results and interpretations.

Ocean Water Distribution

Ocean mass variations are important for diagnosing sea level budgets, the hydrological cycle, the global energy budget, and ocean circulation variability [14]. Seasonal cycles (Figure 5) and decadal trends of ocean mass at global and reginal scales could be determined from GRACE Release-05 data from which global flux of mass into the ocean approaches 2cm per decade in equivalent sea level rise.

Historically, ocean mass variations, whether global or regional, have been difficult to diagnose although Bottom Pressure Recorders (BPRs) have been deployed in regional arrays for short term observations to investigate ocean boundary currents. Although ocean general circulation models have been used to study low-frequency ocean mass variability, results are often suspect owing to the time scale needed for deep ocean conditions to equilibrate.

Large negative trends south of Greenland and north of the West Antarctic Peninsula are likely owing to large ice mass loss signals leaking into the ocean even after attempting to correct with estimates based on GRACE [15].

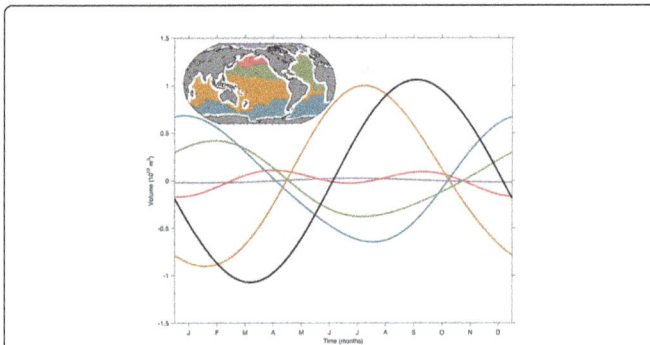

Figure 5: Seasonal mass redistribution among various areas with similar phases for monthly GRACE Release-05 500 km smoothed maps. The seasonal cycles are shown here for the different latitude defined zones (inset map) with the black curve being the integrated Global curve [14].

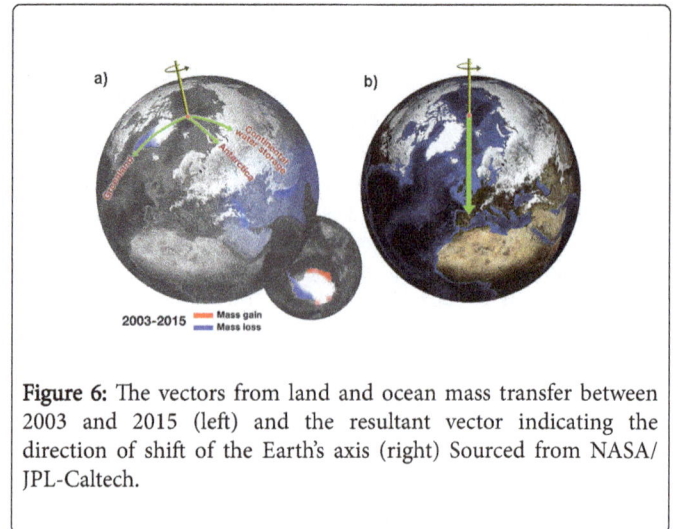

Figure 6: The vectors from land and ocean mass transfer between 2003 and 2015 (left) and the resultant vector indicating the direction of shift of the Earth's axis (right) Sourced from NASA/JPL-Caltech.

Earthquakes and Gravity

Song and Simons [16] has reported that within the regions where one of the Earth's plates slips below another, areas where the attraction due to gravity is relatively high are less likely to experience large earthquakes than areas where the gravitational force is relatively low. The gravity potential signature of an earthquake at a location would be mainly associated with the vertical displacement of the Earth's surface between earthquake events. This vertical displacement triggers isostatic adjustment of the crust with the potential of inducing sea level change. Earthquakes are isolated single events information for which is available from USGS. Should it be possible for segregating the gravitational field variations due to mass changes owning to climatic factors, the remnant field could be attributed to vertical displacement resulting from earthquakes. De Viron et al. [17], provided insights into retrieving signature of gravity variations from GRACE from seismic events and concluded that the empirical orthogonal function (EOF) decomposition is key to separate the geopotential variations emanating from a seismic event. GRACE data can greatly help in understanding how earthquakes influence the earth's gravity field changes and this would help improve the sensitivity of predicting sea level changes, isostatic compensation and sub-crustal mass addition, thereby improving climate prediction models.

Polar Wandering

The slow erratic movement of the Earth's poles relative to the continents throughout geological time is known as polar wandering. This movement was considered to be very slow, largely due to drifting of the Earth's crustal plates. For the first time, Adhikari [18] demonstrated that the strong deviation in linear drift since 2000 was influenced by the decadal variability in global continent ocean mass transport, with changes in terrestrial water storage (TWS) playing a larger role as continental drift is insignificant in this period. Study of GRACE gravity based land and ocean mass movement patterns, it is now possible to forecast the direction of this polar tilts (Figure 6), thereby predicting changes in climate pattern shifts.

Future Needs

The scientific goal of the GRACE satellite mission is to measure the gravitational field of the earth and its changes over time on a global scale with unprecedented accuracy. This has been achieved with significant concessions to spatial resolution and smoothed measurements which has triggered applications that address variations rather than absolute value of the gravity field. A global gravity map is completed approximately every 30 days and therefore the gravity field can be assumed as a monthly average of values [19-21]. This is better than all sources which were used before GRACE and also more accurate by a factor of 100 but at the same time limits the data to a monthly temporal resolution and also impacts the original values (since they are processed, aggregated and smoothed for a seamless monthly global dataset). Future satellite platforms need to address this issue of spatial resolution and sensor scale noise reduction.

Since GRACE does not measure gravity by direct means, this essentially means that the gravity calculations needs to be done, and validated before they are released for any meaningful interpretation. This introduces a latency of weeks for the gravity field data and more than a month for applications based products like the Land mass and Ocean mass products. Automated processing for reliable timely data product generation and validation are needed to support timely studies and for the information to be used for updating predictive models.

Another key challenge is that GRACE gravity is a sum of all contributing factors at an area and isolation of the factors resulting from each competent in this vertical column are based on sparse ground based sensors and observations. This induces reliability issues and uncertainty in the results and their interpretations.

The utility of GRACE, as discussed above, is therefore primarily aimed improving our understanding on the recent past environment and climate and on development of algorithms to isolate the various contributing components for the overall gravity field of an area.

Conclusion

The mass balance study of the continental water content, which is ultimately a sum of precipitation, evaporation, runoff and storage. Water is connected directly to weather and climate and with GRACE data its movement can be monitored on land and on oceans at a global scale. Season-dependent changes in the major river basins could therefore be studied coupled with anthropogenic use of surface water and ground water. Quantification of the increase or decrease of the ice and snow masses in the polar or large glacier areas from GRACE data have been well correlated with climatic phenomenon like shift in patterns of the rainfall patterns in West Antarctica and the reduction of ice mass there. GRACE also contributes to better understanding of global ocean circulation and heat transport from central latitudes to the Polar Regions though temporal observations of surface and deep currents. It was probably for the first time scientists were able to separate mass from ice melt or temperature from global warming to address sea level changes. Changes inside the solid earth and isostatic compensation of the region were also addressed using GRACE data through studying earthquake events since its launch.

GRACE gravity, through its continuous mapping of gravity variation, has largely improved our understanding on earth's surface processes and interactions between hydrosphere, lithosphere, cryosphere and biosphere and the impact of climate change to an extent where new requirements have emerged to define future mission to enhance the gravity signals for objective discourses. Future mission could incorporate the issue of latency, inherent harmonic noise and data resolution issues to advance the application of gravity variations for climate change studies and even into climate forecasting models.

References

1. Wouters B, Bonin JA, Chambers DP, Riva REM, Sasgen I, et al. (2014) GRACE, time-varying gravity, Earth system dynamics and climate change. Rep Prog Phys 77: 1-41.

2. Ray RD, Luthcke SB (2006) Tide model errors and GRACE gravimetry: towards a more realistic assessment. Geophys J Int 167: 1055-1059.

3. Schrama EJO, Visser PNAM (2007) Accuracy assessment of the monthly GRACE geoids based upon a simulation. J Geod 81: 67-80.

4. Rodell M, Houser PR, Jambor U, Gottschalck J, Mitchell K, et al. (2004) The Global Land Data Assimilation System. Bull Am Meteorol Soc 85: 381-894.

5. Watkins MM, Wiese DN, Yuan DN, Boening C, Landerer FW (2015) Improved methods for observing Earth's time variable mass distribution with GRACE using spherical cap mascons. J Geophys Res Solid Earth 120: 2648.

6. Famiglietti JS (2004) Remote sensing of terrestrial water storage, soil moisture and surface waters. American Geophysical Union 150: 197-207.

7. Ramillien G, Famiglietti JS, Wahr J (2008) Detection of Continental Hydrology and Glaciology Signals from GRACE: A Review. Surv Geophys 29: 361-374.

8. Hirschi M, Seneviratne SI, Schaer C (2006) Seasonal variations in terrestrial water storage for major midlatitude river basins. J Meteorol 7: 39-60.

9. Hall AC, Schumann GJ, Bamber JL, Bates PD (2011) Tracking water level changes of the Amazon Basin with space-borne remote sensing and integration with large scale hydrodynamic modelling: A review. Phys Chem Ear 36: 223-231.

10. Xiang L, Wang H, Steffen H, Wu P, Jia L, et al. (2016) Groundwater storage changes in the Tibetan Plateau and adjacent areas revealed from GRACE satellite gravity data. Earth Plane Sci Lett 449: 228-239.

11. Schmidt R, Petrovic S, Guntner A, Barthelmes F, Wunsch J, et al. (2008) Periodic components of water storage changes from GRACE and global hydrology models. J Geophys Res 113: B08419.

12. Cazenave A, Chen J (2010) Time-variable gravity from space and present-day mass redistribution in the Earth system. Earth Plan Sci Lett 298: 263-274.

13. Velicogna I, Wahr J (2013) Timevariable gravity observations of ice sheet mass balance: Precision and limitations of the GRACE satellite data. Geophys Res Lett 40: 3055-3063.

14. Johnson GC, Chambers DP (2013) Ocean bottom pressure seasonal cycles and decadal trends from GRACE Release05: Ocean circulation implications. J Geophys Res Oceans 118: 4228-4240.

15. Chambers DP, Bonin JA (2012) Evaluation of Release-05 GRACE time-variable gravity coefficients over the ocean. Ocean Sci 8: 859-868.

16. Song TA, Simons M (2003) Large Trench-Parallel Gravity Variations Predict Seismogenic Behavior in Subduction Zones. Science 301: 630-633.

17. De Viron O, Panet I, Mikhailov V, Van Camp M, Diament M (2008) Retrieving earthquake signature in grace gravity solutions. Geophys J Int 174: 14-20.

18. Adhikari I (2016) Climate-driven polar motion: 2003-2015, American Association for the Advancement of Science.

19. Karpik AP, Kanushin VF, Ganagina IG, Goldobin DN, Mazurova EM (2015) Analyzing spectral characteristics of the global Earth gravity field models obtained from the CHAMP, GRACE and GOCE space missions. Gyroscopy Navig 6: 101-108.

20. Luthcke SB, Zwally HJ, Abdalati W, Rowlands DD, Ray RD, et al. (2006) Recent Greenland ice mass loss by drainage system from satellite gravity observations. Science 314: 1286-1289.

21. Nerem RS, Jekeli C, Kaula WM (1995) Gravity field determination and characteristics: Retrospective and prospective. J Geophys Res 100: 15053-15074.

26

Analysis of Temperature Variability over Desert and Urban Areas of Northern China

Isaac Mugume[1], Shuanghe Shen[2]*, Sulin Tao[2] and Godfrey Mujuni[3]

[1]Department of Geography, Geo informatics and Climatic Sciences, Makerere University, Uganda

[2]Collaborative Innovation Center on Forecast and Evaluation of Meteorological Disasters, Nanjing University of Information Science and Technology, China

[3]Department of Applied Meteorology, Data and Climate Services, Uganda National Meteorological Authority, Uganda

*Corresponding author: Shuanghe Shen, Collaborative Innovation Center on Forecast and Evaluation of Meteorological Disasters, Nanjing University of Information Science and Technology, China, E-mail: yqzhr@nuist.edu.cn

Abstract

Although many studies carried out have shown evidence of regional temperature variability along with global climate changes, it is important to compare the trends over different regions considering urbanization and levels of development. The study investigated temperature variability over urban and desert areas of Northern China using Mann-Kendall trend test, ranking temperatures and regression analysis for the data from 20 stations. The results show decreasing diurnal temperature range (DTR) over both deserts and cities in spring but decreasing (increasing) for cities (deserts) in summer (cities: -0.140°C/decade and deserts: 0.068°C/decade). The DTR over cities is decreasing faster in spring over deserts (cities: -0.307°C/decade, desert: 0.023°C/decade). The maximum temperature over desert areas is increasing at a higher rate (annual: 0.510°C/decade, spring: 0.540°C/decade and summer: 0.550°C/decade) than over cities (annual: 0.325°C/decade, spring: 0.252°C/decade and summer: 0.389°C/decade). The high temperature days and high temperature extremes for both areas are increasing while the frost days and low temperature extremes for both areas are decreasing. The spring minimum temperatures are also increasing over both areas and increasing at a higher rate over deserts (0.536°C/decade) than over cities (0.529°C/decade).

Keywords: Climate response; Mann-kendall trend test; Urban and desert areas of China

Introduction

The air temperature is one of the key parameters used in classification of climatological zones, for instance: temperate, humid, desert and semi-desert [1]. The variation of daily temperature is controlled mainly by incoming solar energy and outgoing long-wave surface radiation [2]. The air temperate does not vary in isolation and one of the cases is explained by Bryan et al. [3] of high temperatures and high pollution concentrations being associated with strength of high pressure systems and sunny (with fast photolysis rates) conditions.

Over a large area and over a long time, the temperatures can be averaged and thus assisting to describe the climate of area in terms of hotness/coldness. While weather elements vary from day to day or even place to place [4], climate too exhibits variability inter or intra season as well as inter or intra annual.

The changes in climate may manifest as changes in the mean state or in variability of their cycles [3]. Boko [5] illustrates these changes graphically as (Figures 1-3).

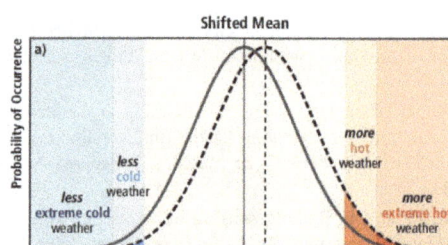

Figure 1: Shift in mean temperature.

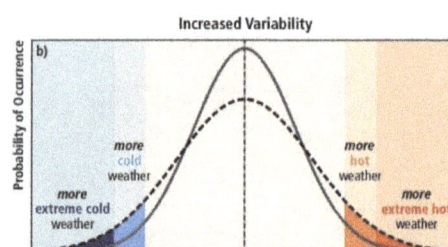

Figure 2: Change in temperature variability.

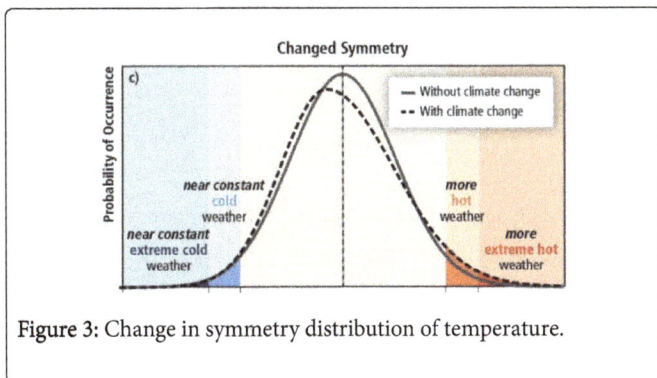

Figure 3: Change in symmetry distribution of temperature.

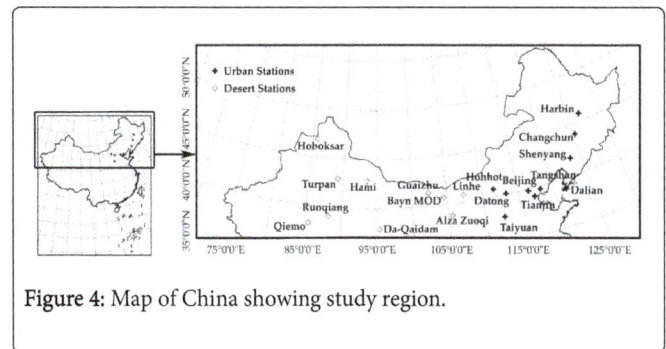

Figure 4: Map of China showing study region.

Climate change and variability has remained a research problem over the years. The previous studies demonstrated changes in precipitation, temperature, and other climatic parameters as well as vegetation cover [6] and these changes have been linked to economic development and urbanization [7]. Additionally, climate extremes have increased both in frequency and magnitude [8] but other areas have seen notable decline [9].

One of the indicators of climate change is changes in temperature and according to Roy et al. [10]; local temperature change is influenced by urbanization. A change in average temperature, can affect the amount of cloudiness as well as the type and amount of precipitation that Occurs [11] and is attributed to changes in radiative forcing [8,12,13] such as changes in landscape leading to changes in surface reflectivity to solar radiation; surface moisture variability; modification of vegetation cover as well as anthropogenic heat release which combine and cause temperature rise.

The increase in temperature has impacts such as: increased incidence and severity of heat-waves, droughts [13,14]; shrinking of glaciers ice caps, mountain glaciers, and permafrost regions of the world [15] and can affect growing practices such as sowing dates and cultivars [16]. The global temperatures are estimated to have increased by 0.5-0.6°C over the last century [9,13] and are estimated to increase by 0.3-0.7°C by 2035 [17].

Due to differential rates of development and urbanization as well as differences in land surface characteristics, we can expect different trends of temperature over deserts and cities. This is because, the economic development of a region is directly correlated with urbanization [18,19] and urbanization affects surface greenness and leads to changes in surface albedo [12]. Thus these changes affect receipt of solar radiation on the earth and influence air temperature. Additionally urbanization and industrialization have led to increase in greenhouse gases (GHGs) which has also influenced temperature [20]. In recent years, China has experienced massive industrial growth and economic development as well as increased urbanization [21]. The growth of cities introduces the need for constant monitoring of weather and climate over different regions of China.

Although temperature trends have been studied in many regions of the world [17], studies considering differential responses of temperature over desert and cities are limited. We compare the temperature variability including their extremes over cities and desert areas of Northern China for the period 1981-2010. The study areas are presented in Figure 4; section 2 describes the data sources, section 3 presents the study methods, section 4 presents results and discussion while section 5 gives summary and conclusion.

Data

The daily maximum and minimum temperature data is obtained from the School of Atmospheric Science of Nanjing University of Information Science and Technology (NUIST). This data is regularly updated and quality controlled to take care of the on-going research in the university. The data is then organized into temporal scales (annual, spring and summer). The diurnal temperature range (DTR) is computed from eqn. (1) as the difference between daily maximum temperature (T_{max}) and daily minimum temperature (T_{min}) and is also organized in terms of annual and seasonal temporal scales.

$$DTR = T_{max} - T_{min} \qquad (1)$$

The high temperature days are computed according to Thomas et al. [13] as days that have T_{max} exceeding 32.0°C and the number of days with extreme high temperature is those that exceed 90th percentile of daily observed T_{max}. Our study used a higher threshold of 35.0°C for high temperature days and the number of days with extreme high temperatures as those with T_{max} over the 95th percentile. The extreme low temperature as those days with T_{min} below the 5th percentiles of daily T_{min} and the frost days as the days with T_{min} below 0°C.

Methods

In order to study the trends of temperature over regions of different climatology, the relative deficit and or surplus (anomalies) of temperature $\hat{T}$ are used and calculated using the eqn (2).

$$\hat{T} = \frac{T_i - \bar{T}}{\bar{T}} \qquad (2)$$

Ti is temperature in question and $\bar{T}$ is respective long-term mean temperature over the study period. The $\bar{T}$ of a temperature data set, {T1, T2, · · ·, Tn} is computed using eqn (3):

$$\bar{T} = \frac{1}{n} \sum_{i=1}^{n} T_i \qquad (3)$$

The Mann-Kendal trend test

The Mann-Kendall (MK) trend test is used to analyze the trends of: T_{max}, T_{min} and DTR. The MK is recommended by Qiang et al. [22] because it is (1) a rank-based nonparametric test, able to test trends without requiring normality or linearity; (2) less sensitive to outliers and (3) recommended by the World Meteorological Organization. Additionally Jagannathan and Parthasarathy [23] have identified the MK test as powerful for testing trends that are linear or non-linear. For a time ordered temperature data-set, we define MK trend test statistic, S (eqn. 4):

$$s = \sum_{i=1}^{n-1} \sum_{j+i=1}^{n} \text{sgn}\left(T_j - T_i\right) \qquad (4)$$

where $\text{sgn}(T_j - T_i)$ is:

$$\text{sgn}\left(T_j - T_i\right) = \begin{cases} +1 : if\, T_j - T_i > 0 \\ 0 : if\, T_j - T_i = 0 \\ -1 : if\, T_j - T_i < 0 \end{cases} \qquad (5)$$

for non-tied values of T_i, the variance $\delta^2(S)$ of the distribution of S is computed using:

$$\delta^2(s) = \frac{n(n-1)(2n+5)}{18} \qquad (6)$$

for tied values of T_i, the variance is given by:

$$\delta^2(s) = \frac{n(n-1)(2n+5) - \sum t_i(i)(i-1)(2i+5)}{18} \qquad (7)$$

t_i is number of ties of extent i. The MK test statistic is then given by the standard Gaussian value, $M\,K_z$ defined as:

$$MK_Z = \begin{cases} \frac{s-1}{\delta(s)} : if\, S > 0 \\ 0 : if\, S = 0 \\ \frac{s+1}{\delta(s)} : if\, S < 0 \end{cases} \qquad (8)$$

The computation of temperature extremes

We use Jenkinson [24] formula (eqn. 9) for computing extreme temperatures. This formula is discussed by Chris and Anderson [25] in comparison with other formula for studying extremes and is recommended for the study of climate extremes [25,26]. According to the Jenkinson formula, the probability, p that a random value is less than or equal to the rank of that value, T_i is given by:

$$p = \frac{m - 0.31}{n + 0.38} \qquad (9)$$

where m is the position of the value and n is the number of values in the data set. The T_i for example summer season which has 92 days, is arranged in ascending order: {T1, T2, · · ·, T91, T92}. The T_i representing the 95th percentile is linearly interpolated between the 88th ranked value (giving: p=94.9%) and 89th ranked value (p=96.0%). The 95th percentile is thus interpolated.

We considered high temperature days, frost days, high temperature extremes and low temperature extremes. The high temperature days are the days with T_{max} greater than 35.0°C, frost days are the days with T_{min} below 0°C, high temperature extreme days are the number of days with T_{max} over the 95th percentiles of daily T_{max}, while low temperature extreme days are the days with T_{min} below the 5th percentiles of daily T_{min}. The number of days with extreme high temperature and extreme low temperature are obtained using the Jenkinson formula (eqn. 9).

Regression method

We use regression to obtain the decadal (10 year) rate of changes of temperature, described as under. Given an n time-ordered temperature

dataset: {T_1, T_2, · · ·, T_{n-1}, T_n}, ordered in time, t the linear regression equation is given as:

$$T_i = \alpha t_i + \epsilon \qquad (10)$$

where

i=1, 2, · · ·, n

α is the rate of change, is the error and the decadal rate of change of temperature (α_{10}) is computed as:

$$\alpha_{10} = \frac{T_{i+10} - T_i}{10} \qquad (11)$$

Results and Discussion

Annual temperature trends

The annual temperature trends over deserts (Table 1) and over cities (Table 2) are obtained using MK_z at 99% confidence level. Over deserts, we find an increasing trend for both T_{max} (MK_z=0.431) and T_{min} (MK_z=0.407) and a decreasing trend for DTR (MK_z=-0.091). The annual DTR shows high variability (Figure 5). It peaked during the period 1990-2000 and declining over the period 2001-2010. This variability probably explains the weak decreasing trend in Table 1. The annual trends of DTR for individual stations were in the range of: -0.35°C/decade to -0.04°C/decade. The T_{max} has been increasing since 1985 (Figure 6) in the range of 0.36-0.70°C/decade and on average at 0.51°C/decade. The T_{min} shows a high variable increasing trend (Figure 7) in the range of 0.06-0.8°C/decade and on average at 0.52°C/decade. This rate is slightly greater than that of T_{max}. We thus argue that T_{min} increase faster than T_{max} which could be accounting for the rate of decrease of DTR over deserts in (Table 1).

Station	T_{max}	T_{min}	DTR
Alza Zuoqi	0.370	0.527	-0.457
Bayn MOD	0.448	0.037	0.269
Da-Qaidam	0.591	0.497	-0.067
Guaizhu	0.301	0.467	-0.223
Hami	0.444	0.144	0.301
Hoboksar	0.264	0.385	-0.269
Linhe	0.315	0.480	-0.545
Qiemo	0.522	0.406	0.278
Ruoqiang	0.545	0.531	0.177
Turpan	0.508	0.596	-0.375
Average	0.431	0.407	-0.091

Table 1: Annual temperature variation for desert areas.

Station	T_{max}	T_{min}	DTR
Beijing	0.260	0.545	-0.384
Changchun	0.269	0.416	-0.320
Dalian	0.228	0.370	-0.315

Datong	0.402	0.554	-0.292
Harbin	0.260	0.582	-0.683
Hohhot	0.384	0.577	-0.503
Shenyang	0.140	-0.145	0.214
Taiyuan	0.407	0.697	-0.393
Tangshan	0.343	0.485	-0.398
Tianjin	0.214	0.016	0.145
Average	0.291	0.410	-0.293

Table 2: Annual temperature variation for cities.

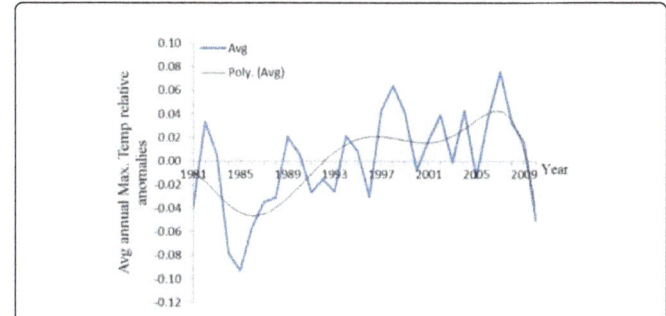

Figure 5: Annual DTR anomalies for selected desert areas.

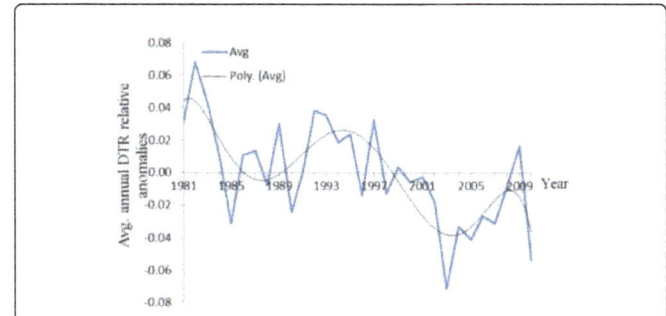

Figure 6: Annual maximum temperature anomalies for selected desert areas.

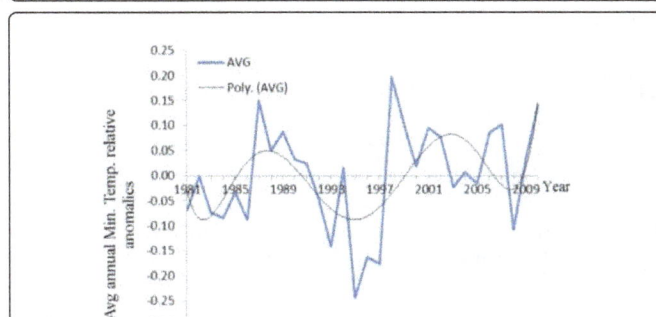

Figure 7: Annual minimum temperature anomalies for selected desert areas.

Over cities, we find both T_{max} and T_{min} increasing annually with exception of Shenyang and a decreasing DTR ($MK_z=-0.293$). The annual DTR for cities decrease sharply (Figure 8) in the range of -0.659 to -0.146°C/decade and on average -0.238°C/decade. The T_{max} is increasing (Figure 9) in the range of 0.11 to 0.501°C/decade and an average of 0.325°C/decade.

The T_{min} (Figure 10) does not present plausible results but we can infer that T_{min} was decreasing slightly over the period 1995-2010. With exception of Shenyang, the annual T_{min} trends for individual stations were increasing in the range of 0.022 to 1.015°C/decade and an average of 0.566°C/decade. Thus the annual rate of increase of T_{min} is greater than that of T_{max} over cities.

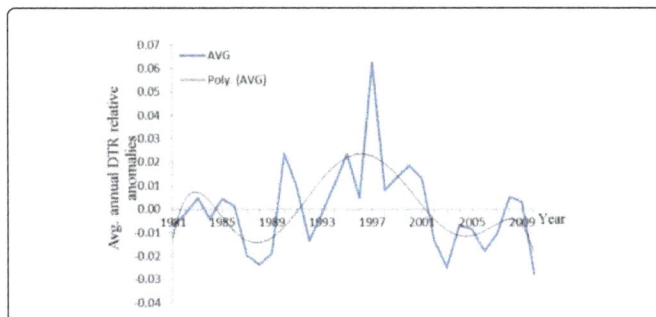

Figure 8: Annual DTR anomalies for selected cities.

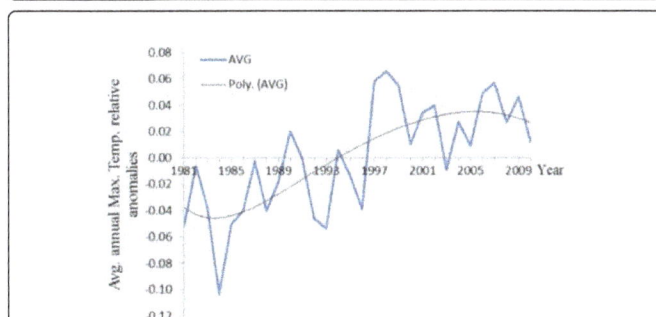

Figure 9: Annual maximum temperature anomalies for selected cities.

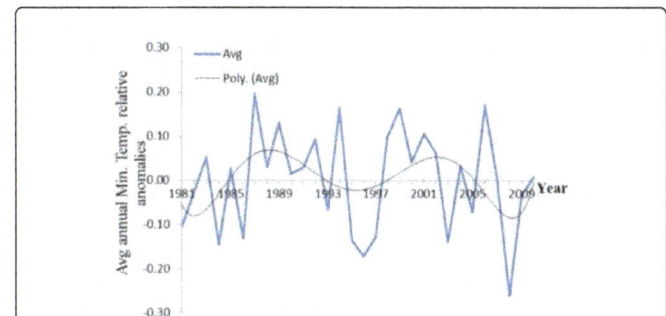

Figure 10: Annual minimum temperature anomalies for selected cities.

Spring temperature trends

The spring temperature trends over deserts (Table 3) and over cities (Table 4) are obtained using MK_z at 99% confidence level.

Station	T_{max}	T_{min}	DTR
Alza Zuoqi	0.274	0.416	-0.218
Bayn MOD	0.329	0.195	0.149
Da-Qaidam	0.467	0.301	0.122
Guaizhu	0.191	0.375	-0.260
Hami	0.255	0.269	0.113
Hoboksar	0.149	0.324	-0.209
Linhe	0.195	0.499	-0.370
Qiemo	0.411	0.425	0.297
Ruoqiang	0.343	0.508	0.090
Turpan	0.177	0.425	-0.209
Average	0.279	0.374	-0.050

Table 3: Spring temperature variation for desert areas.

Station	T_{max}	T_{min}	DTR
Beijing	0.154	0.338	-0.195
Changchun	0.113	0.264	-0.278
Dalian	0.214	0.320	-0.175
Datong	0.186	0.377	-0.195
Harbin	-0.002	0.324	-0.511
Hohhot	0.246	0.457	-0.297
Shenyang	0.117	-0.039	0.039
Taiyuan	0.315	0.476	-0.228
Tangshan	0.163	0.384	-0.338
Tianjin	0.195	0.149	0.090
Average	0.170	0.410	-0.293

Table 4: Spring temperature variation for cities.

Over deserts, we find both spring T_{max} and T_{min} increasing. The DTR over five stations is increasing while the rest are decreasing. On average, DTR is decreasing (MK_z=-0.050) and it increased over the period 1990-2010 (Figure 11). Both T_{max} (Figure 12) and T_{min} (Figure 13) are increasing at MK_z=0.279 and MK_z=0.374 respectively and increased over the period 1987-2007. The differential increases in trends of T_{max} and T_{min}, can in part explain the moderate increase of DTR over the same period. The rate of decrease of DTR of individual stations is in the range of -0.547 to -0.18°C/decade except for the stations that presented an increasing DTR trend.

In general, the rate of decrease of DTR over 1981-2010 is -0.023°C/decade (negative sign is maintained to emphasize the decrease and

differentiate it from increasing trend, where increase is shown with a positive sign). The spring T_{max} is in general increasing in the range of 0.3-0.8°C/decade and on average at 0.54°C/decade. The spring T_{min}, (Figure 13) is increasing too in range of 0.23-0.85°C/decade with an average of 0.56°C/decade.

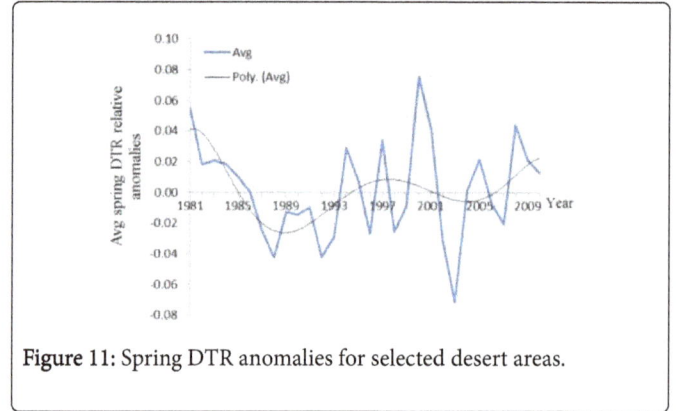

Figure 11: Spring DTR anomalies for selected desert areas.

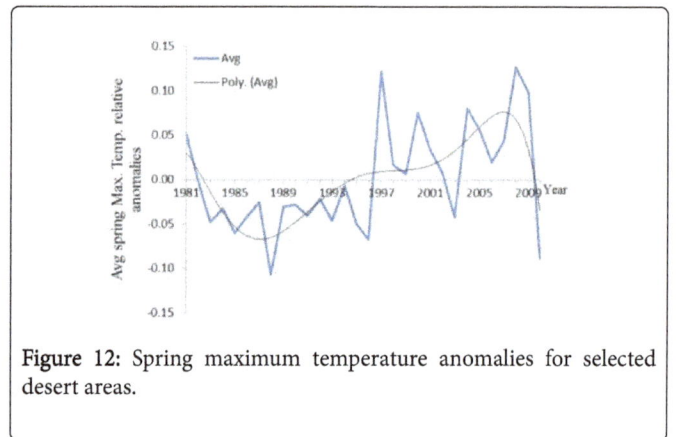

Figure 12: Spring maximum temperature anomalies for selected desert areas.

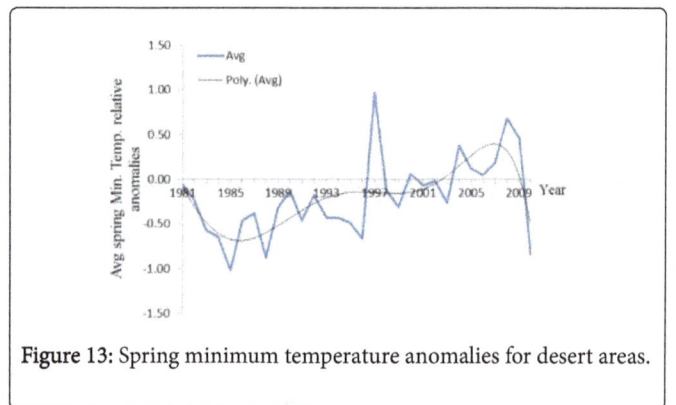

Figure 13: Spring minimum temperature anomalies for desert areas.

Over cities, we find an increasing trend of both spring T_{max} and T_{min} with exception of Harbin whose T_{max} is decreasing and Shenyang's T_{min}. The DTR is decreasing (MK_z=-0.209) with the exception of Shenyang (Figure 14) in the range of -0.829 to -0.157°C/decade and on average at -0.307°C/decade.

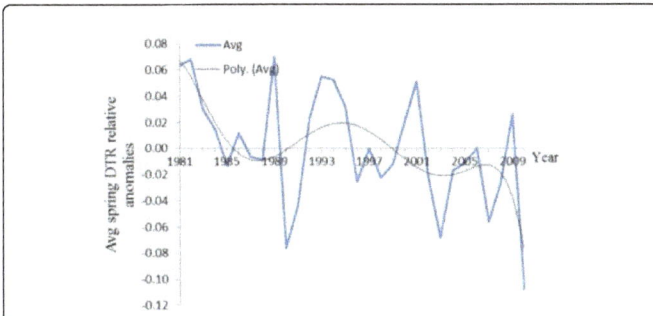

Figure 14: Spring DTR anomalies for selected cities.

The spring T_{max} (Figure 15) is increasing with an average MK_z of 0.170 and in the range of 0.052 to 0.533°C/decade and overall rate of 0.252°C/decade. The spring T_{min} (Figure 16) is increasing as well with exception of Shenyang in the range of 0.137 to 0.928°C/decade and overall increasing at 0.560°C/decade.

The rate at which spring T_{min} is increasing (0.560°C/decade) is greater than the rate of spring T_{max} (0.252°C/decade) which probably explains the decreasing trend of spring DTR (-0.307°C/decade).

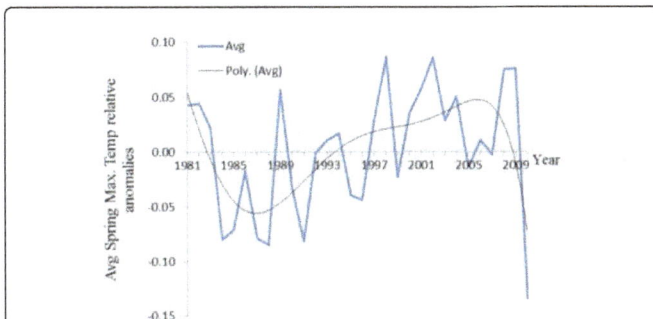

Figure 15: Spring maximum temperature anomalies for selected cities.

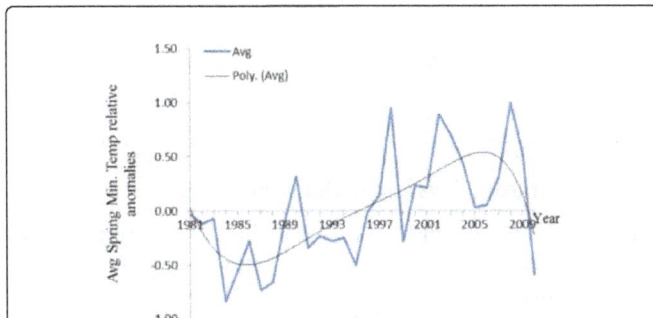

Figure 16: Spring minimum temperature anomalies for selected cities.

Summer temperature trends

The summer temperature trends over deserts (Table 5) and over cities (Table 6) are obtained using MK_z at 99% confidence level.

Station	T_{max}	T_{min}	DTR
Alza Zuoqi	0.287	0.444	-0.301
Bayn MOD	0.343	0.292	0.278
Da-Qaidam	0.439	0.513	-0.232
Guaizhu	0.434	0.410	0.062
Hami	0.485	0.195	0.315
Hoboksar	0.324	0.434	-0.090
Linhe	0.278	0.526	-0.315
Qiemo	0.494	0.343	0.159
Ruoqiang	0.522	0.462	-0.039
Turpan	0.499	0.333	0.163
Average	0.411	0.395	-2.8×10^{-18}

Table 5: Summer temperature variation for desert areas.

Station	T_{max}	T_{min}	DTR
Beijing	0.228	0.494	-0.255
Changchun	0.287	0.352	-0.136
Dalian	0.085	0.186	-0.145
Datong	0.352	0.526	-0.011
Harbin	0.267	0.494	-0.287
Hohhot	0.324	0.536	-0.067
Shenyang	0.071	-0.057	0.182
Taiyuan	0.379	0.568	-0.140
Tangshan	0.237	0.384	-0.329
Tianjin	0.154	0.186	0.094
Average	0.238	0.367	-0.109

Table 6: Summer temperature variation for cities.

Over deserts, the summer T_{max} and T_{min} are increasing with DTR for five stations increasing while the rest decreasing. We find a moderate decreasing trend for summer DTR over the deserts (Figure 17) in the range of 0.026 to 0.382°C/decade. The summer T_{max} is increasing (Figure 18) as well in the range of 0.369 to 0.723°C/decade and on average 0.55°C/decade. The summer T_{min} is also increasing sharply over the period 1985-2010 in the range of 0.28 to 0.94°C/decade and on average 0.536°C/decade (Figure 19).

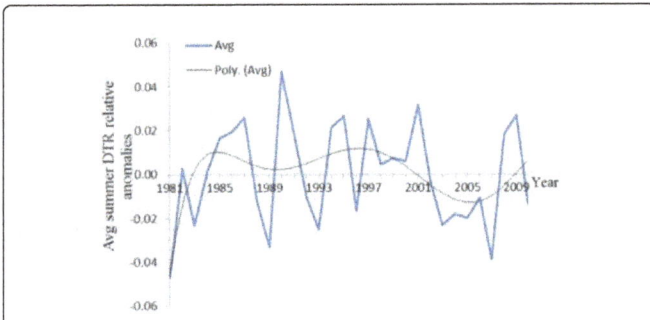

Figure 17: Summer DTR anomalies for selected desert areas.

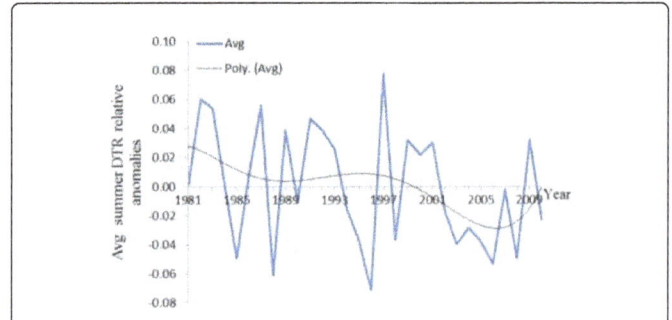

Figure 18: Summer maximum temperature anomalies for selected desert areas.

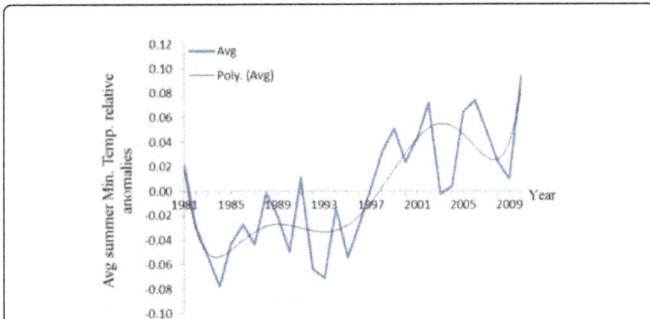

Figure 19: Summer minimum temperature anomalies for selected desert areas.

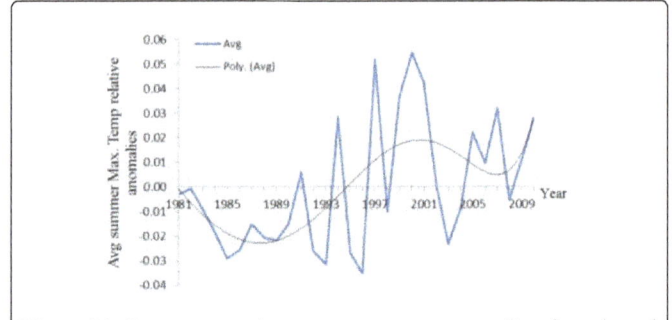

Figure 20: Summer DTR anomalies for selected cities.

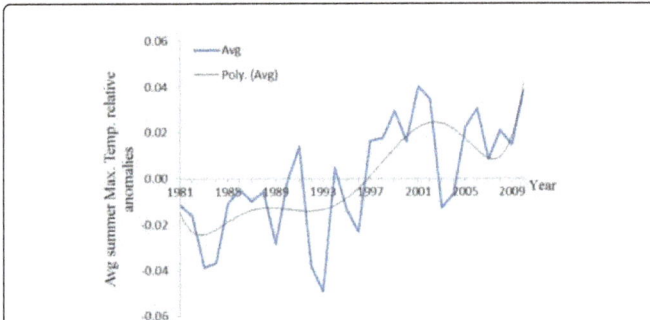

Figure 21: Summer maximum temperature anomalies for selected cities.

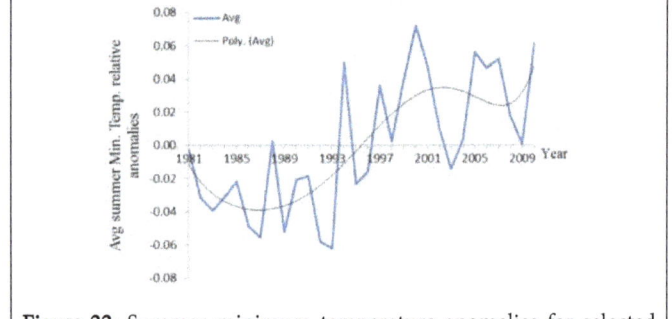

Figure 22: Summer minimum temperature anomalies for selected cities.

Over cities, we find increasing trends of summer T_{max} and T_{min} except Shenyang's T_{min} and decreasing trend of DTR (MK_z=-0.109) (Figure 20). The trend of summer DTR is in the range of -0.419 to -0.023°C/decade and on average decreasing at a rate of -0.140°C/ decade. The summer T_{max} (Figure 21) is increasing over the period 1991-2010 in the range of 0.107 to 0.684°C/decade and in general, at a rate of 0.389°C/decade. From 1993, the summer T_{min} is also increasing (Figure 22) with exception of Shenyang in the range of 0.144 to 0.864°C/decade and in general increasing at 0.529°C/decade. Thus the rate of increase of T_{min} (0.529°C/decade) is greater than that of T_{max} (0.389°C/decade) explaining the decreasing trend of DTR.

High temperature days and high temperature extremes

The Table 7 presents the trend of high temperature days and high temperature extreme days over desert areas during summer. The number of high temperature days, with exception of DaQaidam is increasing for all the stations (MK_z=0.151 to 0.492) as well as the days having high temperature extremes (MK_z=0.159 to 0.493). The highest temperature for Da-Qaidam was below 35°C and thus has the smallest rate of high temperature extremes.

Area	High Temp. Days	High Temp. Extremes
Alza Zuoqi	0.250	0.493
Bayn MOD	0.338	0.440
Da-Qaidam	N/A	0.159

Guaizhu	0.464	0.426
Hami	0.492	0.445
Hoboksar	0.151	0.308
Linhe	0.189	0.328
Qiemo	0.363	0.300
Ruoqiang	0.509	0.339
Turpan	0.386	0.353
Average	0.349	0.326

Table 7: Trend of high temperature days and high temperature extremes for desert areas.

The Table 8 shows the trend of high temperature days and high temperature extremes for selected cities. Like for the desert areas (Table 7), both the number of high temperature days is increasing for all the cities (MK_z=0.140 to 0.511) and the number of days with high temperature extremes (MK_z=0.105 to 0.442).

Area	High Temp. Days	High Temp. Extremes
Beijing	0.283	0.343
Changchun	0.314	0.138
Dalian	0.179	0.105
Datong	0.505	0.321
Harbin	0.269	0.307
Hohhot	0.289	0.303
Shenyang	0.14	0.02
Taiyuan	0.511	0.442
Tangshan	0.242	0.307
Tianjin	0.252	0.327
Average	0.298	0.261

Table 8: Trend of high temperature days and high temperature extremes for cities.

Frost days and low temperature extremes

The Table 9 shows the trend of frost days and days for low temperature extreme over spring for desert areas. We find both the frost days (MK_z=-0.414 to -0.128) and the number of days with extreme low temperatures (MK_z=-0.446 to -0.010) decreasing with a positive correlation of 0.614. This decreasing trend means that few and fewer days have minimum temperature below zero degrees Celsius (0°C). The Table 10 shows the trend of frost days and days for low temperature extreme over spring for cities.

Area	Frost Days	Low Temp. Extremes
Alza Zuoqi	-0.335	-0.209
Bayan MOD	-0.128	-0.01

Da-Qaidam	-0.355	-0.05
Guaizhu	-0.375	-0.222
Hami	-0.215	-0.107
Hoboksar	-0.379	-0.094
Linhe	-0.383	-0.446
Qiemo	-0.155	-0.054
Ruoqiang	-0.414	-0.242
Turpan	-0.339	-0.125
Average	0.308	0.156

Table 9: Trend of frost days and low temperature extremes for desert areas.

Area	Frost Days	Low Temp. Extremes
Beijing	-0.125	-0.304
Changchun	-0.133	-0.381
Dalian	-0.231	-0.148
Datong	-0.334	-0.311
Harbin	-0.272	-0.442
Hohhot	-0.324	-0.29
Shenyang	0.021	0.319
Taiyuan	-0.362	-0.389
Tangshan	-0.272	-0.184
Tianjin	-0.08	0.01
Average	-0.237	-0.308

Table 10: Trend of frost days and low temperature extremes for cities.

Like deserts, the cities show a decreasing trend of both frost days (MK_z=-0.362 to -0.080) and the number of days having extreme low temperature (MK_z=-0.442 to -0.148) with a positive correlation of 0.730. The decreasing trend means that few and fewer days have T_{min} below zero degrees Celsius (0°C.) and it also explains the increase of spring T_{min}. We find a greater decrease of frost days for desert greater than the rate for the cities which probably indicates that spring night are becoming warmer for deserts at a faster rate compared to cities.

Summary and Conclusion

In the present study, we investigated the variability of temperature over desert and cities of Northern China for the period 1981-2010. We found DTR decreasing for both desert and cities. It is evident that the DTR for cities is decreasing faster than the desert's. The T_{max} is increasing for both desert and cities and that T_{max} for the deserts is increasing faster than that for the cities. The T_{min} is increasing as well for both deserts and cities and the rate of increase of T_{min} for the desert greater than the one for cities.

We also find both the deserts and cities exhibiting an increasing trend of high temperature days and high temperature extremes. The rate of increase of high temperature extremes over deserts is slightly higher than the one over cities. In general, the summer T_{max} for all the areas show increasing trend. The frost days and low temperature extremes over both the deserts and cities are decreasing. The frost days over deserts decrease slightly higher than over cities and in general, the spring T_{min} for all the stations is increasing explaining the decreasing trend of the frost days and low temperature extremes.

Acknowledgement

This study is supported by the Research Fund for the Public Sector of China (GYHY201506018) and the Climate Change Specific Foundation of China Meteorological Administration (CCSF201318). The authors also acknowledge expert opinion from reviewers in quest of improving our manuscript.

References

1. Zabeltitz CV, Baudoin WO (1999) Greenhouses and shelter structures for tropical regions. Food and Agriculture Org 154.

2. Ahrens DC (2011) Essentials of meteorology: an invitation to the atmosphere. Cengage Learning.

3. Bryan JB, Konstantin YV, Russell RD (2010) Changes in seasonal and diurnal cycles of ozone and temperature in the eastern US. Atmospheric Environment 44: 2543-2551.

4. Walter JS (1989) Principles of meteorological analysis. Courier Corporation.

5. Boko M (2012) Managing the Risks of Extreme Events and Disasters to Advance Climate Change Adaptation. A Special Report of Working Groups I and II of the Intergovernmental Panel on Climate Change.

6. Wang K, Ye H, Chen F, Xiong Y, Wang C (2012) Urbanization effect on the diurnal temperature range: different roles under solar dimming and brightening. Journal of Climate 25: 1022-1027.

7. Schmal H (1981) Patterns of European urbanisation since 1500 309.

8. Richard B, Arnell NW, Adger WN, Thomas D (2013) Migration, immobility and displacement out- comes following extreme events. Environmental Science and Policy 27: S32-43.

9. Easterling DR, Evans JL (2000) Observed variability and trends in extreme climate events: A brief review. Bulletin of the American Meteorological Society 81: 417-425.

10. Roy S, Shouraseni, Yuan F (2009) Trends in extreme temperatures in relation to urbanization in the Twin Cities Metropolitan Area, Minnesota. Journal of Applied Meteorology and Climatology 48: 669-679.

11. Selvaraju, Ramamasy, Baas S (2007) Climate Variability and Change: Adaptation to Drought in Bangladesh: a Resource Book and Training Guide 9.

12. Sailor, David J (1993) Role of surface characteristics in urban meteorology and air quality. Lawrence Berkeley Lab, CA.

13. Thomas PC, Stott PA, Stephanie H (2012) Explaining extreme events of 2011 from a climate perspective. Bulletin of the American Meteorological Society 93: 1041-1067.

14. Oerlemans, Johannes (2001) Glaciers and climate change. CRC Press.

15. Timothy MK (2010) Climate change: shifting glaciers, deserts, and climate belts. Infobase Publishing.

16. Tao S, Shen S, Li Y, Wang Q, Gao P, Mugume I (2016) Projected Crop Production under Regional Climate Change Using Scenario Data and Modeling: Sensitivity to Chosen Sowing Date and Cultivar. Sustainability 8: 214.

17. Jaswal AK, Rao PCS, Singh V (2015) Climatology and trends of summer high temperature days in India during 1969-2013. Journal of Earth System Science 124: 1-15.

18. Watson RT, Zinyowera MC, Moss RH (1998) The regional impacts of climate change: an assessment of vulnerability. Cambridge University Press.

19. Uddin AJ (2007) Industrialisation in North-Eastern Region. Mittal Publications.

20. Rooij V, Benjamin (2006) Regulating land and pollution in China: law making, compliance, and enforcement: theory and cases. Amsterdam University Press.

21. Zhou L, Dickinson RE, Tian Y (2004) Evidence for a significant urbanization effect on climate in China. Proceedings of the National Academy of Sciences of the United States of America 101: 9540-9544.

22. Qiang Z, Vijay PS, Suna P, Chend X (2011) Precipitation and stream flow changes in China: changing patterns, causes and implications. Journal of Hydrology 410: 204-216.

23. Jagannathan P, Parthasarathy B (1973) Trends and periodicities of rainfall over India. Monthly Weather Review 101: 371-375.

24. Jenkinson AF (1977) The analysis of meteorological and other geophysical extremes. Met Office Synoptic Climatology Branch Memo 58: 41.

25. Chris F, Anderson C (2002) Estimating changing extremes using empirical ranking methods. Journal of climate 20: 2954-2960.

26. Bonsal BR, Zhang X, Vincent LA (2001) Characteristics of daily and extreme temperatures over Canada. Journal of Climate 14: 1959-1976.

27

Modeling the Impact of Climate Change on Production of Sesame in Western Zone of Tigray, Northern Ethiopia

Awetahegn Niguse[1]* and Araya Aleme[2]

[1]Department of GIS and Agro meteorology, Mekelle Agricultural Research center, Tigray Agricultural Research Institute, Mekelle, Ethiopia
[2]Mekelle University, Department of Crop and Horticultural Sciences, Mekelle, Ethiopia

Abstract

Sesame is one of the most important cash crops which is mostly grown in the western and north western zone of Tigray region. The impact of climate change on sesame yield were not addressed yet particularly in the study area. Therefore, this study was aimed at assessing the impact of climate change on production of Sesame in the Western lowlands of Tigray, with the specific objective of modeling the impact of climate change on production of sesame. Historical sesame yield was obtained and climate outputs from HadGEM2-ES, ACCESS1-0 and GFDL-ESM2M models were projected for the near (2010-2039), mid (2040-2069) and end (2070-2099) term periods to evaluate future impacts of climate change. In all periods (near, mid and end term) normal sowing date was better than early and late sowing dates in terms of yield. In late sowing date, yield was simulated to reduce from -5.88% to -23.31% in the end term RCP8.5 by GFDL-ESM2M and HadGEM2-EM climate models respectively. However, in the normal sowing date the yield was increased up to 33.1% by GFDL-ESM2M model in the midterm RCP4.5. Generally, higher yields were found in the normal sowing date. The response of sesame cultivars to the future climate changes should be studied under different management options. The impact should also be studied by different crop and climate models so as to capture the possible variability of sesame yield. Sensitivity to carbon dioxide, temperature, rainfall and other different management activities should be undertaken.

Keywords: RCP; GFDL-ESM2M; ACCESS1-0; HadGEM2-ES

Introduction

The major cause for total annual crop losses in agriculture across the world is related to weather and climatic effects such as drought, flash flood, untimely rain, frost, hail and storms [1]. Rising temperature, drought, flood, desertification and weather extremes severely affect agriculture, especially in the developing countries [2]. The Ethiopian economy is more dependent in Agriculture which contributes about 41% of the total GDP, 80% of the employment and the majority of foreign exchange earnings [3]. Many African countries including Ethiopia, their agricultural system is highly sensitive to climate, extreme weather events and climatic conditions which have major impacts on agriculture. Sesame is an important oil seed crop and is grown in tropical zones as well as in temperate zones between latitudes 40°N and 40°S [4]. The third assessment report of the International Panel on Climate Change (IPCC) indicated that developing countries are expected to suffer most from the negative impacts of climate change and climate variability [5].

Materials and Methods

Area description

Kafta Humera is located in the north-western Ethiopia and in the western part of Tigray Regional State Figure 1 and 585 km away from Mekelle and is located at 14°15' latitude and 36°37' longitude. Kafta Humera is bordered with "*Tsegede*" on the south and with Sudan in the west. In the north, the Tekeze River separates the district from Eritrea, in the east "Tahtay-Adiyabo" bordered with the district and in the southeast with "*Wolqayt*". The district covers an area of 632,877.75 ha which is about 23.6 percent of the western zone of Tigray (Figure 1).

Data collection

Climate, crop and management and soil data: Daily rainfall data of the study area were obtained from the National Meteorological Service Agency (NMA) of Ethiopia which is from 1980-2009 with some data gaps. The quality of the datasets was checked using freely available "tamet" software. R software script was later used to generate future climate scenarios for the near (2010-2039), mid (2040-2069) and end (2070-2099) of century using the fifth phase coupled model inter-comparison project (CMIP5) protocol. Even though, there are about 20 GCM outputs in the CMIP5, only three; HadGEM2-ES, GFDL-ESM2M and ACCESS1-0 climate model were used in this study. Moreover, for the Aqua Crop model, climate data (temperature, rainfall, solar radiation, evapotranspiration and relative humidity), Crop and agronomic parameter (sowing date, sowing depth, flowering date, maturity date, days to emergency, days to heading, grain yield, fertilizer rate, type and application time, seed rate and other management activities were collected from Humera agricultural research center.

Climate data preparation: The inputs of Aqua Crop model are climate, crop, field management and soil while the required output is yield. In simulating yield, four files with extension CLI, ETo, PLU and TEMP were used. CLI is a program file which consists ETo file (reference evapotranspiration), PLU file (for rainfall) and TEMP file (for maximum and minimum temperature). ETo was calculated using Hargreaves equation [6] by the formula: $ETo = 0.0023 \, (Tmean + 17.8) \, (Tmax - Tmin)^{0.5} \, Ra$.

*__Corresponding author:__ Awetahegn Niguse, Department of GIS and Agro meteorology, Mekelle Agricultural Research center, Tigray Agricultural Research Institute, Mekelle, Ethiopia, E-mail: nawetahegn@gmail.com

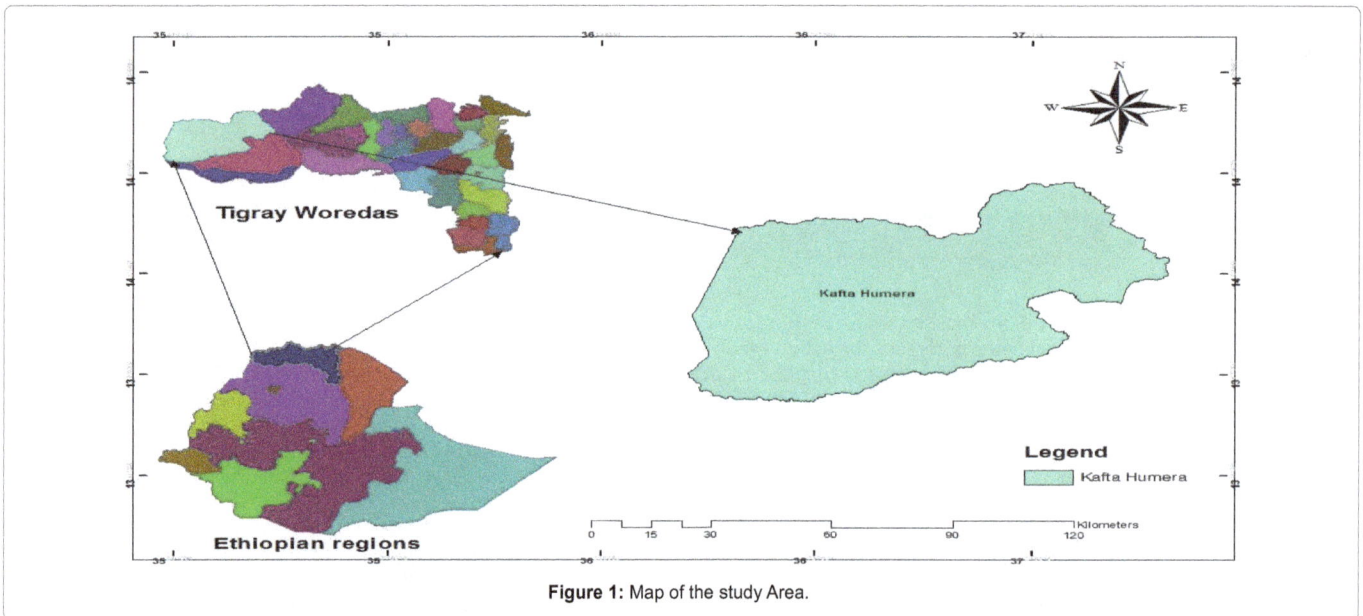

Figure 1: Map of the study Area.

Periods	RCP4.5			RCP8.5		
	ACCESS1-0	GFDL-ESM2M	HadGEM2-ES	ACCESS1-0	GFDL-ESM2M	HadGEM2-ES
Near-term	-2.92	**8.19**	-2.19	2.32	-2.43	0.94
Midterm	-4.11	4.25	0.47	**5.17**	-6.60	**4.16**
End Term	-3.45	-1.27	-2.87	4.39	1.12	-0.37

Table 1: Summarized expected future (near, mid and end term) changes of rainfall (%) compared to the baseline.

Where T=average temperature Tmax=maximum temperature, Tmin=minimum temperature, Ra=radiation.

Method of data analysis

For preparing Aqua Crop simulations, the instructions for running the simulation were modified to accommodate different sets of weather records and soil types. The climatic inputs used for the Sesame yield simulation was examined for four time periods; a baseline of climate (1980-2009) and for three future 30-years' time slices (2010-2039, 2040-2069 and 2070-2099) using RCP 4.5 and RCP 8.5.Three time periods (near term, midterm and end term) and two Representative Concentration Pathways (RCP4.5 and RCP8.5) and three sowing scenarios (Early, Normal and Late) were used to assess the impact of climate change on Sesame production in the study area. A fixed sowing date was set in the model by considering criteria when cumulative rainfall amount of 20 mm occurred over three consecutive days. It was achieved by introducing this condition in the sowing rule of the manager file.

Result and Discussion

Future climate of the study area

Projected rainfall: The GFDL-ESM2M climate model has predicted rainfall to increase by 8.2% under RCP4.5 in the near term. In the midterm, GCM ACCESS1-0 has shown a reduction of rainfall by 4.1% under RCP4. Similarly, GCM ACCESS1-0 and HadGEM2-ES climate models predicted rainfall to increase by 2.3% and 1% under RCP8.5 in the near term, respectively. In the mid-term (30 years) GCM GFDL-ESM2M has shown a reduction of rainfall by 6.6% under RCP8.5. However, in the end term, all models have projected a decreased rainfall under RCP4.5, while an increased rainfall was projected by ACCESS1-0

(4.4%) and GFDL-ESM2M (1.1%) under RCP8.5. Even though the magnitude of the change was small, precipitation was projected to increase under RCP8.5 for all time periods (Table 1).

Projected temperature: The models GFDL-ESM2M and HadGEM2-ES has shown a decreasing of temperature in the near and end term. In the mid-term, almost all models projected an increment of temperature will be occurred. However, the magnitude of increment will not be below 1°C except GCM ACCESS1-0 (1.01°C). This is also very similar with the near term RCP4.5. In the end term RCP8.5, almost all models simulate an increment of temperature by nearly 2°C (Table 2).

Unlike the maximum temperature, the minimum temperature was predicted to increase using all GCM models. In the near term RCP8.5 and end term RCP 4.5, all models projected that the increase in minimum temperature will be below 1°C. Both RCPs of mid-term and near term RCP 4.5 indicated very similar temperature increments. However, in the end term under RCP8.5 all models projected to experience higher temperatures in the study area (Table 3).

Generally, the climate models have shown an increase in temperature under future scenario even though there are slight differences in the magnitudes and extents of changes. The increase was slightly higher for minimum temperature than maximum temperature. In both maximum and minimum temperatures, all models signified that the temperature would reduce under RCP8.5 compared to the RCP4.5 during the near term period. This could be attributed due to the increase in rainfall of this period. The difference in temperature between RCP8.5 and RCP4.5 with in the same period could be due to the elevated concentrations of greenhouse gases. Compared to the east African projections (1.1 to 6.4°C), the study area could experience lower temperature and increments in the amount of rainfall [7]. However,

Periods	RCP4.5			RCP8.5		
	ACCESS1-0	FGDL-ESM2M	HadGEM2-ES	ACCESS1-0	FGDL-ESM2M	HadGEM2-ES
Near-term	1.21	0.98	1.29	0.02	-0.03	0.24
Midterm	1.01	0.37	0.79	0.75	0.93	0.81
EndTerm	0.08	-0.13	-0.02	1.84	1.46	2.27

Table 2: Summarized expected future (near, mid and end) changes of maximum temperature (°c) compared to baseline.

Periods	RCP4.5			RCP8.5		
	ACCESS1-0	GFDL-ESM2M	HadGEM2-ES	ACCESS1-0	GFDL-ESM2M	HadGEM2-ES
Near-term	1.21	1.04	1.40	0.11	0.13	0.29
Midterm	0.91	0.21	1.81	1.04	1.23	0.99
End Term	0.21	0.75	0.11	2.23	2.23	2.88

Table 3: Summarized expected future changes (near, mid and end) of Minimum Temperatures (°c) compared to the baseline.

Periods	Near term						Mid term						End term					
	RCP_4.5 (Rainfall deviation in (%)			RCP_8.5 (Rainfall deviation in (%)			RCP_4.5 (Rainfall deviation in (%)			RCP_8.5 (Rainfall deviation in (%)			RCP_4.5 (Rainfall deviation in (%)			RCP_8.5 (Rainfall deviation in (%)		
Month	ACCESS1-0	GFDL-ESM2M	Had GEM2-ES	ACCESS1-0	GFDL-ESM2M	Had GEM2-ES	ACCESS1-0	GF DL-ESM2M	Had GEM2-ES	ACCESS1-1	GFDL-ESM2M	Had GEM2-ES	ACCESS1-0	GFDL-ESM2M	Had GEM2-ES	ACCESS1-0	GFDL-ESM2M	Had GEM2-ES
Jun	-13.6	10.28	-5.1	-14.1	9.9	-9.8	-2	-17.5	5.1	16.88	-11.3	15	3.5	-15.2	6.3	5.8	-14.2	23.6
Jul	4	-9.73	7.1	4.6	-8.9	14	-3.3	14.7	-9.4	-5.15	-1.3	-12.6	-6	3.8	-7.6	-2.1	-4.4	-12.5
Aug	10.1	-12.69	1	8.1	3.8	7	-9.6	-0.1	-5.5	-3.45	-2.6	-3.6	-6.9	-2.2	-1.8	-0.7	-7.3	-11.3
Sep	10.5	-18.3	0.9	4.1	-20.2	-2.1	-13.3	18.1	10.3	-7.69	2.4	24	-5	7.3	7.5	-12.2	9.7	7.2

Table 4: Deviations of predicted monthly rainfall relative to the baseline under the three periods.

Parameter		Value
Canopy expansion	Very fast to CGC	17.9% per day
	Shape	3
Stomata closure	Moderately sensitive to water stress	
Canopy size		5cm² per plant
Plant density		20 plant per m²
Maximum canopy cover	Very thiny cover	60%
Canopy decline	Very fast	10 days
Date of emergency		4 days
Senescence		71 days
Date of maturity		94 days
Yield build up HI		47 days
Day of flowering		43 days
Harvest index		40%
CWP		14g/m²

Table 5: Input parameters used for Aqua Crop model calibration.

in hotter areas, even a slight change in temperature could have a detrimental effect on crop production. Hence, the impact of climate change could be more in the end term as temperature is projected to increase and precipitation decreases.

Monthly rainfall of the study area: The area is characterized by unimodal rainfall with the highest rainfall in July and Augest months. In the near term RCP4.5, GCM ACCESS1-0 in Jun (-13.6%) and GFDL-ESM2M in Augest (-12.7%) and September (-18.3%) revealed the highest reduction in rainfall. Similarly, in the mid term, GCM ACCESS1-0 in september (-13.3) and GFDL-ESM2M in June (-17.5%) demonstrated a decrease in rainfall. Similarly, in the end term RCP4.5, only GFDL-ESM2M model in June (-15.2%) has shown reduction in rainfall (Table 4).

In the near term RCP8.5, the GCM ACCESS1-0 in June (-14.1%) and GFDL-ESM2M in September (-20.2%) has shown a decrease in rainfall. Moreover, GFDL-ESM2M in June (-11.3%), HadGEM2-ES in July (-12.6%) in the midterm and ACCESS1-0 in September (-12.2%),

GFDL-ESM2M in June (-14.2), HadGEM-ES in July (-12.5%) and August (-11.3%) in the end term revealed the highest reduction in rainfall from the base. To the contrary, HadGEM-ES model has shown the highest increment in rainfall in the midterm in September (24%) and end term June (23.6%) under RCP8.5.

Historical Yield of Sesame

Model calibration

There is no universal model that works everywhere and hence, to capture the variability at the intended site, model calibration is important. The following inputs were used for calibration of the AquaCrop model. In addition to the crop and climate parameters, soil data was also entered into the model (Table 5).

Model performance evaluation

The performance of the Aqua Crop model in predicting the grain yield was evaluated using the root mean square error (RMSE) and the index of agreement. The Root mean square error (RMSE) quantifies the patterns of similarity by measuring the differences between values predicted by the model and the values actually observed while index of agreement (d) is a measure of agreement of the observed and simulated model ($0 \leq d \leq 1$).

For model evaluation, an independent data set of two years (2010 and 2011) which were not used in the model calibration was used. Hence, the index of agreement (d index=0.98) was able to reproduce the observed data with a very good agreement. In the same way, the root mean square error (RMSE=0.16) which indicates the pattern of similarity between the observed and simulated values indicates a promising result. Therefore, both statistical tools revealed that the Aqua Crop model can be used for further simulations.

Impact of climate change on sesame production by sowing date during the baseline period

Accordingly, the historical yield of sesame was obtained by providing the historical (baseline) climate data and other management factors to the model (Aqua Crop). The three sowing dates was evaluated using a statistical test and no significant difference (p=0.34) was found between groups (the sowing dates) (Figure 2).

Apart from the statistical tests, cumulative density function was used to see the probability of sesame yield. Accordingly, normal sowing date which is currently being practiced by the farmers was found to yield better compared to late sowing. The probability of getting 1 t/ha of sesame yield was nearly 60% for normal sowing, while about 50% and 45% for early and late sowing respectively. With certain adaptation this yield could be attained, however, the probability of getting 1.5 t/ha

of yield was very unlikely (20%) which needs more cost of adaptation to get this yield.

Impact of climate change on sesame production by GCM, sowing date and time period

From the (30 years) historical data, predicted climate data were also used to simulate grain yield and accordingly evaluate the impact of climate change on yield of each sowing dates. The likely future deviation of yield from the base year yield of each sowing date was summarized in the following Table 6.

In all periods and both RCPs yield has increased between 22.29 and 33.09% in the normal sowing date (Table 7). Similarly, in the early sowing date, an increment of yield between 13.67 and 25.44% was projected by the GFDL-ESM2M climate model. However, in the late sowing date, yield was reduced in the midterm (-6.87%) and end term (-5.88%) under RCP8.5. The reason why the yield deviation obtained on late sowing was very low could be due to the inadequate amount of precipitation to meet the crop water demand as late sowing exposes the crop to an extended dry spells. Extended dry spells occur during the mid-season stage of a late sown crop due to cessation of rainfall in early to mid of September. Generally, use of late sowing might experience more impacts of climate change as rainfall ceases during the critical growth stages of the crop. In mid and end term periods and all sowing dates (normal, early and late) higher yield was projected under RCP4.5 compared to RCP8.5. However, in the near term period, the yield was higher under RCP8.5 in all sowing dates.

In the normal sowing date, yield has increased between the range of 8.5% and 26% using the Had-GEM2-ES model (Table 7), however, a reduction of yield was observed in all sowing dates under RCP8.5.

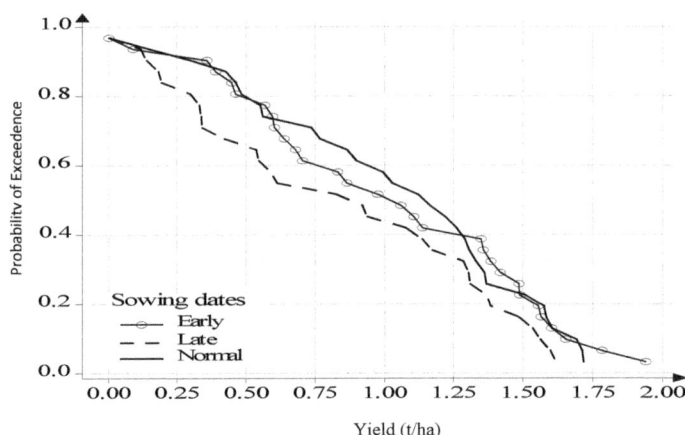

Figure 2: Historical yield (t/ha) of Sesame under different sowing dates.

Periods & RCPs	GFDL-ESM2M			HadGEM2-ES			ACCESS1-0		
	Normal	Early	Late	Normal	Early	Late	Normal	Early	Late
NT_rcp4.5	22.29	13.67	1.79	14.68	4.14	-11.25	27.39	14.31	-19.22
NT_rcp8.5	24.05	18.95	1.68	8.49	2.11	-16.63	22.92	17.66	-13.48
MT_rcp4.5	33.09	25.44	2.81	18.63	14.3	-1.94	32.9	25.12	-17.95
MT_rcp8.5	22.61	19.17	-6.87	16.81	15.28	4.16	15.99	21.22	-10.96
ET_rcp4.5	32.32	22.14	2.11	25.95	19.58	3.78	32.31	22.14	-8.79
ET_rcp8.5	23.58	16.55	-5.88	-3.34	-2.07	-23.31	25.01	16.53	-21.97

NB: NT, MT, ET and RCP represents the near term, midterm, end term and representative concentration pathways respectively.

Table 6: Changes of Sesame yield (%) relative to the baseline by sowing date, time period (near, mid and end term) and GCM scenarios.

Similarly, the yield was projected to increase between 2.11% and 19.58% in the early sowing date under both RCPs. In the late sowing date, yield was only increased in the midterm RCP8.5 (4.16%) and end term RCP4.5 (3.78%). However, the highest yield reduction was projected in the end term RCP8.5 (-23.31%) followed by near term RCP8.5 (-16.63%).

Using the ACCESS1-0 climate model, the yield of sesame was projected to increase between the range of 16% and 32.9% in the normal sowing date in all periods and both RCPs. Even in the early sowing date, the yield was increased between 14.31% and 25.12%. However, in the late sowing date, yield was simulated to decrease between the range of -8.8% and -21.97% in all periods and RCPs.

Sesame yield probabilities by RCP and planting date during the near term period (2010–2039)

The temporal impact of climate change was evaluated based on the yield difference relative to the baseline (1980–2009). Graphically probability of exceedence was drawn to show the difference of simulated yields based on the three sowing dates and three GCMs and two RCPs in comparison with the baseline scenario.

In the early sowing date, the probability of getting 1t/ha was indicated to be 60% by the three climate models under RCP4.5, while below 40% for 1.5 t/ha of yield. In the normal sowing date, the probability of getting 1t/ha was about 65% by the ACCESS1-0 GCM, while the remaining two demonstrated about 75% of probability under RCP4.5. Here, all models agreed with the probability of getting 1.5 t/ha was 40%. In the late sowing date, the highest probability of obtaining yield was observed by ACCESS1-0 followed by GFDL-ESM2M. However, Had GEM2-ES model has been similar with the baseline yield (Figure 3). In the near term RCP8.5, the probability of getting 1t/ha were about 50% by HadGEM2-ES, 60% (ACCESS1-0) and 65% (GFDL-ES M2M) in the early sowing date. Similar to the RCP4.5 of this sowing date, the probability of getting 1.5t/ha was below 40% by all models. In the normal sowing date, almost all models indicated that the probability of getting 1t/ha was about 70% under RCP8.5. Similarly, all models demonstrated below 40% probability of getting 1.5 t/ha of yield. In the late sowing of RCP8.5 the GFDL-ESM2M revealed better yield, while the remaining two models coincides with the base year yield. Generally, in the near term, the intermodal variability was low in the early and normal sowing dates under both RCPs. However, the variability between models was observed to be higher in the late sowing date of both RCPs. Unlike the two models, ACCESS1-0 GCM demonstrated an increasing probability of yield. In RCP8.5 for this plantation date, both HadGEM2-ES and ACCESS1-0 indicated almost the same yield comparing the base year yield which clearly indicated that the yield has greatly reduced here. In the contrary, GFDL-ES M2M has confirmed a higher yield than the other models in RCP8.5 of late sowing. In summary, in the near term using the normal planting date was found better as models consistently simulate higher yield than early and late sowings.

Sesame yield probabilities by RCP and planting date during the midterm period (2040–2069)

Similar to the near term, in the early sowing all models simulate better yield than the base year. Not only this but also, the simulated yield between RCP4.5 and RCP8.5 of this sowing date were almost the same. This was even similar to the near term early sowing date of both RCPs. In the normal sowing date (Figure 4).

Almost all models simulate similar probability of getting 1t/

ha of yield. However, to get 1.5 t/ha GFDL-ESM2M indicated 60% probability in the RCP4.5 while the remaining models indicate 40% probability in both RCPs. In the late sowing date, GCM ACCESS1-0 revealed 30% while the others about 50% probability of getting 1 t/ha of yield in RCP4.5. Similarly, ACCESS1-0 has shown a reduced yield (25%) and the remaining 35% probability to get 1.5 t/ha in both RCPs. In RCP8.5 the same probability (40%) was indicated by ACCESS1-0 and GFDL-ESM2M models with the base year to get 1t/ha of yield, while 50% by HadGEM2-ES. Generally, even in this period normal sowing date reveals a better yield than early and late sowing dates.

Sesame yield probabilities by RCP and planting date during the end term period (2070–2099)

In the end term RCP4.5 of early sowing date, all climate models simulate similar yield probabilities like in the near and midterm periods. Nevertheless, in RCP8.5 HadGEM2-ES and ACCESS1-0 has shown 50% of probability to obtain 1t/ha, while GFDL-ESM2M predicted 60% of probabilities? The possibilities of getting 1.5t/ha was about 35% in both RCPs (Figure 5).

In the normal sowing date RCP4.5, the probability of getting 1t/ha was 70% by all models, while below 50% of probability to achieve

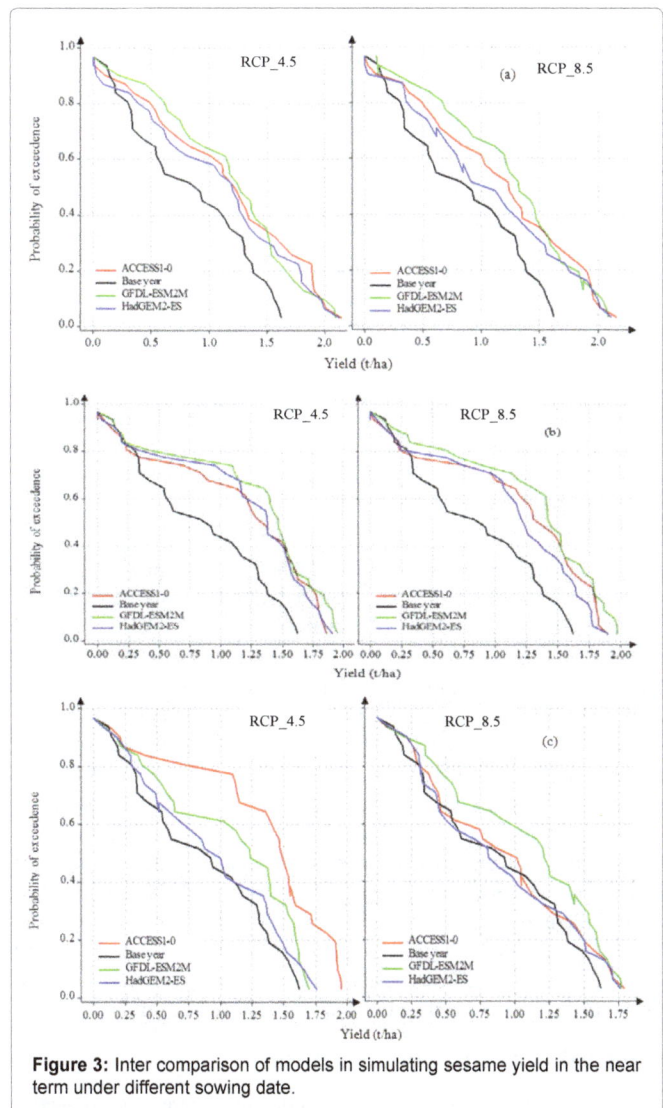

Figure 3: Inter comparison of models in simulating sesame yield in the near term under different sowing date.

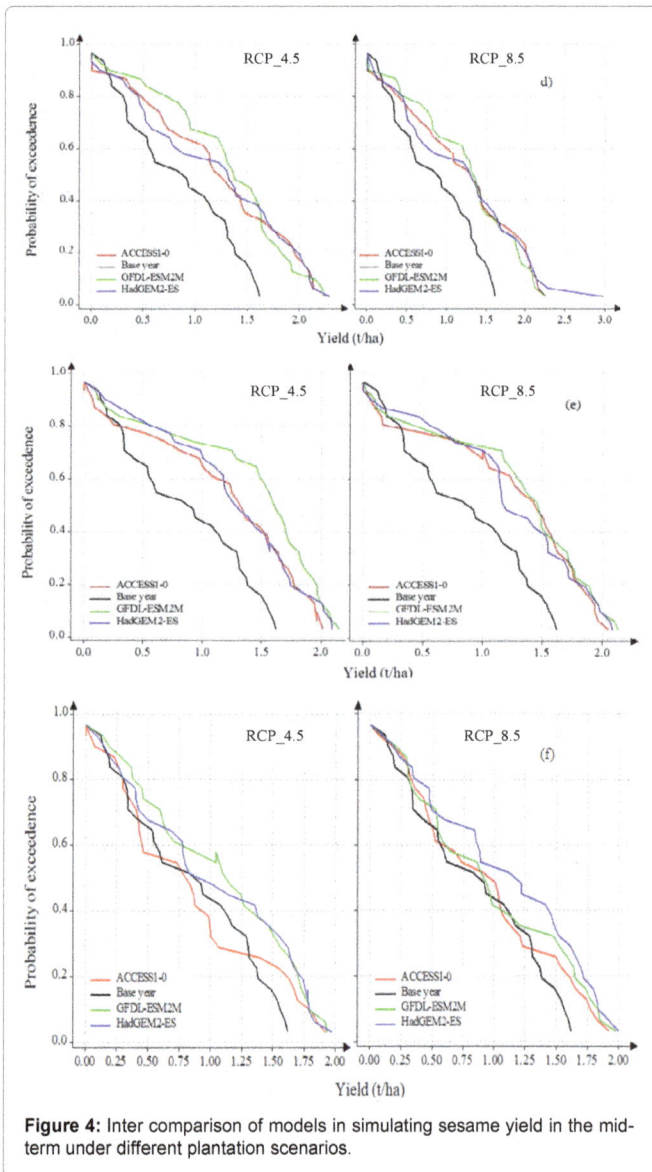

Figure 4: Inter comparison of models in simulating sesame yield in the mid-term under different plantation scenarios.

1.5 t/ha of yield. However, in the RCP8.5, there is a model variability to attain 1 t/ha. HadGEM2-ES simulated 45% while ACCESS1-0 and GFDL-ESM2M 60% and 70% of probability, respectively. To obtain 1.5 t/ha of yield, all models agreed with the probability of 40%. In the end term or late sowing, almost all models suggested a very similar yield in comparison to the base year. In the RCP4.5, the probability of getting 1t/ha and 1.5 t/ha was about 50% and 35, respectively. However, in RCP8.5, the probability of getting 1 t/ha has reduced to 40% while it remains the same in achieving 1.5 t/ha. In conclusion, yield has decreased in the end term in all plantation scenarios and similarly temperature was projected to increase and precipitation to decrease in this period. However, in all periods (near, mid and end term) normal plantation date was better than early and late sowing dates and there was less variability of yield between models in both RCPs and sowing scenarios, This can be evidenced by the occurrence of almost insignificant changes of the rainfall over the historic records.

Conclusion and Recommendation

Ethiopia's economy is highly dependent on climate sensitive sector,

rain fed agriculture and hence variability or change of climate might have major impacts on future livelihood of the people. the climate models revealed that the increase in temperature will be slightly higher in the minimum than maximum temperature. In both maximum and minimum temperatures, all models indicated that temperature will be dropped under RCP4.5 compared to the RCP8.5 in the near term. However, in already hotter areas even a slight change in temperature could have a detrimental effect on crop production. Hence, the impact of climate change could be more severe during the end term period as temperature and precipitation are projected to increase and decreases, respectively. The increase in temperature in these periods could induce a significant impact on crop production.

Modeling of sesame production using Aqua Crop has sufficiently reproduced the observed data and hence, was found convincing in using this model for predicting the impact of climate change. This was approved by measuring the performance of the model based on root mean square error (RMSE=0.16) and index of agreement (d index=0.98). In the study area early sowing date was the most common

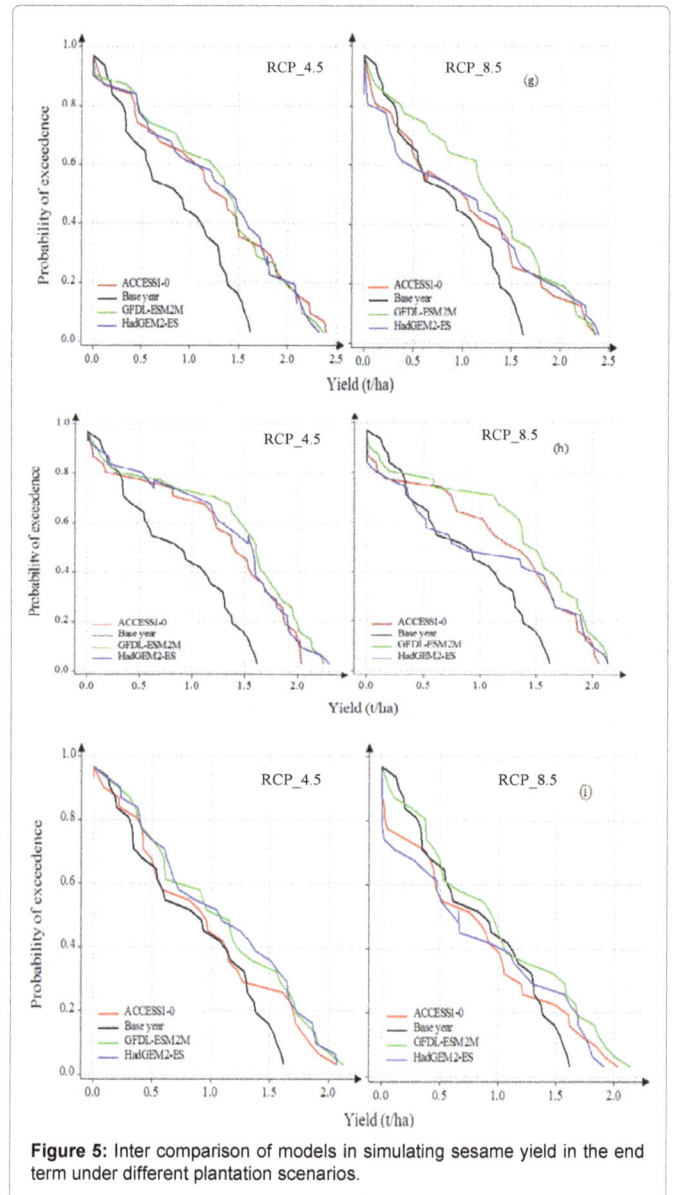

Figure 5: Inter comparison of models in simulating sesame yield in the end term under different plantation scenarios.

sowing date of sesame used by most farmers. This study indicated thatseasame perform well when normal planting was used compared to early or late sowing. Generally, the climate of the study area has significantly been changed putting its consequences on agricultural crops, which is evidenced by the change in annual rainfall totals, rainy days, length of growing period and onset date of the historical years. As a recommendation Climate change will have a negative impact on the selected local sesame cultivar (Hirhir) in all sowings, hence, July 7 sowing date was recommended based on the gate high yield. The response of sesame cultivars to the future climate changes should be studied under different management options. The impact of climate change should also be studied by different crop and climate models so as to capture the possible variability of sesame yield. Sensitivity to carbon dioxide, temperature, rainfall and other different management activities should be undertaken.

References

1. Sivakumar, Mannava S, Raymond M (2007) managing weather and climate risks in agriculture.

2. IPCC (2007). Climate Change 2007: The Physical Science Basis. Contribution of Working Group I to the Fourth Assessment Report of the Intergovernmental Panel on Climate Change, Cambridge University Press, Cambridge, United Kingdom and New York, NY, USA.

3. Gebreegziabher Z, Stage J, Mekonnen A (2011) Climate change and the Ethiopian economy: A computable general equilibrium analysis.Environment for Development Discussion

4. Hajarpoor A, Soltani A, Zeinali E, Sayyedi F (2014) Simulating climate change impacts on production of chickpea underwater-limited conditions 3: 209-217.

5. IPCC (2001) Climate Change 2001: Impacts, Adaptation and Vulnerability, Contribution of Working Group II to the Third Assessment Report of the Intergovernmental Panel on Climate Change (IPCC). Cambridge University Press, Cambridge.

6. Hargreaves GH (1994) Simplified coefficients for estimating monthly solar radiation in North America and Europe. Departmental Paper, Dept. of Biol. And Irrig. Engrg,Utah State University, Logan, Utah.

7. Cooper P, Dimes J, Rao KP, Shapiro B, Shiferaw B, Twomlow S (2008) Coping better with current climatic variability in the rain-fed farming systems of sub-Saharan Africa, An essential first step in adapting to future climate change. Agriculture, Ecosystems and Environment 126: 24-35.

28

The Effects of the Trade Winds on the Distribution of Relative Humidity and the Diurnal Air Temperature Cycle on Oceanic Tropical Islands

Ronalds RT[1], Emiliano A[1] and Rafael Mendez-Tejeda[2]*

[1]School of Natural Science and Technology, Universidad del Turabo, Gurabo, PR, USA
[2]Atmospheric Sciences Laboratory, University of Puerto Rico at Carolina, PO Box 4800, Carolina, PR, USA

Abstract

On many oceanic tropical islands the trade winds blow from the east and as the air passes over the island it loses moisture to rain. The two hypotheses for this study are that the western part of the island is less humid and has a larger diurnal air temperature cycle. Using data available on the Internet, the two hypotheses were assayed 7 times on 6 different islands in the Pacific and Caribbean. The islands used in this study are Puerto Rico, Hispaniola, the Big Island of Hawaii, Maui, Oahu, and Kauai. The first hypothesis was tested only on Hispaniola and is true. The second hypothesis was tested on all the islands and is true on all the islands except on the Big Island of Hawaii and Maui. Using a p-value of 0.05, these hypotheses are as predicted 5 of the 7 times, which has a p-value of 6×10^{-6}. These findings should apply to thousands oceanic tropical islands where data are sparse.

Keywords: Rain shadow; Oceanic tropical Islands; Trade winds

Introduction

The Earth has an estimated 45,000 tropical islands larger than 0.05 km^2 (5 hectares) [1]. The populations of these islands range from zero to 143 million on the Indonesian island of Java. Oceanic tropical islands often have environmental problems that include high population densities, water scarcity, and extinctions [2]. Data are often sparse [1], but because inputs and processes are similar, many oceanic tropical islands have commonalities including climate and in general vegetation. This study will look for patterns on data-rich islands. Nothing in this study will prove that these patterns apply to data-sparse islands but it will generate hypotheses that can help guide future research on other oceanic tropical islands.

This study has two parts. First a simple theoretical approach, that examines how on many oceanic tropical islands, the easterly trade winds shape the distribution of the two variables of this study. And second, searching for available data to test the hypotheses generated from the theoretical approach. This is a pilot study. The vast majority of oceanic tropical islands do not have any available data to verify these hypotheses.

Over the open water, upwind of an oceanic tropical island the relative humidity is typically around 80 percent [3]. The average value of relative humidity is an equilibrium between evaporation which, on a daily basis, is almost constant and frequent but light rain storms that remove moisture out of the air. As a parcel of air moves over land it is disconnected from its source of moisture and from upwind to downwind it loses moisture to rainfall and becomes drier. When the winds are consistent, the interaction between the wind and the land creates a region with a reduced water vapor concentration in the atmosphere. The western part of the island is in the rain shadow of eastern end of the island. This structure in the atmosphere is a permanent feature even as the air in it is being constantly replaced. The exact patterns of climate will be affected by many factors including topography, the shape of the coastline, distance from the coast, and anthropogenic factors such as urban heat islands. The existence of this area of reduced water vapor content can be predicted solely on the basis of longitude and is independent of the specifics of the other factors. The distance from the eastern tip of the island can be measured either in degrees of longitude or kilometers and it would not change the results. The rates of rainfall and runoff are typically much higher in the

humid tropics than in temperate regions [4] and the higher rainfall will increase the contrast between upwind and downwind areas.

There are at least three methods to study the interaction between the prevailing trade winds and the land mass. Large numbers of temporary instruments can be deployed to provide a large amount of data over the span of a few weeks. Such an experiment is described for the Big Island of Hawaii by Chen and Nash [5]. A second approach is to use digital elevation models, complex mathematical algorithms, and expert knowledge to generate detailed maps of climate variables and this is the approach used in the Parameter-elevations Regression on Independent Slope Model (PRISM) [6]. The application of PRISM to the island of Puerto Rico is in Daly et al. [7]. A third approach is used in this study. This study looks at data-rich islands and uses simplified mathematical relationships to look for generalized patterns that probably exist on thousands of oceanic tropical islands where data are sparse. The simplified relationship will require less data than more complex patterns.

On many oceanic tropical islands the trade winds consistently blow from the east. This has been documented for Hawaii [5], Hispaniola [8] and Puerto Rico [9]. The eastern two-thirds of Hispaniola is the Dominican Republic while the western one-third is Haiti. In Puerto Rico, the wind blows between northeast and southeast at least 65 percent of the time [10]. A detailed study of the trade winds in Hawaii is in Giambelluca and Nullet [11]. Data-rich islands have studies but there are few studies that compare the environments of widely separated islands.

The structure of the atmosphere in the Caribbean is important

***Corresponding author:** Rafael Mendez-Tejeda, Atmospheric Sciences Laboratory, University of Puerto Rico at Carolina, Carolina, PR, USA 00984
E-mail: rafael.mendez@upr.edu

locally for the islands and is connected to continental scale patterns. The North Atlantic Oscillation (NAO) is derived from the difference in barometric pressure between Iceland and Portugal. High levels of the NAO index are associated with high precipitation in northern Europe and low precipitation in southern Europe and Puerto Rico [12]. In late summer in the northern Antilles rainfall is associated with a negative NAO index combined with warm conditions in the El Niño Southern Oscillation in the eastern Pacific Ocean [13]. The Caribbean Low Level Jet (CLLJ) transports moisture to Central America and the eastern part of North America. Variability in the CLLJ is associated with variability of precipitation in the downwind areas [14]. Several studies have analyzed the inter-annual variability of Caribbean rainfall [15-18]. The climatological mean, annual, and monthly rainfall in Haiti is a complex function of fixed factors including the topography and shape of the country, as well as the annual cycle of regional-scale oceanic and atmospheric factors. The atmosphere is in the northern Antilles is dominated by the permanent Azores high, which induces permanent easterlies across the Caribbean islands [19].

The variables in this study are affected by multiple factors such as elevation, distance from coast, the shape of mountain ranges, and anthropogenic factors such as urban heat islands. The central question of this study is if longitude is needed to be added to the list to explain the observed data. The basic idea is that the consistency of the trade winds creates a rain shadow in the western parts of oceanic tropical islands. Mountains in Puerto Rico create a rain shadow [7]. At 3098 m, Pico Duarte is the highest mountain in Hispaniola and the Caribbean. The rain shadow of this mountain affects the climate of Jamaica and eastern Cuba [20]. Data are extremely limited for most oceanic tropical islands and this study will use simple patterns with limited data requirements to see if the results are useful and can then be extended to other islands.

The trade winds blow from the east and the air loses water to rain as it passes over the land. This concept generated two hypotheses. The second hypothesis derives from the first. In numerical form, the hypotheses will be stated for the western hemisphere; in the eastern hemisphere the correlations are reversed. Each hypothesis will be stated as a correlation with longitude. Longitude has an arbitrary datum in Greenwich, United Kingdom but it would not affect the correlations of this study if the datum were the eastern tip of the island or any other location. The hypotheses, for oceanic tropical islands, are:

First: the western part of the island is in the rain shadow of the east and there will be an inverse correlation between longitude and relative humidity.

Second: the diurnal air temperature cycle is the average difference between the daily highs and the lows at night and will be correlated with longitude. Dry air is more transmissive of infrared radiation [21], and with the same amount of solar radiation, the days will be hotter and the nights will be colder, as occurs in deserts [22]. There is a published map of the diurnal air temperature cycle in Puerto Rico [10]. The correlation with longitude is visible on the map but the interpretation is based on distance from the coast.

The vast majority of the oceanic tropical islands on Earth have no data to assay these hypotheses. There are other islands that have data but they are not readily available. This study uses data collected by government meteorological agencies and made available on the Internet. At least one hypothesis was tested on 6 islands. The two Caribbean islands in this study are Hispaniola and Puerto Rico. All data from Hispaniola are from the Dominican Republic as no data are available for the Haitian side of the island. The Hawaiian Islands used in this study are the Big Island of Hawaii, Maui, Oahu, and Kauai. Puerto Rico and Hawaii are part of the United States. .

The 6 islands in this study are not representative of the 45,000 tropical islands on Earth. All are in the northern hemisphere and all are in the western hemisphere. All are near the Tropic of Cancer and none are in the deep tropics. None of the islands are in the Indian Ocean. The largest concentration of tropical islands on Earth is between Asia and Australia and none of the islands in this study are in this part of the world. These are the islands for which data are available on the Internet and if the hypotheses work on these islands it can guide the work of future researchers on other islands. The islands of this study are well separated from continental landmasses. Future work will be needed to understand the relationship between distance and the effect of continents on the climate of tropical islands. The large masses of ocean water around the islands of this study buffers the climate and reduces the annual air temperature cycle.

The location of the islands are in Figure 1. The data were collected by the National Weather Service (NWS), except in the Dominican Republic where they were collected by the *Oficina Nacional de*

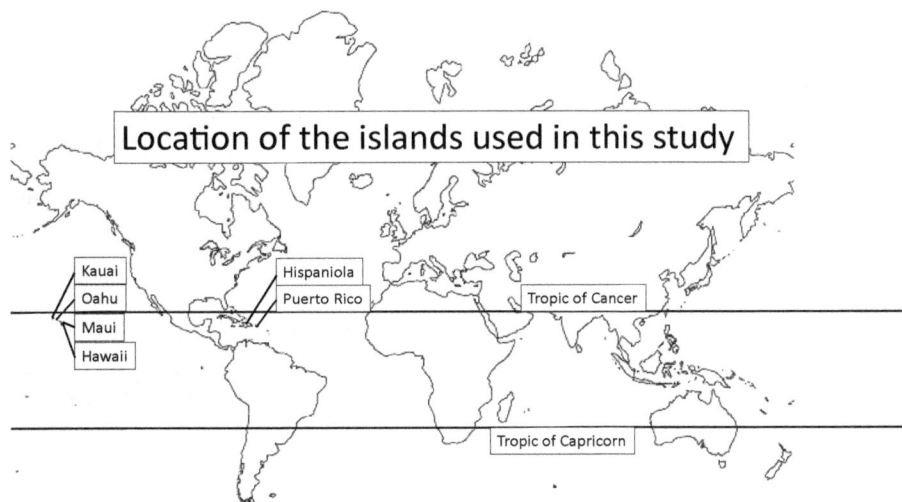

Figure 1: Locations of the oceanic tropical islands used in this study.

Meteorología (ONAMET). NWS data are at the Southeast Regional Climate Center [23] and the Western Regional Climate Center [24]. The data from the Dominican Republic are at World Climate [25] and were submitted to the World Meteorological Organization and edited by the National Climate Data Center (NCDC). Both the NWS and the NCDC are part of the National Oceanic and Atmospheric Administration of the United States.

Material and Methods

The hypotheses were tested for every oceanic tropical island for which data could be found on the Internet. Typically for each station the metadata includes latitude, longitude, and elevation. For each hypothesis assayed, each island has between 15 and 68 data points. The data were plotted versus longitude and visually examined for outliers, which were excluded. The Pearson correlation coefficient was calculated, the p-value used was 0.05.

The hypotheses were assayed with climate stations that had at least 7 years of data and in many cases 30 or more years. The first hypothesis, which states that the western ends of oceanic tropical islands have lower relative humidity than the eastern parts, was assayed only for the island of Hispaniola. The data were the monthly average relative humidity, which were averaged to produce the average relative humidity. The second hypothesis, stating that the diurnal air temperature cycle is larger in the western parts of the islands was assayed for all of the islands. The monthly air temperature cycle is the average high temperature in a month less the average low temperature in the month. In the tropics, the diurnal air temperature cycle is 65 percent of the monthly cycle [26], so the monthly average high and low temperature was used as a proxy for the diurnal cycle. A proxy was needed because data on the diurnal air temperature cycle are not readily available. Each station has a value for the monthly difference between high and low temperature and these were averaged to produce a single value for the station over the course of a year.

The hypotheses were also assayed for north-south differences. If the trade winds are converging on the equator then it would be logical that the low latitude part of the island would be dryer and have a larger diurnal air temperature cycle.

Results

The results are in Table 1. The graph of average relative humidity versus longitude for Hispaniola is shown in Figure 2. There are 15 data points. The 13 data points from stations below 200 m above sea level showed a clear trend from 83 to 63 percent relative humidity. The two stations higher than 400 m above sea level were more humid and were excluded as outliers. The Pearson correlation coefficient was -0.74 which had a p-value of 0.0019.

Hypothesis number two predicts that the western parts of oceanic tropical islands will have a larger diurnal air temperature cycle and this was tested with monthly air temperature data. The average monthly air temperature cycles for Hispaniola and Puerto Rico are shown in Figure 3, while Figure 4 has the same data for the four Hawaiian Islands. All six islands had a larger diurnal temperature cycle in the western parts of the islands but the correlations were not significant in the Big Island of Hawaii and Maui.

The hypotheses of this study were assayed 7 times on 6 oceanic tropical islands in two groups in two oceans. The hypotheses were as predicted 5 of the 7 times. The binomial distribution was used to calculate the p-value for this to occur by random chance and the result is 6 times in a million trials. The testing of the north-south hypotheses produced no significant results.

Discussion

The correlations in this study are based on the consistency of the easterly trade winds. In parts of south Asia and the island-rich region between Asia and Australia, the climate is dominated by the monsoon pattern which can include abrupt changes in wind direction and precipitation [27]. Understanding how the Asia monsoon affects the climatology of tropical islands is crucial in the extension of this study to other areas such as Indonesia and the Philippines; however none of the islands in this study are affected by the monsoon pattern.

In Maui the hypothesis on the correlation between longitude and the monthly air temperature cycle has 20 data points, a Pearson correlation coefficient of 0.37, and a p-value of 0.054. It was classified as insignificant but it is right on the edge and with one more station the correlation would be significant. The Big Island of Hawaii is a different story. With 37 data points, the Pearson correlation coefficient is 0.091, and the p-value is 0.29. To be significant a Pearson correlation coefficient of 0.1 requires more than 250 data points.

The largest and most populated island used in this study is Hispaniola, which has an area of 76,500 km^2, a maximum elevation of 3098 m above sea level, and a population of over 20 million people, split almost evenly between the Dominican Republic and Haiti. The population density of Hispaniola is 259 people per square kilometer. Kauai is the smallest and least populated island in this study. Kauai has an area of 1430 km^2, a maximum elevation of 1598 m above sea level, a population of 67,000, and a population density of 47 people per square kilometer. Oahu is the lowest and most densely populated island in the study. Oahu has an area of 1545 km^2, an elevation of 1220 m, a population of 953,000, and a population density of 617 people per square kilometer. The Big Island of Hawaii is the highest in elevation and has the lowest population density. The Big Island of Hawaii has an area of 10,432 km^2, an elevation of 4205 m, and a population of 185,000 with a population density of 18 people per square kilometer.

| | Relative humidity | | | | | | Diurnal air temperature cycle | | | | |
| | Hypothesis #1 | | | | | | Hypothesis #2 | | | | |
Island	m[a]	b[b]	r[c]	n[d]	p-value		m	b	r	n	p-value
Puerto Rico	[e]						2.47	-154	0.54	44	<0.001
Hispaniola	-4.05	362	-0.74	13	0.0019		1	-60	0.52	68	<0.001
Hawaii							0.379[f]	-50	0.091	37	0.29
Maui							287	-439	0.37	20	0.054
Oahu							7.23	-1130	0.65	30	<0.001
Kauai							9.01	-1430	0.9	16	<0.001

[a]m is slope; [b]b is intercept; [c]r is Pearson correlation coefficient; [d]n is sample size; [e]blank cells represent no data; [f]highlighted in gray are insignificant

Table 1: Results of assaying the hypotheses on oceanic tropical Islands.

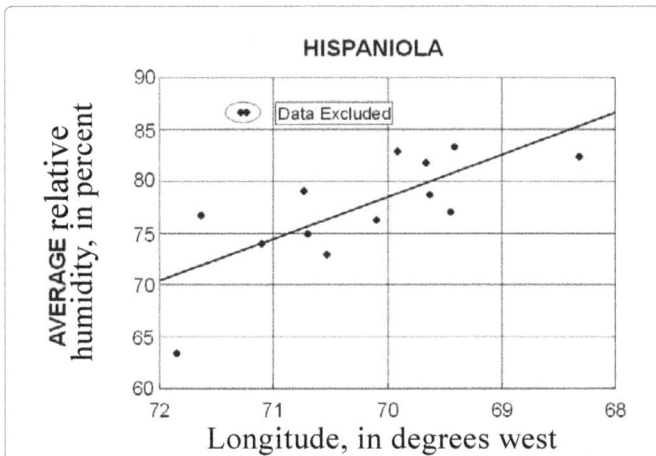

Figure 2: Hypothesis #1- Average Relative Humidity Versus Longitude for Hispaniola. The excluded data are from stations over 400 m in elevation while the rest of the data are from stations within 200m of sea level. Data modified from World Climate.

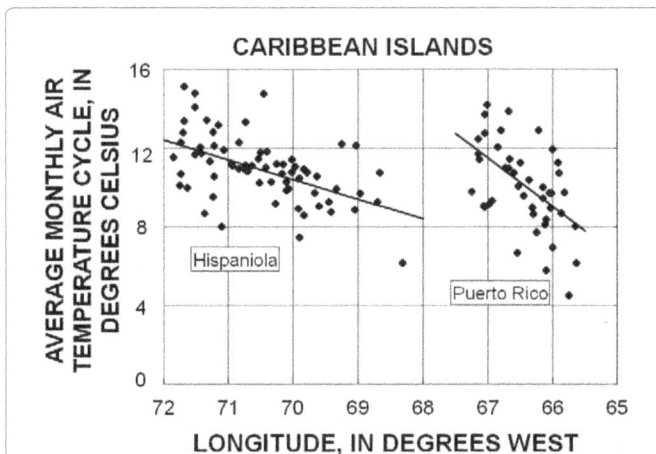

Figure 3: Average monthly air temperature cycle versus longitude for Hispaniola and Puert Rico. Data modified from World Climate and Southeast Regional Climate Center.

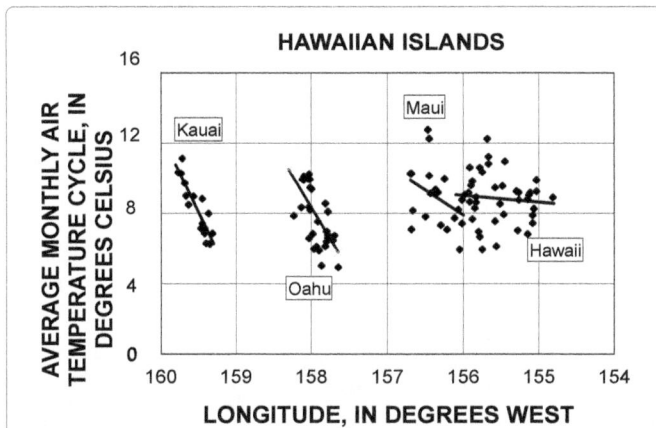

Figure 4: Hypothesis #2--Average monthly air temperature cycle versus longitude for four Hawaiian Islands. The monthly air temperature cycle is a proxy for the diurnal air temperature cycle. Dashed lines are used when the correlation is not significant. The Big Island of Hawaii and Maui slightly overlap in longitude. Data modified from Western Regional Climate Center.

The wind blows from the east, and as the air mass moves over land from east to west it loses moisture to rain and becomes drier. The pattern is simplistic but it produced two hypotheses which could be tested with data that are publically available on the Internet. The hypotheses were successful in predicting observations on widely separated oceanic tropical islands. The variables used in this study are affected by multiple factors like elevation, proximity to the coast, the shape of mountains, and anthropogenic factors like urban heat islands, the consistency of the easterly trade winds adds longitude to the list. A simple correlation with limited data requirements produced useful results even though the distance between Puerto Rico and Hawaii is more than 9,000 km.

Oceanic tropical islands are tiny specks of land spread out over vast distances of ocean. On thousands of these islands, the easterly trade winds shape the environment in predictable ways. This study is the first part of a larger effort to identify these underlying physical processes that can help improve the management of water and other natural resources on these islands. The easterly trade winds affects not only the humidity and diurnal air temperature cycle but probably also the temperatures of sea surface, rivers, and groundwater. These abiotic conditions shape the environment for plants, animals, bacteria and fungi. Puerto Rico is a good place to start because its political relationship with the United States has made it one of the most data-rich places on Earth. The goal is science that can improve the management of water and other natural resources in Puerto Rico while at the same time providing insights that can help in the environmental management of thousands of islands where data are sparse. Puerto Rico is and island but the viewpoint should not be insular but rather one that is inclusive of the tens of thousands oceanic tropical islands with similar climates and environmental problems.

Conclusion

The consistency of the easterly trade winds on oceanic tropical islands leads to two predictions which can be assayed with data that are publically available on the Internet. The hypotheses that were verified on two islands in the Caribbean and four in the Pacific are that on the western end of the island the relative humidity is lower and the diurnal air temperature cycle is larger. These patterns probably exist on thousands of islands for which there are no readily available data. With more data it should be possible to observe these patterns much more widely.

It has been observed that on islands, the topography of the island can act as an obstacle or barrier to the wind causing an unequal distribution of humidity and diurnal air temperature cycle. This study establishes the difference between the east (windward) and the west (leeward) part of the islands. The east side benefit from the moisture of the trade winds, while the west slope receives dryer wind. This generates microclimates that also depend on elevation and large differences can occur across small horizontal distances.

Acknowledgements

This study was possible because of the effort of observers and employees of different agencies who for more than 50 years have collected, processed, and archived the data used in this study. Their names are unknown, but their work is appreciated.

References

1. Arnberger H, Arnberger E (2001) The tropical islands of the Indian and Pacific Oceans. Austrian Academy of Sciences. Vienna pp. 556.

2. Olson SL, James HF (1984) The role of Polynesians in the extinction of the

avifauna of the Hawaiian Islands. In: Quaternary Extinctions: A prehistoric revolution, The University of Arizona Press, Tucson pp. 760-780.

3. Emanuel KA (1987) The dependence of hurricane intensity on climate. Nature 326: 483-485.

4. Bonell M, Hufschmidt MM, Gladwell JS (2005) Hydrology and Water Management in the Humid Tropics: Hydrological Research.

5. Chen YL, Nash AJ (1994) Diurnal variation of surface airflow and rainfall frequencies on the island of Hawaii. Monthly Weather Review 122: 34-56.

6. Daly C, Gibson WP, Taylor GH, Johnson GL, Pasteris P (2002) A knowledge-based approach to the statistical mapping of climate. Climate Research 22: 99103.

7. Daly C, Helmer EH, Quiñones M (2003) Mapping the climate of Puerto Rico, Vieques and Culebra. International Journal of Climatology 23: 1359-1381.

8. Izzo M, Rosskopf CM, Aucelli PPC, Maratea A, Méndez R, et al. (2010) A new climatic map of the Dominican Republic based on the Thornthwaite classification. Physical Geography 31: 455-472.

9. Carter MM, Elsner JB (1996) Collective rainfall regions of Puerto Rico. International Journal of Climatology 16: 1033-1043.

10. Colón Torres JA (2009) Weather In Puerto Rico. Editorial University of Puerto Rico. San Juan.

11. Giambelluca TW, Nullet D (1991) Influence of the trade-wind inversion on the climate of a leeward mountain slope in Hawaii. Climate Research 1: 207-216.

12. Malmgren BA, Winter A, Chen D (1998) El Niño-Southern Oscillation and North Atlantic Oscillation control of climate in Puerto Rico. Journal of Climate 11: 2713-2717.

13. Gouirand I, Jury MR, Sing B (2012) An analysis of low- and high frequency summer climate variability around the Caribbean Antilles. Journal of Climate 25: 3942-3952.

14. Wang C (2007) Variability of the Caribbean Low-Level Jet and its Relations to Climate. Climate Dynamics 29: 411-422.

15. Enfield DB, Alfaro EJ (1999) The dependence of Caribbean rainfall on the interaction of the tropical Atlantic and Pacific Oceans. Journal of Climate 12: 2093-2103.

16. Giannini A, Kushnir Y, Cane MA (2000) Interannual variability of Caribbean rainfall, ENSO, and the Atlantic Ocean. Journal of Climate 13: 297-311.

17. Giannini A, Chiang JCH, Cane MA, Kushnir Y, Seager R (2001) The ENSO teleconnection to the tropical Atlantic Ocean: Contribution of the remote and local SSTs to rainfall variability in the tropical Americas. Journal of Climate 14: 4530-4544.

18. Taylor MA, Enfield DB, Chen AA (2002) Influence of the tropical Atlantic versus the Tropical Pacific on Caribbean rainfall. Journal of Geophysical Research 107: 10-14.

19. Moron V, Frelat, R, Jean-Jeune PK, Gaucherel C (2014) Interannual and intra-annual variability of rainfall in Haiti. Climate Dynamics 45: 915-932.

20. Jury MR (2009) An intercomparison of observational, reanalysis, satellite, and coupled model data on mean rainfall in the Caribbean, Journal of Hydrometeorology 19: 413-430.

21. Pierrehumbert RT (2011) Infrared Radiation and Planetary Temperature. Physics Today 64: 33-38.

22. McGregor GR, Nieuwolt S (1998) Tropical climatology: an introduction to the climate of the low latitudes. John Wiley and Sons, Hoboken, pp. 339.

23. Southeast Regional Climate Center (2014). http://www.sercc.com

24. Western Regional Climate Center (2014). http://www.wrcc.com

25. World Climate (2014). http://www.climate-charts.com

26. Watterson IG (1997) The diurnal cycle of surface air temperature in simulated present and doubled CO2 climates. Climate Dynamics 13: 533-545.

27. Chang C-P, Wang Z, McBride J, Liu CH (2005) Annual cycle of Southeast Asia -Maritime Continent rainfall and the asymmetric monsoon transition. Institutional Archive of the Naval Postgraduate School, J Climate 18: 287-301.

29

Observed and Simulated Changes in Precipitation over Sahel Region of West Africa

Agumagu O*

Department of Geography, University of Sussex, UK

***Corresponding author:** Agumagu O, Department of Geography, University of Sussex, UK, E-mail:obroma4u@yahoo.com

Abstract

There is an increasing need for strategic evaluations of changes in precipitation in current and future conditions. The paper simulates West African Sahel region climatology, describing the major characteristic of precipitation variability and trends in the region. The study assess the changes in precipitation in Sahel region using observational data output from Global Precipitation Climate Centre (GPCC) and Climate Research Unit (CRU) data set together with General Climate Model (GCM) AR4 used by Intergovernmental Panel on Climate Change (IPCC) to investigate precipitation trends over the Sahel region. In West Africa, rainfall has been subject to large decadal and inters decadal variation. A period of (1910-2009, 2050-2080) was chosen to understand the climatic variability across the Sahel region. Mike Hulme in his paper described Sahel climate as the most dramatic example of climatic variability that the world has observed. It is consequently significance to understand if, and to which extent climate trends in the Sahel region exist that influence availability of water in future.

The study demonstrates that the characteristics of rainfall variability in Sahel region differ markedly. The analyses show that the Northern part of the Sahel region is characterizes by low precipitation while the West Coast and Gulf of Guinea rainfall are apparent, although uncertainty from the GCM cannot be over role. The rationale behind the study is the need to understand the nature of climate change in the region over the past 100 years in relation to changes in precipitation together with the spatial pattern of these changes.

Keywords: Sahel region; Precipitation; Climate change; Weather hazards

Introduction

Changes in the climate and climate variability have become important topics raised by relevant scientific communities, most especially climatological branches that assess the influence of climate change. Recent developments in understanding, the influence of climate change in food production, water resource planning and ecosystems, especially in regions with scarce freshwater resources, have brought seasonal-to-inter annual climate predictions into everyday life. The effect of variability on precipitation is of great concern specifically across Africa where rainfall plays an essential role in sustaining livelihoods and economic development. Changes in precipitation are very significant, it is one of the weather elements whose future changes will have a large impact on African nations whose natural resources are dependent on rainfall [1]. A decrease in mean precipitation would lead to increased risk of drought while an increase in mean precipitation would lead to increased risk of flooding. Recent years have seen a number of weather events cause large losses of life as well as a tremendous increase in economic losses from weather hazards.

Sahel region of West Africa has been recognized as an area of global environmental change, but understanding of climatic and anthropogenic-forcing driving the change is not sufficient [2]. The region source of livelihood is dependent on rain-fed agriculture and very strong drought (in the 1970s and 1980s) was a great loss of agricultural production and livestock; loss of human lives to hunger, undernourishment and diseases; displacement of people and devastated economies [3]. The drought incidence really attracted scientific attention to investigate the mechanism responsible for the occurrence [4]. Many researchers have tried to link the drought to tropical factor such as ENSO, Sea Surface Temperature (SSTs) and Inter Tropical Convergence Zone [5]. Despite the controversy regarding the albedo change Xue and Shukla [6] argue that desertification has been occurring in Africa, which may have led to Sahel drought.

Drought is said to occur when a region receives consistent below average precipitation. However Sahel droughts seem to occur a little more frequently due to reduced rainfall. Batterbury and Warren [7] described the short and long-term effects of drought as greatest in the region that is poor, as indeed was much of the Sahel region in the early 1970s and 1980s. Because of the severity of Sahel drought it was estimated that in 1972 and 1982 about 100,000 people in Sahel region died from hunger and in 1974 about 75,000 were depending on food aid [8].

Although many studies have examined the aspects of Sahel region precipitation trends, the model forecasts a decrease of annual rainfall over Sahel in future [9]. Nevertheless, the degrees to which this variability can be termed natural as opposed to anthropogenic remain uncertain. Thus, there has not been a robust finding in the literature and many opinions reflect decrease in precipitation tends [10]. Liebmann et al. [11] argue that understanding the possible effect of climate variability in Sahel region precipitation is difficult because, Africa as a whole has not been studies in a clearly noticeable manner. The future occurrence of Sahel region droughts and the development of its hydrological balance are therefore of great concern.

Observational records are crucial for identifying changes in rainfall patterns. GCM are valuable tools for understanding the physical mechanisms driving past, current and future climate variability and changes through the use of climate change scenarios. And provide the only physically based approach accessible to predict future changes in the climate. These portray the important of GCM in assessing the precipitation change. Despite the progress attained in the last few years in using GCM to estimate the influence of the changing climate on water resource, there are still many unsolved problems. An uncertainty has often been a limiting factor. The paper examine Sahel region of West Africa precipitation variability and trends as portrayed by an range of observational data sets output from GPCC and CRU together with GCM AR4 used by IPCC to characterize and understand vigorous changes in Sahel region rainfall over the past 100 years (1910-2009) and 30 years in the future (2050-2080).

Background

The Sahel is the northern edge of the transition region, south of the desert. In this study the nomenclature Sahel is used is refer to the swath over West Africa located approximately between 10°N and 20°N, it includes much of the countries of Mauritania, Senegal, Mali, Niger, Chad, the Sudan, and the northern fringes of Burkina Faso and Nigeria [4] (Figure 1).

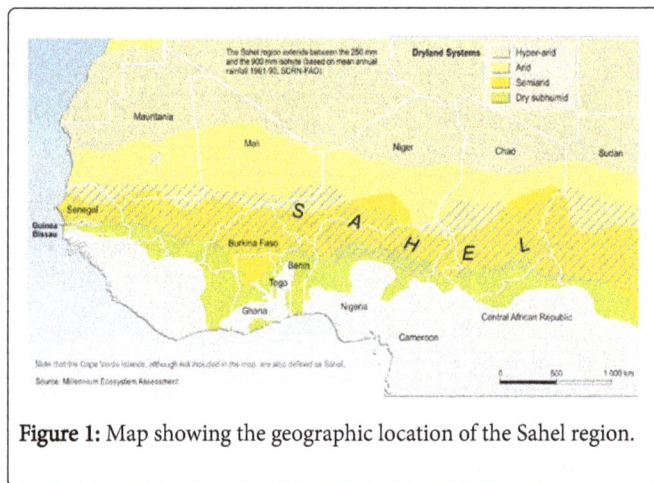

Figure 1: Map showing the geographic location of the Sahel region.

The ecological and geomorphologic diversity is distinctive from other regions in the world. This region is increasingly vulnerable to the impact of climate change; rainfall scarceness, prolonged drought and intensified land use lead to enlarging desertification [3]. The Sahel's climate is confirmed by a dry subtropical climate in the north and a dry tropical climate in the south. This dry trade wind over the Sahel is often named Harmattan [12].

The main climatic feature is the annual cycle of dry and rainy monsoon seasons resulting from the Inter-tropical Discontinuity (ITD) oscillations [13]. The ITD is defined as the interface between the cool, moist south westerly monsoon flow and the warm, dry-laden north easterly Harmattan flow [14]. Drought can be describes by late monsoon onset, it often create severe societal impacts [8]. According to Druyan [8] 95% of the land use in Sahel West Africa is devoted to agriculture and that 65% population is employed with agriculture work.

During boreal winter (October-March), northern Sahel region is dominated by the low level north easterly Harmattan winds. The Harmattan flow is influenced by day to day synoptic scale variability in the strength of the north-south pressure gradient across the region intensifying anti cyclonic conditions notably controlled by ridging of the Libyan high [12]. In boreal summer, atmospheric circulation in the Sahel bound becomes more complex, dominated in a great extent by the West African Monsoon (WAM). The WAM coincides with the crossing of the ITD to its northernmost location.

Desertification from land use change may have contributed to decrease in precipitation in the Sahel region that lead to 1979 and 1980 drought [6].Therefore, changes in future climate is likely to exacerbate the on-going and economic challenges that are already happening in Sahel region because of its dependent on resources that are sensitive in changing climate and combine with additional natural and anthropogenic threats [15].

Xue and Shukla [16] argue that SST anomalies has a strong influence to Sahel drought, both globally and regionals. A prediction was carried out on the regional climate up to 2050 by Paeth and Hense [17]; Wang and Alo [9], to examine the concentration of CO_2 and land use change for A1B and B1 emission scenarios it was discover that West Africa climate is dominated by drought.

The key in climate variability in Sahel region is land use change both natural and man-made. Though, the projections of West Africa precipitation changes using atmospheric oceanic general circulation models (AOGCMs) are highly uncertain this is due to inability of climate models to capture the basic characteristics of the present-day climate variability in Sahel region [18].

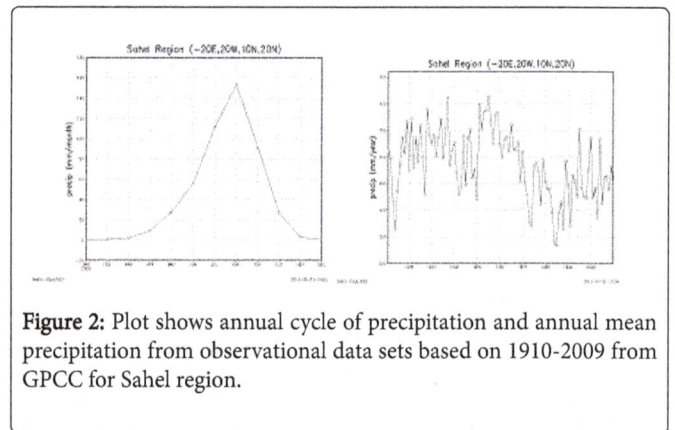

Figure 2: Plot shows annual cycle of precipitation and annual mean precipitation from observational data sets based on 1910-2009 from GPCC for Sahel region.

The Sahel simulated and observes display climatology showing the Sahel region rainfall pattern distribution and the yearly variability in Figure 2 above. The rainy season in Sahel region start around June, July, August and September (JJAS), with the pick at August while the pick of winter season is from November, December, January and February (NDJF).

The region experiences strong variation in climate characterized by the strong seasonality of the climate with a rainy season in the boreal summer and less rain in the winter season. The annual rainfall varies from around 200 mm in the North of Sahel than 600 mm in the South [4]. The main feature of climate of Sahel region is WAM system (Figure 3).

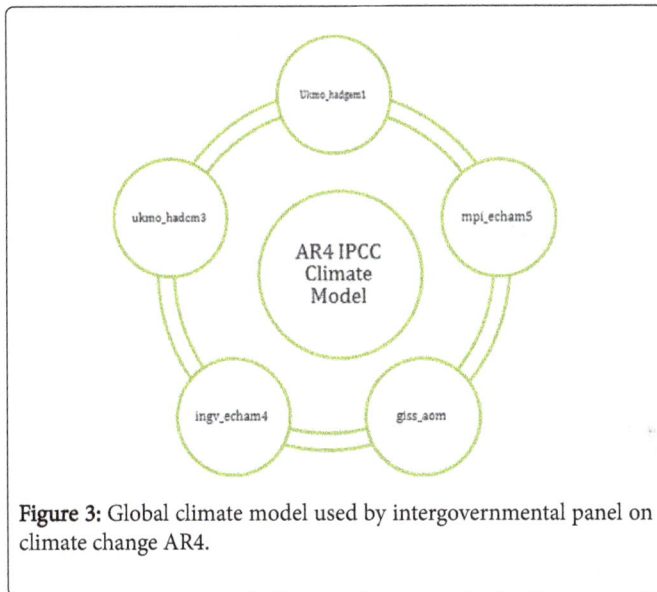

Figure 3: Global climate model used by intergovernmental panel on climate change AR4.

Sahel rainfall is mostly during the movement of the ITCZ. Biasutti [19] and Buontempo [18] argue that the distributions of annual rainfall in Sahel region can be coarsely divided into different similar regions: the highest is along the west coast, the weaker is around Lake Chad and the minimal rainfall distribution is center around Greenwich meridian. The observed feature on inter-annual time scale of Sahel region is dipole structure that associates dry nature in the regions and the wet situations of Guinean coast (south of 10°N) with the presence of warm Gulf of SST anomalies [18,20]. However, the observed dipole is connected to the variability in SST over the Gulf of Guinea that is responsible for dry and wet years in the region [21,22] (Figure 4).

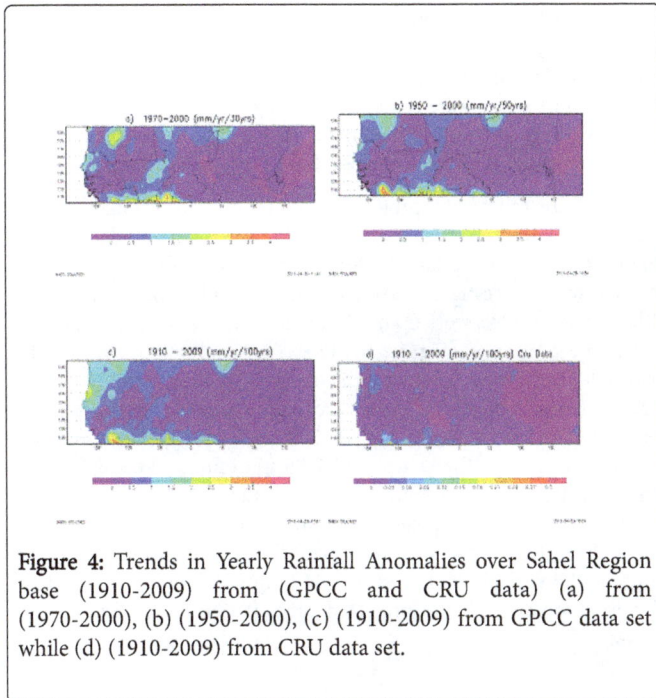

Figure 4: Trends in Yearly Rainfall Anomalies over Sahel Region base (1910-2009) from (GPCC and CRU data) (a) from (1970-2000), (b) (1950-2000), (c) (1910-2009) from GPCC data set while (d) (1910-2009) from CRU data set.

Data and Methods

Data were obtained from the global monthly precipitation data set held by the GPCC and CRU. Data set from CRU consists of historical time series of monthly precipitation from worldwide [1].

Observation

The study analyses the precipitation trends over Sahel region using the observe data set supplied from the GPCC, full data Reanalysis product from World Climate Research Programme (WCRP), as historical observation data. The spatial resolution is 0.5° of Latitude and Longitude [23]. CRU data set was also used to provide the comparison with GPCC data set (Figures 5-7).

The paper analysed the period from 1970 to 2000, 1950-2000 and 1910-2009 for a single time series of area average rainfall over the Sahel region. The rationale behind this period was due to well-documented changes in rainfall climate on a decadal time series in the Sahel region [24]. This region receives about 70% of the annual rainfall from June to September [25] The long-term variation of the area average yearly anomaly precipitation within this region is shown in Figure 7(c).

The region rainfall climatology is determined mainly by seasonal changes in large-scale circulation, part of which involves the seasonal north-south movement of the ITCZ.

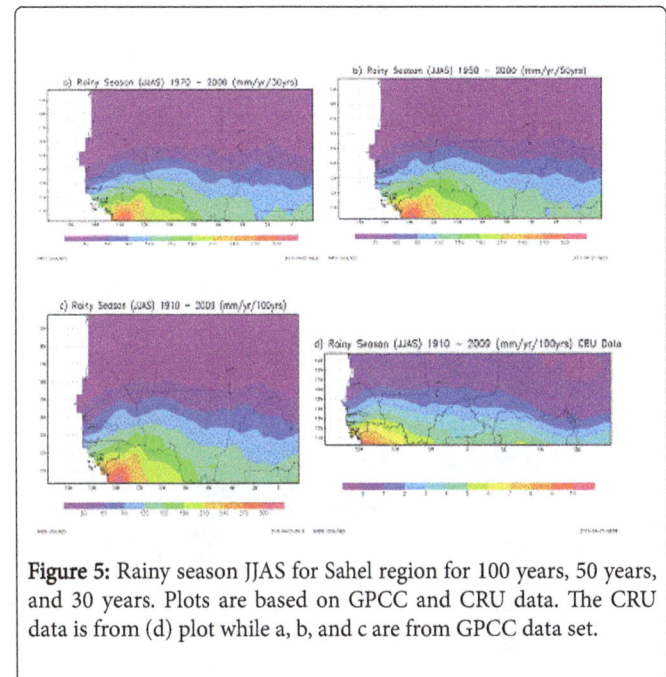

Figure 5: Rainy season JJAS for Sahel region for 100 years, 50 years, and 30 years. Plots are based on GPCC and CRU data. The CRU data is from (d) plot while a, b, and c are from GPCC data set.

Model simulations

When investigating the effects of one selected forcing mechanism in climatological time series against the background of natural variability, [17] argue that, it is necessary to rely on an ensemble of several model simulations, forced by the same mechanism but starting from different independent initial conditions. GCM used by IPCC AR4 were used to compare seasonal and the annual precipitation variability over Sahel region. The study carried out the analyses using the following AR4 IPCC climate model.

Analyses

In order to assess precipitation trends over West Africa Sahel region in term of area average, the yearly variability and seasonality for the period of study 1910-2009 the data were analysed using Exceed on Demand and Climate Data Operators (CDO) software. To visualize climate data the result were display in Linux based software, including the Grid Analysis and Display System (GrADS). The analyses were based on GPCC and CRU precipitation data set to observe the spatial distribution of Sahel region. GPCC contribute to the global observing system (GCOS) and the WCRP. The analysis started by extracting the West Africa Sahel region for the study period that is from 1970-2000, 1950-2000 and 1910-2009. These analyses were carried out using CDO command; this was to get the area average over the region.

Again, the study analysed 100 years climatology for the period of 1910-2009, using CRU data set to compare with the GPCC data. The study needed to find out how the data sets are represented in Sahel region precipitation variability. Precipitation in African continent is strongly seasonal; these reflect the leading role of the migrating ITCZ in determining the rainfall season [1]. Thus, the paper analyses the seasonal variability with the pick of rainy season from June to September and the pick of winter season from November to February to get rainfall distribution during JJAS and NDJF Figures 5 and 6. Again the study analysed the rainfall anomalies; this is to get the anomalies for all the months JJAS and NDJF. The precipitation anomaly time series were represented in Figure 7. The analyses carried out above were to identify the character and amount of precipitation trends over Sahel West Africa region for 30, 50 and 100 years.

GCMs are reliable to project the future climate change; it has a spatial resolution that is too coarse to accurately model spatially inhomogeneous precipitation [26]. IPCC AR4 models were used in this study to investigate precipitation variability on Sahel region, although the models were used to compare with historical observations and future projection (Figures 8 and 9). The study used, five-member ensemble runs have been conducted corresponding to the above historical observation. The same processes used in observation were also used for model simulation for 20th century. The study used the differences between the wet period JJAS to examine the seasonal and annual variation from the models 20th century. The models were also run for 21st century to assess the future precipitation trends in Sahel region from the period of (2050-2080) for A1B scenario this was to examine the variability of precipitation trends in future over Sahel and also to compare with the projected precipitation variability over the region.

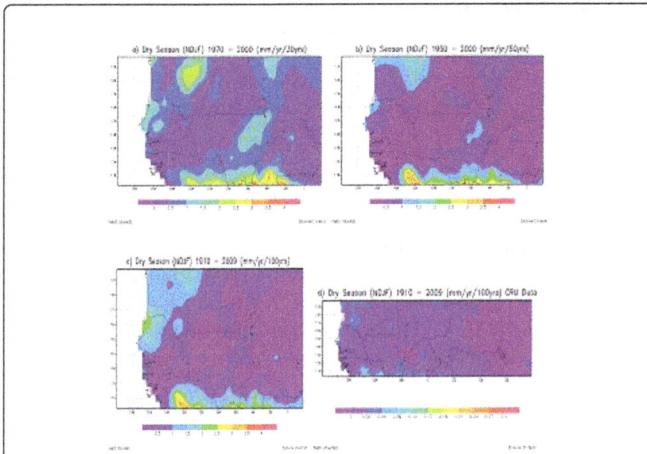

Figure 6: Dry season, NDJF for Sahel region for 100 years, 50 years and 30 years. Plots are based on GPCC and CRU data. The CRU data is plot d while GPCC data is plot a, b and c.

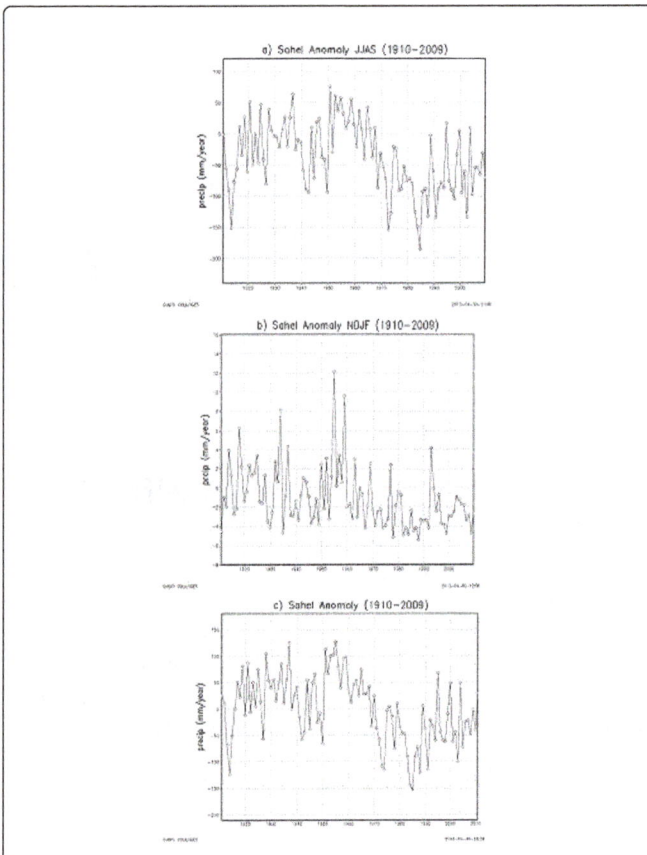

Figure 7: The anomaly of JJAS and NDJF annual average rainfall climatology from 1910-2009 for 100 years, the anomaly plots is based on GPCC data.

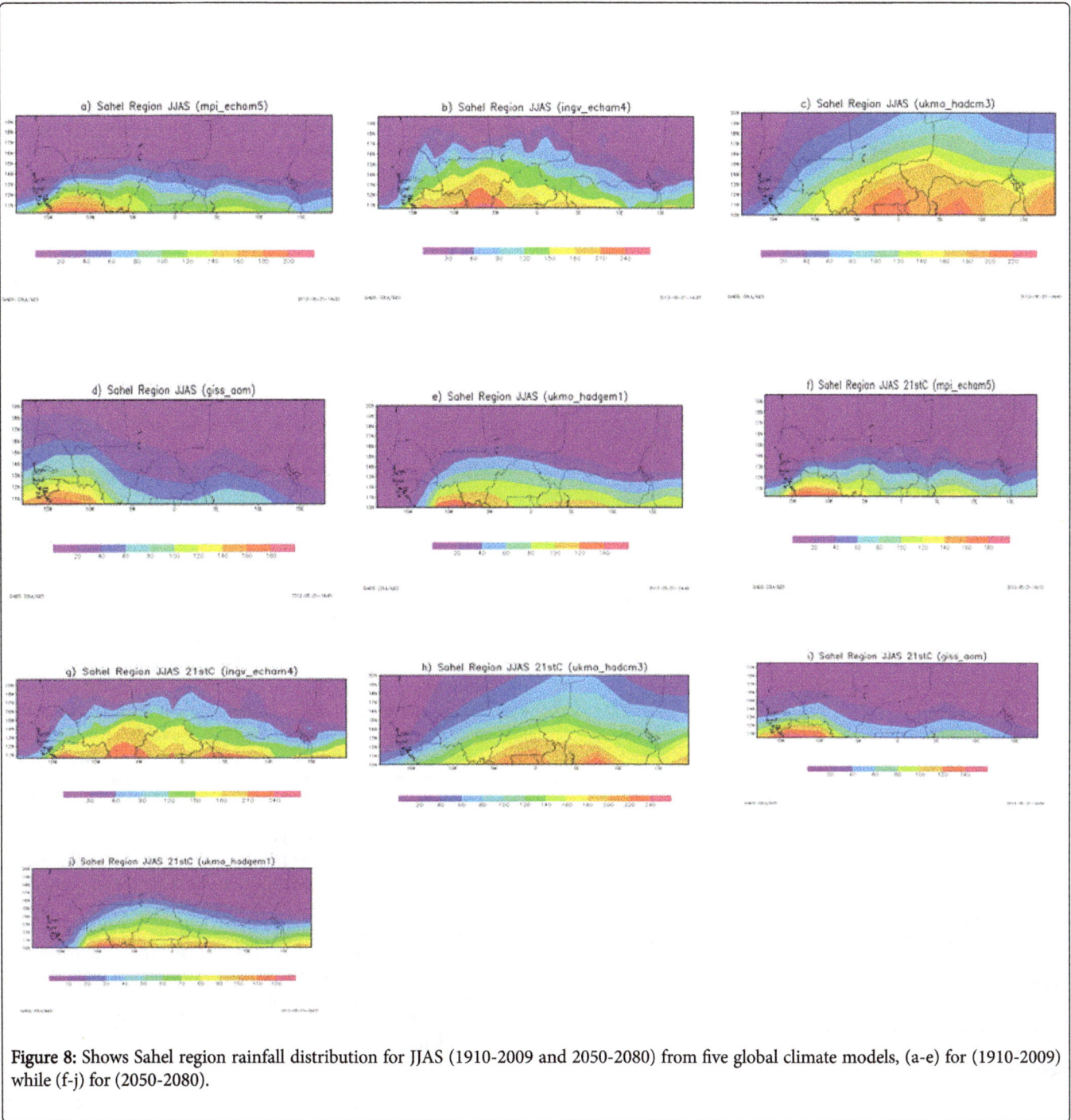

Figure 8: Shows Sahel region rainfall distribution for JJAS (1910-2009 and 2050-2080) from five global climate models, (a-e) for (1910-2009) while (f-j) for (2050-2080).

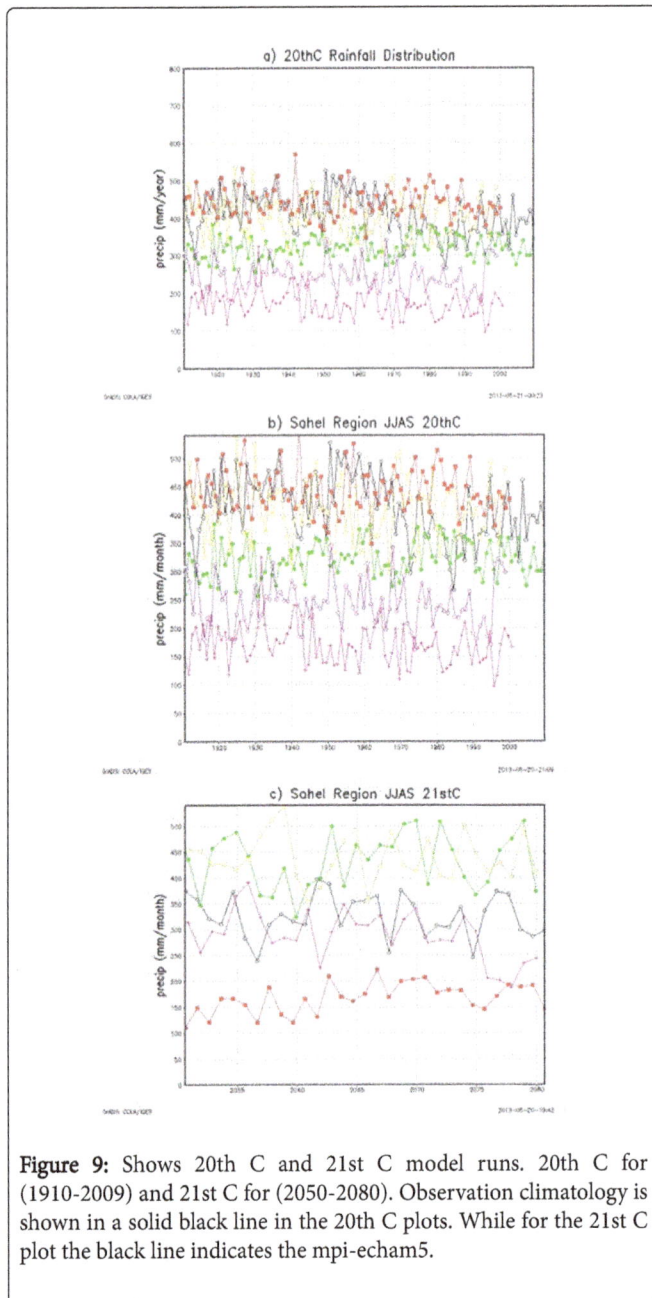

Figure 9: Shows 20th C and 21st C model runs. 20th C for (1910-2009) and 21st C for (2050-2080). Observation climatology is shown in a solid black line in the 20th C plots. While for the 21st C plot the black line indicates the mpi-echam5.

Results

Mean rainfall pattern

The result reported in this paper is the analysed precipitation variability of Sahel region in terms of area average rainfall distribution, annual variability and seasonality. The analyses provide a characterization of the weather dynamics of this region. The observed rainfall distributions for the present study 1970-2000, 1950-2000 and 1910-2009. Figure 4, from GPCC and CRU data set, shows that the two data sets has good agreement to locate and observed the pattern of inter-annual variability and the overall trend in Sahel region between the study periods. The observed changes in rainfall indicate that the Sahel may have undergone a climatic change in the first and last half of

the twentieth century. The major disagreement is that Sahel rainfalls are not zonally oriented as observed. The study observes high intensity of rainfall over the West coastal and Gulf of Guinea (Figures 4 and 5), such that Hunt [22] noticed in his study.

The observation shows that the Northern part of the Sahel region is characterizes by low precipitation. On the inter annual and decadal time scales, Sahel region rainfall is known to be affected by a variety of regional and global SST difference pattern. The study notices that the region yearly mean rainfall distribution patterns during the past 100 years are not consistent (Figure 7(c)). Though the long-term yearly trends are freely apparent in Sahel region, some changes appear to be reasonably robust and consistent within specific regions. The observed rainfall climatology for the Sahel region in Figure 4 for 30, 50 and 100 year shows that significant inters annual variability is apparent. The decadal length of dry periods occurred between around 1910-1930, 1930-1950 and during 1960-1980s. Figure 7(c), this indicates that an interval of 30 years exist apart from 1960-1980s during Sahel rainfall variability. According to Lebel and Ali [27] the key change of rainfall regime during the droughts period is the dramatically smaller occurrence rate of rain events in June to September, however the monthly average of their point intensity did not change much.

The observed significant variability in Sahel region is characterized by dryness in the second half of 20th C and the alternating incidence of multi decadal dry and wet influences in recent history; this is associated with the oscillation of lake level [28]. Xue and Shukla [6] argue that desertification in Sahel region may have caused a decrease of precipitation, which led to drought observed in the 1960s. Although number of literature has endeavoured to link the drought to tropical factors such as ENSO, SSTs and ITCZ, yet there is still extensive divergence about the role each of these played in rainfall variability. The drying trend in Sahel region observed in this study is similar with that highlighted by Nicholson and Palo [5]. The analyses observed that the northern Sahel was drying more than the Gulf of Guinea this is because the north Sahel is dominated by the low level north easterly Harmattan winds.

The change in precipitation during 1968 was very abrupt and very extreme. The Sahel latitudes of (15°N to 20°N) mean JJAS rainfall from 1968 to 1997 were roughly lower than the mean for the 30 year period 1931 to 1960 [4]. The drought was more persistence in the central Sahel. Thus the JJAS precipitation climatology from GPCC and CRU data for 1910-2009. Figures 4(c) and 4(d) observed two maxima over west Africa Sahel, one is along West Coast between 5°N and 12°N and the other is at 10°E and 7°N over the Eastern Guinea Coast. The North of 15°N, rainfall rate fall drastically such that PU and Cook noticed in their study [20].

The study observed that the result display from the two data sets GPCC 100 years (Figure 4(c)) and CRU 100 year's (Figure 4(d)) demonstrates that the variability of rainfall for the data sets does not correspond. The CRU data values are much smaller than GPCC, this show that the observational records for Sahel region are not totally precise. The different trend by GPCC and CRU data set shows that the findings for Sahel region are relatively robust.

The study focus on JJAS rainfall patterns because this four-month accounts for the highest levels of rainfall along the Sahel region Figure 2 above. This four month rainfall patterns determine the intensity or the duration of dry trends that usually occur in the Sahel region. It was observed from the build climatology (Figure 2) that during JJAS the precipitation trends increased. It is deficit in rainfall during JJAS that affects seasonal and annual rainfall amount in Sahel region. Although rainy season in Sahel region varies with latitude, fluctuating from about 100 to 200 mm in the northern extreme to 500 to 600 mm in the southern Sahel region [4]. The West African Sahel region rainfall is connected to the south-west monsoon movement during the summer; this is because of the specific geometry of the West African continent, which increases the strong sea-land contrasts westwards of about 20"E [13]. During the boreal summer, the atmospheric circulation in the Sahel region bound becomes more complex, dominated in a great extent by the West African Monsoon (WAM). Peyrille et al. [29] argue that the seasonal cycle of monsoon is linked to the energy balance in the Sahel region.

The study observed from the build climatology the yearly average rainfall over different season in Sahel region. The Four month contributes to 50% to 60% of the rainfall in the Sahel region. The northern Sahel rainfall is also relatively high in JJAS. There was a wetting trend observed apparently in the West coastline and in the southern part of the Sahel, i.e., Guinea, southern and southern Burkino Faso (Figure 5). This trend is reflected in the 100 year and with comparable with GPCC and CRU data for JJAS (Figures 5c and 5d). The overall seasonal rainfall trends observed indicate that the Sahel region rainfall distribution during JJAS assume to be sufficiently not well produced for the purpose of this study.

The observed trends of rainfall distribution during NDJF. Figure 2 above are regarded as boreal winter period in Sahel region. Rainfall distribution during this period is less or no rain. The north Sahel during this period is dominated by the low level north easterly Harmattan winds. The observed rainfall distribution over Sahel region display decreasing in annual rainfall. Figure 6, this is because this season produces minimal or no rainfall in Sahel region. Thus the dry season NDJF plot also shows wetting in the West coastline and in the southern part of the Sahel, i.e., Guinea, southern and southern Burkino Faso area over the 100 year period, but the numbers are very small and so indicate that this wetting trend during the dry season was quite minimal. The study notice that the recent years irregular dry rainy seasons and irregular dry months of JJAS were more recurrent than wet ones in the north of 10°N [13] (Figure 6).

Series of annual mean rainfall anomalies

Anomalies show deviation from the normal either higher or lower. The observed rainfall trends in the Sahel region for JJAS and NDJF annual mean rainfall anomalies for the period of (1910-2009) displayed in Figure 7 were more homogenous. The rainy season JJAS over Sahel is not normally distributed (Figure 7(a)). The basic features of the simulated rainfall changes in the Sahel region are in agreement with observations [22]. The observation shows that the Sahel region is characterizes by drier conditions during the first and second half of 20th century. The droughts that occur during 1984-1985, the JJAS season was anomalously drier than usual due to a large decrease in the trend with an approximate value of 180mm.This was very dry season, which may coincide with a drought season [30].

Another important features that the paper notice about Sahel drought was, during 1910s and 1940s, the spread about 30 year interval before 1970s drought, this shows that a cycle of 30 year exist in Sahel region drought (Figure 7) [31]. The study notice that the period of 1950 to 1960s was wet (positive values) (Figure 7). The drought in Sahel region, which is as a result of rainfall deficit, is connected with a general reduction of the occurrence rate of rain events during the period. The noisy spatial features of drought in Sahel region suggest

that the seasonal shift in the ITD might not likely be the most contributing mechanism of drought in the region [32], El Nino event might have also play a part in it [21].

Model simulations of climatology

The sensitivity of the Sahel region rainfall to the changing climate has important significance on the climate models to simulate properly the current mean state of the monsoon system. The simulated patterns of rainfall climatology for Sahel region are display in Figure 8 below. The study uses five global climate models for the 20th C to compare with seasonality from the observational data set. The models were also run for 21st C (2050-2080) to examine the projected scenario over Sahel rainfall variability. The study displays the analyses for each of the five models for the 20th and 21st C for JJAS rainy season (Figure 8). The year-to-year rainfall anomaly of the region is a robust feature across the models simulated, given huge dissimilarities in model formulation, and the difficulties of GCM in simulating tropical rainfall. Although despite the dissimilarities in the model, the models captured the large scales drying of Sahel region during the first and second half of 20th C (Figure 8(a)) [19]. The analyses show slightly similar features to observations across Sahel region. The specific location of anomalies varies from the five-difference model. For example, the decrease in Sahel rainfall during 20th C is simulated west (east) of 0° for mpi_echam5.

However, one important feature of the ukmo_hadcm3 model observed in this study is that the model produces more rainfall Figures 8(c) and 8(h) than the other four models. And the ukmo_hadcm3 simulation display shows close trends with the observation. Again ingv_echam4 also shows more rainfall but not to the extent of ukmo_hadcm3. The simulated rainfall distribution over Sahel appears to be weaker than the observed.

The five models were also used to simulate the 21st C for A1B scenarios (2050-2080) climatology with mean of the rainy season JJAS showing that southern Sahel region will get wetter (i.e., Burkina Faso, north Ghana, southern Mali, Guinea, northern Nigeria) Druyan [8], but the ukmo_hadcm3 has a bigger magnitude of increased rainfall (Figure 8). There is a robust drying trend reflected by the five models over Sahel region with reductions being most significant during the past 50 year. The models observe that the southern areas are wetter compared to central Sahara. Thus the five models indicated a drier Sahel in the future, due primarily to increasing greenhouse.

Discussion

Implication of the result to changes in precipitation

The changes in precipitation in Sahel region of West Africa in light of the synthesis of the observational data sets evidences as well as the model simulation has shown that Sahel region is characterized by dryness during the first and second half of the 20th C. The simulation shows that Sahel rainfall has changed substantially over the last 100 years in the 20th C. This change has been most notable over Northern part of the region that is describes by low precipitation. The Intensity of rainfall was observed over the West coastline and Gulf of Guinea areas (Figure 4). According to Dennett [10] these differences in Sahel rainfall trends have been reported by Gregory who has studied the spatial variability of dissimilarity between rainfall regimes in Sahel region.

The climatology in Figure 1 indicates that JJAS is the peak of rainy season while NDJF is the peak of winter season. From the anomalies it can be seen that Sahel rainfall distributions during JJAS were below normal rainfall for the period of 100 years. The time series provided evidence of drought clustering in the region. But the causes of the drought remain uncertain; investigation suggests the existence of a strong correlation between the moisture flux, heat and variations in precipitation within the Sahel [33]. However, despite the models difference in simulating Sahel region the results still indicates drying trends over the region, although Ukmo_hadcm3 indicate a greater magnitude of increase in rainfall. Furthermore, the changes in precipitation trend within GCM influences on the simulation, and the uncertainty due to the model is noticeable.

In general rainfall variability in JJAS is predominantly liable for the tendency for rainfall anomalies of the sign prevail through the analyses. This four-month is responsible for the long down ward trend, the high inter annual persistence of rainfall variation through Sahel region. Therefore, Sahel's region climate in the future is disturbing. It is apparent that Sahel precipitation variability is inextricably tied up to global climate variability, both in the inter-annual and the inter-daedal timescales. Hence, it is likely to say that Sahel's region rainfall responds to El Niño/La Niña events [34].

Conclusion

The study showed that the change in precipitation over Sahel region is a characterization of the weather dynamics of this region. From the prospective of climate variability based on the scientific evident presented in this paper, it is increasingly observed that Sahel region is characterized by dryness during the first and second half of the 20th C. Rainfall deficit in the region is linked with a general reduction of the incidence rate of rain events during the period. According to Adger et al. [15] climate change is likely to intensify the already felt economic challenges that are experiencing in the region because of the region dependent on rain-fell for its livelihoods.

However, this inescapable climate effect is not a problem of Africa's making, yet Sahel region stand to be predominantly more effected because of their geography, their agricultural dependence, and because of low level of adaptation this region face. It is therefore paramount for policy makers to strategies for facilitating adaptation need to be developed, also ensure investment and promotion of more energy efficient infrastructure in order to cope with these extreme weather and climate events.

Acknowledgement

The author acknowledges the support by Prof. Martin Todd of University of Sussex United Kingdom during my MSc programme. I am also thankful to National Environmental Standards and Regulations Enforcement Agency (NESREA) for cooperation and assistance from them.

References

1. Hulme M (1992) Rainfall Changes in Africa: 1931-1960 to 1961-1990. International Journal of Climatology 12: 685-699.

2. Hickler T, Eklundh L, Seaquist J, Smith B, Ardo J, et al. (2005) Precipitation controls Sahel greening trend. Geophysical Research Letters by the American Geophysical Union 32.

3. UNEP and ICRAF (2006) Climate Change and Variability in the Sahel Region, Impacts and Adaptation Strategies in the Agricultural Sector.

4. Nicholson SE (2013) The West African Sahel: A Review of Recent Studies on the Rainfall Regime and Its Inter annual Variability. ISRN Meteorology 2013: 1-32.

5. Nicholson SE and Palo MI (1993) A Re-Revolution of Rainfall Variability in Sahel part1. Characteristics of Rainfall Fluctuation. International Journal of Climatology 13: 371-389.

6. Xue Y, Shukla J (1993) The influence of land surface properties on Sahel climate. Part I: desertification. Journal of Climate 6: 2232-2245.

7. Batterbury S and Warren A (2001) The African Sahel 25 years after the great drought, Assessing progress and moving towards new agendas and approaches. Global Environmental Change 11: 1-8.

8. Druyan LM (2010) Review Studies of 21st century precipitation trends over West Africa. International Journal of Climatology 31: 1415-1424.

9. Wang G, Alo CA (2011) Changes in Precipitation Seasonality in West Africa Predicted by RegCM3 and the Impact of Dynamic Vegetation Feedback. International Journal of Geophysics 2012: 1-10.

10. Dennett MD, Elston J, Rodgers JA (1984) A Reappraisal of Rainfall Trends in the Sahel. Journal of Climatology 5: 353-361.

11. Liebmann B, Blade I, Kiladis GN, Carvalho LMV, Senay GB, et al. (2011) Seasonality of African Precipitation from 1996 to 2009. Journal of Climate 25.

12. Kalu AE (1979) The African dust plume: its characteristics and propagation across West Africa in winter, in Saharan Dust: Mobilization, Transport, Deposition 95-118.

13. Fontaine B, Bigot S (1993) West African Rainfall Deficits and Sea-Surface Temperatures. International Journal of Climatology 13: 271-285.

14. Hayward D and Oguntoyinbo JS (1987) Climatology of West Africa, Rowman and Littlefield.

15. Adger WN, Huq S, Brown K, Conwaya D, Hulme M (2003) Adaptation to climate change in the developing world. Progress in Development Studies 3: 179-195.

16. Xue Y, Shukla J (1998) Model Simulation of the Influence of Global SST Anomalies on Sahel Rainfall 126: 2782-2792.

17. Paeth H, Hense A (2004) SST versus Climate Climatic: Change 65: 179-208.

18. Buontempo C (2010) Sahelian climate: past, current, projections: Sahel and club west Africa Secretariat.

19. Biasutti M (2013) Forced Sahel rainfall trends in the CMIP5 archive. Journal of Geophysical Research 118: 1613-1623.

20. Pu B, Cook KH (2012) Role of the West African Westerly Jet in Sahel Rainfall Variations. Journal of Climate volume 25.

21. Folland CK, Palmer TN, Parker DE (1986) Sahel Rainfall and worldwide, Sea surface temperature 1901-85. Nature 320: 602-607.

22. Hunt BG (2000) Natural climatic variability and Sahelian rainfall trends. Global and Planetary Change 24: 107-131.

23. Munemoto M, Tachibana Y (2010) The recent trend of increasing precipitation in Sahel and the associated inter-hemispheric dipole of global SST. Climate and Ecosystem Dynamics Division 32: 1346-1353.

24. Lucio PS, Molion LCB, Conde FC, Melo MLD (2011) A study on the West Sahel rainfall variability: The role of the inter-tropical convergence zone (ITCZ). African journal of Agricultural Research 7: 2096-2113.

25. Hameed S, Riemer N (2012) Relationship of Sahel Precipitation and Atmospheric Centers of Action. Advances in Meteorology 1-8.

26. Sapiano MRS (2004) Trends and Variability in Observations of Winter Precipitation, A thesis submitted for the degree of Doctor of Philosophy Department of Meteorology.

27. Lebel T, Ali A (2009) Recent trends in the Central and Western Sahel rainfall regime (1990-2007). Journal of Hydrology 375: 52-64.

28. Wang G and Eltahir EAB (2002) Role of vegetation dynamics in enhancing the low-frequency variability of the Sahel rainfall. Water Resources Research 36: 1013-1021.

29. Peyrille P, Lafore JP, Redelsperger JL (2007) An idealized two-dimensional framework to study the West African monsoon. Part I: Validation and key controlling factors. Journal of the Atmospheric Sciences 64: 2765-2782.

30. Reardon T, Matlon P (1989) Seasonal food insecurity and vulnerability in drought-affected regions of Burkina Faso. Seasonal variability in Third World agriculture: The consequences for food security 118-136.

31. Nicholson SE, Enetekhabi D (1986) The Quasi-Periodic Behavior of Rainfall Variability in Africa and Its Relationship to the Southern Oscillation. Arch Met Geoph Biocl Ser A 34: 311-348.

32. Oladipo EO (1993) Some Aspects of the Spatial Characteristics of Drought in Northern Nigeria. Natural Hazards 8: 171-188.

33. Stevenson S, Fox-Kemper BM, Neal R, Deser C, Meehl G (2011) Will there be a significant change to El Nino in the Twenty-First Century. Community Climate System Model CCSM4, 2129-2145.

34. Paulo SL, Luiz CBM, Cati EAV, Fábio CC, Andrea MR, et al. (2012) Dynamical Outlines of the Rainfall Variability and the ITCZ Role over the West Sahel 2: 1-14.

30

Diagnostic Evaluation of September 29, 2012 Heavy Rainfall Event over Nigeria

Akinsanola AA* and Aroninuola BA

Department of Meteorology and Climate Science, Federal University of Technology Akure, Nigeria

*Corresponding author: Akinsanola AA, Department of Meteorology and Climate Science, Federal University of Technology Akure, Ondo State, Nigeria
E-mail: mictomi@yahoo.com

Abstract

The study diagnostically evaluate the atmospheric conditions that led to the heavy rainfall event of September 29, 2012 in Nigeria using data from Tropical Rainfall Measurement Mission Multi-Platform Analysis (TMPA 3B43v7) and European Centre for Medium-range Weather Forecast (ECMWF), Era-Interim reanalysis. The spatial pattern of precipitation, horizontal divergence of moisture flux and vertical structure of the zonal wind speed were investigated over the study area. Results show that the heavy rainfall event was only limited to the south of latitude 9°N within the study domain. Furthermore, strong moisture convergence (divergence) was observed in the lower (middle) troposphere during the pre-event days of September 27 and 28, 2012, implying strong upward motion of convectively unstable moist air over majority of the study area. This observation may be responsible for the heavy rainfall delivery that led to the flooding. Furthermore, a well-organized African Easterly Jet (AEJ) at 700 hpa and Tropical Easterly Jet (TEJ) at 200 hpa were respectively observed South of 9°N. It is worthy to note that during the pre-event days, the intensity of the AEJ was greater in magnitude. Disappearing TEJ and a well-organized AEJ was observed North of 9°N. This observation may be the reason why the heavy rainfall was majorly over South of 9°N. The distinctive appearance of the two summer easterly Jets South of 9°N and the observed moisture convergence at the lower troposphere may have impacted to a very large extent the development of deep moist convection that led to the heavy precipitation experienced over the region. The aforementioned findings may be responsible for the extreme rainfall over the study area.

Keywords: Moisture flux; Divergence; African easterly jet; Tropical easterly jet; Heavy rainfall

Introduction

One of the most crucial and dynamic phenomena of the West African climate system during the summer is the West Africa Monsoon (WAM) [1,2] a phenomenon caused by the seasonal reversal of winds due to differential heating between land and ocean [3,4] it plays a vital role in producing majority of the annual rainfall in the region [5-7]. In general, summer monsoon rainfall in Nigeria usually last between the first week of July until late September [8]. On 29th September, 2012, a heavy rainfall event occurred in major part of Nigeria. Historically, the rainfall recorded was of higher intensity. The Nigeria Meteorological Agency (NIMET) and National Emergency Management Agency (NEMA) reported that the event led to substantial damages (e.g. displacement of thousands of people, loss of crops, homes, farms, belongings, and livelihoods). The effects of the calamity enormously destabilized the economic situation of the Country. Presently, no studies examined the meteorological conditions which produced the September 29, 2012 heavy rainfall. However, such a heavy downpour leading to catastrophic flood disaster deserves a closer look at the atmospheric environment which led to the extreme phenomenon. Such incidences of heavy rainfall need to be monitored regularly in order to avoid the heavy damages that can be caused by subsequent flash flooding. Hence, this study aimed at diagnosing the atmospheric conditions that led to the heavy rainfall event of 29th September, 2012.

Study Area

Nigeria which lies between latitude 4° and 14°N and longitude 4°to 14°E with elevation ranging from 6 to 1290 m above, it is bounded on the North by the Republic of Niger, East by Cameroon and West by Benin Republic while the Southern boundary is Gulf of Guinea which is an arm of the Atlantic ocean. There are two main seasons which are wet (June-September) and dry (December-January). These periods vary from the Northern part to the Southern part of the country.

Data and Methodology

Daily rainfall data used for this study was obtained from the Tropical Rainfall Measurement Mission Multi-Platform Analysis (TMPA 3B43v7) [9,10] from September 27-30, 2012. Also daily zonal and meridional wind components and specific humidity were obtained from the European Centre for Medium-range Weather Forecast (ECMWF), Era- Interim reanalysis dataset for four pressure levels (1000, 850, 700, and 200 hpa) for the same time period. It is important to state that all the dataset used were at a spatial resolution of 0.25° by 0.25°.

Furthermore, the horizontal divergence of moisture flux (HFD) was computed for all the days under study over Nigeria using equation 1.

$$HFD = -\left(\left(u\frac{\partial q}{\partial x} + v\frac{\partial q}{\partial y}\right) + \left(q\frac{\partial u}{\partial x} + q\frac{\partial v}{\partial y}\right)\right)(1)$$

where q is the specific humidity (g/kg), u and v are respectively the zonal and meridional wind speed components (m/s). Negative HFD means convergence while positive means divergence.

Finally the vertical profile of the zonal wind was plotted to investigate the position and role of the Jet streams.

Results and Discussion

Before diagnostically evaluating the atmospheric condition that led to the heavy rainfall event. There is need to examine the spatial pattern of rainfall over Nigeria before, during and after the event. September 27 and 28 were used as the pre event days, September, 29 was the event day while September 30 was regarded as the post event day. During the pre-event days there was little or no rainfall in the major part of the country (Figure 1). However, on the day of the event, it was observed that majority of the country experienced rainfall above the daily normal threshold. It is worth mentioning that South of latitude 9°N experience rainfall above 120 mm. The intensity of the rainfall had been reduced drastically on the post event day of September 30.

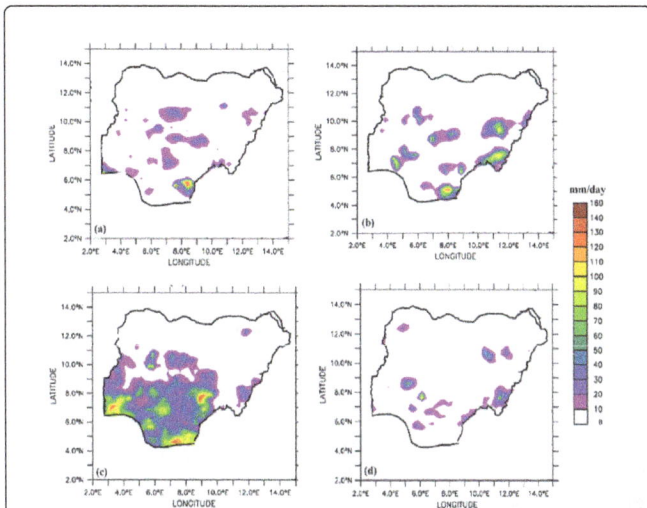

Figure 1: Daily precipitation pattern over Nigeria for the pre event days (a) 27th September, 2012, (b) 28th September, 2012, during the event (c) 29th September, 2012 and post event day (d) 30th September, 2012.

The characteristics of moisture flux at three different tropospheric layers are presented in Figures 2-4. Here the practical usage of equation 1 to diagnose the heavy rainfall event that led to flooding had been displayed. At the surface (1000 hpa), strong moisture convergence was observed on the pre and post event days while dominance of diverging wind was observed in major part of the country on the event day. However, in the middle troposphere (700 hpa), diverging wind with high humidity loadings dominated days 27, 28 and 29. The observed moisture convergence in the lower troposphere and divergence in the middle troposphere during pre-event days may imply strong upward motion of moist air; release of latent heat due to condensation process (this available heat energy might be the source of low-level latent heat instability) and in turn intense rainfall delivery of September 29, 2012. Furthermore, at the upper troposphere, the HFD distribution depicts a wider spread of converging wind over the country in all the days considered. Thus high loading of upper-level moisture transport may have influenced the heavy rainfall South of 9°N. The results presented here are in mutual agreement with the assumption of Dine's compensation which says "convergence at the surface will lead to divergence at upper level and vice versa".

The dynamics of the atmosphere associated with rainfall over Nigeria is further assessed taking into consideration the zonal wind profile average for two latitudinal domains over Nigeria; South of 9°N, and North of 9°N as shown in Figure 5.

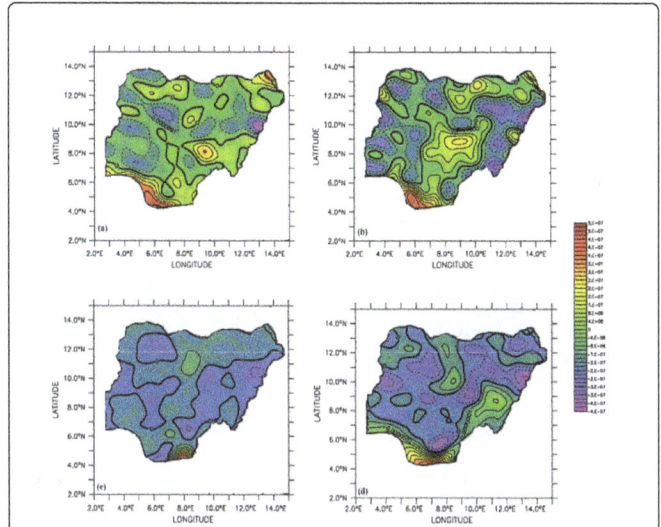

Figure 2: Horizontal Divergence of Moisture flux at 1000 hpa over Nigeria for the pre event days (a) 27th September, 2012, (b) 28th September, 2012, during the event (c) 29th September, 2012 and post event day (d) 30th September, 2012.

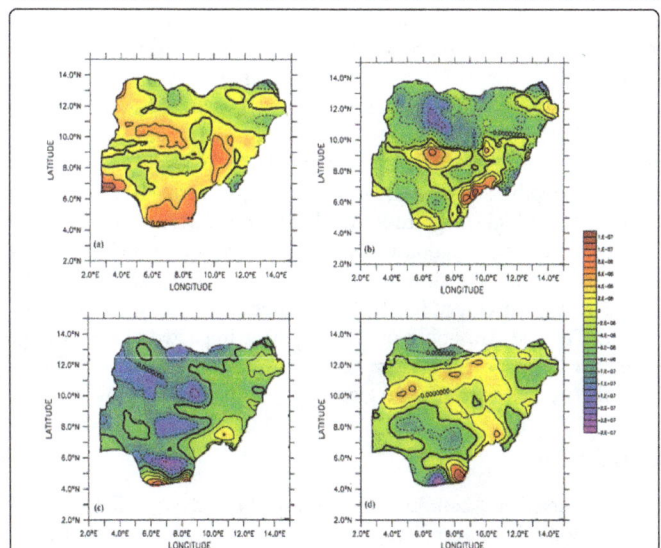

Figure 3: Horizontal Divergence of Moisture flux at 700 hpa over Nigeria for the pre event days (a) 27th September, 2012, (b) 28th September, 2012, during the event (c) 29th September, 2012 and post event day (d) 30th September, 2012.

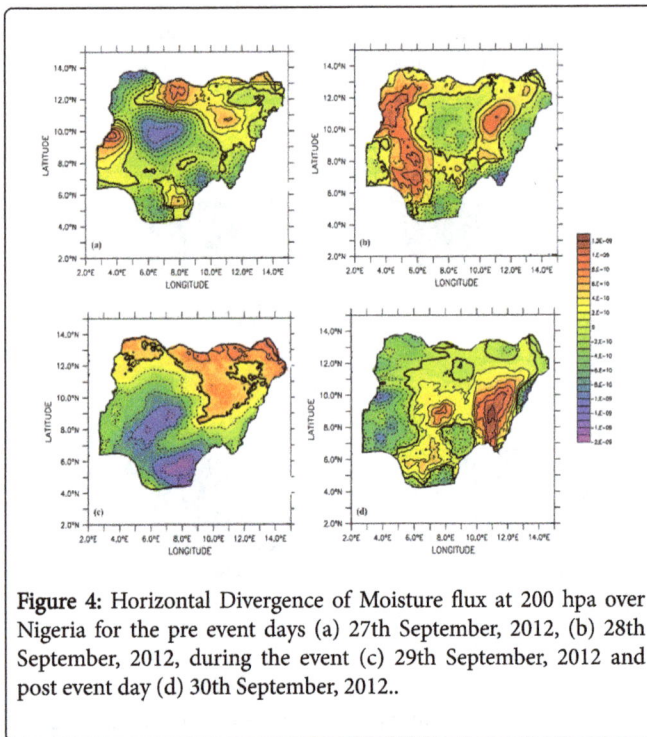

Figure 4: Horizontal Divergence of Moisture flux at 200 hpa over Nigeria for the pre event days (a) 27th September, 2012, (b) 28th September, 2012, during the event (c) 29th September, 2012 and post event day (d) 30th September, 2012..

The diagram illustrates the main dynamical large scale features associated with convective activity and rainfall delivery over Nigeria [11,12]. There was an appearance of the African Easterly Jet (AEJ) and Tropical Easterly Jet (TEJ) at 700 hpa and 200 hpa respectively in all five days considered South of 9°N. However, it is worthy to note that the intensity of the AEJ in the pre event days especially September 28, 2012 was stronger in magnitude (11 m/s), with observation of a reasonably distinct TEJ except for day 27. The zonal wind profile average at North of 9°N as shown in Figure 5b indicates a deeper moisture depth during the pre-event days with a considerable organised and high magnitude AEJ but with a disappearing TEJ. The disappearance of the Tropical Easterly Jet North of Latitude 9°N may be the reason for the little or no rainfall observed over the region. Hence it can be concluded that the combine appearance of a well distinct TEJ and AEJ South of 9°N may have impacted to a very large extent the development of deep moist convection that led to the heavy rainfall experienced over the region.

Conclusion

The atmospheric condition that led to the heavy rainfall event of September 29, 2012 over Nigeria has been diagnosed in this study. The spatial pattern of rainfall over Nigeria indicated that majority of the country experienced rainfall above the daily threshold especially South of 9°N. Furthermore, the analysis of the horizontal divergence of moisture flux revealed very important characteristics of the heavy rainfall over study region. The observed moisture convergence in the lower troposphere and divergence in the middle troposphere during pre-event days implies strong upward motion of moist air which may be responsible for producing the observed heavy rainfall. Similarly, the occurrence and dominance of the heavy rainfall event South of 9°N on 29th September, 2012 may be attributed to a very large extent the appearance of a well-organized and distinct AEJ and TEJ which impacted to the development of deep, moist convection.

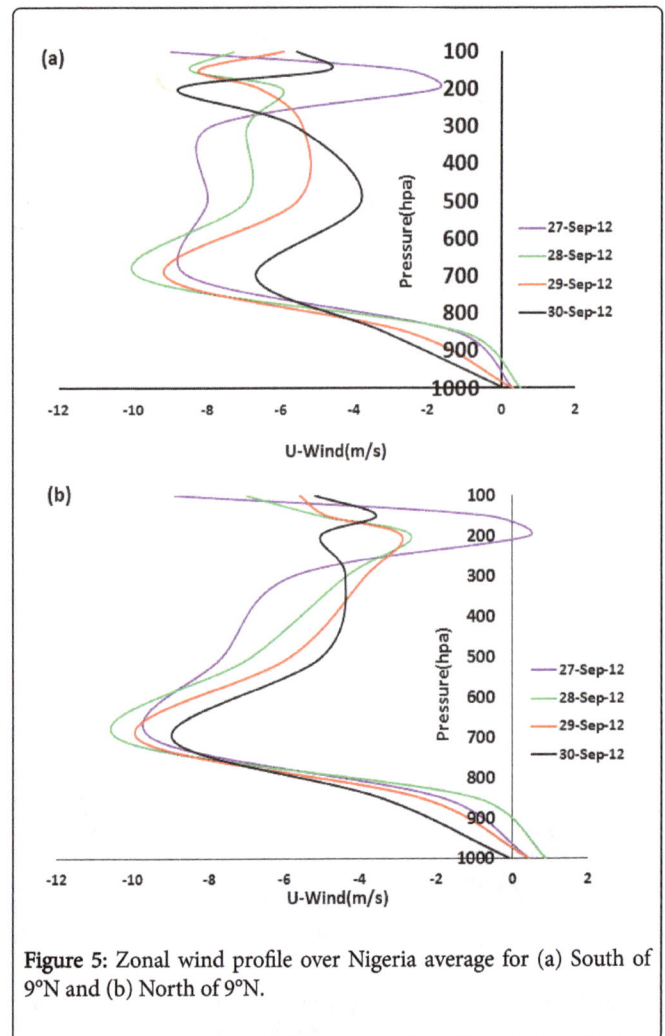

Figure 5: Zonal wind profile over Nigeria average for (a) South of 9°N and (b) North of 9°N.

References

1. Janicot S, Caniaux G, Chauvin F, de Coetlogon G, Fontaine B, et al. (2010) Intraseasonal variability of the West African monsoon. Atmospheric Science Letters 12: 58-66.

2. Akinsanola AA, Ogunjobi KO, Gbode IE, Ajayi VO (2015) Assessing the Capabilities of three Regional Climate Models over CORDEX Africa in Simulating West African Summer Monsoon Precipitation. Advances in Meteorology 2015: 1-13.

3. Sultan B, Janicot S (2000) Abrupt shift of the ITCZ over West Africa and intraseasonal variability. Geophys Res. Lett 27: 3353-3356.

4. Le Barbe´ L, Lebel T (1997) Rainfall climatology of the HAPEX Sahel region during the years 1950-1990. J Hydrol 189: 43-73.

5. Omotosho JB, Abiodun BJ (2007) A numerical study of moisture build-up and rainfall over West Africa. Meteorol Appl 14: 209-225.

6. Omotosho JB (1992) Long-range prediction of the onset and end of the rainy season in the West African Sahel. Int J of Climatology 12: 369-382.

7. Omotosho JB (1990) Onset of thunderstorms and precipitation over Northern Nigeria. International Journal of Climatology 10: 849-860.

8. Akinsanola AA and Ogunjobi KO (2015) Recent Homogeneity Analysis and Long Term Spatio-temporal Rainfall Trends in Nigeria. Theoretical and Applied Climatology 1-15.

9. Kummerow C, Hong Y, Olson WS, Yang S, Adler RF et al. (2001) The evolution of the Goddard profiling algorithm (GPROF) for rainfall

estimation from passive microwave sensors. Journal of Appl Met 40: 1801-1840.

10. Huffman GJ, Adler RF, Bolvin DT (2010) The TRMM multisatellite precipitation analysis (TMPA). In satellite rainfall applications for surface hydrology 3-22.

11. Cook KH (1999) Generation of the African Easterly Jet and its role in determining West African precipitation. J Clim 12: 1165-1184.

12. Thorncroft CD, Blackburn M (1999) Maintenance of the African Easterly Jet. Q J R Meteorol Soc 125: 763-786.

PERMISSIONS

LIST OF CONTRIBUTORS

Boon Allwin, Nishit S Gokarn and Serma S Pandian
Madras Veterinary College, Chennai, TANUVAS, Tamil Nadu, India

Stalin Vedamanickam, Sathish Gopal and Bharath Jothi S
Department of Animal Husbandry, Tamil Nadu, India

Manoj K
Forest College and Research institute, TNAU, Tamil Nadu, India

Nazia Hassan Choudhury and Ataur Rahman and Sara Ferdousi
Department of Water Resources Engineering, Bangladesh University of Engineering and Technology, Dhaka-1000, Bangladesh

Mahdi Dehghan Tezerjani, Kamal Omidvar and Ahmad Mazidi
Department of Geography, Yazd University, Iran

Rediat Takele and Solomon Gebretsidik
Jigjiga University, Jigjiga, Somali regional state, Ethiopia

Asmelash T. Reda
Mekelle University, Mekelle Institute of Technology,Mekelle Tigray Ethiopia, Ethiopia

Amitabh Chandra Dwivedi and Priyanka Mayank
Regional Centre, ICAR-Central Inland Fisheries Research Institute, 24 Panna Lal Road, Allahabad 211002, India

Sheeba Imran
Department of Biological Sciences, SHIATS, Allahabad, India

Hussain Alsarraf
Kuwait Meteorology Department, Kuwait

Matthew Van Den Broeke
Department of Earth and Atmospheric Sciences, University of Nebraska-Lincoln, USA

Luigi Cimorelli
Department of Civil, Architectural and Environmental Engineering, University of Naples Federico II, Via Claudio n.21 – 80125, Napoli, Italy

Luca Cozzolino, Andrea D'Aniello, Francesco Morlando and Domenico Pianese
Department of Engineering, Parthenope University of Naples - Centro Direzionale di Napoli, 80142 Napoli, Italy

Wales Singini, Mavuto Tembo and Chikondi Banda
Mzuzu University, Private Bag 201 Mzuzu 2, Malawi

Farzana Raihan and Guangqi Li
Department of Biological Sciences, Macquarie University, North Ryde, NSW 2109, Australia

Sandy P. Harrison
Department of Biological Sciences, Macquarie University, North Ryde, NSW 2109, Australia

School of Archaeology, Geography & Environmental Sciences, Reading University, Whiteknights, Reading, UK

Amin Donyaei and Alireze Pourkhabbaz
Department of Environmental Sciences, University of Birjand, Birjand, Iran

Shashank Shekhar Singh and Singh SK and Shuchita Garg
Environmental Engineering Department, Delhi Technological University, Delhi, India

Mahamad B Pathan
Department of Statistics, Poona College of Arts, Science and Commerce, Savitribai Phule Pune University, Pune, India

Amaury de Souza and Flavio Aristones
Federal University of Mato Grosso do Sul, Institute of Physics, PO Box 549, CEP 79070-900, Campo Grande Mato Grosso do Sul, Brazil

Fabio Verissimo Goncalves
Federal University of Mato Grosso do Sul, Faculty of Engineering, Architecture and Geography, Graduate Program in Environmental Technologies, PO Box 549, CEP 79070-900, Campo Grande Mato Grosso do Sul, Brazil

Christopher G A Harrison
Rosenstiel School of Marine and Atmospheric Science, University of Miami, 4600 Rickenbacker Causeway, Miami FL-33149, USA

Oyenike Mary Eludoyin
Department of Geography and Planning Sciences, Adekunle Ajasin University, Akungba-Akoko, Nigeria

Yan T, Pietrafesa LJ, Gayes PT and Bao S
School of Coastal & Marine Systems Science, Coastal Carolina University, Conway, SC 29528, USA

Jae-Won Choi, Yumi Cha and Jeoung-Yun Kim
National Institute of Meteorological Sciences, 33, Seohobuk-ro, Jeju 63568, Korea,

Singh SK, Anunay G, Rohit G, Shivangi G and Vipul V
Department of Environmental Engineering, Delhi Technological University, Delhi, India

Shafiq MU, Bhat MS, Rasool R, Ahmed P, Singh H and Hassan H
Department of Geography and Regional Development, University of Kashmir, J & K India

Nazari P, Kardavany H, Farajirad P and Abdolreza A
Department of Geography, Science and Research Branch, Islamic Azad University, Tehran, Iran

Alinune Musopole
University of Malawi, The Polytechnic Blantyre, Malawi

Anila Kumary KS
Kuriakose Gregorios College, Pampady, Kottayam, Kerala, India

El Sakka S
Department of management, School of Business, Future University in Egypt, 5th Settlement, New Cairo, Egypt

Sandeep N Kundu
Department of Geography, National University of Singapore, Singapore

Isaac Mugume
Department of Geography, Geo informatics and Climatic Sciences, Makerere University, Uganda

Shuanghe Shen and Sulin Tao
Collaborative Innovation Center on Forecast and Evaluation of Meteorological Disasters, Nanjing University of Information Science and Technology, China

Godfrey Mujuni
Department of Applied Meteorology, Data and Climate Services, Uganda National Meteorological Authority, Uganda

Awetahegn Niguse
Department of GIS and Agro meteorology, Mekelle Agricultural Research center, Tigray Agricultural Research Institute, Mekelle, Ethiopia

Araya Aleme
Mekelle University, Department of Crop and Horticultural Sciences, Mekelle, Ethiopia

Ronalds RT and Emiliano A
School of Natural Science and Technology, Universidad del Turabo, Gurabo, PR, USA

Rafael Mendez-Tejeda
Atmospheric Sciences Laboratory, University of Puerto Rico at Carolina, PO Box 4800, Carolina, PR, USA

Agumagu O
Department of Geography, University of Sussex, UK

Akinsanola AA and Aroninuola BA
Department of Meteorology and Climate Science, Federal University of Technology Akure, Nigeria

Index